An Introduction to Statistical Thermodynamics

An Introduction to Statistical Thermodynamics

Terrell L. Hill
National Institutes of Health
Bethesda, Maryland

Dover Publications, Inc.
New York

Copyright

Bibliographical Note

This Dover edition, first published in 1986, is an unabridged, corrected republication of the second printing (1962) of the work first published by the Addison-Wesley Publishing Company, Inc., Reading, Mass., 1960. (A volume in the Addison-Wesley Series in Chemistry, ed. by Francis T. Bonner and George C. Pimentel.)

Library of Congress Cataloging-in-Publication Data

Hill, Terrell L. An introduction to statistical thermodynamics.
Reprint. Originally published: Reading, Mass. : Addison. : Wesley, 1960. Originally published in series: Addison-Wesley series in chemistry.
Includes bibliographical references and index.
ISBN-13: 978-0-486-65242-9
ISBN-10: 0-486-65242-4
1. Thermodynamics. I. Title.
QD501.H573 1986 541.3'69 86-13468

Manufactured in the United States by RR Donnelley
65242416 2015
www.doverpublications.com

To Laura

PREFACE

This is an introductory textbook on equilibrium statistical mechanics. It is called *Statistical Thermodynamics* to distinguish it from the author's earlier work, *Statistical Mechanics* (McGraw-Hill, 1956). The latter is a treatise covering selected topics in detail and at a relatively advanced level; it is neither introductory nor primarily a textbook (except for a second or specialized course). The two volumes in fact complement each other. *Statistical Thermodynamics* (or some other introductory textbook, such as Rushbrooke's) is prerequisite to *Statistical Mechanics.* Where there is overlapping (e.g., imperfect gases, distribution functions, Ising problem, etc.), the discussion given in the present work is more elementary, less general, and should serve to smooth the way for a more detailed study in *Statistical Mechanics.*

Although introductory, *Statistical Thermodynamics* provides a quite extensive coverage of topics of current interest in equilibrium statistical mechanics. This is its principal justification for existence. However, nonequilibrium statistical mechanics has been omitted because: (a) the foundations are not yet established; (b) it is difficult to discuss this subject on the elementary level used in the rest of the book; and (c) the author is not expert enough in this field, at least at the present time, to do it justice. The reader will find treatments of nonequilibrium problems in Kittel (see the list of references at the end of this Preface) and, much more extensively, in a book by I. Prigogine.*

Up to, say, 1945, the usual course in statistical thermodynamics in chemistry departments was concerned primarily or entirely with how to calculate thermodynamic functions of ideal gases from spectroscopic data. Such a choice of subject matter was not inappropriate, for this was the principal area of interest and research in statistical thermodynamics in the 1930's. However, the author feels that a modern introductory course in statistical thermodynamics should reflect the developments of the 1940's and 1950's. For this reason, the more traditional material (Chapters 4 and 8 through 10) referred to above is given a rather condensed treatment to provide room for a survey of more recent advances.

The book is divided into four parts. Part I (Chapters 1 and 2) is concerned with the principles or postulates of statistical mechanics. The argument is based on elementary quantum-mechanical ideas such as

* I. Prigogine, *Nonequilibrium Statistical Mechanics.* New York: Interscience, 1961.

energy levels, states and eigenfunctions, degeneracy, etc. The principles of classical statistical mechanics are almost entirely omitted because these follow as a limiting form of quantum statistics and also because the classical theory is more sophisticated and difficult in many ways than the quantum approach (at least in the elementary form given here). The interested reader should consult the very elegant exposition of the principles of classical statistical mechanics given by Tolman (see the reference list below). Certainly any student who intends to pursue statistical mechanics beyond a first course must turn to a study of Tolman sooner or later, preferably sooner.

Part II (Chapters 3–13) contains applications of the principles developed in Part I to systems of independent molecules (e.g., an ideal gas) or of other independent subsystems. The more complicated but also more interesting problems which arise when molecules can no longer be treated as independent of each other (because of intermolecular forces) provide the subject matter of Part III (Chapters 14–21).

Most of the applications in Parts II and III have to do with the classical (high-temperature) limit of quantum statistics. Part IV (Chapter 22) is concerned with problems for which the classical limit is *not* valid (e.g., helium gas at low temperatures). The sections of Chapter 22 could have been distributed among the appropriate chapters of Parts II and III (some instructors may in fact prefer to use this order), but the present arrangement has the advantage (for many students) that only the most rudimentary quantum-mechanical background is necessary in Chapters 1–21. Somewhat more quantum mechanics is required in Chapter 22 but still not very much.

The level of the first 16 chapters is fairly uniform. Of these, Chapter 1 may seem the most difficult, at least on first reading, because it is necessarily the most general. It may be wise for the average student to return to this chapter for rereading after having acquired some familiarity with applications in later chapters. Parts of Chapters 17–22 are somewhat more advanced than the earlier chapters. For this reason, and because more recent contributions are involved, there are rather more references to the research literature in the last part of the book.

Throughout the text, the intent, with each topic treated, is *not* to give a complete discussion that brings the reader up to the research frontier on the subject, but rather to give an *introduction* only. Usually this means that a somewhat approximate first-order theory is outlined. The most recent details and refinements are intentionally omitted to keep the book within reasonable compass and also because the level would otherwise be too advanced for a beginning textbook. As a next step in pursuing any given subject to a more advanced level, the student should consult the works listed at the end of each chapter and the literature references.

An author index has been omitted because many of the more advanced and important references have not been included in the text—for reasons just mentioned. In this connection an apology should be made for the disproportionate number of references to the author's own papers. This is a consequence not of the relative importance of these papers but rather of their relative simplicity and of the author's familiarity with them.

The arrangement of chapters is such that a great deal of flexibility is possible in adopting the book as a text. As examples, we list below chapters and sections which might be used for (a) a one-semester graduate course in chemistry; (b) a two-semester graduate course in chemistry; and (c) a one-semester senior or graduate course in physics.

Chemistry *One Semester* (3 *hours*)	Chemistry *Two Semesters* (3 *hours*)	Physics *One Semester* (3 *hours*)
Chapter 1	Chapters 1 through 16	Chapters 1–4
Sections 2–3 and 2–4	Sections 17–1, 17–2, and 17–4	Sections 22–1 through 22–4
Chapters 3 through 6	Section 18–1	Chapters 5 and 6
Sections 7–1 and 7–2	Section 19–1	Section 7–1
Chapters 8 through 11	Section 20–1	Chapter 8
Sections 14–1, 14–4, and 14–6	Sections 21–1 and 21–2	Section 22–8
Sections 15–1 and 15–2	Sections 22–1 through 22–4, and 22–8	Chapters 9 and 10
Chapter 16		Chapter 12
Section 17–1		Chapters 14 through 16
Section 18–1		Sections 17–1 and 17–2
Section 20–1		Section 18–1
Sections 22–1 and 22–8		

It will be obvious from the choice of subjects that the author has had physical chemists (and physical biochemists) particularly in mind in writing the book. However, as just indicated above, by suitable omissions and rearrangements one can easily use the text for an introductory course in statistical mechanics in a physics department.

The reader is assumed to have studied thermodynamics, calculus, elementary differential equations, and elementary quantum mechanics.

The problems vary widely in difficulty. They range from simple numerical exercises to small-scale "research" problems. The first problems listed in each chapter are referred to, in passing, in the text. Many of these contain details that the author feels should not be spelled out in the text but that the reader should verify.

The most important sources of reference for the student are the following:

Band, W., *Introduction to Quantum Statistics*. New York: Van Nostrand, 1955.
Flory, P. J., *Principles of Polymer Chemistry*. Ithaca, N. Y.: Cornell, 1953.

Fowler, R. H., and Guggenheim, E. A., *Statistical Thermodynamics.* Cambridge: 1939.

Guggenheim, E. A., *Mixtures.* Oxford: 1952.

Hildebrand, J. H., and Scott, R. L., *Solubility of Nonelectrolytes.* Third Edition. New York: Reinhold, 1950.

Hill, T. L., *Statistical Mechanics.* New York: McGraw-Hill, 1956.

Hirschfelder, J. O., Curtiss, C. F., and Bird, R. B., *Molecular Theory of Gases and Liquids.* New York: Wiley, 1954.

Kittel, C., *Elementary Statistical Physics.* New York: Wiley, 1958.

Landau, L. D., and Lifshitz, E. M., *Statistical Physics.* Reading, Mass.: Addison-Wesley, 1958.

Mayer, J. E., and Mayer, M. G., *Statistical Mechanics.* New York: Wiley, 1940.

Prigogine, I., *Molecular Theory of Solutions.* Amsterdam: North-Holland, 1957.

Rushbrooke, G. S., *Introduction to Statistical Mechanics.* Oxford: 1949.

Schrödinger, E., *Statistical Thermodynamics.* Cambridge: 1948.

Slater, J. C., *Introduction to Chemical Physics.* New York: McGraw-Hill, 1939.

ter Haar, D., *Elements of Statistical Mechanics.* New York: Rinehart, 1954.

Tolman, R. C., *Principles of Statistical Mechanics.* Oxford: 1938.

Wilson, A. H., *Thermodynamics and Statistical Mechanics.* Cambridge: 1957.

Note. A supplementary reading list is given at the end of each chapter. To avoid repetition, the above works are listed by authors only. The present writer's earlier book is referred to throughout the present text as S. M.

The author is greatly indebted to Dr. Dirk Stigter and Mr. Robert E. Salomon for reading the entire manuscript and making many helpful suggestions. Parts of the manuscript were also read and valuable criticism given by Professors George Pimentel, Nobuhiko Saito, and Tsunenobu Yamamoto. While writing the book, the author had the benefit of many stimulating discussions with Professors Saito and Yamamoto. Finally, the author wishes to express his appreciation for partial support from the Alfred P. Sloan Foundation during the period in which the book was written.

February 1960 T. L. H.

CONTENTS

PART III. SYSTEMS OF INTERACTING MOLECULES

Part I

Principles of Quantum Statistical Mechanics

CHAPTER 1

STATISTICAL-MECHANICAL ENSEMBLES AND THERMODYNAMICS

1-1 Introduction. The object of thermodynamics is to derive mathematical relations which connect different experimental properties of macroscopic systems in equilibrium—systems containing many molecules, of the order of, say, 10^{20} or more. However useful, these interconnections of thermodynamics give us no information at all concerning the interpretation or explanation, on a molecular level, of the observed experimental properties. For example, from thermodynamics we know that experimental values of the two heat capacities C_p and C_V for a given system must be interrelated by an exact and well-known equation, but thermodynamics is unable to furnish any explanation of why particular experimental values of either C_p or C_V, taken separately, should be observed. Such an explanation falls rather within the province of statistical mechanics or statistical thermodynamics, terms which we regard in this book as synonymous. That is, *the object of statistical mechanics is to provide the molecular theory or interpretation of equilibrium properties of macroscopic systems.* Thus the fields covered by statistical mechanics and thermodynamics coincide. Whenever the question "why?" is raised in thermodynamics—why, for example, a given equilibrium constant, Henry's law constant, equation of state, etc., is observed—we are presented with a problem in statistical mechanics.

Although thermodynamics itself does not provide a molecular picture of nature, this is not always a disadvantage. Thus there are many complicated systems for which a molecular theory is not yet possible; but regardless of complications on the molecular level, thermodynamics can still be applied to such systems with confidence and exactness.

In recent years both thermodynamics and statistical mechanics have been extended somewhat into the nonequilibrium domain. However, the subject is new and changing, and the foundations are still a little shaky; hence we omit this area from our consideration. An exception is the theory of absolute reaction rates, which we discuss in Chapter 11. This approximate theory is based on a quasi-equilibrium approach which makes it possible to include the theory within the framework of equilibrium statistical mechanics.

Aside from the postulates of statistical mechanics themselves, to be introduced in the next section, the foundation on which our subject is based is quantum mechanics. If we seek a molecular interpretation of the properties of a system containing many molecules, as a starting point we must certainly be provided with knowledge of the properties of the individual molecules making up the system and of the nature of the interactions between these molecules. This is information which can in principle be furnished by quantum mechanics but which in practice is usually obtained from experiments based on the behavior of individual molecules (e.g., spectroscopy), pairs of molecules (e.g., the second virial coefficient of an imperfect gas), etc.

Although quantum mechanics is prerequisite to statistical mechanics, fortunately a reasonably satisfactory version of statistical mechanics can be presented without using any quantum-mechanical concepts other than those of quantum-mechanical states, energy levels, and intermolecular forces. Only in Part IV of the book is it necessary to go beyond this stage.

Another very helpful simplification is that the classical limit of quantum mechanics can be used, without appreciable error, in most problems involving significant intermolecular interactions. Problems of this type are very difficult without this simplification (Part IV).

Despite our extensive use of classical statistical mechanics in the applications of Parts II and III, we introduce the principles of statistical mechanics, beginning in the next section, in quantum-mechanical language because the argument is not only more general but is actually much simpler this way.

1–2 Ensembles and postulates. As mentioned above, our problem is to calculate macroscopic properties from molecular properties. Our general approach is to set up postulates which allow us to proceed directly with this task insofar as "mechanical" thermodynamic properties are concerned; the "nonmechanical" properties are then handled indirectly by an appeal to thermodynamics. By "mechanical" properties we mean, for example, pressure, energy, volume, number of molecules, etc., all of which can be defined in purely mechanical terms (quantum or classical) without, for example, introducing the concept of temperature. Examples of "nonmechanical" thermodynamic variables are temperature, entropy, free energy (Gibbs or Helmholtz), chemical potential, etc.

Let us consider the pressure as a typical mechanical variable. In principle, if we wished to calculate the pressure in a thermodynamic system from molecular considerations, we would have to calculate (by quantum or possibly classical mechanics) the force per unit area exerted on a wall of the system, taking into account the change in the state of the whole system with time. The force itself would be a function of time. What we would need, therefore, is a time average of the force over a period of time sufficiently long to smooth out fluctuations, i.e., sufficiently long to give a time average which is independent, say, of the starting time, $t = t_0$, in the averaging. Because of the tremendous number of molecules in a typical system, and the fact that they interact with each other, such a hypothetical calculation is of course completely out of the question in either quantum or classical mechanics.

Therefore we are forced to turn to an alternative procedure, the ensemble method of Gibbs, based on postulates connecting the desired time average of a mechanical variable with the ensemble average (defined below) of the same variable. The validity of these postulates rests on the agreement between experiment and deductions (such as those in this book) made from the postulates. So far, there is no experimental evidence available that casts doubt on the correctness of the postulates of statistical mechanics.

Before stating the postulates, we must introduce the concept of an ensemble of systems. An ensemble is simply a (mental) collection of a very large number $\mathfrak{N}$ of systems, each constructed to be a replica on a thermodynamic (macroscopic) level of the actual thermodynamic system whose properties we are investigating. For example, suppose the system of interest has a volume V, contains N molecules of a single component, and is immersed in a large heat bath at temperature T. The assigned values of N, V, and T are sufficient to determine the thermodynamic state of the system. In this case, the ensemble would consist of $\mathfrak{N}$ systems, all of which are constructed to duplicate the thermodynamic state (N, V, T) and environment (closed system immersed in a heat bath) of the original system. Although all systems in the ensemble are identical from a thermodynamic point of view, they are not all identical on the molecular level. In fact, in general, there is an *extremely* large number of quantum (or classical) states consistent with a given thermodynamic state. This is to be expected, of course, since three numbers, say the values of N, V, and T, are quite inadequate to specify the detailed molecular (or "microscopic") state of a system containing something in the order of 10^{20} molecules.

Incidentally, when the term "quantum state" is used here, it will be understood that we refer specifically to *energy* states (i.e., energy eigenstates, or stationary states).

At any instant of time, in an ensemble constructed by replication of a given thermodynamic system in a given environment, many different quantum states are represented in the various systems of the ensemble. In the example mentioned above, the calculated instantaneous pressure would in general be different in these different quantum states. The "ensemble average" of the pressure is then the average over these instantaneous values of the pressure, giving the same weight to *each system* in the ensemble in calculating the average. A similar ensemble average can be calculated for any mechanical variable which may have different values (i.e., which is not held constant) in the different systems of the ensemble.

We now state our *first postulate: the (long) time average of a mechanical variable M in the thermodynamic system of interest is equal to the ensemble average of M, in the limit as $\mathfrak{N} \to \infty$, provided that the systems of the ensemble replicate the thermodynamic state and environment of the actual system of interest.* That is, this postulate tells us that we may replace a time average on the one actual system by an instantaneous average over a large number of systems "representative" of the actual system. The first postulate by itself is not really helpful; we need in addition, in order to actually compute an ensemble average, some information about the relative probability of occurrence of different quantum states in the systems of the ensemble. This information must be provided in a second postulate.

Note that the ensemble average of M in the limit as $\mathfrak{N} \to \infty$, referred to above, must be independent of time. Otherwise the original system which the ensemble "represents" is not in equilibrium.

We shall work out details in this chapter for the three most important thermodynamic environments: (a) *an isolated system* (N, V, and E given, where E = energy); (b) *a closed, isothermal system* (N, V, and T given); and (c) *an open, isothermal system* (μ, V, and T given, where μ = chemical potential). N and μ stand for the sets $N_1, N_2, \ldots$ and $\mu_1, \mu_2, \ldots$ if the system contains more than one component. Also, V might stand for a set of "external variables"* if there are more than one. The representative ensembles in the above three cases are usually called *microcanonical*, *canonical*, and *grand canonical*, respectively. The first postulate is applicable to all these cases and to other ensembles which will be introduced in Section 1–7. The second postulate, however, can be limited to a statement concerning only the microcanonical ensemble. The corresponding statement for other ensembles can then be deduced (as in Section 1–3, for example) from this limited second postulate without any further assumptions.

Our *second postulate* is: *in an ensemble ($\mathfrak{N} \to \infty$) representative of an isolated thermodynamic system, the systems of the ensemble are distributed uniformly, that is, with equal probability or frequency, over the possible quantum states consistent with the specified values of N, V, and E.* In other words, each quantum state is represented by the same number of systems in the ensemble; or, if a system is selected at random from the ensemble, the probability that it will be found in a particular quantum state is the same for all the possible quantum states. A related implication of this postulate, when combined with the first postulate, is that the single isolated system of actual interest (which serves as the prototype for the

---

* There is one "external variable" for each thermodynamic work term, e.g., volume, area, length, etc.

systems of the ensemble) spends equal amounts of time, over a long period of time, in each of the available quantum states. This last statement is often referred to as the quantum "ergodic hypothesis," while the second postulate by itself is usually called the "principle of equal *a priori* probabilities." The ergodic hypothesis in classical statistical mechanics is mentioned at the end of Section 6–3. For a more detailed discussion, see Tolman, pp. 63–70 and 356–361. (For full identification of works referred to by only the author's last name, see Preface.)

The value of E in the second postulate must be one of the energy levels of the quantum-mechanical system defined by N and V. Since N is extremely large, the energy levels for such a system will be so close together as to be practically continuous, and furthermore, each of these levels will have an extremely high degeneracy. We shall in general denote the number of quantum states (i.e., the degeneracy) associated with the energy level E for a quantum-mechanical system with N and V by $\Omega(N, V, E)$. Thus the number of "possible quantum states" referred to in the second postulate is Ω.

A complication in the above discussion is the fact that, from an operational point of view, E cannot be known precisely; there will always be a small uncertainty δE in the value of E. For all thermodynamic purposes this complication is completely inconsequential.* Hence for the sake of simplicity we ignore it.

It should also be mentioned that the point of view in the above statement of the second postulate is not so general as it might be. If the energy level E for the system N, V has a degeneracy Ω, there are Ω orthogonal (and therefore linearly independent) wave functions ψ which satisfy the Schrödinger equation $\mathcal{H}\psi = E\psi$. The particular choice of the Ω ψ's is somewhat arbitrary, since other possible choices can always be set up by forming suitable linear combinations of the ψ's in the first choice. In any case, the "Ω quantum states" mentioned in connection with the second postulate refers to some set of orthogonal ψ's all "belonging" to the same E. But regardless of the set of ψ's chosen, the wave function representing the actual quantum-mechanical state of any system selected from the ensemble will in general *not* be one of the chosen set of ψ's, but will be some linear combination of all of them. The contrary is really implied in the above statement of the second postulate. Fortunately, this simplification in our statement of the postulate makes no difference† in any deductions we shall make that can be compared with experiment.

---

* See S. M. (the present author's earlier work identified in the Preface), p. 113, and Mayer and Mayer, pp. 55–56, 100–102.

† See S. M., pp. 50–55, 79.

We turn now to a derivation from the above two postulates of the essential properties of the canonical and grand ensembles.

1-3 Canonical ensemble. The experimental system of interest here has a fixed volume V, fixed numbers of molecules N (which stands for $N_1, N_2, \ldots$ in a multicomponent system), and is immersed in a very large heat bath at temperature T. The heat bath is assumed "very large" to be consistent with the use of the limit $\mathfrak{N} \to \infty$ below. Our first objective is to set up the machinery necessary for calculating the average value of mechanical variables, such as energy and pressure, in the system. In view of the first postulate, this means that we need to be able to calculate the *ensemble* average of such variables. This, in turn, can be done if we know the value of the particular variable in question in a given quantum state and the fraction of systems in the ensemble which are in this quantum state. It might be noted that because the thermodynamic system here is not isolated but is in contact with a heat bath, the energy of the system can fluctuate; therefore quantum states belonging to *different* energy levels E will have to be reckoned with. Since mechanical variables have well-defined values in a given quantum state (in fact we can use this property as the definition of a "mechanical variable"), the task that remains is to determine the fraction of systems in the ensemble in a given quantum state (or the probability that a system selected arbitrarily from the ensemble is in a given quantum state). This is the problem we now consider.

The experimental, or prototype, system is in a very large heat bath at temperature T. Therefore each system in the ensemble representative of the experimental system must also be in a very large heat bath at T. Specifically, we contemplate the following arrangement, which satisfies this requirement. We imagine $\mathfrak{N}$ macroscopic systems as our ensemble, each with N and V (duplicating the values in the experimental system), stacked together in a lattice (Fig. 1-1). The walls between the different systems in the ensemble are heat conducting, but impermeable to all molecules. To establish the temperature T, we imagine further that the entire stack of systems (i.e., the ensemble) is placed in a sufficiently large heat bath at T. After equilibrium is reached, thermal insulation (represented schematically by the double lines in Fig. 1-1) is placed on the outside walls of the ensemble, and the ensemble is removed from the heat bath. The entire ensemble itself is now an isolated system with volume $\mathfrak{N}V$, numbers of molecules $\mathfrak{N}N$, and a total energy which we shall denote by E_t (t = total). The relation between E_t and the temperature T will emerge later. Observe that each system in the ensemble is immersed in a large (we shall later use the limit $\mathfrak{N} \to \infty$) heat bath at temperature T, as is required if the ensemble is to be representative of the original thermo-

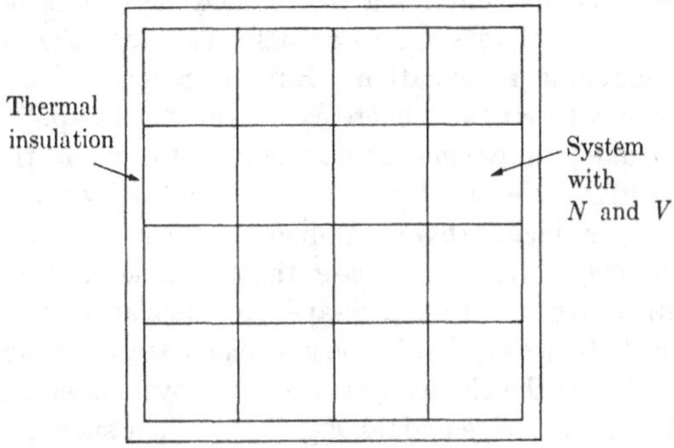

FIG. 1-1. Canonical ensemble of $\mathfrak{N}$ systems, each with N and V.

dynamic system. That is, the remaining $\mathfrak{N} - 1$ systems in the ensemble serve as the heat bath for any one selected system.

At this point we come to the essential step in the argument, which is to note that since the ensemble itself is an isolated system, *we can apply the second postulate to the whole ensemble.* Thus the entire canonical ensemble shown in Fig. 1-1 is now regarded as a prototype thermodynamic system, characterized by the variables $\mathfrak{N}V$, $\mathfrak{N}N$, and E_t. We shall refer to this system as a "supersystem" in order to avoid confusion with the original experimental closed, isothermal system. The second postulate tells us, then, that every possible quantum state of this supersystem (canonical ensemble) is equally probable and hence should be given equal weight in the calculation of average values of interest. As we show next, it is possible to take advantage of this observation in order to find the required probability of occurrence of a given quantum state in the systems of a canonical ensemble.

We return now to a single system in the canonical ensemble. As a quantum-mechanical system, it is characterized by N and V. Let us list all possible energy states for such a system in increasing order of the energy eigenvalue, $E_1, E_2, \ldots, E_j, \ldots$. Here, for later convenience, each state is listed separately so that when degeneracy occurs several successive E_j's will have the same value. For example, in the notation used above and to which we shall return later, the energy value E occurs Ω successive times in the list.

Each E_j is a function of N and V. If V is changed infinitely slowly, each E_j changes in a continuous manner. However, the number of molecules of any one of the components can be changed only discontinuously—one molecule at a time. Hence the energy levels must jump discontinuously if N changes.

For most systems containing many molecules, it is not possible, for purely mathematical reasons, to actually calculate the energies E_1, E_2, ... from Schrödinger's equation. But for generality, we assume in the present argument that these energies are known. In applications, we shall be able to make progress in each case only to the extent that we can overcome this difficulty either by approximation, or by use of classical mechanics, or by reducing this many-body problem to a one-, two-, three-, ... body problem, etc. In any case, the ensemble method has the following advantage over a direct time-average calculation on a single system (see Section 1–2): we need only the stationary states of the system and do not have to follow the change in state of the system with time.

The list of energy eigenvalues E_1, E_2, ... is assumed, then, to be the correct list for any given problem. The argument that follows is valid irrespective of such complications as intermolecular forces, symmetry restrictions on wave functions, etc.

Since each system in the canonical ensemble has the same N and V, all systems have the same set of energy states, represented by $E_1, E_2, \ldots,$ $E_j, \ldots$. Now suppose we observe, simultaneously, the energy state of each system in the ensemble, and count the number of systems found in each of the listed states. We let n_1 be the number of systems found in state $E_1, \ldots, n_j$ in state E_j, etc. The set of numbers $n_1, n_2, \ldots$ is called a "distribution." There are, of course, many possible distributions that might be observed, but obviously all must satisfy the relations

$$\sum_j n_j = \mathfrak{N}, \tag{1–1}$$

$$\sum_j n_j E_j = E_t. \tag{1–2}$$

The individual systems in the supersystem (canonical ensemble) are macroscopic in size, are arranged in a certain order, and can be separately labeled. Then the energy state of the whole supersystem would be completely specified if we indicated the system energy state (i.e., $E_1, E_2, \ldots$) for each of the (labeled) systems in the supersystem. To take a simple example, suppose there are four systems (A, B, C, D) in the supersystem ($\mathfrak{N} = 4$) and the possible energy states for each system are E_1, E_2, and E_3. Then one possible energy state for the supersystem would be, say,

A	B	C	D
E_2	E_3	E_2	E_1

provided that (compare Eq. 1–2)

$$E_1 + 2E_2 + E_3 = E_t \text{ (preassigned)}.$$

Here $n_1 = 1$, $n_2 = 2$, $n_3 = 1$. Actually, there are 12 possible states of the supersystem consistent with this distribution. Three of them are:

A	B	C	D
E_2	E_2	E_3	E_1
E_2	E_3	E_2	E_1
E_3	E_2	E_2	E_1

But there are four sets of this type, corresponding to the four possible assignments of E_1. In general, the number of states of the supersystem, $\Omega_t(n)$, consistent with a given distribution $n_1, n_2, \ldots$ (n represents the entire set $n_1, n_2, \ldots$) is given by the well-known combinatorial formula

$$\Omega_t(n) = \frac{(n_1 + n_2 + \cdots)!}{n_1!n_2!\cdots} = \frac{\mathfrak{N}!}{\prod_j n_j!}. \tag{1-3}$$

Recall that we are attempting to find the probability of observing a given quantum state (say E_j) in a system selected from a canonical ensemble (or the fraction of systems in the ensemble in the state E_j). For a particular distribution $n_1, n_2, \ldots,$ this probability or fraction is just $n_j/\mathfrak{N}$ for state E_j. But, in general, there are very many possible distributions for given N, V, $\mathfrak{N}$, and E_t. What we need is the over-all probability; that is, an average of $n_j/\mathfrak{N}$ over these distributions, based on an assignment of equal weight to each state of the supersystem. Assignment of equal weights to supersystem states implies immediately that the weight assigned to each distribution, in calculating an average over different distributions, should be proportional to $\Omega_t(n)$ for the distribution.

Now consider the numerical example above, and suppose further that there are just two distributions which satisfy the conditions of Eqs. (1-1) and (1-2), namely,

$$n_1 = 1,\ n_2 = 2,\ n_3 = 1, \qquad \Omega_t = 12,$$
$$n_1 = 2,\ n_2 = 0,\ n_3 = 2, \qquad \Omega_t = 6.$$

The probability of observing E_3 is $\frac{1}{4}$ in the first distribution and $\frac{1}{2}$ in the second distribution, while the over-all probability is $\frac{1}{3}$:

$$\overline{n}_3 = \frac{1 \times 12 + 2 \times 6}{12 + 6} = \frac{4}{3}, \qquad \frac{\overline{n}_3}{\mathfrak{N}} = \frac{1}{3}.$$

In general, the required probability of observing a given quantum state E_j in an arbitrary system of a canonical ensemble is

$$P_j = \frac{\overline{n}_j}{\mathfrak{N}} = \frac{1}{\mathfrak{N}} \frac{\sum_n \Omega_t(n) n_j(n)}{\sum_n \Omega_t(n)}, \tag{1-4}$$

where $n_j(n)$ means the value of n_j in the distribution n. The sum is over all distributions satisfying Eqs. (1–1) and (1–2). Of course, by definition, $\sum_j P_j = 1$.

Then the desired ensemble averages of, for example, the energy and pressure are

$$\overline{E} = \sum_j P_j E_j \tag{1–5}$$

and

$$\overline{p} = \sum_j P_j p_j, \tag{1–6}$$

where p_j is the pressure in state E_j, defined by

$$p_j = -\left(\frac{\partial E_j}{\partial V}\right)_N. \tag{1–7}$$

That is, $-p_j\,dV = dE_j$ is the work that has to be done on the system, when in the state E_j, in order to increase the volume by dV.

In principle, Eq. (1–4) for P_j tells us all we need to know to calculate canonical ensemble averages of mechanical variables. But in practice, a much more explicit expression for P_j is necessary. We must now face this problem.

The most elegant way to proceed is to employ the Darwin-Fowler technique,* based on the mathematical method of steepest descents. However, in the present discussion, since we can take $\mathfrak{N} \to \infty$, the so-called maximum-term method, which involves the use of undetermined multipliers, is equally rigorous though not so elegant. The latter method, which we shall use, has the important advantage of requiring much less of the reader in the way of mathematical background.

In any particular case we are given $\mathfrak{N}$, the E_j (determined by N and V), and E_t (determined by $\mathfrak{N}$, N, V, and T). There are then many possible distributions n consistent with the restrictions of Eqs. (1–1) and (1–2). For each of these distributions we can calculate from Eq. (1–3) the weight $\Omega_t(n)$ to be used in obtaining averages, as already explained. The situation here parallels exactly that illustrated in Appendix II. That is, because of the large numbers involved (the present example is ideal in this respect because we can take the limit $\mathfrak{N} \to \infty$), the most probable distribution, and distributions which differ only negligibly from the most probable distribution, completely dominate the computation of the average in Eq. (1–4). By the most probable distribution, denoted by n^*, we mean of course that distribution to which the largest $\Omega_t(n)$ belongs. In effect this means that, in the limit as $\mathfrak{N} \to \infty$, we can regard all other weights $\Omega_t(n)$

* See, for example, Schrödinger, Chapter 6.

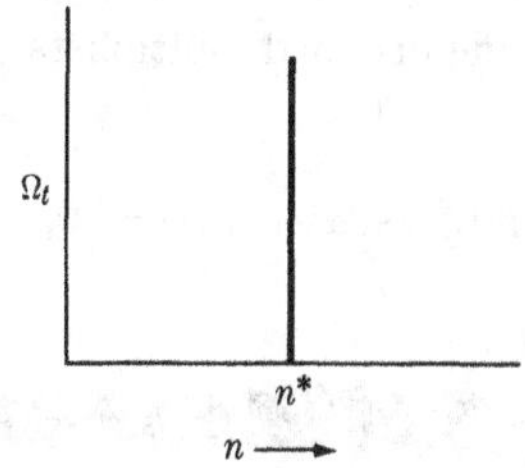

FIG. 1-2. Number of states Ω_t as a function of the distribution n (schematic).

as negligible compared with $\Omega_t(n^*)$. This is illustrated diagrammatically in Fig. 1-2. With $\mathfrak{N}$ large but finite, there would be a narrow gaussian distribution centered about $n = n^*$. But in the limit as $\mathfrak{N} \to \infty$, this distribution becomes completely sharp (a Dirac δ-function).

Naturally, as we let $\mathfrak{N} \to \infty$ (i.e., increase the size of the ensemble), holding N, V, and T fixed, each $n_j \to \infty$ also. But all ensemble averages depend only on the ratio $n_j/\mathfrak{N}$, which remains finite.

Equation (1-4) becomes, then,

$$P_j = \frac{\bar{n}_j}{\mathfrak{N}} = \frac{1}{\mathfrak{N}} \frac{\Omega_t(n^*)n_j^*}{\Omega_t(n^*)} = \frac{n_j^*}{\mathfrak{N}}, \tag{1-8}$$

where n_j^* is the value of n_j in the most probable distribution, n^*. Equation (1-8) tells us that in the computation of P_j we can replace the mean value of n_j by the value of n_j in the most probable (largest Ω_t) distribution. This leads us to a purely mathematical question: Which of all possible sets of n_j's satisfying Eqs. (1-1) and (1-2) gives us the largest Ω_t?

We solve this problem by the method of undetermined multipliers (see Appendix III). The distribution giving the largest Ω_t is also the distribution giving the largest $\ln \Omega_t$, since $\ln x$ increases monotonically with x. We work with $\ln \Omega_t$ instead of Ω_t because it is more convenient. From Eq. (1-3),

$$\ln \Omega_t(n) = \left(\sum_i n_i\right) \ln \left(\sum_i n_i\right) - \sum_i n_i \ln n_i,$$

where we have used Stirling's approximation (Appendix II) and changed the running index from j to i. This "approximation" is in fact exact here because we are interested in the limit $\mathfrak{N}, n_i \to \infty$. According to the method of undetermined multipliers, the set of n_j's which leads to the maximum value of $\ln \Omega_t(n)$, subject to the conditions (1-1) and (1-2), is found from the equations

$$\frac{\partial}{\partial n_j}\left[\ln \Omega_t(n) - \alpha \sum_i n_i - \beta \sum_i n_i E_i\right] = 0, \qquad j = 1, 2, \ldots,$$

where α and β are the undetermined multipliers. On carrying out the differentiation, we find

$$\ln\left(\sum_i n_i\right) - \ln n_j^* - \alpha - \beta E_j = 0, \qquad j = 1, 2, \ldots,$$

or

$$n_j^* = \mathfrak{N} e^{-\alpha} e^{-\beta E_j}, \qquad j = 1, 2, \ldots. \tag{1-9}$$

This is the most probable distribution, expressed in terms of α and β. If desired, $\mathfrak{N}$ may be substituted for $\sum_i n_i$ in $\ln \Omega_t(n)$ at the outset, and treated as a constant in the differentiation. This will change the meaning of α, but not any physical results.

The straightforward procedure here is to substitute the distribution (1–9) into Eqs. (1–1) and (1–2) in order to determine α and β as functions of $\mathfrak{N}$ and E_t, or of $\mathfrak{N}$ and $\overline{E}$ (since obviously $E_t = \mathfrak{N}\overline{E}$). The result is

$$e^{\alpha} = \sum_j e^{-\beta E_j}, \tag{1-10}$$

$$\overline{E} = \frac{\sum_j E_j e^{-\beta E_j}}{\sum_j e^{-\beta E_j}}, \tag{1-11}$$

where $\mathfrak{N}$ has dropped out of both equations. Equation (1–11) provides β as an implicit function of $\overline{E}$ (and also of N and V, since the energies E_j are functions of N and V). Equation (1–10) then gives α in terms of β (and N, V). However, the independent variables of real interest here are N, V, T rather than N, V, $\overline{E}$, and we have no information yet about the dependence of $\overline{E}$ on T. Hence we do not pursue the above approach any further (see Problem 1–2, however), but turn instead, in Section 1–4, to a thermodynamic argument which provides a direct connection between β and T.

We note in passing that elimination of $e^{-\alpha}$ in Eq. (1–9) by use of Eq. (1–10) (or comparison of Eqs. 1–5 and 1–11) gives us P_j as a function of β, N, and V:

$$P_j = \frac{n_j^*}{\mathfrak{N}} = \frac{e^{-\beta E_j(N,V)}}{\sum_i e^{-\beta E_i(N,V)}}, \qquad j = 1, 2, \ldots. \tag{1-12}$$

Anticipating the fact that β turns out to be a positive number, we deduce from this equation that the probability of observing a given quantum state in a canonical ensemble decreases exponentially with the energy of the quantum state.

1–4 Canonical ensemble and thermodynamics. To bring nonmechanical thermodynamic variables such as temperature and entropy into our discussion, we now combine the above "mechanical" considerations with thermodynamics. In the first place, by virtue of the first postulate, we can associate the thermodynamic pressure p and energy E with the statistical-mechanical ensemble averages $\bar{p}$ and $\bar{E}$. Let us take the differential of $\bar{E}$ in Eq. (1–5), holding N constant (the system being closed):

$$d\bar{E} = \sum_j E_j \, dP_j + \sum_j P_j \, dE_j$$

$$= -\frac{1}{\beta} \sum_j (\ln P_j + \ln Q) \, dP_j + \sum_j P_j \left(\frac{\partial E_j}{\partial V}\right)_N dV, \quad (1\text{–}13)$$

where we have defined

$$Q = \sum_j e^{-\beta E_j}, \quad (1\text{–}14)$$

used Eq. (1–12) in the first sum, and have recognized in the second sum that $E_j(N, V)$ can vary only with V if N is fixed. The first sum simplifies further in view of the relations

$$\sum_j P_j = 1, \qquad \sum_j dP_j = 0,$$

and

$$d\left(\sum_j P_j \ln P_j\right) = \sum_j \ln P_j \, dP_j.$$

Thus, using Eq. (1–6), we can write

$$-\frac{1}{\beta} d\left(\sum_j P_j \ln P_j\right) = d\bar{E} + \bar{p} \, dV. \quad (1\text{–}15)$$

Since we already have the associations with thermodynamics $E \leftrightarrow \bar{E}$ and $p \leftrightarrow \bar{p}$, and since in thermodynamics (N constant)

$$T \, dS = dE + p \, dV,$$

we can deduce from Eq. (1–15) the further association

$$T \, dS \leftrightarrow -\frac{1}{\beta} d\left(\sum_j P_j \ln P_j\right). \quad (1\text{–}16)$$

With these associations established, let us digress to note that from Eq. (1–13) and

$$dE = DQ^* - DW,$$

we have

$$DQ^* = T\,dS \leftrightarrow \sum_j E_j\,dP_j, \tag{1-17}$$

$$DW = p\,dV \leftrightarrow \sum_j P_j\,dE_j, \tag{1-18}$$

where Q^* and W are heat absorbed and work done by the system, respectively. These relations provide us, in a general way, with the molecular interpretation of the thermodynamic concepts of heat and work. We see that when a closed thermodynamic system increases its energy infinitesimally by the absorption of heat from its surroundings, this is accomplished not by changing the energy levels of the system but rather by a shift in the fraction of time the system spends in the various energy states. The converse statement can be made about the work term.

We now return to the main argument, the purpose of which is to relate S to the P_j. From Eq. (1–16),

$$dS \leftrightarrow \frac{1}{\beta T}\,dG, \tag{1-19}$$

where G is defined by

$$G = -\sum_j P_j \ln P_j.$$

From thermodynamics we know that the left side of Eq. (1–19) is an exact differential. Hence the right side must be also. This condition will be met provided that $1/\beta T$ is any function of G, say $\varphi(G)$. That is,

$$dS \leftrightarrow \varphi(G)\,dG = df(G), \tag{1-20}$$

where

$$f(G) = \int \varphi(G)\,dG, \qquad \varphi(G) = \frac{df(G)}{dG}.$$

From Eq. (1–20),

$$S \leftrightarrow f(G) + c, \tag{1-21}$$

where c is an integration constant independent of G and therefore inde-

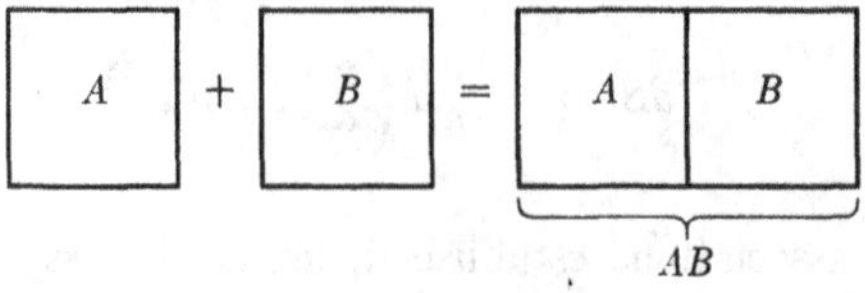

FIG. 1–3. Systems A and B combined to form AB. All systems are at same temperature.

pendent of the variables on which G depends (e.g., β and V, with N constant). In thermodynamic language, c is independent of the thermodynamic state of a closed system. Experimental information about the entropy always involves a *difference* in entropy between two states (e.g., the entropy change ΔS between T_1 and T_2 at constant N and V), never an absolute value. The constant c in Eq. (1-21) always cancels on taking such a difference. Hence its value is completely arbitrary from an operational point of view. But for convenience and simplicity, we adopt the particular choice $c = 0$ from now on. The connection between this choice and the third law of thermodynamics will be discussed in Section 2-4.

Up to this point we have that $S \leftrightarrow f(G)$, but we do not know the function f. To settle this matter we make use of a thermodynamic property of the entropy, namely its additivity. Specifically, suppose we have two thermodynamic systems A and B at the same temperature and with entropies S_A and S_B. Then if we regard the combined systems (Fig. 1-3) as a new system AB, we have $S_{AB} = S_A + S_B$. This relationship suffices to determine f, as we now show.

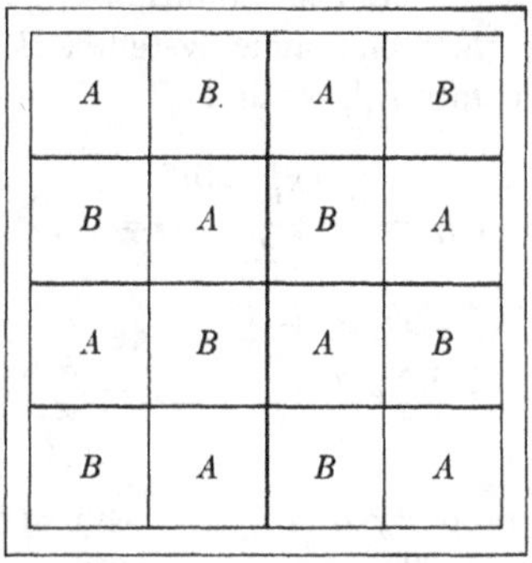

FIG. 1-4. Canonical ensemble of $\mathfrak{N}$ systems, each of type AB.

We first investigate whether the statistical-mechanical quantity G is additive in the above sense. For this purpose we form a canonical ensemble of $\mathfrak{N}$ systems AB (as shown in Fig. 1-4) representative of a thermodynamic (prototype) AB system at temperature T. Heat can flow through all interior walls of the ensemble. The A part of the thermodynamic system is characterized further by N^A and V^A, and the B part by N^B and V^B (A and B are not exponents). In general, the types of molecules may be different in A and B. We have two sets of energy states for the separate systems, $E_1^A, E_2^A, \ldots$ and $E_1^B, E_2^B, \ldots$. If n_j^A stands for the number of A systems in the ensemble in state E_j^A, with a similar meaning for n_j^B, then the number of states of the whole ensemble (Fig. 1-4), or supersystem, consistent with given distributions n^A and n^B is

$$\Omega_t(n^A, n^B) = \frac{(\sum_j n_j^A)!}{\prod_j n_j^A!} \times \frac{(\sum_j n_j^B)!}{\prod_j n_j^B!}, \tag{1–22}$$

since the A and B systems are independent of each other (except for energy exchange through the walls). The distributions of interest must satisfy the equations

$$\sum_j n_j^A = \mathfrak{N}, \qquad \sum_j n_j^B = \mathfrak{N},$$

$$\sum_j (n_j^A E_j^A + n_j^B E_j^B) = E_t.$$

The argument from here on is essentially the same as before, so we omit details (Problem 1–3). The three restrictions above require three undetermined multipliers, α_A, α_B, and β, respectively. We note in particular that because of energy exchange between the A and B systems, only one energy equation and one multiplier β are necessary. For the probability that the thermodynamic system AB has its A part in state E_i^A and its B part in state E_j^B, we find

$$P_{ij} = \frac{e^{-\beta E_i^A} e^{-\beta E_j^B}}{Q_A Q_B} = P_i^A P_j^B, \tag{1–23}$$

where

$$Q_A = \sum_j e^{-\beta E_j^A}, \qquad Q_B = \sum_j e^{-\beta E_j^B}.$$

This multiplicative property of P_{ij} is of course what we should expect from the form of Eq. (1–22). We deduce from Eq. (1–23) that if two systems are in thermal contact with each other (and therefore have the same temperature), they have the same β. This suggests a close connection between β and T, which we verify below.

For the combined system AB,

$$\begin{aligned} G_{AB} &= -\sum_{i,j} P_{ij} \ln P_{ij} \\ &= -\sum_{i,j} P_i^A P_j^B (\ln P_i^A + \ln P_j^B) \\ &= -\sum_i P_i^A \ln P_i^A - \sum_j P_j^B \ln P_j^B \\ &= G_A + G_B. \end{aligned} \tag{1–24}$$

That is, G is additive. Also, since $S_{AB} = S_A + S_B$, we have

$$f(G_{AB}) = f(G_A) + f(G_B).$$

Then, from Eq. (1–24),

$$f(G_A + G_B) = f(G_A) + f(G_B).$$

The question before us becomes, then: Given that

$$f(x + y) = f(x) + f(y), \tag{1–25}$$

what is the function f? Let us differentiate* Eq. (1–25) with respect to x and y:

$$\frac{df(x+y)}{d(x+y)} \frac{\partial(x+y)}{\partial x} = \frac{df(x+y)}{d(x+y)} = \frac{df(x)}{dx},$$

$$\frac{df(x+y)}{d(x+y)} \frac{\partial(x+y)}{\partial y} = \frac{df(x+y)}{d(x+y)} = \frac{df(y)}{dy}.$$

Hence

$$\frac{df(x)}{dx} = \frac{df(y)}{dy}.$$

This says that a certain function of x is equal to the same function of y. But this is only possible if the function is a constant, say k. Then

$$\frac{df(x)}{dx} = k, \qquad f(x) = kx + a,$$

where a is another constant. But we have to choose $a = 0$ in order to satisfy Eq. (1–25). Therefore, finally, we have found that $f(x) = kx$, and that

$$S \leftrightarrow f(G) = kG$$

$$\leftrightarrow -k \sum_j P_j \ln P_j. \tag{1–26}$$

Also, from Eq. (1–20),

$$\frac{1}{\beta T} = \varphi(G) = \frac{df(G)}{dG} = k,$$

or

$$\frac{1}{T} \leftrightarrow \beta k, \qquad \frac{1}{kT} \leftrightarrow \beta. \tag{1–27}$$

---

* This argument is from Schrödinger, p. 13.

The constant k is still unevaluated at this stage. We have seen that if *any* two systems are in thermal contact, they have the same β and T. Therefore they have the same k. What is more, k is a *universal* constant, since one system of the pair, say A, can be retained and B can be varied over all other possible systems, C, D, E, ... The value of k can thus be obtained once and for all by comparing statistical-mechanical and experimental values of the same property, on any convenient system (A, above). The pressure of an ideal gas is usually used. The numerical value of k depends, of course, on the absolute temperature scale employed. We anticipate from our treatment of an ideal gas in Chapter 4 that $k = +1.38044 \times 10^{-16}\ \mathrm{erg \cdot deg^{-1}}$, with the conventional kelvin temperature scale. However, the important fact that k is a positive number can easily be checked here in several ways. For example, if we put $\beta = 1/kT$ in Eq. (1–11), differentiate with respect to T, and use the experimental thermodynamic fact that $C_V = (\partial E/\partial T)_{N,V}$ is always positive, we find that k must be positive (Problem 1–4).

We are now in a position to summarize the basic statistical-mechanical equations that can be used to calculate the thermodynamic properties of a closed, isothermal system. In the first place, the probability that the system is in any particular energy state E_j is

$$P_j(N, V, T) = \frac{e^{-E_j(N,V)/kT}}{Q(N, V, T)}, \tag{1–28}$$

where

$$Q(N, V, T) = \sum_j e^{-E_j(N,V)/kT}. \tag{1–29}$$

We call Q the "canonical ensemble partition function." Because of the association (1–27), the independent thermodynamic variables here turn out to be N, V, and T, which is just the desired set for a closed, isothermal system (see Section 1–3). The entropy is

$$S(N, V, T) = -k \sum_j P_j \ln P_j, \tag{1–30}$$

where P_j is given by Eq. (1–28). If we substitute Eq. (1–28) into Eq. (1–30), we find

$$S = \frac{E}{T} + k \ln Q = \frac{E}{T} - \frac{A}{T},$$

where the last expression is a thermodynamic one (A is the Helmholtz free energy). Therefore

$$A(N, V, T) = -kT \ln Q(N, V, T). \tag{1–31}$$

This equation is particularly useful because A is the "characteristic function" in thermodynamics for the independent variables N, V, T:

$$dA = -S\,dT - p\,dV + \sum_{\alpha} \mu_\alpha\, dN_\alpha. \tag{1–32}$$

Thus,

$$S = -\left(\frac{\partial A}{\partial T}\right)_{V,N} = kT\left(\frac{\partial \ln Q}{\partial T}\right)_{V,N} + k \ln Q, \tag{1–33}$$

$$p = -\left(\frac{\partial A}{\partial V}\right)_{T,N} = kT\left(\frac{\partial \ln Q}{\partial V}\right)_{T,N}, \tag{1–34}$$

$$E = -T^2\left(\frac{\partial A/T}{\partial T}\right)_{V,N} = kT^2\left(\frac{\partial \ln Q}{\partial T}\right)_{V,N}. \tag{1–35}$$

Hence, if the function $Q(N, V, T)$ is available from Eq. (1–29), differentiation of Q yields S, p, and E. Furthermore, despite the fact that Eq. (1–31) was derived from the study of a closed system, we can make use of the thermodynamic equation (1–32) and $Q(N, V, T)$ to deduce the chemical potential of any component, say i, from

$$\mu_i = \left(\frac{\partial A}{\partial N_i}\right)_{T,V,N_{\alpha\neq i}} = -kT\left(\frac{\partial \ln Q}{\partial N_i}\right)_{T,V,N_{\alpha\neq i}}. \tag{1–36}$$

Thus we have a complete set of thermodynamic functions (from which all others can be derived): N, V T; A, E, S, p, μ. Incidentally, whether the averaging bars over E and p in the above equations are dropped or not is optional; it depends on whether one has in mind primarily the thermodynamic or the statistical-mechanical aspect of the equation in question.

The above equations, which allow us to deduce all the thermodynamic properties from Eq. (1–29) for the partition function Q, are general but quite formal. In fact, the reader may feel that these relations are rather useless since, in general, the E_j must be expected to be very difficult to calculate for a system with many molecules. While such an attitude is perhaps justified in complicated cases, there are many systems for which considerable progress of one kind or another can be made. Much of the rest of this book will be devoted to such examples.

For many purposes it is convenient to group together all energy states belonging to the same energy level. Let $\Omega_i(N, V)$ be the number of such states (that is, the degeneracy) for an energy level $E_i(N, V)$. In other words, in the list of energy states $E_1, E_2, \ldots,$ the same value E_i occurs Ω_i times. Then,

$$Q(N, V, T) = \sum_{\substack{j \\ \text{(states)}}} e^{-E_j(N,V)/kT} = \sum_{\substack{i \\ \text{(levels)}}} \Omega_i(N, V)e^{-E_i(N,V)/kT}. \tag{1–37}$$

Also,

$$P\text{ (level)} = \Omega P\text{ (state)} = \frac{\Omega e^{-E/kT}}{Q} \tag{1-38}$$

is the probability that the system exists in the energy level E. We have dropped subscripts here to avoid confusion between i and j. Whether a sum such as one of those occurring in Eq. (1–37) is over "states" or "levels" can always be judged by noticing whether or not degeneracies are included as weights for the so-called Boltzmann factors ($e^{-E_i/kT}$).

We have already mentioned that P_j, being proportional to the Boltzmann factor $e^{-E_j/kT}$, falls off exponentially with increasing E_j. We shall discuss essentially this point in more detail in Chapter 3, but in anticipation we should mention here two important extreme cases:

(a) If $T \to 0$ and the lowest level E_1 is nondegenerate, then

$$Q \to e^{-E_1/kT}[1 + \Omega_2 e^{-(E_2-E_1)/kT} + \cdots] \to e^{-E_1/kT}$$

and

$$P_1 \to 1, \qquad P_j \to 0, \qquad j = 2, 3, \ldots .$$

That is, in the limit as $T \to 0$, the system is certain to be found in the lowest energy state. From Eq. (1–30), $S \to 0$.

(b) If $T \to \infty$, the relative effect of different E_j's on the Boltzmann factors is washed out, and P_j (state) $\to$ constant (independent of j); that is, the probability distribution over states becomes uniform. Then $S \to \infty$, assuming that there is an infinite number of energy states (Problem 1–5).

1–5 Grand canonical ensemble. In this section we suppose that the thermodynamic system of volume V, whose properties we wish to calculate from molecular considerations, is in a large heat bath and is "open" with respect to the molecules in the system. That is, both heat and matter (molecules) can be transported across the walls of the system. The bath provides a reservoir of heat at temperature T and of molecules at chemical potentials $\mu_1, \mu_2, \ldots$. The system is thus characterized by the thermodymamic variables $V, T, \mu_1, \mu_2, \ldots$. The numbers of molecules $N_1, N_2, \ldots$ do not have fixed values, as they do in a closed system, but fluctuate about mean values $\overline{N}_1, \overline{N}_2, \ldots$.

We employ here the same type of argument as for the canonical ensemble: (a) the first postulate permits us to use ensemble averages over mechanical variables in place of time averages on the actual system; (b) by regarding the entire ensemble as an isolated supersystem, we can deduce ensemble-average weighting (probability) factors from the second postulate, in terms of undetermined multipliers; and (c) the significance of the undetermined multipliers, as nonmechanical variables, can then be

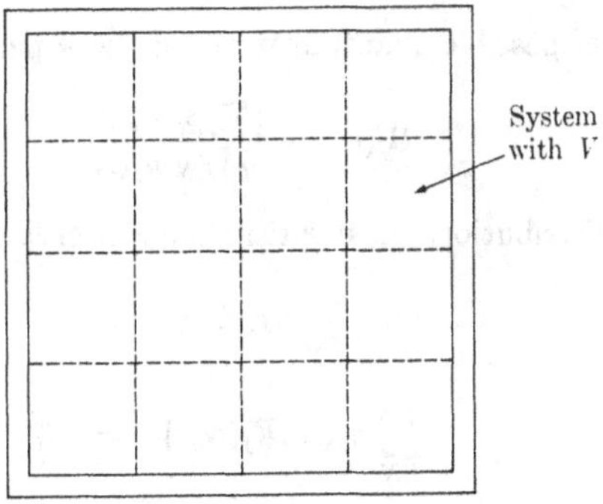

FIG. 1-5. Grand canonical ensemble of $\mathfrak{N}$ systems, each with V.

established by comparing statistical-mechanical and thermodynamic expressions for mechanical variables.

For simplicity, we consider now a one-component system, with V, T, and μ given. As an ensemble (which we call a "grand canonical ensemble") representative of this system, we introduce a lattice (Fig. 1-5) of $\mathfrak{N}$ systems, each with volume V and with walls permeable to molecules (indicated by the dashed lines) and to heat. To establish the desired values of T and μ in each of the $\mathfrak{N}$ systems, we imagine that the whole ensemble is immersed in a giant reservoir at T and μ until equilibrium is reached. We then place walls around the ensemble (solid double lines in Fig. 1-5) that are impermeable to both heat and molecules, and finally remove the ensemble from the reservoir. The ensemble itself is then an isolated supersystem to which the second postulate can be applied. The volume of the supersystem is $\mathfrak{N}V$, and we let E_t and N_t be its total energy and number of molecules.

Since the subsequent details are very similar to those in Sections 1-3 and 1-4, we condense the discussion here. For each value of N, there will be a different set of energy states $E_j(N, V)$. The quantum-mechanical state of the supersystem (ensemble) is specified when we give the value of N and the state $E_j(N, V)$ for each system in the supersystem. In a given state of the supersystem, let $n_j(N)$ be the number of systems which contain N molecules *and* are in the particular energy state $E_j(N, V)$. N can range from zero to infinity (unless there is some upper limit set by the model being used). For a given distribution n, that is, for a set of numbers

$$\begin{aligned} &n_1(0), \\ &n_1(1), n_2(1), n_3(1), \ldots, \\ &n_1(2), n_2(2), n_3(2), \ldots, \\ &\vdots \end{aligned}$$

the number of possible quantum states of the supersystem is

$$\Omega_t(n) = \frac{[\sum_{j,N} n_j(N)]!}{\prod_{j,N} n_j(N)!}. \tag{1-39}$$

Acceptable distributions must satisfy the conservation relations

$$\sum_{j,N} n_j(N) = \mathfrak{N}, \tag{1-40}$$

$$\sum_{j,N} n_j(N)E_j(N, V) = E_t, \tag{1-41}$$

$$\sum_{j,N} n_j(N)N = N_t. \tag{1-42}$$

If we let α, β, and γ be the respective undetermined multipliers, the most probable distribution turns out to be (Problem 1-6)

$$n_j^*(N) = \mathfrak{N}e^{-\alpha}e^{-\beta E_j(N,V)}e^{-\gamma N}. \tag{1-43}$$

Again, in principle we can substitute Eq. (1-43) into Eqs. (1-40) through (1-42) and find α, β, and γ as functions of $\mathfrak{N}$, E_t, and N_t. But, instead, we follow a procedure analogous to that used for the canonical ensemble.

From Eqs. (1-40) and (1-43),

$$e^{\alpha} = \sum_{j,N} e^{-\beta E_j(N,V)}e^{-\gamma N} \tag{1-44}$$

and

$$P_j(N) = \frac{\overline{n_j(N)}}{\mathfrak{N}} = \frac{n_j^*(N)}{\mathfrak{N}} = \frac{e^{-\beta E_j(N,V)}e^{-\gamma N}}{\sum_{i,N'} e^{-\beta E_i(N',V)}e^{-\gamma N'}}, \tag{1-45}$$

where $P_j(N)$ is the probability that a system selected at random from the grand ensemble will contain N molecules and be in the energy state $E_j(N, V)$; or, $P_j(N)$ is the probability that the single prototype thermodynamic system contains exactly N molecules and is in the energy state $E_j(N, V)$. We note that $P_j(N)$ has an exponential dependence on both $E_j(N, V)$ and N. An open system has a definite volume, but both the energy and number of molecules in the system fluctuate. In a closed, isothermal system (canonical ensemble), N is fixed but the energy fluctuates. The magnitude of these fluctuations will be examined in Chapter 2.

From the first postulate, we have the associations

$$E \leftrightarrow \overline{E}(= E_t/\mathfrak{N}) = \sum_{j,N} P_j(N)E_j(N, V), \tag{1-46}$$

$$N \leftrightarrow \overline{N}(= N_t/\mathfrak{N}) = \sum_{j,N} P_j(N)N, \tag{1-47}$$

$$p \leftrightarrow \overline{p} = \sum_{j,N} P_j(N) \left\{ -\left[\frac{\partial E_j(N, V)}{\partial V} \right]_N \right\}. \tag{1-48}$$

These are mechanical variables. To include nonmechanical variables and to evaluate β and γ, we utilize the expression

$$d\overline{E} = \sum_{j,N} E_j(N, V)\, dP_j(N) + \sum_{j,N} P_j(N)\, dE_j(N, V). \tag{1-49}$$

Since we are summing over all values of j and N, $E_j(N, V)$ is in effect a function of V only in the second term on the right. In the first term, we substitute for $E_j(N, V)$ from Eq. (1-45). Then,

$$d\overline{E} = -\frac{1}{\beta} \sum_{j,N} [\gamma N + \ln P_j(N) + \ln \Xi]\, dP_j(N) + \sum_{j,N} P_j(N) \frac{\partial E_j(N, V)}{\partial V}\, dV, \tag{1-50}$$

where

$$\Xi = \sum_{j,N} e^{-\beta E_j(N,V)} e^{-\gamma N}. \tag{1-51}$$

Using [Eq. (1-47)]

$$d\overline{N} = \sum_{j,N} N\, dP_j(N),$$

Eq. (1-50) simplifies to

$$-\frac{1}{\beta} d\left[\sum_{j,N} P_j(N) \ln P_j(N) \right] = d\overline{E} + \overline{p}\, dV + \frac{\gamma}{\beta} d\overline{N}. \tag{1-52}$$

We compare this with the thermodynamic equation

$$T\, dS = dE + p\, dV - \mu\, dN, \tag{1-53}$$

and conclude that

$$\mu \leftrightarrow -\frac{\gamma}{\beta}, \tag{1-54}$$

$$T\, dS \leftrightarrow -\frac{1}{\beta} d\left[\sum_{j,N} P_j(N) \ln P_j(N) \right]. \tag{1-55}$$

By the same kind of lengthy argument already employed for the canonical ensemble, we arrive at the further results

$$S \leftrightarrow -k \sum_{j,N} P_j(N) \ln P_j(N), \tag{1-56}$$

$$\frac{1}{kT} \leftrightarrow \beta, \tag{1-57}$$

and therefore, from (1–54),

$$\frac{\mu}{kT} \leftrightarrow -\gamma. \tag{1-58}$$

The relation (1–56) has the same formal appearance as Eq. (1–30) in the canonical ensemble. In fact, this form for S is quite general (see Problem 1–8, for example).

According to Eq. (1–45), there is only one β (and therefore k) for all values of N. Furthermore, this is the same β as in Section 1–4 for a closed, isothermal system, since a grand ensemble is just an aggregate of canonical ensembles. That is, we can imagine "freezing" the composition of the systems in a grand ensemble by suddenly inserting, between the systems, walls which are heat conducting but impermeable to molecules. Then the original grand canonical ensemble becomes simply a collection of canonical ensembles (in fact, this is the significance of the word "grand") in thermal contact with each other, each characterized by a definite N.

Let us now summarize results for an open, isothermal system whose thermodynamic state is specified by the variables V, T, and μ. The probability that such a system contains N molecules and is in the energy state $E_j(N, V)$ is

$$P_j(N; V, T, \mu) = \frac{e^{-E_j(N,V)/kT} e^{N\mu/kT}}{\Xi(V, T, \mu)}, \tag{1-59}$$

where

$$\Xi(V, T, \mu) = \sum_{j,N} e^{-E_j(N,V)/kT} e^{N\mu/kT}. \tag{1-60}$$

We call Ξ the "grand partition function." The notation used for P in Eq. (1–59) means that N and j are essentially running indices (the notation P_{Nj} might have been used), while V, T, and μ are independent thermodynamic variables. An alternative form for Ξ is

$$\begin{aligned} \Xi(V, T, \mu) &= \sum_N \left[e^{N\mu/kT} \sum_j e^{-E_j(N,V)/kT} \right] \\ &= \sum_N Q(N, V, T) e^{N\mu/kT}. \end{aligned} \tag{1-61}$$

The probability that the system has N molecules, irrespective of the energy state, is

$$P(N; V, T, \mu) = \sum_j P_j(N) = \frac{Q(N, V, T)e^{N\mu/kT}}{\Xi(V, T, \mu)}. \tag{1-62}$$

Thus, for example, the average value of N is

$$\overline{N}(V, T, \mu) = \frac{\sum_N NQ(N, V, T)e^{N\mu/kT}}{\Xi(V, T, \mu)}. \tag{1-63}$$

If we substitute Eq. (1–59) into Eq. (1–56), we find

$$S = \frac{\overline{E}}{T} - \frac{\overline{N}\mu}{T} + k \ln \Xi = \frac{E}{T} - \frac{N\mu}{T} + \frac{pV}{T},$$

where the last expression is thermodynamic in origin. Hence

$$pV = kT \ln \Xi \; (V, T, \mu). \tag{1-64}$$

Now pV is the thermodynamic characteristic function for the variables V, T, and μ:

$$d(pV) = S\,dT + N\,d\mu + p\,dV. \tag{1-65}$$

Therefore, from Eq. (1–64), we have the following relations which, together with Eq. (1–64), permit us to calculate all the thermodynamic properties of a system if Ξ is known as a function of V, T, and μ:

$$S = kT\left(\frac{\partial \ln \Xi}{\partial T}\right)_{V,\mu} + k \ln \Xi, \tag{1-66}$$

$$N = kT\left(\frac{\partial \ln \Xi}{\partial \mu}\right)_{V,T}, \tag{1-67}$$

$$p = kT\left(\frac{\partial \ln \Xi}{\partial V}\right)_{\mu,T} = kT\frac{\ln \Xi}{V}. \tag{1-68}$$

The last form of Eq. (1–68) follows from Eq. (1–64) or, on thermodynamic grounds, from the fact that the variables held constant in the derivative are both intensive.

We shall see in Chapter 2 that one can choose an ensemble from which to calculate thermodynamic functions on the basis of convenience, and irrespective of the actual environment of a system (heat bath, constant pressure, etc.). In many problems the grand ensemble is easier to use than the canonical ensemble. When this is the case, the reason is usually either (a) that a mathematically awkward restraint of constant N in the canonical ensemble can be avoided by summing over N (Eq. 1–61), or (b) that a many-body problem can be reduced to a one-body, two-body, etc., problem by viewing Eq. (1–61) as a power series in the "absolute activity" $\lambda = e^{\mu/kT}$:

$$\Xi\,(V, T, \mu) = Q(0, V, T) + Q(1, V, T)\lambda + Q(2, V, T)\lambda^2 + \cdots. \tag{1–69}$$

This is the preferable method in treating an imperfect gas, for example (Chapter 15).

The above discussion is limited to a one-component system, but it can easily be extended to any number of components (Problem 1–7). For example, for two components there will be two equations like (1–42) and two undetermined multipliers, $\gamma_1 = -\mu_1/kT$ and $\gamma_2 = -\mu_2/kT$. Equation (1–59) becomes

$$P_j(N_1, N_2; V, T, \mu_1, \mu_2) = \frac{e^{-E_j(N_1,N_2,V)/kT}\lambda_1^{N_1}\lambda_2^{N_2}}{\Xi(V, T, \mu_1, \mu_2)}, \tag{1–70}$$

with

$$\Xi = \sum_{N_1,N_2} Q(N_1, N_2, V, T)\lambda_1^{N_1}\lambda_2^{N_2}, \tag{1–71}$$

where

$$\lambda_1 = e^{\mu_1/kT}, \qquad \lambda_2 = e^{\mu_2/kT}.$$

Also

$$pV = kT \ln \Xi\,(V, T, \mu_1, \mu_2), \tag{1–72}$$

$$d(pV) = S\,dT + N_1\,d\mu_1 + N_2\,d\mu_2 + p\,dV, \tag{1–73}$$

from which we can immediately write the extensions of Eqs. (1–66) through (1–68).

1–6 Microcanonical ensemble. Here we are concerned with an isolated system with given E, V, and N (N again represents a set N_1, N_2, ... if the system is multicomponent). The representative ensemble is called a microcanonical ensemble, as stated in Section 1–2. For an isolated system it is difficult to achieve a direct connection between our two postulates and thermodynamics (e.g., we have used variations in $\bar{E}$ for this purpose in Sections 1–4 and 1–5, but here E is constant). The most common procedure for avoiding this difficulty is to introduce, essentially as a new postulate, the equation $S = k \ln \Omega$, where $\Omega(N, V, E)$ is the degeneracy of the energy level E (see Section 1–2). However, a new postulate is not really needed; its introduction is therefore unsatisfactory from a logical point of view. Instead, we derive the properties of a microcanonical ensemble from either the canonical ensemble or the grand ensemble.

First, consider a canonical ensemble. A microcanonical ensemble, as the name is meant to imply, is a degenerate canonical ensemble in which all systems have (virtually) the same energy. Thus, suppose we start with a canonical ensemble, pick out just those systems with an energy level E, place thermal insulation around each of them, and then remove these systems from the other systems in the canonical ensemble (with energies different from E). As a result of this operation, we have a collection of isolated systems, all with the same N, V, *and* E (a microcanonical ensemble). This degenerate canonical ensemble* may be thought of as being representative of a hypothetical closed, isothermal system that is somehow restrained from having values of E other than $E = \bar{E}$. Another way of saying this is that the only quantum states accessible to the system are those with energy E. In this new ensemble, according to Eq. (1–28), the fraction of systems P_j in a given quantum state (energy E) is proportional to $e^{-E/kT}$. But E is the same for all quantum states, $\Omega(N, V, E)$ in number. Hence P_j is the same for all Ω quantum states. Since $\sum_j P_j = 1$, $P_j = 1/\Omega$. Then, from Eq. (1–30),

$$S(N, V, E) = -k\sum_j P_j \ln P_j = -k\Omega\left(\frac{1}{\Omega}\ln\frac{1}{\Omega}\right)$$

$$= k \ln \Omega(N, V, E). \tag{1–74}$$

This relation between the thermodynamic S and statistical-mechanical (actually, quantum-mechanical) Ω can then be employed to derive all thermodynamic functions of interest, as we shall see below.

A microcanonical ensemble is also a degenerate grand ensemble: we can pick out of a grand ensemble only those systems with certain prescribed values of N and E. But there is a different, in fact, complementary, way in which Eq. (1–74) can be deduced from a grand ensemble.

In what follows, we have to make use of Section 1–5, which was restricted for simplicity to a one-component system, but the method and result are independent of the number of components. In Section 1–5 we applied the second postulate to the whole grand ensemble, or supersystem (Fig. 1–5). That is, the supersystem itself is an example of an isolated system. The point of view we adopt here is that the supersystem may be regarded, for present purposes, not as an imaginary construct but as a single, very large, *real isolated system*. In this case, the dashed lines in Fig. 1–5, dividing the "supersystem" into "systems," represent *mathematical* rather than physical planes. A "system" (Fig. 1–6) is then an imaginary macroscopic portion, between mathematical planes, of the total or "supersystem." Each "system" is *open* and *isothermal*. Note that

---

* We shall see in Section 2–2 that, because relative fluctuations in E about $\bar{E}$ in a canonical ensemble are virtually negligible in magnitude, this somewhat artificial way of forming a microcanonical ensemble is really unnecessary: canonical and microcanonical ensembles are essentially indistinguishable in any case.

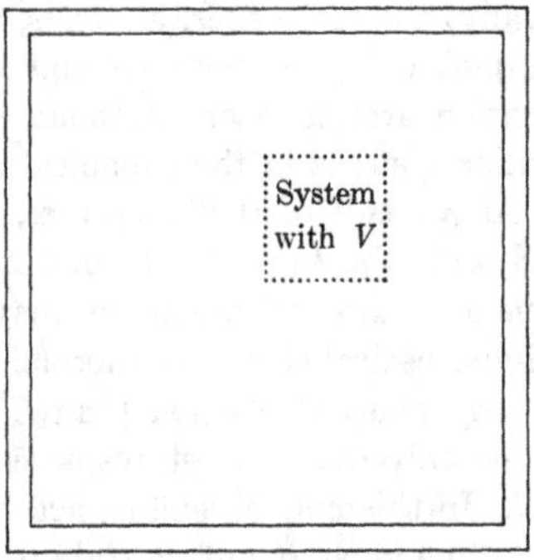

FIG. 1-6. Portion of a larger system forming a smaller, open, isothermal system of volume V.

we cannot use this point of view in connection with a canonical ensemble, because this ensemble requires that each "system" be *closed*. The way we proceed is to employ Eq. (1-39) to find the total number of quantum states Ω_t of the "supersystem" (specified E_t, N_t, $V_t = \mathfrak{N}V$) and relate this number to the entropy S_t of the "supersystem" using (a) the connections with thermodynamics already found for an open, isothermal system (Eqs. 1-59 through 1-68), and (b) the additive property of the entropy, $S_t = \mathfrak{N}S$, where S is the entropy of one "system" in the "supersystem." To evaluate $\ln \Omega_t$, we start with

$$\ln \Omega_t = \ln \sum_n \Omega_t(n) = \ln \Omega_t(n^*) \tag{1-75}$$

and then substitute

$$n_j^*(N) = \frac{\mathfrak{N}e^{-E_j(N,V)/kT}e^{N\mu/kT}}{\Xi}, \tag{1-76}$$

which follows from Eqs. (1-45) and (1-59), into $\ln \Omega_t(n)$ from Eq. (1-39), to give $\ln \Omega_t(n^*)$. We find

$$\begin{aligned}\ln \Omega_t(n^*) &= \mathfrak{N} \ln \mathfrak{N} - \sum_{j,N} n_j^*(N) \ln n_j^*(N)\\ &= \frac{\mathfrak{N}\bar{E}}{kT} - \frac{\mathfrak{N}\bar{N}\mu}{kT} + \mathfrak{N} \ln \Xi\\ &= \mathfrak{N}\left(\frac{E}{kT} - \frac{N\mu}{kT} + \frac{pV}{kT}\right) = \frac{\mathfrak{N}S}{k},\end{aligned}$$

so that

$$S_t = k \ln \Omega_t, \tag{1-77}$$

in agreement with Eq. (1-74).

Equation (1–74), due to Boltzmann, is possibly the best-known equation in statistical mechanics, mainly for historical reasons. We shall discuss it further in Chapter 2. But let us note here that: (a) it applies to an isolated system; (b) if the ground state of the system is nondegenerate, $\Omega = 1$ and $S = 0$ $(T \to 0)$; and (c) for any isolated system whatever, the more quantum states available to the system, the higher the entropy. This is the origin of qualitative statements which correlate the entropy with "probability," "randomness," "disorder," etc. We shall encounter many examples of such a correlation in the present book. Incidentally, we can appeal to thermodynamics for a quick estimate of the magnitude of Ω to be expected in statistical mechanics in general. That is, since it is found experimentally that* $S = O(Nk)$, $\ln \Omega = O(N)$, and $\Omega = O(e^N) = O(10^{10^{20}})$, an impressively large number.

Let us assume that, from quantum mechanics, we have $\Omega(N, V, E)$ for the system of interest. Our next problem is to calculate all thermodynamic functions, not just the entropy S. But S is the characteristic function for the variables N, V, E:

$$dS = \frac{1}{T}\,dE + \frac{p}{T}\,dV - \sum_{\alpha} \frac{\mu_\alpha}{T}\,dN_\alpha. \tag{1–78}$$

Hence

$$\frac{1}{kT} = \left(\frac{\partial \ln \Omega}{\partial E}\right)_{V,N}, \tag{1–79}$$

$$\frac{p}{kT} = \left(\frac{\partial \ln \Omega}{\partial V}\right)_{E,N}, \tag{1–80}$$

$$-\frac{\mu_i}{kT} = \left(\frac{\partial \ln \Omega}{\partial N_i}\right)_{E,V,N_{\alpha \neq i}} \tag{1–81}$$

The temperature is determined by the dependence of Ω on E. Since we know from thermodynamics that T is positive, we can anticipate that Ω will increase with E for any macroscopic quantum-mechanical system. Clearly, the same statement can also be made about $\Omega(V)$.

In practice, except in very simple systems, $\Omega(N, V, E)$ is not available, and the microcanonical equations (1–74) and (1–79) through (1–81) cannot be utilized. In particular, the restriction to constant energy E is usually a difficulty. This can be avoided by passing to the canonical ensemble.

---

* The notation $O(x)$ means "of order x."

1–7 Other ensembles. Many other ensembles and partition functions are possible and are often useful. For example, for a one-component system, if we start with $\Omega(N, V, E)$ and then sum one at a time over E, N, and V, we obtain* the four relations (we write $1/kT = \beta$ here for convenience),

$$\ln \Omega(N, V, E) = \beta TS = \frac{S}{k}, \tag{1–82}$$

$$\ln \sum_E \Omega(N, V, E)e^{-\beta E} = \text{function of } N, V, \beta = -\beta A, \tag{1–83}$$

$$\ln \sum_N \Omega(N, V, E)e^{\beta\mu N} = \text{function of } V, E, \beta\mu = \beta H, \tag{1–84}$$

$$\ln \sum_V \Omega(N, V, E)e^{-\beta pV} = \text{function of } N, E, \beta p = \beta(TS - pV). \tag{1–85}$$

In Eq. (1–84), H is the "heat content," $E + pV$. The summations over possible values of E and V may be replaced by integrations in most problems, as we shall see. But here we use the present simple notation for convenience. The sum in Eq. (1–83) is the same, except for notation, as the sum over energy levels in Eq. (1–37). Previously we established only Eqs. (1–82) and (1–83), but the other two cases, and those below, can be worked out in detail by the same general methods already used (Problem 1–8).

Continuing, we can also sum two at a time over E, N, and V:

$$\ln \sum_{E,N} \Omega(N, V, E)e^{-\beta E}e^{\beta\mu N} = \text{function of } V, \beta, \mu = \beta pV, \tag{1–86}$$

$$\begin{aligned}\ln \sum_{E,V} \Omega(N, V, E)e^{-\beta E}e^{-\beta pV} &= \text{function of } N, \beta, p \\ &= -\beta N\mu = -\beta F,\end{aligned} \tag{1–87}$$

$$\ln \sum_{N,V} \Omega(N, V, E)e^{\beta\mu N}e^{-\beta pV} = \text{function of } E, \beta\mu, \beta p = \beta E. \tag{1–88}$$

Equation (1–86) is the logarithm of the grand partition function, already encountered. The other two equations are new. Equation (1–87) is particularly important† because it is applicable to a system with the familiar

---

* T. L. HILL, *J. Chem. Phys.* **29**, 1423 (1958). See also A. MÜNSTER (Supplementary Reading list); W. B. BROWN, *Mol. Phys.* **1**, 68 (1958); A. MÜNSTER, *ibid.* **2**, 1 (1959); R. A. SACK, *ibid.* **2**, 8 (1959).

† W. B. BROWN, *Mol. Phys.* **1**, 68 (1958).

set of independent variables N, T, and p. In Eq. (1–87), F is the Gibbs free energy, $A + pV$, equal to $N\mu$ for a one-component system. Finally, we can sum over all of E, N, and V:

$$\ln \sum_{E,N,V} \Omega(N, V, E) e^{-\beta E} e^{\beta \mu N} e^{-\beta p V} = \text{function of } \beta, \mu, p = 0. \tag{1–89}$$

This is an exceptional case, since T, μ, and p appear to be independent variables, whereas we know from thermodynamics that at most two of these variables can be independent. The special treatment necessary for this partition function is provided elsewhere (S. M., Chapters 2 and 3).

The characteristic functions are redefined here, in a more systematic way, as dimensionless quantities. In every case there is an appropriate thermodynamic equation which permits calculation of other thermodynamic functions from knowledge of the partition function. For example, for Eq. (1–83),

$$d(-\beta A) = -E\, d\beta + \beta p\, dV - \beta\mu\, dN, \tag{1–90}$$

or for Eq. (1–87),

$$d(-\beta F) = -E\, d\beta - V\, d(\beta p) - \beta\mu\, dN. \tag{1–91}$$

Equation (1–90) is, of course, just a rearrangement of Eq. (1–32) for a closed, isothermal system.

The reader has perhaps noticed that the characteristic function can be written immediately on inspection of the partition function. The rule is: if we replace Ω by $e^{\beta TS}$, then the characteristic function is the sum of the exponents in the partition function. For example, for Eqs. (1–83) and (1–87), respectively:

$$\begin{aligned} \beta TS - \beta E &= -\beta A, \\ \beta TS - \beta E - \beta p V &= -\beta F. \end{aligned} \tag{1–92}$$

The reason for the existence of this rule will be obvious from our discussion of the thermodynamic equivalence of ensembles in Chapter 2. It depends on the legitimacy of replacing the logarithm of a partition function by the logarithm of its maximum term.

There are two further types of ensembles or partition functions we should mention, since they will be encountered in the applications. First, there are some problems in which the "external variable" V is replaced by another external variable (e.g., length or area) or is supplemented by additional external variables (e.g., volume and area are both external variables). Second, in multicomponent systems, there are cases in which it is helpful or necessary to regard the system as open with respect to some components, but not all.

Problems

1-1. Modify the derivation, in Section 1-3, of the canonical ensemble probability distribution (1-12) to obtain Eq. (1-38) directly. That is, use "levels" instead of "states."

1-2. Obtain the derivatives $(\partial \overline{E}/\partial V)_{\beta,N}$ and $(\partial \overline{p}/\partial \beta)_{N,V}$ from Eqs. (1-5), (1-6), and (1-12), and show that

$$\left(\frac{\partial \overline{E}}{\partial V}\right)_{\beta,N} + \beta\left(\frac{\partial \overline{p}}{\partial \beta}\right)_{N,V} = -\overline{p}.$$

This result is of interest because comparison with the thermodynamic equation,

$$\left(\frac{\partial E}{\partial V}\right)_{T,N} + \left(\frac{1}{T}\right)\left[\frac{\partial p}{\partial(1/T)}\right]_{N,V} = -p,$$

suggests that β = constant/T, as we find in Section 1-4. (Page 12.)

1-3. Derive the canonical ensemble probability distribution (1-23) for a combined system, starting with Eq. (1-22) and the associated restraints. (Page 16.)

1-4. Make use of Eq. (1-11) for $\overline{E}$, the relation $\beta = 1/kT$, and the experimental fact that C_V is always positive to prove that k is a positive constant. (Page 18.)

1-5. Use the method of undetermined multipliers to show that $-\sum_j P_j \ln P_j$, subject to the condition $\sum_j P_j = 1$, is a maximum when P_j = constant. (Page 20.)

1-6. Use the method of undetermined multipliers to find the most probable distribution in a grand ensemble, Eq. (1-43). (Page 22.)

1-7. Derive the basic equations (1-70) and (1-72) for an open system with two components. (Page 26.)

1-8. Derive Eq. (1-84) for a system of fixed volume in thermal and material contact with a reservoir at T and μ, but with the transports of heat and molecules across the walls of the system coupled in such a way as to maintain E constant. (Page 30.)

1-9. Verify that Eqs. (1-34) and (1-35) for p and E in terms of Q are equivalent to Eqs. (1-5), (1-6), and (1-28) in terms of ensemble averages.

Supplementary Reading

KITTEL, Part I.

MAYER and MAYER, Chapters 3, 4, and 10.

MÜNSTER, A., *Handbuch der Physik*, Vol. III/2, pp. 176–412. BERLIN: SPRINGER, 1959.

MÜNSTER, A., *Statistische Thermodynamik*. BERLIN: SPRINGER, 1956.

SCHRÖDINGER, Chapters 1, 2, and 6.

S. M., Chapters 2 and 3.

TER HAAR, D., *Revs. Mod. Phys.* **27**, 289 (1955).

TOLMAN, Chapters 13 and 14.

WILSON, Chapter 5.

CHAPTER 2

FURTHER DISCUSSION OF ENSEMBLES AND THERMODYNAMICS

2–1 Fluctuations. The general approach in Chapter 1 was, first, to set up postulates from which a procedure could be worked out for calculating the mean values of mechanical thermodynamic variables, such as energy and pressure; then, to extend the treatment to include nonmechanical variables, such as temperature and entropy, by comparing corresponding statistical-mechanical and thermodynamic equations. Thus we now have at our disposal a set of general equations that can be employed to accomplish our primary task, namely, the calculation of thermodynamic functions from molecular properties. Beginning with the next chapter, we shall encounter many illustrations of such calculations.

For the strictly thermodynamic purposes referred to above, only the *mean* values of mechanical variables need be considered. Yet any student of statistical mechanics will or should be curious about the extent to which mechanical variables fluctuate around their mean values. For example, in a closed, isothermal system (canonical ensemble), the system has a certain probability of being found in any given energy state E_j, though the average value of the energy is $\overline{E}$. What is the dispersion or spread of the energy probability distribution about $\overline{E}$? However, curiosity does not provide the only motive for investigating fluctuations. The matter is of some practical interest as well. For example, (a) a statistical-mechanical investigation of fluctuations justifies or explains, theoretically, why it is possible and legitimate to ignore fluctuations in thermodynamics; (b) a study of fluctuations leads us to the important conclusion [essentially equivalent to (a)] that in the calculation of thermodynamic properties of macroscopic systems, we can choose a statistical-mechanical ensemble to work with strictly on grounds of convenience, ignoring the actual kind of environment (isolated; closed, isothermal; etc.) the system of interest is in; (c) some important topics, such as light scattering and the theory of solutions, can be analyzed explicitly in terms of fluctuations; and finally, (d) fluctuation theory is very important in nonequilibrium statistical mechanics.

Note that we have not mentioned fluctuations in nonmechanical variables. The reason for this is the following. To define the mean value (or examine the extent of fluctuations about the mean value) of a property of a system which can exist in various states j with probabilities P_j, the property itself must be well defined in each state j. As was mentioned at the beginning of Section 1–3, a property that meets this criterion is "mechanical," by definition. For example, E_j, $p_j = -\partial E_j/\partial V$, N, and V are all well defined for a single quantum state, but S and T are not. Hence we can discuss fluctuations in E, N, etc., but not in S, T, etc.*

We have remarked that thermodynamic functions calculated in statistical mechanics turn out to be independent of the ensemble used in the calculation. This is not true, however, of fluctuations. For each environment (and therefore for each ensemble) the problem is different; in fact, even the variables which fluctuate are different. Hence, in a study of fluctuations, the actual environment of the system of interest must be taken note of and the appropriate ensemble used. For example, if the system is closed and isothermal, there will be fluctuations in p and E, but not in N and V (since values of N and V are prescribed and fixed), and these fluctuations must be investigated using a canonical ensemble.

Each environment has a corresponding ensemble and partition function: eight possibilities are listed in Section 1–7 for a one-component system with a single external variable (V). It is clear then that there must be a great many fluctuation formulas which might be derived and discussed (see Problem 2–1, for example). We confine ourselves here to perhaps the three most important special cases (for a one-component system).

Let us first consider the fluctuation in the energy of a closed system immersed in a large heat bath (N, V, T given and fixed). Because V and N are fixed, fluctuations in energy must be associated with heat exchange between system and bath. We use the canonical ensemble, of course. Now, as we shall soon verify, these energy fluctuations turn out to be very small indeed, so the probability distribution function for different energies is gaussian in form about the mean value $\overline{E}$ (Fig. 2–1). The dispersion or spread in this probability distribution may therefore be characterized completely by the standard deviation σ_E, that is, the root mean-square deviation from the mean:

$$\sigma_E = \left[\overline{(E - \overline{E})^2}\right]^{1/2}. \tag{2–1}$$

Before evaluating σ_E explicitly, we note that

$$\overline{(E - \overline{E})^2} = \overline{E^2 - 2E\overline{E} + (\overline{E})^2} = \overline{E^2} - (\overline{E})^2. \tag{2–2}$$

This is necessarily a positive quantity.

If we differentiate

$$\overline{E} \sum_j e^{-E_j(N,V)/kT} = \sum_j E_j(N, V) e^{-E_j(N,V)/kT} \tag{2–3}$$

* A less restricted point of view is taken, for example, by Landau and Lifshitz, Chapter 12.

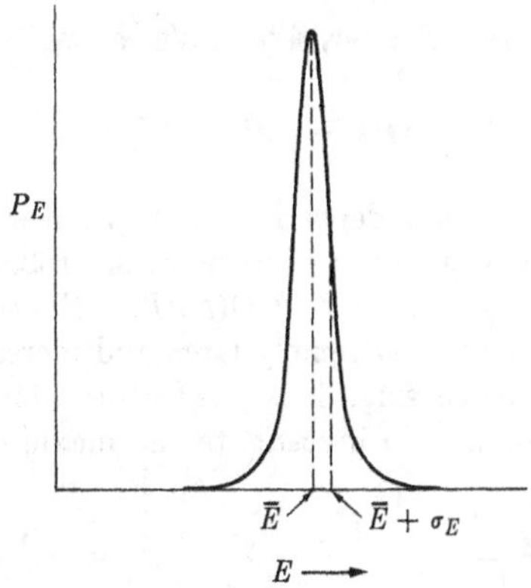

FIG. 2–1. Probability distribution in energy for a closed, isothermal system (schematic).

with respect to T, and then divide by Q, we find

$$\left(\frac{\partial \overline{E}}{\partial T}\right)_{V,N} + \frac{\overline{E}}{QkT^2}\sum_j E_j e^{-E_j/kT} = \frac{1}{QkT^2}\sum_j E_j^2 e^{-E_j/kT},$$

or

$$\overline{E^2} - (\overline{E})^2 = \overline{(E - \overline{E})^2} = \sigma_E^2 = kT^2 C_V. \tag{2–4}$$

From thermodynamics, we know that in general $C_V = O(Nk)$ and $\overline{E} = O(NkT)$. Therefore

$$\frac{\sigma_E}{\overline{E}} = \frac{(kT^2 C_V)^{1/2}}{\overline{E}} = O(N^{-1/2}). \tag{2–5}$$

Thus we have found that in a typical closed, isothermal system the standard deviation of the energy probability distribution is of the order of $10^{-10}\overline{E}$, an *extremely* small quantity. The probability distribution is therefore practically a Dirac δ-function at $E = \overline{E}$. It is for this reason that in thermodynamics all values of E other than $\overline{E}$ can, in effect, be ignored.

It may be helpful to look at these energy fluctuations from a more explicit and slightly different point of view, in a special case. According to Eq. (1–38), the probability P_E (Fig. 2–1) of observing a certain energy E is proportional to $\Omega(N, V, E)e^{-E/kT}$. In order to have a concrete example before us, let us anticipate from Eq. (4–39) that the energy dependent factor in Ω for an ideal, classical, monatomic gas is $E^{3N/2}$. As mentioned in Section 1–6, the fact that Ω always increases with E can be seen from Eq. (1–79) (T is positive). We can write, in our special case,

$$P_E = cx^{3N/2}e^{-x},$$

where $x = E/kT$ and c is independent of x (we need not specify c more closely here). In the neighborhood of the most probable energy $E^*(=\overline{E})$, x is clearly of order N, since $E^* = O(NkT)$. Hence one of the factors involving x in P_E above is extremely large and increasing, and the other extremely small and decreasing. It is this feature which causes P_E to have such a sharp maximum. To investigate the maximum, we need

$$\frac{\partial \ln P_E}{\partial x} = 0 = \frac{3N}{2x} - 1, \qquad x^* = \frac{3N}{2},$$

$$\frac{\partial^2 \ln P_E}{\partial x^2} = -\frac{3N}{2x^2} = -\frac{2}{3N}, \qquad \text{when } x = x^*.$$

Then from

$$\ln P_E = \ln P_{E^*} + \frac{1}{2}\left(\frac{\partial^2 \ln P_E}{\partial x^2}\right)_{x=x^*}(x - x^*)^2 + \cdots,$$

we find

$$P_E = P_{E^*}e^{-(E-E^*)^2/2\sigma_E^2},$$

where

$$\sigma_E^2 = \frac{3N(kT)^2}{2}.$$

This is the conventional form for a gaussian distribution in terms of the standard deviation of the distribution. This result for σ_E agrees with Eq. (2–4), since $C_V = 3Nk/2$ in this special case (Chapter 4). We also note that $E^* = x^*kT = 3NkT/2$, which is correct.

Next we examine the fluctuation in N in an open, isothermal system. Here V, T, and μ are fixed. We differentiate

$$\overline{N}\sum_N Q(N, V, T)e^{N\mu/kT} = \sum_N NQ(N, V, T)e^{N\mu/kT} \tag{2–6}$$

with respect to μ, divide by Ξ, and obtain

$$\left(\frac{\partial \overline{N}}{\partial \mu}\right)_{V,T} + \frac{\overline{N}}{\Xi kT}\sum_N NQe^{N\mu/kT} = \frac{1}{\Xi kT}\sum_N N^2Qe^{N\mu/kT},$$

or

$$\overline{N^2} - (\overline{N})^2 = \sigma_N^2 = kT\left(\frac{\partial \overline{N}}{\partial \mu}\right)_{V,T}. \tag{2–7}$$

Since, in thermodynamics, $\mu = O(kT)$, $\sigma_N^2 = O(\overline{N})$ and $\sigma_N/\overline{N} = O(\overline{N}^{-1/2})$. This is the same order of magnitude for the relative fluctuation as we found above for the energy of a closed, isothermal system. Indeed, it is the standard result in statistical-mechanical fluctuation formulas. We conclude that even an open system contains virtually a fixed number of molecules, $\overline{N}$, for given V, T, and μ: the probability distribution in N is practically a δ-function.

The right side of Eq. (2–7) can be put in more familiar thermodynamic form. From

$$d\mu = v\,dp \qquad (T \text{ constant}),$$

we have

$$\left(\frac{\partial \mu}{\partial \rho}\right)_T = v\left(\frac{\partial p}{\partial \rho}\right)_T,$$

where $v = V/N$ and $\rho = N/V$. Then

$$\left(\frac{\partial \mu}{\partial N}\right)_{V,T} V = -\frac{V^3}{N^2}\left(\frac{\partial p}{\partial V}\right)_{N,T}. \tag{2–8}$$

Now we put Eq. (2–8) in Eq. (2–7) and find

$$\left(\frac{\sigma_N}{\overline{N}}\right)^2 = \frac{kT\kappa}{V}, \tag{2–9}$$

where κ, the compressibility, is defined by

$$\kappa = -\frac{1}{V}\left(\frac{\partial V}{\partial p}\right)_{N,T}.$$

For an ideal gas, $\kappa = 1/p$ and $kT\kappa/V = 1/\overline{N}$. In view of the fact that V is constant, Eq. (2–9) is also a formula for the fluctuation in number density ($\rho = N/V$):

$$\frac{\overline{(N - \overline{N})^2}}{(\overline{N})^2} = \frac{\overline{(\rho - \bar{\rho})^2}}{(\bar{\rho})^2} = \left(\frac{\sigma_\rho}{\bar{\rho}}\right)^2 = \frac{kT\kappa}{V}. \tag{2–10}$$

As our third case, let us investigate a system with N, p, and T fixed, but with fluctuations in volume. According to Eqs. (1–87) and (1–91), we have as the appropriate starting point,

$$\overline{V}\Delta = \sum_V VQ(N, V, T)e^{-pV/kT}, \tag{2–11}$$

where

$$\begin{aligned}\Delta &= \sum_{E,V} \Omega(N, V, E)e^{-E/kT}e^{-pV/kT} \\ &= \sum_{V} Q(N, V, T)e^{-pV/kT}.\end{aligned} \tag{2-12}$$

That is, the probability of observing the volume V is

$$P_V = \frac{Q(N, V, T)e^{-pV/kT}}{\Delta}. \tag{2-13}$$

On differentiation of Eq. (2–11) with respect to p, and division by Δ, there results

$$\left(\frac{\partial \overline{V}}{\partial p}\right)_{N,T} - \frac{\overline{V}}{\Delta kT}\sum_{V} VQe^{-pV/kT} = -\frac{1}{\Delta kT}\sum_{V} V^2Qe^{-pV/kT},$$

or

$$\sigma_V^2 = \overline{V^2} - (\overline{V})^2 = -kT\left(\frac{\partial \overline{V}}{\partial p}\right)_{N,T}. \tag{2-14}$$

Hence the formula here for $\sigma_V/\overline{V}$ is the same as that for $\sigma_N/\overline{N}$ in an open system (Eq. 2–9).

Finally, we mention an exception to the negligible fluctuations we have been encountering above. Consider Eq. (2–9), for example. At a critical point or when two phases exist together in the system, $(\partial p/\partial V)_{N,T}$ is essentially zero and κ infinite. Hence the fluctuations are large rather than negligible. To be a little more specific and exact, in a two-phase system with the two number densities ρ_1 and ρ_2, ρ in the expression $\overline{(\rho - \bar{\rho})^2}$ can range from ρ_1 to ρ_2 (rather than being centered very closely around $\bar{\rho}$), so that $\overline{(\rho - \bar{\rho})^2}$ is now of the order of $(\bar{\rho})^2$ itself. Hence $\sigma_\rho/\bar{\rho}$ in Eq. (2–10) is of order unity rather than $\overline{N}^{-1/2}$. That is, the standard deviation is of the same magnitude as the mean value itself.

Fluctuations in density at the critical point are responsible for the well-known critical opalescence phenomenon.

2–2 Thermodynamic equivalence of ensembles. The formulas derived in the preceding section are typical in showing that fluctuations about mean values are so small they can be ignored in thermodynamics except under very special circumstances (even in a two-phase system, fluctuations within each phase separately are normal, i.e., thermodynamically negligible). For example, in the grand partition function

$$\Xi = \sum_{N} Q(N, V, T)e^{N\mu/kT} \tag{2-15}$$

for a one-component open system, the only values of N which have an appreciable probability of being observed experimentally are those that deviate only negligibly from the mean value $\overline{N}$. Therefore, just as we found in Appendix II, we can replace $\ln \Xi$ by the logarithm of the largest term in the above sum, without making a detectable error to terms of thermodynamic order of magnitude. To be more specific: both $\ln \Xi\ (=pV/kT)$ and the logarithm of the largest term in the sum are of order $\overline{N}$, and Eq. (2–7) tells us that there are $O(\overline{N}^{1/2})$ terms in the sum of the same order of magnitude as the maximum term; hence the situation is completely analogous to that in Appendix II [see Eq. (II–11)].

Let us proceed to find the maximum term in Ξ, for the result is rather interesting. Let

$$t_N(V, T, \mu) = Q(N, V, T)e^{N\mu/kT}.$$

Then

$$\left(\frac{\partial \ln t_N}{\partial N}\right)_{V,T,\mu} = 0 = \left(\frac{\partial \ln Q}{\partial N}\right)_{V,T} + \frac{\mu}{kT}. \tag{2–16}$$

If we denote by N^* the value of N satisfying Eq. (2–16), t_{N^*} is the maximum term in Ξ. Then

$$\ln \Xi = \frac{pV}{kT} = \ln t_{N^*} = \ln Q(N^*, V, T) + \frac{N^*\mu}{kT},$$

or

$$N^*\mu - pV = A(N^*, V, T) = -kT \ln Q(N^*, V, T). \tag{2–17}$$

Equation (2–16) determines $N^*\ (=\overline{N})$ as a function of V, T, and μ. But if we take the alternative point of view that the independent variables are N^*, V, and T, and that Eq. (2–16) gives μ as a function of N^*, V, and T, then Eqs. (2–16) and (2–17) are just the canonical ensemble equations (1–36) and (1–31), respectively. In other words, application of the maximum-term procedure, which is legitimate because of the small fluctuations in N, causes the grand ensemble to degenerate into the canonical ensemble. This conclusion can be verified by noting further that the grand ensemble equations for S and p [(1–66) and (1–68)] also go over into the corresponding canonical ensemble equations [(1–33) and (1–34)] when $\ln \Xi$ is replaced by $\ln t_{N^*}$ (Problem 2–2).

For practical thermodynamic purposes, then, there is no distinction between a grand ensemble and a canonical ensemble. In a given problem one can choose between them simply on the basis of mathematical convenience.

Let us digress to note that we can write, as in Eq. (II–8),

$$t_N = Q(N, V, T)e^{N\mu/kT} = t_{N^*}e^{-(N-N^*)^2/2\sigma_N^2},$$

where σ_N is given by Eq. (2–7). The grand partition function sum can then be replaced by an integral:

$$\ln \Xi = \ln t_{N^*} + \ln \int_{-\infty}^{+\infty} e^{-x^2/2\sigma_N^2}\, dx$$

$$= \ln t_{N^*} + \ln [(2\pi)^{1/2}\sigma_N],$$

where $x = N - N^*$. The second term is of order $\ln \overline{N}$, which is negligible compared with the first term.

The argument above can be applied to any of the partition functions, (1–83) through (1–88). As one further example, observe the degeneration of the canonical ensemble into the microcanonical ensemble:

$$\ln Q = \ln \sum_E t_E = \ln t_{E^*},$$

where

$$t_E = \Omega(N, V, E)e^{-E/kT}.$$

Then $E^*(N, V, T)(=\overline{E})$ is found from

$$\frac{\partial \ln t_E}{\partial E} = 0 = \left(\frac{\partial \ln \Omega}{\partial E}\right)_{N,V} - \frac{1}{kT}. \tag{2–18}$$

Also,

$$\ln Q = -\frac{A}{kT} = \ln \Omega(N, V, E^*) - \frac{E^*}{kT},$$

or

$$S = k \ln \Omega(N, V, E^*). \tag{2–19}$$

Equations (2–18) and (2–19) will be recognized as microcanonical ensemble relations.

In thermodynamics the functional relations between the thermodynamic variables of a system are independent of the environment (open, closed, isobaric, isothermal, etc.). Another way of saying this is that the choice of independent thermodynamic variables is arbitrary and not prescribed by the environment. In statistical mechanics, just as we should expect, we have now come to the same conclusion: regardless of environment, we can select whichever ensemble or partition function (and therefore independent thermodynamic variables) we wish in calculating thermodynamic properties; the results must be independent of the choice.

Since the entropy is a particularly important and interesting thermodynamic function, we add a few comments concerning it. The canonical ensemble expression for S, Eq. (1–30), is

$$S = -k \sum_j P_j \ln P_j. \tag{2–20}$$

Now we have found that in a closed, isothermal system, the probability that the system has an energy departing appreciably from $\bar{E}$ is virtually zero. Hence we may assume that the only P_j's in Eq. (2-20) which make a significant contribution to the sum are those associated with $\bar{E}$ $(=E^*$, above). The number of quantum states with energy $\bar{E}$ is $\Omega(N, V, \bar{E})$, so that each $P_j = 1/\Omega(N, V, \bar{E})$ on this assumption. Then

$$S = -k\Omega\left(\frac{1}{\Omega}\ln\frac{1}{\Omega}\right)$$

$$= k\ln\Omega(N, V, \bar{E}), \tag{2-21}$$

in agreement with Eq. (2-19). This confirms our assumption. The reader will have noticed that this argument is practically the same as that used in deducing Eq. (1-74) as the starting point for a discussion of the microcanonical ensemble. But the point of view here is quite different. In Section 1-6 we had to either mentally pick out those systems in a canonical ensemble with a certain E, or artificially limit the accessibility of quantum states to those with energy E. But we have found from our study of fluctuations in this chapter that these restricting devices are really unnecessary: a canonical ensemble (in fact, any ensemble) is, so to speak, *by its own choosing* virtually a microcanonical ensemble.

Similarly, in the grand ensemble relation (Eq. 1-56),

$$S = -k\sum_{j,N} P_j(N)\ln P_j(N), \tag{2-22}$$

the only important $P_j(N)$'s are those associated with $\bar{E}$ and $\bar{N}$. The number of these states is $\Omega(\bar{N}, V, \bar{E})$. Then

$$S = k\ln\Omega(\bar{N}, V, \bar{E}). \tag{2-23}$$

Thus the relation $S = k\ln\Omega(N, V, E)$ has quite general validity. It is restricted in principle but not in practice to isolated systems. But we must take for N, V, and E in $\Omega(N, V, E)$ the mean or most probable values (when fluctuations are possible), which are themselves functions of the independent variables of the system. With this understanding we can make a general correlation between the magnitude of S and the number of available quantum states Ω. The greater the degree of randomness or disorder in a system, the larger Ω and therefore S. We shall encounter many explicit examples in later chapters, but we can anticipate, on qualitative grounds, that, for example: Ω and the entropy increase in the sequence crystal → liquid → gas for the same substance at the same pressure; the entropy of a group of molecules is less than the entropy of the dissociated atoms from which the molecules are made; molecules with small vibrational force constants ("weak interatomic springs") will have higher entropies than those with large force constants; a group of gas molecules occupying a large volume has a higher entropy than when occupying a small volume; etc.

2-3 Second law of thermodynamics. In this section we discuss the second law of thermodynamics from the standpoint of statistical mechanics. First we consider an isolated system.

If an isolated system changes its thermodynamic state infinitesimally, the change in entropy in the process satisfies the relation $dS \geq 0$. This is one way to state the second law of thermodynamics. The inequality holds for any spontaneous (irreversible) change. After the isolated system has exhausted all possible spontaneous changes (each with $dS > 0$) available to it, the entropy will have reached a maximum value. A further infinitesimal change in state will now be reversible, and the equality will hold, $dS = 0$. Similarly, in any finite spontaneous change in thermodynamic state (i.e., both initial and final states are equilibrium states) in an isolated system, $\Delta S > 0$. It is this inequality in particular which we examine.

Probably no system is ever in complete equilibrium with respect to all possible processes. In practice, we require in thermodynamics only that the system be in equilibrium with respect to those rate processes with half-lives short compared with the time available for an experiment. Processes so slow as to proceed to a negligible extent during a thermodynamic measurement present no complication and can be ignored, even though the system is not in equilibrium with respect to such processes. For example, in the absence of a catalyst, the equilibrium properties of a mixture of nitrogen and hydrogen gases can be studied without taking into account the possibility of formation of ammonia molecules, because the rate of formation of these molecules from nitrogen and hydrogen is extremely slow. In quantum-mechanical language, we would say that only those quantum states associated with the presence of only nitrogen and hydrogen are, in fact, *accessible* states; states corresponding to the existence of some ammonia molecules in the system are *inaccessible*. In effect, there is a *restraint* (of chemical kinetic origin, in this case—an activation energy or, rather, activation free energy) in operation which limits the accessibility of quantum states.

However, if we start in this example with a mixture of nitrogen and hydrogen gases at equilibrium in an isolated system and then add a small amount of catalyst (not considered part of the system, for simplicity): (a) some ammonia molecules will be formed spontaneously; (b) the states accessible to the system are now extended, by removal of

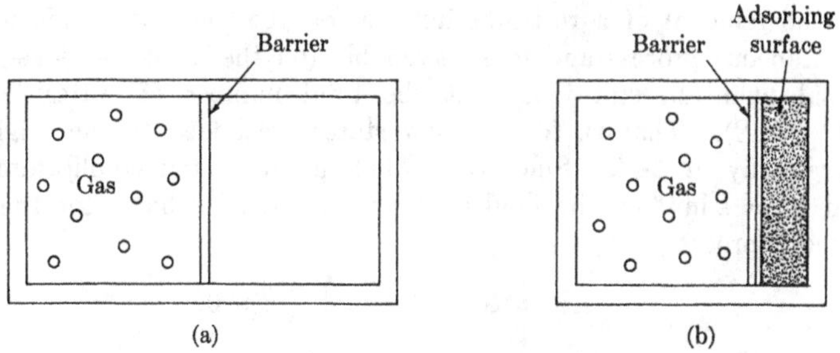

FIG. 2-2. (a) Gas confined to one side of container by a removable barrier. (b) Gas restrained from contact with adsorbing surface by removable barrier.

the restraint,* to include a group of formerly inaccessible states as well as the originally accessible ones; and (c) after the new equilibrium point is reached, now including the chemical equilibrium, we must have $\Delta S > 0$ for this spontaneous process, according to the second law.

Quite generally, any spontaneous thermodynamic process in an isolated system can be viewed in this way: the initial equilibrium state of the system includes a certain set of accessible quantum states but excludes other (inaccessible) quantum states; the spontaneous process leading to the final equilibrium state occurs because of, or is made possible by, the removal of one or more restraints, which makes an additional group of quantum states accessible to the system. Further examples are: (a) a gas is confined to one side of a barrier (Fig. 2-2a), and the barrier (restraint) is then removed; and (b) gas molecules are initially prevented by a barrier from coming in contact with an adsorbing surface (Fig. 2-2b), but then the barrier is removed.

The removal of a restraint can probably always be analyzed in terms of the reduction or obliteration of an activation energy (i.e., potential barrier) or free energy. Thus the function of a catalyst for a chemical reaction is to reduce the activation free energy of the reaction (see Chapter 11). In cases (a) and (b) in the previous paragraph, the physical barrier presents an essentially infinite potential barrier to the passage of molecules, which becomes zero when the physical barrier is taken out.

Suppose an isolated system is characterized by N, V, E (N might be a set of numbers of molecules and V a set of external parameters) and in its initial state has accessible to it $\Omega(N, V, E)$ quantum states. Further,

* We have not defined the word "restraint" because we are using its conventional meaning. "Removal of restraint" always implies that the number of accessible quantum states *increases*. Mayer and Mayer (pp. 81-85) use the word "inhibition" instead of "restraint."

let the removal of a restraint initiate (on the thermodynamic level) a spontaneous process and make available (on the molecular level) additional quantum states so that the total number of states becomes $\Omega'(N, V, E)$. That is, $\Omega' - \Omega$ new states are added to the original Ω. Necessarily, $\Omega' > \Omega$. Since $S = k \ln \Omega$ in the initial equilibrium state and $S' = k \ln \Omega'$ in the final equilibrium state, we have, for the spontaneous process,

$$\Delta S = S' - S = k \ln \frac{\Omega'}{\Omega} > 0. \tag{2-24}$$

This is the statistical-mechanical version of the second law of thermodynamics for an isolated system. It is beautifully simple, following directly from the ideas of the number and accessibility of quantum states.

It is interesting to calculate from the above considerations the probability of the spontaneous occurrence in an isolated system of a process with a negative entropy change, in contradiction to the second law of thermodynamics. For example, suppose we have gas occupying the whole container (Ω' quantum states) in Fig. 2–2(a) and we wish to calculate the probability of observing the gas occupying (spontaneously) only the left half of the container (Ω quantum states). From thermodynamics we know that the entropy change in this process for an ideal gas of N molecules is $-Nk \ln 2$. Now in accordance with the notation of the preceding paragraph, of the total Ω' quantum states accessible when the whole container is available to the gas molecules, Ω of these states correspond to molecules actually being present only in the left half of the container. Thus, since all Ω' quantum states have equal probability of occurrence, the probability that all molecules will be found in the left half is Ω/Ω'. If ΔS represents the (positive) entropy change for the process "left half" → "whole," then (Eq. 2–24)

$$\frac{\Omega}{\Omega'} = e^{-\Delta S/k}, \qquad \Delta S > 0. \tag{2-25}$$

For an ideal gas $\Delta S = Nk \ln 2$, and the probability of observing a violation of the second law in this case is $\Omega/\Omega' = 1/2^N$, a *very* small number of order $1/10^{10^{20}}$. This result can also be deduced from simple probability considerations: the probability that any one of the N molecules will be found on the left is 1/2, so the probability that all will be found on the left is $(1/2)^N$ (ideal gas).

Equation (2–25) is, of course, a general equation applicable to other cases in which the second law would be violated. But as significant experimental values of ΔS are always of order Nk, the probability Ω/Ω' is always very small, of order e^{-N}. That is, a nonnegligible entropy increase always corresponds to a tremendous increase in the number of accessible quantum states: $\Omega' \gg \Omega$. Thus, we conclude that while it is not rigorously impossible for a spontaneous process with a negative entropy change to occur in an isolated system, the probability of such an occurrence is so slight that the process can be considered impossible in actual practice (i.e., operationally), and this is all that thermodynamics requires.

We turn now to a statistical-mechanical analysis of the second law of thermodynamics in closed, isothermal systems. The situation here is a little more complicated, but of more practical interest since isolated systems are exceptional in experimental thermodynamics. Again we have to consider a spontaneous process initiated by the removal of a restraint, but since now the system is immersed in a heat bath, initial and final equilibrium states have the same temperature rather than the same energy (as in an isolated system). The three examples above (chemical reaction, diffusion, adsorption) also serve as examples here, with the understanding that in each case the system is put in thermal contact with a large constant-temperature bath. In all these cases and in general, as is well known, the second law of thermodynamics takes the form $\Delta A < 0$ for a finite, spontaneous process in a closed, isothermal system. It is this inequality that we now wish to deduce from statistical mechanics.

A system in a heat bath has fluctuations in E, as we have seen in Section 2-1. For given N and V, many energy levels E have to be considered. Let $\Omega(N, V, E)$ be the number of quantum states accessible to the system, in the initial thermodynamic state, for a particular energy level E. When the restraint is removed, causing the spontaneous process in question to occur, the number of accessible quantum states for energy E may increase, but cannot decrease since all the Ω original states are still available to the system. That is, $\Omega'(N, V, E) \geq \Omega(N, V, E)$, and *this will be true for every E*. Of course, in principle, some values of E may be energy eigenvalues (i.e., energy levels) in the final state but not in the initial state. If so, $\Omega = 0$ and $\Omega' > 0$ for such an E. In the initial state,

$$Q = \sum_E \Omega(N, V, E)e^{-E/kT}, \tag{2-26}$$

and in the final state,

$$Q' = \sum_E \Omega'(N, V, E)e^{-E/kT}, \tag{2-27}$$

where every possible E in the final state is included in the two sums. Now no term in the sum in Eq. (2-26) can be larger than the corresponding term (same E) in Eq. (2-27), in view of the relation $\Omega' \geq \Omega$ just mentioned above. Hence, since $A = -kT \ln Q$,

$$Q' > Q \qquad \text{and} \qquad \Delta A < 0. \tag{2-28}$$

The above argument provides a general statistical-mechanical justification of the equilibrium criterion A = minimum, $dA = 0$, for a closed, isothermal system (i.e., the system runs through all possible spontaneous processes—each with $\Delta A < 0$—and finally settles, at equilibrium, on the lowest available value of A). A frequent theoretical application of essentially this principle might be mentioned. Suppose that in a statistical-mechanical model of a thermodynamic system a continuously variable parameter ξ enters, in addition to the independent thermodynamic variables. If we are able to find, from the model, $Q(N, V, T, \xi)$ and therefore $A(N, V, T, \xi)$, then the appropriate value of ξ to assign to the real system is that value which minimizes A (maximizes Q) for given N, V, T. An example would be the fraction ξ of holes in a lattice model for a crystal or liquid, with N, V, T fixed. Or, ξ might be the "degree of advancement" of a chemical reaction, or the fraction of molecules in one or the other phase of a two-phase system.

It may be instructive to examine ΔE and ΔS separately for spontaneous isothermal processes with $\Delta A < 0$. Consider first the case where $\Delta E = \overline{E}' - \overline{E}$ is positive. An example: initial state = a gas of hydrogen molecules at T; final state = an equilibrium mixture of hydrogen molecules and atoms at T. The spontaneous process here is the dissociation of some hydrogen molecules into atoms, a process assumed to have a negligible rate in the initial state. $\overline{E}' > \overline{E}$ because the newly accessible quantum states (belonging to a mixture of hydrogen molecules and atoms) in the final thermodynamic state lie in general at higher energies than the original quantum states (belonging to hydrogen molecules only). The bond energy of the hydrogen molecule is of course primarily responsible for this. From Eq. (2–21), in the initial state

$$S = k \ln \Omega(N, V, \overline{E}), \tag{2–29}$$

and in the final state

$$S' = k \ln \Omega'(N, V, \overline{E}'). \tag{2–30}$$

In the example, N is the number of hydrogen molecules started with (i.e., in the initial state). Now the function $\Omega(N, V, E)$ always increases with increasing E (holding N and V constant), as we have already seen (Eq. 1–79). Therefore, since $\overline{E}' > \overline{E}$ in this case,

$$\Omega(N, V, \overline{E}') > \Omega(N, V, \overline{E}).$$

But also, as already emphasized for any E,

$$\Omega'(N, V, \overline{E}') \geq \Omega(N, V, \overline{E}').$$

Hence

$$\Omega'(N, V, \overline{E}') > \Omega(N, V, \overline{E}),$$

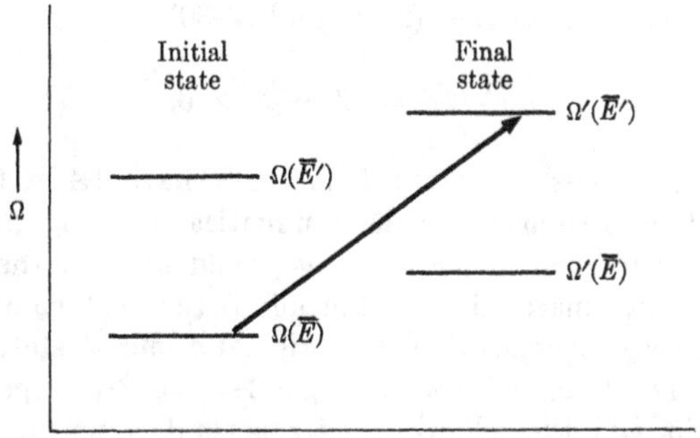

FIG. 2-3. Spontaneous isothermal process with $\Delta A < 0$, $\Delta E > 0$, and $\Delta S > 0$.

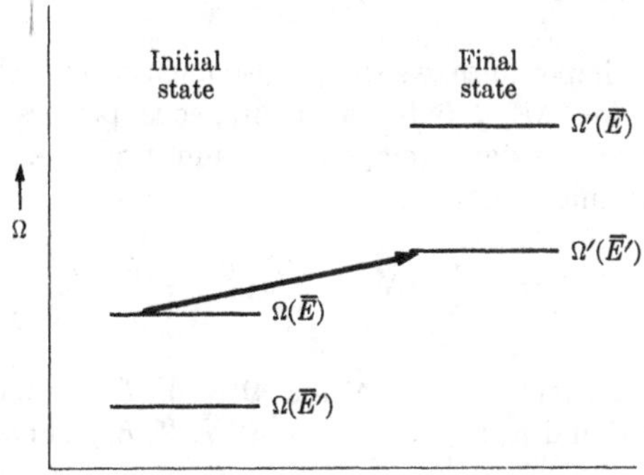

FIG. 2-4. Spontaneous isothermal process with $\Delta A < 0$, $\Delta E < 0$, and $\Delta S > 0$.

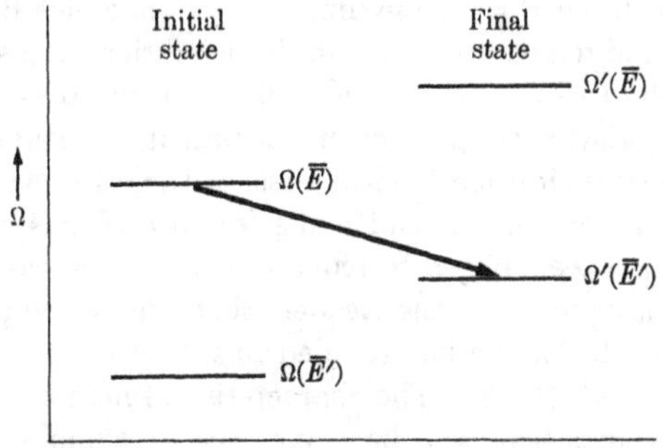

FIG. 2-5. Spontaneous isothermal process with $\Delta A < 0$, $\Delta E < 0$, and $\Delta S < 0$.

and consequently, from Eqs. (2–29) and (2–30),

$$\Delta S = S' - S > 0, \tag{2–31}$$

as is in fact obviously required if we are to have $\Delta A < 0$ and $\Delta E > 0$. This situation is summarized diagrammatically in Fig. 2–3.

On the other hand, if $\Delta E < 0$ (as would occur if the group of new quantum states, inaccessible in the initial thermodynamic state, lies in general at lower energies than the original group of states), ΔS may be positive or negative, as shown in Figs. 2–4 and 2–5 respectively.

We might add the following quite general comment on closed, isothermal systems. Such a system will be thermodynamically more stable the lower A, the larger Q, and therefore, in view of the form of Eq. (2–26) for Q (a sum of terms, all positive), the lower the energy levels E available to the system and the denser (Ω) the quantum states, especially at low energies.

Because of its importance, we should also mention explicitly the thermodynamic condition $\Delta F < 0$ for a spontaneous process occurring in a closed system at constant temperature and pressure. The partition function in the initial state is

$$\Delta = \sum_{E,V} \Omega(N, V, E) e^{-E/kT} e^{-pV/kT}.$$

In the final state, replace Δ by Δ' and $\Omega(N, V, E)$ by $\Omega'(N, V, E)$. For any particular V and E, $\Omega'(N, V, E) \geq \Omega(N, V, E)$, as before. Therefore

$$\Delta' > \Delta \quad \text{and} \quad \Delta F < 0, \tag{2–32}$$

since $F = -kT \ln \Delta$. This is the statistical-mechanical basis for the equilibrium condition $F =$ minimum, $dF = 0$, in a closed system at constant pressure and temperature. From this condition, well-known (thermodynamic) results follow, such as $\Delta\mu = 0$ for chemical (see Section 10–1) or phase equilibria, where $\mu =$ chemical potential. Having shown once and for all the statistical-mechanical basis of the criterion $F =$ minimum (and therefore $\Delta\mu = 0$), we shall make free use of this condition in applications and not feel obliged to rederive it in special cases.

It will be apparent to the reader that the argument leading to Eqs. (2–28) and (2–32) can be extended to any of the partition functions, Eqs. (1–83) through (1–88). The characteristic function on the right side of any of these equations will be maximized at equilibrium if the independent variables are held constant (e.g., $-\beta A$ is a maximum at constant N, V, β).

2-4 Third law of thermodynamics. Our object here is to present the statistical-mechanical basis of the third law of thermodynamics. But first we review very briefly the status of the third law in pure thermodynamics, and its relation to certain thermodynamic integration constants. As usual, there are some fine points that we shall omit or pass over lightly in this introductory discussion. The reader is referred to the Supplementary Reading list at the end of the chapter for fuller discussion.

It has been found experimentally that $\Delta S \rightarrow 0$ as $T \rightarrow 0$ for a large number of isothermal processes involving pure phases. Examples are: (a) a phase transition at temperature T between two different crystalline forms of the same substance (e.g., tin or silicon dioxide); (b) the phase transition between liquid and solid helium at T; (c) a chemical reaction between pure crystals at p and T (e.g., $Pb + I_2 \rightarrow PbI_2$); and (d) a crystal at T and $x \rightarrow$ crystal at T and x', where x represents one or more independent thermodynamic variables other than T (e.g., $x =$ pressure, volume, or magnetic field strength). The experimental result, $\Delta S \rightarrow 0$, for systems of this type is presumed to be general (with some understandable exceptions) and is not a consequence of the first two laws of thermodynamics. Hence the *third law of thermodynamics* has been proposed. It can be stated as follows: for any isothermal process involving only pure phases in internal equilibrium (or including frozen metastable phases if the process does not disturb the frozen equilibrium),

$$\lim_{T \rightarrow 0} \Delta S = 0. \tag{2-33}$$

This statement does not exclude, for example, processes in which metastable atomic nuclei are present but unchanged in the process, or in which negligibly slow chemical reactions are thermodynamically possible but do not occur. But it does exclude, say, the process glass $\rightarrow$ crystal. Also, the word "pure" does not exclude isotopic mixtures if the mixing is unaffected by the process.

The term "metastable" is closely related to the terms "accessible" and "inaccessible" quantum states. Metastability implies the possibility of another, more stable, thermodynamic state which, however, is unavailable because of the inaccessibility of the associated quantum states (see Section 2-3).

An essentially equivalent but slightly more general statement of the third law* is: it is impossible by any procedure, no matter how idealized, to reduce any thermodynamic system to the absolute zero of temperature in a finite number of operations. This is called "the principle of the un-

* See Fowler and Guggenheim, pp. 224–227.

attainability of absolute zero." We shall not use the third law in this form, or establish the connection with Eq. (2–33).

To illustrate the use of the third law in thermodynamics, consider, as an example, the simple gaseous chemical reaction

$$A \rightarrow B$$

occurring at p and T, and meeting the metastability restrictions of Eq. (2–33). From experimental heat capacities (extrapolated to $T = 0$) and heats of phase transitions, we have the well-known equations

$$H_A(T) = H_A(0) + h_A(T), \tag{2–34}$$

$$S_A(T) = S_A(0) + g_A(T), \tag{2–35}$$

where

$$h_A(T) = \sum \Delta H^A_{\text{trans}} + \int_0^T C^A_p(T')\, dT',$$

$$g_A(T) = \sum \Delta S^A_{\text{trans}} + \int_0^T \frac{C^A_p(T')}{T'}\, dT'.$$

$H_A(0)$ and $S_A(0)$ are integration constants, and the sums are over the phase transitions encountered at p between $T' = 0$ and $T' = T$. Similar equations can be written for B. Integration of $\partial F_A/\partial T = -S_A$ gives, further,

$$F_A(T) = H_A(0) - TS_A(0) - \int_0^T g_A(T')\, dT'. \tag{2–36}$$

The integration constant $F_A(0)$ is equal to $H_A(0)$, since the relation $F = H - TS$ becomes $F = H$ at $T = 0$. Then for the reaction $A \rightarrow B$,

$$\Delta H(T) = H_B(T) - H_A(T) = \Delta H(0) + \Delta h(T), \tag{2–37}$$

$$\Delta S(T) = \Delta S(0) + \Delta g(T), \tag{2–38}$$

$$\Delta F(T) = \Delta H(0) - T\,\Delta S(0) - \int_0^T \Delta g(T')\, dT'. \tag{2–39}$$

The value of the constant $H_A(0)$ depends on the location of the zero of energy, and this is arbitrary for any single substance. But in a process such as $A \rightarrow B$, the choices of $H_A(0)$ and $H_B(0)$ have to be self-consistent. That is, $\Delta H(0)$ is not arbitrary, but is the limit ($T \rightarrow 0$) of an experimental quantity. In general, $\Delta H(0) \neq 0$. The third law simplifies Eqs. (2–38) and (2–39) considerably by asserting that $\Delta S(0) = 0$ in these equations. There is no operational (experimental) justification—since thermodynamics gives us information about entropy *differences* only—for going beyond this and simplifying Eqs. (2–35) and (2–36) by setting $S_A(0) = 0$ as well. But this step is obviously convenient and is consistent with the requirement $\Delta S(0) = 0$, so it is adopted as a convention for pure phases (crystals, except for helium). That is, $S_A(T)$ is set equal to the experimental (including an extrapolation) quantity $g_A(T)$.

With this condensed thermodynamic survey as background, we turn now to the statistical-mechanical side. The first point that requires comment is the integration constant c in Eq. (1–21), where we remarked that c is independent of the thermodynamic state of a closed system. This is a sufficiently inclusive observation for our present purposes, provided we understand "state" in a general way. For example, any of the isothermal processes (a) through (d) mentioned at the beginning of this section may be regarded as involving changes in the thermodynamic state of a closed system. In every case, one can in principle proceed reversibly from the initial state to the final state by infinitely slow changes in one or more parameters on which the state depends (volume, pressure, degree of advancement, etc.). Thus for all isothermal processes embraced by the third law of thermodynamics (Eq. 2–33), the constant c is independent of these parameters of state, and hence for such a process (the prime refers to the final state),

$$\begin{aligned} \Delta S = S' - S &= (k \ln \Omega' + c') - (k \ln \Omega + c) \\ &= k \,\Delta \ln \Omega, \qquad \Delta c = 0. \end{aligned} \tag{2–40}$$

The value assigned to c clearly has no operational significance at all, for it always cancels on taking differences. Hence we can put $c = 0$ if we wish, as in Chapter 1.

Having thus disposed of Δc, the remaining question is whether statistical mechanics predicts, as is required for agreement with Eq. (2–33), that

$$\lim_{T \to 0} \Delta \ln \Omega \equiv \Delta \ln \Omega_0 = 0 \tag{2–41}$$

for the processes included in our above statement of the third law. The subscript zero refers to the ground state, the quantum state of interest, since $T \to 0$. If we include only the degrees of freedom associated with the electrons and centers of mass of the nuclei (electronic, vibrational, rotational, translational), we can say that quantum mechanics predicts or experiment indicates in known cases that the ground state of a pure crystal or liquid helium is nondegenerate.* For these degrees of freedom, then, $\Omega_0 = \Omega_0' = 1$. Now, fortunately, this is a much stronger statement

* See Fowler and Guggenheim, pp. 199–205.

than Eq. (2–41) requires. For one thing, as far as thermodynamic consequences (such as the third law) are concerned, a value of Ω_0 in the range $1 \ll \Omega_0 \ll e^N$ is indistinguishable from $\Omega_0 = 1$, as can be seen from Eq. (2–40) $[\Delta S = O(Nk)]$. This is a considerable safety factor. More important, if we include intranuclear degrees of freedom, about which there is still a vast amount of ignorance, we have to expect in general that $\ln \Omega_0$ and $\ln \Omega_0'$ are not thermodynamically negligible, owing to degeneracy of the nuclear ground state or possibly of a metastable state. But the third law is saved despite this because we expect $\Omega_0 = \Omega_0'$ (ordinary thermodynamic changes do not affect the nuclear state), and hence $\Delta \ln \Omega_0 = 0$. Similarly, if we include the entropy of isotopic mixing, which is present in most real thermodynamic systems, $\ln \Omega_0$ and $\ln \Omega_0'$ will not be negligible, but still $\Omega_0 = \Omega_0'$. The same remarks can be made yet again concerning any type of metastability, allowed in our statement of the third law, which introduces nonnegligible degeneracy at 0°K.

We come to the conclusion, then, that for the same systems which obey Eq. (2–33) thermodynamically, we anticipate on molecular grounds that Eq. (2–41) will also be obeyed. This is all that we need for a statistical-mechanical understanding of the third law.

But we also conclude that $\ln \Omega_0$ and $\ln \Omega_0'$ separately are in general unknown and may not be thermodynamically negligible. This conclusion destroys any hope of setting up a scale of truly "absolute entropies" for single substances (in addition, there is the arbitrariness of the additive constant c).

In spite of the above remarks, it is conventional to introduce, for pure substances, what might be called "practical absolute entropies." If we exclude any substance which possesses at 0°K frozen metastability that "thaws out" between 0°K and T°K (we shall encounter some of these exceptions in Chapter 9), we have

$$S(T) = c + k \ln \frac{\Omega(T)}{\Omega_0} + k \ln \Omega_0 = S(0) + g(T),$$

or

$$S(T) - c - k \ln \Omega_0 = k \ln \frac{\Omega(T)}{\Omega_0} = g(T). \qquad (2\text{–}42)$$

In these equations one might just as well set $c = 0$. In Eq. (2–42), the expression on the left is the definition of the practical absolute entropy, the second expression is the statistical form of it, and the third is the thermodynamic form (Eq. 2–35). In other words: (a) $c + k \ln \Omega_0$ has the same significance as $S(0)$ in Eq. (2–35), and both are nonoperational quantities; (b) Ω/Ω_0 is the number of quantum states associated with the electrons and centers of mass of the nuclei (i.e., with the electronic, vibrational, rotational, and translational degrees of freedom); and (c) the Ω/Ω_0 quantum states referred to in (b) are the ones that are "excited" in ordinary thermodynamic systems when the system absorbs heat ($DQ^* = T\,dS$) to increase its temperature from 0°K to T°K.

For simplicity, in the rest of the book (except in Chapter 9) we shall write Eq. (2–42) simply as

$$S(T) = k \ln \Omega(T) = g(T), \tag{2–43}$$

with the understanding that we are referring to the "operational" part of the entropy only: we count in Ω only those quantum states which are actually "excitable" and exclude those which are not (nuclear ground state, isotope mixing, any metastability frozen over the whole temperature range). As we pointed out following Eq. (2–41), this $\Omega = 1$ in the ground state, and hence this $S = 0$ at $T = 0$°K.

The entropy calculated as $k \ln \Omega$ in Eq. (2–43) is often called the "spectroscopic entropy," since energy levels from spectroscopy are used in practice; and the entropy measured as g (with an extrapolation of C_p to 0°K) is called the "calorimetric entropy." We shall compare these quantities in Chapter 9, and discuss some exceptions (owing to "thawed" metastability) to the expected equality.

PROBLEMS

2-1. Note that Eqs. (1-83) through (1-85) can be written in the form

$$\ln \sum_{G_1} \Omega(G_1, G_2, G_3)e^{g_1G_1} = \varphi_{\text{I}}(g_1, G_2, G_3) = \frac{S}{k} + g_1\overline{G}_1,$$

where g is intensive and G extensive. Show that

$$\overline{G_1^2} - (\overline{G}_1)^2 = \left(\frac{\partial^2\varphi_{\text{I}}}{\partial g_1^2}\right)_{G_2,G_3} = \left(\frac{\partial \overline{G}_1}{\partial g_1}\right)_{G_2,G_3}.$$

Also, Eqs. (1-86) through (1-88) can be written

$$\ln \sum_{G_1,G_2} \Omega(G_1, G_2, G_3)e^{g_1G_1}e^{g_2G_2} = \varphi_{\text{II}}(g_1, g_2, G_3) = \frac{S}{k} + g_1\overline{G}_1 + g_2\overline{G}_2.$$

Show that

$$\overline{G_1^2} - (\overline{G}_1)^2 = \left(\frac{\partial^2\varphi_{\text{II}}}{\partial g_1^2}\right)_{g_2,G_3} = \left(\frac{\partial \overline{G}_1}{\partial g_1}\right)_{g_2,G_3},$$

$$\overline{G_1G_2} - \overline{G}_1\overline{G}_2 = \left(\frac{\partial^2\varphi_{\text{II}}}{\partial g_1\,\partial g_2}\right)_{G_3} = \left(\frac{\partial \overline{G}_1}{\partial g_2}\right)_{g_1,G_3} = \left(\frac{\partial \overline{G}_2}{\partial g_1}\right)_{g_2,G_3}.$$

(Page 34.)

2-2. Show that the grand ensemble equations for S and p go over into the canonical ensemble equations for the same variables when $\ln \Xi$ is replaced by the logarithm of the largest term in Ξ. (Page 39.)

2-3. Prove the following: if the probability distribution for G is gaussian about $\overline{G}$ with $\sigma_G/\overline{G}$ very small, then the probability distribution for $1/G$ about $1/\overline{G}$ is gaussian with

$$\frac{\sigma_{1/G}}{1/\overline{G}} = \frac{\sigma_G}{\overline{G}}.$$

Show that this result and Eq. (2-14) lead to the density fluctuation formula (2-10).

2-4. Show that in the N, p, T ensemble (Eq. 2-12),

$$\overline{H^2} - (\overline{H})^2 = kT^2C_p.$$

2-5. Show that in a two-component, open, isothermal system,

$$\overline{N_1N_2} - \overline{N}_1\overline{N}_2 = kT\left(\frac{\partial \overline{N}_1}{\partial \mu_2}\right)_{V,T,\mu_1} = kT\left(\frac{\partial \overline{N}_2}{\partial \mu_1}\right)_{V,T,\mu_2}.$$

2-6. Prove that in the canonical ensemble (see Problem 1-2),

$$\overline{(E - \overline{E})(p - \overline{p})} = kT\left[\left(\frac{\partial E}{\partial V}\right)_{N,T} + p\right] = kT^2\left(\frac{\partial p}{\partial T}\right)_{N,V}.$$

2–7. Show that the canonical ensemble equations for p and μ go over into the microcanonical ensemble equations for the same variables when $\ln Q$ is replaced by the logarithm of the largest term in Q.

2–8. If one combines Eqs. (2–34) and (2–35) to obtain $F_A(T) = H_A(T) - TS_A(T)$, the resulting expression appears to differ from Eq. (2–36). Prove that the two equations for $F_A(T)$ are equivalent.

2–9. Justify the statement above Eq. (2–1) that the energy distribution is gaussian.

Supplementary Reading

Fowler, R. H., *Statistical Mechanics*, Cambridge, 1936 (Second Edition), Chapter 20.

Fowler and Guggenheim, Chapter 5.

Greene, R. F., and Callen, H. B., *Phys. Rev.* **83**, 1231 (1951).

Kittel, Part 2.

Landau and Lifshitz, Chapter 12.

Mayer and Mayer, Chapters 3 and 4.

Schrödinger, Chapter 3.

S. M., Chapters 3 and 4.

Tolman, Chapter 14.

Wilson, Chapter 7.

Part II

Systems Composed of Independent Molecules or Subsystems

CHAPTER 3

GENERAL RELATIONS FOR INDEPENDENT DISTINGUISHABLE AND INDISTINGUISHABLE MOLECULES OR SUBSYSTEMS

The simplest problems in statistical mechanics are those concerned with systems composed of molecules, groups of molecules, or degrees of freedom which are effectively independent of each other. Part II of this book is devoted to problems of this type, except for Chapter 6, which returns to more general considerations. A lack of such independence is usually associated with two causes: (a) intermolecular forces (treated primarily in Part III), and (b) symmetry restrictions on quantum-mechanical wave functions (discussed in Part IV). Of course, no statistical-mechanical system can consist of *strictly* noninteracting molecules (or subsystems), for then the system could not achieve internal equilibrium. By "independent" we shall imply "weak interaction" only: the molecules or degrees of freedom interact sufficiently to maintain thermal equilibrium by energy exchange, but not to such an extent as to require taking into account intermolecular forces, etc. The "weak interaction" may be direct (e.g., by collisions) or indirect (e.g., via the walls or a heat bath). The most obvious example is an ideal gas; the density is low enough so that intermolecular forces make a negligible contribution to thermodynamic properties (e.g., the equation of state), but equilibrium is still maintained by intermolecular collisions and/or by collisions with the walls.

The "subsystems" of a system, other than molecules, referred to in the chapter title and above might be, for example: (a) different degrees of freedom in the same molecule (translation, rotation, etc.); (b) independent vibrational modes in a monatomic crystal; (c) molecules adsorbed on independent groups of adsorption sites on a solid surface; etc.

Before turning in the next chapter to our first specific example of a system of independent molecules, we set down in the present chapter certain relationships of quite general validity for systems of this type. A few brief comments will be made about possible applications, but details in every case will be reserved for the appropriate chapter later in the book.

3–1 Independent and distinguishable molecules or subsystems. Let H be the classical Hamiltonian function for the macroscopic system under consideration. If the system is composed of independent molecules or subsystems, by definition H will be given by a sum of independent contributions,

$$H = H_a + H_b + \cdots, \tag{3-1}$$

where H_a, H_b, etc., refer to individual molecules, degrees of freedom, etc. Similarly, for the Hamiltonian operator,

$$\mathcal{H} = \mathcal{H}_a + \mathcal{H}_b + \cdots. \tag{3-2}$$

Let the eigenvalues and eigenfunctions of $\mathcal{H}_a$, $\mathcal{H}_b$, etc., be denoted by $\epsilon_a, \epsilon_b, \ldots$ and $\psi_a, \psi_b, \ldots$, respectively. Then we try, as a solution of $\mathcal{H}\psi = E\psi$, the product $\psi = \psi_a \psi_b \ldots$, and find

$$\begin{aligned}\mathcal{H}\psi &= (\mathcal{H}_a + \mathcal{H}_b + \cdots)\psi_a\psi_b \cdots \\ &= \psi_b\psi_c \cdots \mathcal{H}_a\psi_a + \psi_a\psi_c \cdots \mathcal{H}_b\psi_b + \cdots \\ &= \psi_b\psi_c \cdots \epsilon_a\psi_a + \psi_a\psi_c \cdots \epsilon_b\psi_b + \cdots \\ &= (\epsilon_a + \epsilon_b + \cdots)\psi = E\psi.\end{aligned}$$

That is, the possible energy eigenvalues for the whole system are of the form

$$E = \epsilon_a + \epsilon_b + \cdots, \tag{3-3}$$

just the sum of the separate energies $\epsilon_a, \epsilon_b, \ldots$, as we might expect. The essential practical point here is the following: for a system of independent molecules or subsystems, we do not have to solve the complete Schrödinger equation $\mathcal{H}\psi = E\psi$ with $O(10^{20})$ coordinates, but only the separate equations $\mathcal{H}_a\psi_a = \epsilon_a\psi_a$, etc., each with one or relatively few coordinates. Thus we have *a reduction of a many-body problem to a one- (or few-) body problem,* because of the form of Eq. (3–1).

For one molecule or subsystem at a time, let us define partition functions of the canonical ensemble type,

$$q_a = \sum_j e^{-\epsilon_{aj}/kT}, \qquad q_b = \sum_j e^{-\epsilon_{bj}/kT}, \qquad \text{etc.,} \tag{3-4}$$

where the sums are over molecular or subsystem energy states j (an energy level with degeneracy is represented by several terms in the sum). Next we notice that the product of all the q's generates all possible values of E. For example, suppose there are just two subsystems, one with three energy states and one with two states. Then

$$\begin{aligned}q_aq_b &= (e^{-\epsilon_{a1}/kT} + e^{-\epsilon_{a2}/kT} + e^{-\epsilon_{a3}/kT})(e^{-\epsilon_{b1}/kT} + e^{-\epsilon_{b2}/kT}) \\ &= e^{-(\epsilon_{a1}+\epsilon_{b1})/kT} + e^{-(\epsilon_{a1}+\epsilon_{b2})/kT} + e^{-(\epsilon_{a2}+\epsilon_{b1})/kT} \\ &\quad + e^{-(\epsilon_{a2}+\epsilon_{b2})/kT} + e^{-(\epsilon_{a3}+\epsilon_{b1})/kT} + e^{-(\epsilon_{a3}+\epsilon_{b2})/kT}.\end{aligned}$$

The sums of ϵ's in the exponents exhaust the possible energy states E for the system. Thus we have in this simple example, and in general,

$$Q = \sum_j e^{-E_j/kT} = \left(\sum_j e^{-\epsilon_{aj}/kT}\right)\left(\sum_j e^{-\epsilon_{bj}/kT}\right)\cdots = q_a q_b \cdots. \qquad (3\text{–}5)$$

That is, the canonical ensemble partition function Q for the whole system is a product of separate q's for the independent molecules or subsystems.

In Chapter 5 we shall apply Eq. (3–5) to the small vibrations of identical atoms about their equilibrium positions in a monatomic crystal. The model adopted is restricted: each atom in the crystal is confined in its motion to the immediate neighborhood of a particular site (equilibrium position) in the crystal lattice. Of course, in real crystals the atoms can jump occasionally from one site to another, but this is not allowed in the model. Now suppose we number the sites in the crystal lattice in some regular order. Then these labels can be considered as belonging as well to the particular atoms confined to the various sites. In this sense, and for this model, we can regard the atoms of the system to be identical and *distinguishable*. This is an exception (owing to the idealized model) to the general quantum-mechanical principle that identical molecules are *indistinguishable*, i.e., that no experiment can distinguish between the order *ab* and the order *ba* for two identical molecules (see Section 22–1).

Note that Eq. (3–5) has been derived on the basis that the molecules or subsystems are independent and distinguishable (labels a, b, . . .). The assumption of distinguishability is obviously correct if the molecules or subsystems are all different. If, however, the molecules are independent and identical, then Eq. (3–5) is correct only if a model artificially introduces molecular distinguishability. In this case, the energy states for each of the identical molecules will be the same, and

$$q_a = q_b = \cdots \equiv q = \sum_j e^{-\epsilon_j/kT},$$

$$Q = q^N. \qquad (3\text{–}6)$$

The Einstein model of a crystal (Chapter 5) is the most obvious example, but not the only one.

Incidentally, the independent subsystems might be *macroscopic* (and therefore distinguishable) parts of the complete system in thermal contact with each other. Then we would write Eq. (3–5) in the notation

$$Q = Q_A Q_B \ldots. \qquad (3\text{–}7)$$

We have already encountered an example of this in Eq. (1–23). Equation (3–7) implies that A, S, E, etc., are additive; that is,

$$A = A_A + A_B + \cdots, \quad \text{etc.}$$

3–2 Independent and indistinguishable molecules. In a gas of N identical and independent molecules, or in a simplified model of a liquid (Chapter 16) in which the molecules are assumed to be independent of each other, we are faced with the problem of handling independent and *indistinguishable* molecules. Let us return to the simple example following Eq. (3–4), where now a and b are identical and indistinguishable. Here, from the quantum-mechanical point of view, the arrangement $\epsilon_{a1} + \epsilon_{b2}$ is operationally indistinguishable from the arrangement $\epsilon_{a2} + \epsilon_{b1}$. That is, these two arrangements belong to a single quantum state* and should count only once instead of twice in the sum $Q = \sum_j e^{-E_j/kT}$ (Eq. 3–5), which is supposed to be over all quantum states of the system, with one term for each state.

In general, a term in Q of the form

$$e^{-(\epsilon_{ai}+\epsilon_{bj}+\cdots)/kT}, \tag{3–8}$$

where $a, b, \ldots$ are identical and indistinguishable molecules, each in a *different* molecular quantum state ($i \neq j \neq \cdots$) as in the example above, will occur $N!$ times if we use Eq. (3–6) for Q. This follows because the N states $i, j, \ldots$ can be permuted in $N!$ ways among the N molecules $a, b, \ldots$, and all of these $N!$ ways will appear in the product q^N. The reader may wish to verify this by forming the product $q_a q_b q_c$ for three identical molecules and three different molecular states, 1, 2, 3. If Eq. (3–6) for Q contained terms of this type only, i.e., with each molecule in a different molecular quantum state, the correction for indistinguishability would be easy. We would simply divide q^N by $N!$ so as not to weight each quantum state of the system $N!$ times in the sum $\sum_j e^{-E_j/kT}$ instead of only once.

Actually, many terms in Eq. (3–6) are not of this sort. For example,

$$e^{-(\epsilon_{ai}+\epsilon_{bi}+\epsilon_{cj}+\cdots)/kT}, \tag{3–9}$$

where at least two molecules are in the *same* molecular quantum state i. These terms lead to complications: (a) such states of the system are allowed in Bose-Einstein statistics (Chapter 22), but $N!$ is no longer the proper correction factor; and (b) in Fermi-Dirac statistics (Chapter 22), quantum states of the system in which two identical molecules are in the same molecular quantum state are not allowed at all, i.e., such states should not appear in the sum $\sum_j e^{-E_j/kT}$.

In a quantum-statistical treatment of identical and indistinguishable molecules, the two possibilities, Bose-Einstein and Fermi-Dirac statistics, have to be discussed as separate cases. We postpone such a discussion to Chapter 22, because it involves more quantum mechanics than we wish to use in Parts I, II, and III.

---

* This point is amplified considerably in Chapter 22.

We can make further progress at this point only if we restrict ourselves to an important special case, which is a limiting or asymptotic form of both Bose-Einstein and Fermi-Dirac statistics: the number of molecular states "available" (i.e., with energies between the molecular ground state and the ground state plus, say, $10kT$, more or less) is *very large* compared with N, the number of molecules in the system. In this case, in the product $Q = q^N$, the number of terms of type (3–9) will be negligibly small [virtually all terms will be of type (3–8)] simply because so many choices of molecular states are available to the molecules that two molecules only rarely find themselves in the same state. The Bose-Einstein and Fermi-Dirac complications above then disappear. Thus, the correction $N!$ is now appropriate for all terms making a significant contribution to Q, and we have

$$Q = \frac{1}{N!}\, q^N \tag{3–10}$$

for a system of identical, indistinguishable molecules satisfying the condition that the number of available molecular states is much greater than N. We shall see in Chapter 4 that for a monatomic ideal gas (or for the translational quantum states of a diatomic or polyatomic gas), this condition is satisfied at ordinary temperatures and densities.

The limiting form of Bose-Einstein and Fermi-Dirac statistics, represented by Eq. (3–10), is called *classical* or *Boltzmann* statistics. The interconnections between the three types of statistics will be examined in Chapter 22.

In the expression $q = \sum_j e^{-\epsilon_j/kT}$ for a molecule in a gas of independent molecules, the ϵ_j depend only on the volume V, besides molecular parameters such as mass, moments of inertia, force constants, etc. We shall see this in detail in Chapters 4, 8, and 9. In other words, $q = q(V, T)$, a function of the thermodynamic variables V and T only. It is interesting that this conclusion alone [i.e., without the exact form of $q(V, T)$], together with Eq. (3–10), determines the ideal gas equation of state (Problem 3–1), for we have

$$\Xi = \sum_{N\geq 0} Q(N, V, T)\lambda^N = \sum_{N\geq 0} \frac{[q(V, T)\lambda]^N}{N!} = e^{q\lambda} = e^{pV/kT} \tag{3–11}$$

and

$$\overline{N} = \lambda\left(\frac{\partial \ln \Xi}{\partial \lambda}\right)_{V,T} = q\lambda = \frac{pV}{kT}. \tag{3–12}$$

Of course, we still have to investigate more closely, in Chapter 4, the conditions under which Eq. (3–10), which we have used here, is valid. Incidentally, we cannot deduce Eq. (3–12) for an approximate model (mentioned at the beginning of this section) of a liquid in which the molecules are assumed to be independent, because in such a model q will be a function not of V and T but of V/N and T (Chapter 16).

Returning to the relation $Q = q^N/N!$ for a one-component system, where $q = \sum_j e^{-\epsilon_j/kT}$, an important simplification or approximation (Chapters 4, 8, and 9) is that each ϵ is itself expressible as a sum of separate contributions, because the Hamiltonian for one molecule is at least approximately separable. For example,

$$\begin{aligned} H &= H_t + H_r + H_v + H_e, \\ \epsilon &= \epsilon_t + \epsilon_r + \epsilon_v + \epsilon_e, \end{aligned} \tag{3-13}$$

where t = translation, r = rotation, v = vibration, and e = electronic In this case (Chapters 8 and 9), just as in Eq. (3–5),

$$q = q_t q_r q_v q_e$$

$$Q = \frac{1}{N!}(q_t q_r q_v q_e)^N. \tag{3-14}$$

Equation (3–10) is easily generalized for systems with several components. For example, for a binary mixture (Problem 3–2),

$$Q = \frac{q_1^{N_1} q_2^{N_2}}{N_1! N_2!}. \tag{3-15}$$

3–3 Energy distribution among independent molecules. Consider first a system of N independent, *distinguishable* molecules, with $q = \sum_j e^{-\epsilon_j/kT}$. (At the end of this section we shall see that the final equations derived here for independent, distinguishable molecules are also valid for independent, indistinguishable molecules, provided that the use of Boltzmann statistics is justified.) The question arises: What fraction of these molecules will be in some particular molecular quantum state, say state i (energy ϵ_i)? An equivalent question is: What is the probability that one particular molecule, say molecule 1, will be observed, in an experiment, to be in state i? Or, what fraction of the time does molecule 1 (or any molecule) spend in state i? To answer, we use Eq. (1–28) as a starting point. According to this equation, the probability that molecule 1 is in state i, molecule 2 in state j, molecule 3 in state l, etc., is

$$P_{ijl\cdots} = \frac{e^{-E_{ijl\cdots}/kT}}{Q} = \frac{e^{-(\epsilon_i+\epsilon_j+\epsilon_l+\cdots)/kT}}{q^N}, \tag{3-16}$$

since an assignment of a molecular state to each molecule specifies a definite quantum state of the whole system. Let η_i be the probability that molecule 1 is in state i *irrespective* of the states of the other molecules. This is just the probability asked for above. Then

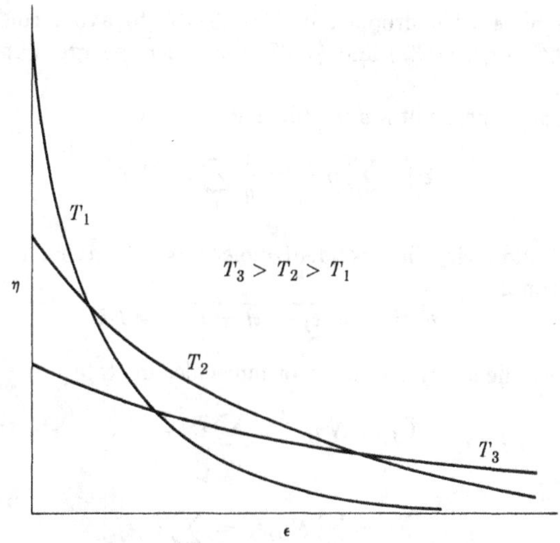

FIG. 3-1. Probability η of a molecule having an energy ϵ, at three different temperatures (schematic).

$$\eta_i = \sum_{j,l,\ldots} P_{ijl\ldots} = \frac{e^{-\epsilon_i/kT}(\sum_j e^{-\epsilon_j/kT})(\sum_l e^{-\epsilon_l/kT})\ldots}{q^N}$$

$$= \frac{e^{-\epsilon_i/kT}}{q}. \tag{3-17}$$

This has precisely the form of Eq. (1-28) itself, but molecular states replace system states. Of course, $\sum_i \eta_i = 1$. The probability η_i decreases exponentially with increasing ϵ_i, and the fall-off with ϵ_i is more rapid at lower temperatures (Fig. 3-1). At very low temperatures, practically all molecules will be in the ground state (ϵ_1): $\eta_1 \to 1$ as $T \to 0$. At high temperatures, $\eta_i \to$ constant (independent of ϵ_i). We shall encounter several numerical examples of Eq. (3-17) in later chapters.

For simplicity of notation we shall continue in this chapter to use the language of molecular *states*. But for many purposes, it is more convenient to use molecular energy *levels*. If ω is the degeneracy of the level ϵ, then

$$\eta\ (\text{level}) = \omega\eta\ (\text{state}) = \frac{\omega e^{-\epsilon/kT}}{q}, \tag{3-18}$$

$$q = \sum_{\substack{j \\ (\text{states})}} e^{-\epsilon_j/kT} = \sum_{\substack{i \\ (\text{levels})}} \omega_i e^{-\epsilon_i/kT}. \tag{3-19}$$

Subscripts have been dropped in Eq. (3–18) to avoid confusion. The analogy with Eqs. (1–37) and (1–38) for *system* energy states and levels is obvious.

The average energy of a molecule is

$$\bar{\epsilon} = \sum_j \eta_j \epsilon_j = \frac{1}{q} \sum_j \epsilon_j e^{-\epsilon_j/kT}. \tag{3-20}$$

Because of additivity (independent molecules), the average energy of the system is simply

$$\bar{E} = \overline{\epsilon_i + \epsilon_j + \epsilon_l + \cdots} = N\bar{\epsilon}. \tag{3-21}$$

If we denote the average number of molecules in state j by $\bar{C}_j$, then

$$\bar{C}_j = N\eta_j, \qquad \sum_j \bar{C}_j = N, \tag{3-22}$$

so that

$$\bar{E} = \sum_j N\eta_j\epsilon_j = \sum_j \bar{C}_j\epsilon_j. \tag{3-23}$$

Fluctuations in ϵ are of some interest. If we differentiate the equation

$$q\bar{\epsilon} = \sum_j \epsilon_j e^{-\epsilon_j/kT}$$

with respect to temperature, we find

$$\overline{\epsilon^2} - (\bar{\epsilon})^2 = kT^2 \frac{\partial \bar{\epsilon}}{\partial T} = \frac{kT^2 C_V}{N}. \tag{3-24}$$

The last relation follows from Eq. (3–21). Since $C_V = O(Nk)$ and $\bar{\epsilon} = O(kT)$, we see that $[\overline{\epsilon^2} - (\bar{\epsilon})^2]^{1/2}$ is of the same order of magnitude as $\bar{\epsilon}$ itself. That is, the energy probability distribution function for a single molecule is broad and *not* sharp, in contrast to the total energy distribution for a system of many molecules (Eq. 2–5). The order of magnitude we find here for the molecular energy fluctuations is consistent with the order predicted by Eq. (2–5) if we put $N = 1$. A sharp energy probability distribution is a many-molecule effect: the large relative fluctuations in the energies of individual molecules practically cancel each other, leaving only an extremely small relative fluctuation in the total energy.

A broad probability distribution is in general not gaussian, and the shape of the curve is not completely characterized by the quantity $\overline{\epsilon^2} - (\bar{\epsilon})^2$. The most familiar example of a broad distribution is the Maxwell-Boltzmann distribution in velocity or translational kinetic energy for a classical one-component system (Chapter 6), usually chosen as an ideal monatomic gas.

At this point let us take advantage of the relative simplicity of a system of independent molecules to look again [see Eqs. (1-17) and (1-18)] at the molecular interpretation of thermodynamic work (W) and heat (Q^*). From Eq. (3-23), holding N constant,

$$d\overline{E} = \sum_j \epsilon_j \, d\overline{C}_j + \sum_j \overline{C}_j \, d\epsilon_j = DQ^* - DW. \tag{3-25}$$

Then

$$DW = p\,dV = \sum_j \overline{C}_j \left(- \frac{\partial \epsilon_j}{\partial V} \right) dV, \tag{3-26}$$

or

$$p = \sum_j \overline{C}_j \left(- \frac{\partial \epsilon_j}{\partial V} \right). \tag{3-27}$$

Each molecule makes an additive contribution to the pressure, in an amount depending on the change of its energy ϵ_j with volume. When work DW is done by the system, the (average) population of different energy states remains fixed, but the energies (ϵ_j) themselves shift. The negative of the thermodynamic work, $-DW$, is the total work necessary to shift the energy levels of all the molecules in the system when the volume is varied by an amount dV.

Also, from Eq. (3-25),

$$DQ^* = \sum_j \epsilon_j \, d\overline{C}_j, \tag{3-28}$$

which says that heat DQ^* is absorbed by shifting the population of energy states, holding the energy levels fixed. Consider a special case: heat is absorbed by the system as a result of a temperature increase dT, with N and V both held constant. Then

$$d\overline{E} = C_V \, dT = DQ^* = T\,dS = \sum_j \epsilon_j \frac{\partial \overline{C}_j}{\partial T} \, dT. \tag{3-29}$$

The change in population of state j is, from Eqs. (3-17) and (3-22),

$$d\overline{C}_j = \frac{\partial \overline{C}_j}{\partial T} \, dT = \frac{\overline{C}_j(\epsilon_j - \bar{\epsilon}) \, dT}{kT^2}. \tag{3-30}$$

That is, if dT is positive, the population of states with energy greater than $\bar{\epsilon}$ increases at the expense of those states with energy less than $\bar{\epsilon}$. For a state with energy larger than $\bar{\epsilon}$ by an amount of order kT, the fractional population increase ($d\overline{C}_j/\overline{C}_j$) is of the same order as the fractional temperature increase (dT/T).

If we substitute Eq. (3–30) into Eq. (3–29), we recover Eq. (3–24) for C_V (Problem 3–3). In the special case ($T \to 0$) that all molecules are in the ground state ($j = 1$), $\epsilon_j - \bar{\epsilon} = 0$ for $j = 1$ and $\overline{C}_j = 0$ for $j > 1$; hence, by Eq. (3–30), $\partial\overline{C}_j/\partial T = 0$ at $T = 0$ for all j and $C_V = 0$ (Problem 3–3).

If a molecule has separable or approximately separable "internal" degrees of freedom, as for example in Eq. (3–13), the energy distribution can be broken down one step further. Thus, in this example, Eq. (3–17) becomes

$$\eta_{ijlm} = \frac{e^{-(\epsilon_{ti}+\epsilon_{rj}+\epsilon_{vl}+\epsilon_{em}/kT}}{q_t q_r q_v q_e}, \tag{3–31}$$

where the molecular quantum state ($ijlm$) is characterized by the translational state (i), rotational state (j), etc. Then the probability that, say, the rotational state is j, irrespective of the translational, vibrational, and electronic states (or the fraction of all molecules in rotational state j), is

$$\eta_j^{\text{rot}} = \sum_{i,l,m} \eta_{ijlm} = \frac{e^{-\epsilon_{rj}/kT}}{q_r}, \tag{3–32}$$

just as in passing from Eq. (3–16) to Eq. (3–17). Applications of this equation will appear in Chapters 4, 8, and 9.

Finally, we have to discuss *indistinguishable* molecules. So far the argument in this section has been concerned with the distinguishable case. We recall first that we are limited to situations in which the number of available molecular states greatly exceeds the number of molecules. In the notation used above, the average population (number of molecules) per molecular state must be much less than one, $\overline{C}_j \ll 1$: every molecule in the system is in a different molecular state (with negligible exceptions), or, an equivalent statement is that each molecular state is either unoccupied or occupied by just one molecule.

Equation (1–28) gives the probability of observing a particular quantum state of the system of N identical molecules. With indistinguishable molecules, a single quantum state of the system is specified if we say that one molecule (any one) is, say, in molecular state i, another (any other) is in state j, another in state l, etc. Which molecule is in which state is immaterial since the molecules cannot be distinguished from each other. Equation (1–28) then reads

$$P_{ijl\cdots} = \frac{e^{-(\epsilon_i+\epsilon_j+\epsilon_l+\cdots)/kT}}{(q^N/N!)}. \tag{3–33}$$

Note that this probability is $N!$ times larger than the corresponding probability (distinguishable molecules) for a single quantum state of the system in Eq. (3-16). This is because, in the case of indistinguishable molecules (when $\overline{C}_j \ll 1$), the system has a total of $N!$ times fewer quantum states from which to choose. To find the probability that any one molecule (*not* one *particular* molecule, such as molecule 1) in the system is, say, in state i, we should sum $P_{ijl} \ldots$ above over all values of $j, l, m, \ldots$. However, different permutations of the values of $j, l, m, \ldots$ should not be included in the sum, because such permutations do not generate new quantum states of the system. For example, if $N = 4$, in the sum over j, l, and m: $j = 4$, $l = 6$, and $m = 17$, say, should be included; but then $j = 6$, $l = 4$, and $m = 17$, or $j = 17$, $l = 4$, and $m = 6$, etc., should be omitted. On the other hand, it is actually simpler to include *all* values of $j, l, m, \ldots$ and correct for the permutations referred to above by dividing by $(N - 1)!$. With a sufficiently large number of molecular quantum states ($\overline{C}_j \ll 1$), we can ignore the rare terms in the sum in which $j = l$, or $l = m$, etc.

In summary: if we sum $P_{ijl} \ldots$ over all values of $j, l, \ldots$ and divide by $(N - 1)!$, the result is the probability that any one molecule in the system will be in state i. But this is just the quantity $\overline{C}_i$. Therefore

$$\overline{C}_i = \frac{1}{(N-1)!} \sum_{j,l,\ldots} P_{ijl} \ldots = \frac{e^{-\epsilon_i/kT} q^{N-1}}{(N-1)!(q^N/N!)}$$

$$= \frac{Ne^{-\epsilon_i/kT}}{q}, \qquad \overline{C}_i \ll 1. \tag{3-34}$$

As a check, we note that if we also sum over all values of i and divide by $N!$ instead of $(N - 1)!$, we get

$$\frac{1}{N!} \sum_{i,j,l,\ldots} P_{ijl} \ldots = \frac{q^N/N!}{q^N/N!} = 1,$$

as we should. If we define η_i by $\eta_i = \overline{C}_i/N$, as in Eq. (3-22), then

$$\eta_i = \frac{e^{-\epsilon_i/kT}}{q}, \qquad \eta_i \ll \frac{1}{N}, \tag{3-35}$$

which is the same as Eq. (3-17) for distinguishable molecules. That is, η_i is the probability that one *particular* molecule is in state i. Or, more properly here, since a "particular" molecule is a nonoperational concept, η_i is the fraction of all molecules in state i. This is a very small number, since $\overline{C}_i$ itself is already a small fraction.

Thus, as one would expect intuitively and as we anticipated at the beginning of the section, the molecular energy distribution for a given set of energy states ϵ_j is the same for distinguishable and indistinguishable independent molecules. There is the important distinction, however, that

Eq. (3–34) is valid for indistinguishable molecules only if $\overline{C}_i \ll 1$, whereas no such restriction is operative for distinguishable molecules (e.g., as $T \rightarrow 0$, $\overline{C}_1 \rightarrow N$).

The high density of molecular states necessary to satisfy $\overline{C}_j \ll 1$ is guaranteed in most applications (as we shall see in the next chapter) just by the translational degrees of freedom. Then if there are internal degrees of freedom, as for example in Eq. (3–31), $\eta_{ijlm} \ll 1/N$ for a single molecular state. But if we now sum η_{ijlm} over all translational states i, the result is the fraction of molecules in a given rotational-vibrational-electronic state jlm—a number which is *not* required to be small compared with unity. In other words, the internal distribution of indistinguishable molecules among nontranslational energy states is ordinarily unrestricted, just as in the case of distinguishable molecules. Thus, Eq. (3–32) gives the distribution over rotational states; the distribution is the same for distinguishable and indistinguishable molecules; and there is no restriction on η_j^{rot} [for example, at low temperatures, $\eta_1^{\text{rot}} \rightarrow 1$ (Chapter 8)].

3–4 "Ensembles" of small, independent "systems." In the Einstein model of a crystal, each of the N molecules carries out its own independent vibrational (and possibly also internal) motion around a definite lattice point, and hence is "distinguishable." Furthermore, the molecules are in thermal contact ("weak interaction") with each other. The reader will recognize that this situation is rather closely analogous to that of the canonical ensemble of Section 1–3. There we were dealing with an ensemble, a collection of $\mathfrak{N}$ independent, distinguishable macroscopic systems in thermal contact with each other. Here a single molecule is a "system," and the crystal (system) is the "ensemble." If we re-examine the argument in Section 1–3, we see that it is still valid even when the systems of the ensemble are very small. In the crystal example, the system is a single molecule, but other cases are possible in which the system contains more than one but still only a relatively small number of molecules. To avoid confusion, let us return now, as in the first part of this chapter, to the use of the term "subsystem" in referring to one of these small systems. If we let C_j be the number of molecules (or subsystems, generally) in the molecular state j with energy ϵ_j, then a set of numbers $C_1, C_2, \ldots$ is a "distribution," and we have, in place of Eqs. (1–1) through (1–3),

$$\sum_j C_j = N, \tag{3–36}$$

$$\sum_j C_j \epsilon_j = E, \tag{3–37}$$

$$\Omega(C) = \frac{N!}{\prod_j C_j!}, \tag{3–38}$$

where $\Omega(C)$ is the number of quantum states of the macroscopic system (e.g., crystal) for a given distribution C. The macroscopic system is isolated (i.e., N, V, and E are constant). As in Eq. (1–12), we find that

$$\eta_j = \frac{\overline{C}_j}{N} = \frac{C_j^*}{N} = \frac{e^{-\beta\epsilon_j}}{\sum_i e^{-\beta\epsilon_i}}, \tag{3–39}$$

which is consistent with Eq. (3–17). The macroscopic system has to be very large (i.e., $N \to \infty$) in order fully to justify the replacement of $\overline{C}_j$ by C_j^* and the use of Stirling's approximation on $\ln C_j!$. The connection with thermodynamics (e.g., $\beta = 1/kT$) is most easily established through the relation $S = k \ln \Omega(C^*)$ and the thermodynamic equation (1–78) (Problem 3–4). The most probable or "equilibrium" distribution C^* is the one which makes $\Omega(C)$, and therefore S, a maximum in an isolated system, in agreement with the second law of thermodynamics (Section 2–3).

The temperature of a single molecule (or subsystem) is well defined. This is the temperature which appears in the Boltzmann energy probability distribution for a single molecule, Eq. (3–39) with $\beta = 1/kT$. But other functions, such as the energy, are not well defined, in the *thermodynamic* sense, for the individual molecules (or subsystems) because of large fluctuations about mean values [e.g., $\bar{\epsilon}$ in Eq. (3–24): a measurement of ϵ will not almost certainly yield $\bar{\epsilon}$, as would be the case for $\overline{E}$ in a macroscopic system].

Of course, no new results are obtained by adopting the above "subsystem" point of view. Furthermore, the argument is restricted to macroscopic systems made up of independent, distinguishable molecules or subsystems; it is a special case of the more general argument of Section 1–3. However, when possible (independent molecules or subsystems), there is a considerable intuitive or conceptual advantage to thinking in terms of a system containing only one or a few molecules rather than a macroscopic number.

We have seen in the preceding section that when a system of N independent, indistinguishable molecules satisfies the condition $\overline{C}_j \ll 1$, the energy distribution is the same as for distinguishable molecules but the total number of quantum states of the macroscopic system is reduced by a factor $N!$. We therefore anticipate that the (molecule, system) $\leftrightarrow$ (system, ensemble) analogy, argument, and results above for distinguishable molecules will still hold for indistinguishable molecules, provided only that we use $\Omega(C)/N!$ in place of Ω. There is, however, a serious complication: since $\overline{C}_j = C_j^* \ll 1$, we have to use Stirling's approximation on very small numbers. Actually, this procedure happens to give correct results, but the method can hardly be considered satisfactory. The rigorous way to handle this problem, with the same results, is to use the Darwin-Fowler application of the method of steepest descents.

It should be emphasized that our derivation in the preceding section of Eq. (3–34) for $\overline{C}_j$ in a system of independent, indistinguishable molecules is *not* open to the above objection. That derivation is based, ultimately, on the use of a canonical ensemble of $\mathfrak{N}$ distinguishable, macroscopic systems, with $\mathfrak{N} \to \infty$ (Section 1–3).

In a number of applications it is very helpful to use an extension to the grand ensemble of the "subsystem" point of view brought out above for the canonical ensemble. Since these applications are somewhat specialized, we shall postpone (see, however, Problem 3–5) our discussion of this subject to Chapter 7 ("binding" or adsorption on sites) and Chapter 22 (Bose-Einstein and Fermi-Dirac statistics).

Problems

3–1. For a gas of independent molecules, $\ln \Xi = q(V, T)\lambda = \overline{N}$ (Eq. 3–12). Use Eq. (1–68), $p = kT\,(\partial \ln \Xi/\partial V)_{\mu,T}$, to deduce that q must have the form $q = Vf(T)$, where $f(T)$ is a function of temperature only. (Page 63.)

3–2. Extend the derivation of Eq. (3–10), $Q = q^N/N!$, for a one-component system of independent, indistinguishable molecules to the two-component case, Eq. (3–15). (Page 64.)

3–3. Use Eq. (3–30) to verify that

$$C_V = \sum_j \epsilon_j \frac{\partial C_j}{\partial T} = \frac{N[\overline{\epsilon^2} - (\overline{\epsilon})^2]}{kT^2}$$

for a system of independent molecules. Note that C_V is never negative. (Page 68.)

3–4. Show that substitution of the most probable distribution (3–39) into $S = k \ln \Omega(C^*)$ leads to

$$A = -NkT \ln \sum_i e^{-\beta \epsilon_i}.$$

Also, with the aid of $(\partial S/\partial E)_{V,N} = 1/T$ (Eq. 1–78), prove that $\beta = 1/kT$. (Page 71.)

3–5. Consider a one-component macroscopic system which consists of M independent, equivalent, and distinguishable subsystems. Here M plays the role that V does in most systems. Each subsystem can contain any number from zero up to m molecules, where $m \geq 1$. Modify the notation used in Section 1–5 (the system here corresponds to the *grand ensemble* in Section 1–5) to obtain the equivalent of Eq. (1–45) for the probability distribution. Use $S = k \ln \Omega$ (most probable distribution) to establish connections with thermodynamics. Note the analogy with Eqs. (1–75) through (1–77). (Page 72.)

3–6. For a gas of independent molecules, $Q = q(V, T)^N/N!$. From the canonical ensemble equations (1–31) through (1–36) prove, as in Problem 3–1, that $q = Vf(T)$.

3–7. Use Eq. (3–15) for a two-component system, where q_1 and q_2 are functions of V and T only, to deduce by the method of Eqs. (3–11) and (3–12) that

$$\overline{N}_1 + \overline{N}_2 = \frac{pV}{kT}.$$

3–8. Show from Eqs. (2–4) and (3–24) that

$$\frac{\sigma_\epsilon}{\overline{\epsilon}} = N^{1/2}\frac{\sigma_E}{\overline{E}}.$$

This emphasizes that the ϵ probability distribution is much broader than the E distribution for N large.

Supplementary Reading

Gurney, R. W., *Introduction to Statistical Mechanics.* New York: McGraw-Hill, 1949. Chapters 1 and 2.

Rushbrooke, Chapters 2 and 3.

Tolman, Chapter 14.

CHAPTER 4

IDEAL MONATOMIC GAS

As our first specific example we consider a one-component monatomic gas dilute enough so that intermolecular forces can be ignored. At the outset, each atom is treated as a mass point (internal degrees of freedom will be discussed briefly in Section 4–4) with three translational degrees of freedom. This discussion will also be applicable, in Chapters 8 and 9, to the translational degrees of freedom of dilute diatomic and polyatomic gases, because of the additivity of the Hamiltonian function [as in Eq. (3–1)].

4–1 Energy levels and canonical ensemble partition function. By virtue of the high dilution, the molecules of the gas are independent of each other. They are also indistinguishable. Therefore, the canonical ensemble partition function is (Eq. 3–10)

$$Q = \frac{1}{N!} q^N, \tag{4–1}$$

provided that the number of available molecular quantum states is large compared with N. Our first task is to investigate q.

Let the volume V be in the shape of a cube of edge L; then $V = L^3$. We choose a cube merely for convenience; thermodynamic properties of the gas do not depend on the shape of the container. In the sum $q = \sum_j e^{-\epsilon_j/kT}$, the energies ϵ_j in this case are those associated with one molecule possessing three translational degrees of freedom only and confined to a cubical box. This is a standard problem in quantum mechanics. The possible energy states are

$$\epsilon_{l_x l_y l_z} = \frac{h^2(l_x^2 + l_y^2 + l_z^2)}{8mL^2}, \tag{4–2}$$

$$l_x, l_y, l_z = 1, 2, 3, \ldots,$$

where l_x, l_y, and l_z are the three quantum numbers.

In the three-dimensional $l_x l_y l_z$ space (Fig. 4–1 shows a two-dimensional version of this space), there is a one-to-one correspondence between possible molecular quantum states and points in the space with positive integers as coordinates. Therefore, there is one quantum state per unit volume of this space. The equation

$$l_x^2 + l_y^2 + l_z^2 = R^2, \tag{4–3}$$

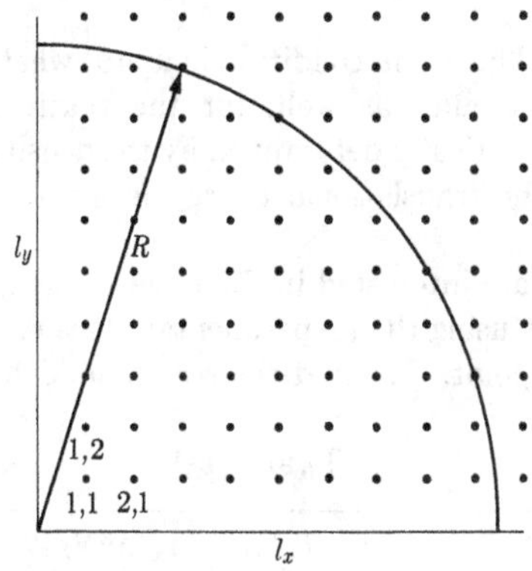

FIG. 4–1. Quantum number (l_x, l_y) space in two dimensions.

with

$$R^2 = \frac{8mV^{2/3}\epsilon}{h^2}, \tag{4–4}$$

is the equation of a sphere of radius R in $l_x l_y l_z$ space. The volume of the positive (l_x, l_y, l_z all positive) octant of a sphere of radius R is $\pi R^3/6$. This is also the number of quantum states $\Phi(\epsilon)$ with energy less than ϵ, where ϵ and R are related by Eq. (4–4), provided that $\Phi(\epsilon)$ is very large so that edge effects are negligible. That is,

$$\Phi(\epsilon) = \frac{\pi R^3}{6} = \frac{\pi}{6}\left(\frac{8m\epsilon}{h^2}\right)^{3/2} V. \tag{4–5}$$

For Eq. (4–1) to be legitimate, we have to have $\Phi \gg N$ when we put $\epsilon = O(kT)$ in Eq. (4–5), since states with ϵ very much larger than kT will not be appreciably populated and are therefore effectively unavailable. Later it proves convenient to choose the unimportant numerical coefficients, of order unity, so that this condition reads

$$\Phi(kT) \cong \left(\frac{2\pi mkT}{h^2}\right)^{3/2} V \gg N$$

or

$$\frac{\Lambda^3 N}{V} \ll 1, \tag{4–6}$$

where

$$\Lambda = \frac{h}{(2\pi mkT)^{1/2}}. \tag{4–7}$$

We see that the condition (4–6) is favored by low density, large mass, and high temperature. This same condition is useful when applied to diatomic and polyatomic molecules as well, for the total density of molecular quantum states is practically determined by the density of the translational states alone (i.e., the translational energy levels are relatively *very* close together).

Even though we are interested in dilute gases here, let us make a stringent test of (4–6) by using the (experimental) density of the *liquid* phase at the normal boiling point. We find the results in Table 4–1, which will be

TABLE 4–1

	T, °K	$\Lambda^3 N/V$
He	4.2	1.5
H_2	20.4	0.44
Ne	27.2	0.015
A	87.4	0.00054

referred to again in Section 6–2 and Chapter 22. In the gas phase at the boiling point, the values of $\Lambda^3 N/V$ would be reduced further by a factor of the order of 100. The conclusion we reach, then, is that, except for light molecules at appreciable densities and very low temperatures, the condition (4–6) is easily satisfied (Problem 4–1).

It is interesting to note that Λ is a length of just the order of magnitude of the de Broglie wavelength h/mv ($v =$ velocity) of a particle with kinetic energy of order kT. Since $(V/N)^{1/3}$ is a distance of the order of the average nearest-neighbor distance between molecules, Eq. (4–6) asserts that quantum effects (requiring the use of Bose-Einstein or Fermi-Dirac statistics) will be absent if neighboring molecules are usually far apart relative to the "thermal" de Broglie wavelength Λ.

We return now to

$$q = \sum_{l_x, l_y, l_z = 1}^{\infty} e^{-\epsilon_{l_x l_y l_z}/kT}. \tag{4–8}$$

We can replace the sum by an integral if, in passing from one energy level to the next higher level, $\Delta\epsilon \ll kT$, for in this case the summand will change its value essentially continuously with l_x, l_y, and l_z. From Eq. (4–2), $\Delta\epsilon = O(h^2/mV^{2/3})$ and

$$\frac{\Delta\epsilon}{kT} = O\left(\frac{h^2}{mkTV^{2/3}}\right) = O\left(\frac{\Lambda^2}{V^{2/3}}\right).$$

But we already have the restriction (4–6), which can be written

$$\frac{\Lambda^2}{V^{2/3}} \ll \frac{1}{N^{2/3}} = O(10^{-14}).$$

Thus the requirement $\Delta\epsilon/kT \ll 1$ is overwhelmingly satisfied.

To integrate Eq. (4–8) (see also Problem 4–2), we note from Eq. (4–5) that the number of molecular quantum states between ϵ and $\epsilon + d\epsilon$ is

$$\omega(\epsilon)\, d\epsilon = \frac{d\Phi}{d\epsilon}\, d\epsilon = \frac{\pi}{4}\left(\frac{8m}{h^2}\right)^{3/2} V\epsilon^{1/2}\, d\epsilon. \tag{4–9}$$

Therefore [compare Eq. (3–19)],

$$\begin{aligned} q(V, T) &= \int_0^\infty \omega(\epsilon) e^{-\epsilon/kT}\, d\epsilon = \frac{\pi}{4}\left(\frac{8mkT}{h^2}\right)^{3/2} V \int_0^\infty u^{1/2} e^{-u}\, du \\ &= \left(\frac{2\pi mkT}{h^2}\right)^{3/2} V = \frac{V}{\Lambda^3}, \end{aligned} \tag{4–10}$$

where $u = \epsilon/kT$. This confirms Problem 3–1. That is, q is proportional to V. We see that q is dimensionless and has a very simple expression in terms of the thermal de Broglie wavelength Λ.

The fraction of molecules with energies between ϵ and $\epsilon + d\epsilon$ is [compare Eq. (3–35)]

$$P(\epsilon)\, d\epsilon = \frac{\omega(\epsilon) e^{-\epsilon/kT}\, d\epsilon}{q}. \tag{4–11}$$

This is essentially the well-known Maxwell-Boltzmann distribution (Section 6–4), since $\omega \propto \epsilon^{1/2}$.

We have for the canonical ensemble partition function

$$Q = \frac{1}{N!}\, q^N = \frac{1}{N!}\left(\frac{V}{\Lambda^3}\right)^N, \tag{4–12}$$

or

$$\ln Q = -N \ln N + N + N \ln q = N \ln\left[\left(\frac{2\pi mkT}{h^2}\right)^{3/2} \frac{Ve}{N}\right]. \tag{4–13}$$

We can now derive all the thermodynamic properties of the gas from this equation.

4–2 Thermodynamic functions. First, the Helmholtz free energy A is simply

$$A(N, V, T) = -kT \ln Q = -NkT \ln\left[\left(\frac{2\pi mkT}{h^2}\right)^{3/2} \frac{Ve}{N}\right]. \tag{4–14}$$

We can confirm that A is an extensive property. That is, A/N is seen to be a function of the intensive variables V/N and T only. We should emphasize that in this equation and others below: (a) the zero of energy for each molecule is implicit in Eq. (4–2) (i.e., the zero is at the bottom of the potential well or box of volume V—classically, this corresponds to a particle at rest); and (b) the entropy $S = k \ln \Omega$ is based on the number Ω of translational quantum states of the system of N molecules (i.e., the zero of entropy is associated with $\Omega = 1$, although this situation is unrealizable for a dilute classical gas).

The pressure is given by

$$p = kT\left(\frac{\partial \ln Q}{\partial V}\right)_{T,N} = NkT\left(\frac{\partial \ln q}{\partial V}\right)_T = \frac{NkT}{V}. \tag{4–15}$$

It was pointed out, following Eq. (1–27), that k is a universal constant whose value can be obtained from any convenient experimental system. We see from Eq. (4–15) that

$$\lim_{N/V \to 0} \frac{pV}{NT}$$

for a real gas of known molecular weight will give k. The value found is $k = 1.38044 \times 10^{-16}$ erg·deg^{-1}.

The energy is

$$E = kT^2\left(\frac{\partial \ln Q}{\partial T}\right)_{V,N} = NkT^2 \frac{d \ln T^{3/2}}{dT} = \frac{3}{2} NkT. \tag{4–16}$$

This is all kinetic energy, $kT/2$ per degree of freedom per molecule. The potential energy is zero because of the absence of appreciable intermolecular forces and by virtue of our choice of the zero of energy. Also,

$$C_V = \left(\frac{\partial E}{\partial T}\right)_{V,N} = \frac{3}{2} Nk, \tag{4–17}$$

as is found experimentally. From Eqs. (4–15) and (4–17),

$$pV = \tfrac{2}{3}E. \tag{4–18}$$

Equations (4–15) through (4–18) are often called "classical" results because they can be derived directly from classical statistical mechanics without any use of quantum theory (Chapter 6). The use of Boltzmann statistics (Eq. 4–1) and the replacement of a sum over quantum states by an integral, as in Eq. (4–10), lead to classical results from a quantum-mechanical starting point. This statistical-mechanical correspondence principle is explored in much more detail in Chapters 6 and 22.

Equation (4–18) is interesting because it is obtained, for a dilute monatomic gas, not only from Boltzmann statistics but also from Bose-Einstein and Fermi-Dirac statistics (Chapter 22). We can anticipate this result by an argument based on the energy-level expression (4–2) alone, without use of Eq. (4–1) (Boltzmann statistics). We have to assume that [see

Eqs. (3–21) and (3–27)], for a dilute enough gas in a volume V, p and E are proportional to N. That is, each molecule makes an independent or additive average contribution to p and E. For a molecule in state j ($j = l_x, l_y, l_z$), the energy, according to Eq. (4–2), has the form $\epsilon_j = a_j/V^{2/3}$, and the contribution of this molecule to the pressure is [compare Eq. (1–7)]

$$-\frac{\partial \epsilon_j}{\partial V} = \frac{2}{3} a_j V^{-5/3} = \frac{2}{3}\frac{\epsilon_j}{V}.$$

Now we average over all states j, multiply by N, and find

$$p = N\left(-\overline{\frac{\partial \epsilon_j}{\partial V}}\right) = \frac{2}{3}\frac{N\bar{\epsilon}}{V} = \frac{2}{3}\frac{E}{V}, \tag{4–19}$$

which is the same as Eq. (4–18).

The entropy follows from Eqs. (4–14) and (4–16) for A and E:

$$S = \frac{E - A}{T} = Nk \ln\left[\left(\frac{2\pi mkT}{h^2}\right)^{3/2} \frac{Ve^{5/2}}{N}\right]. \tag{4–20}$$

From (4–6), we see that the quantity in brackets is necessarily large compared with unity, say 10^3 or more, but the logarithm of the quantity will be of order unity. Hence $S = O(Nk)$, as expected from thermodynamics. In terms of the pressure instead of the number density N/V, Eq. (4–20) becomes

$$S = Nk \ln\left[\left(\frac{2\pi mkT}{h^2}\right)^{3/2} \frac{kTe^{5/2}}{p}\right]. \tag{4–21}$$

Equation (4–21) is in agreement, as it should be, with the well-known thermodynamic expression for the entropy increase in an isothermal expansion of an ideal gas,

$$\Delta S = S(p_2, T) - S(p_1, T) = Nk \ln \frac{p_1}{p_2}.$$

For the chemical potential and Gibbs free energy, we have

$$\mu = \frac{F}{N} = -kT\left(\frac{\partial \ln Q}{\partial N}\right)_{T,V} = -kT \ln \frac{q}{N} \tag{4–22}$$

$$= -kT \ln\left[\left(\frac{2\pi mkT}{h^2}\right)^{3/2} \frac{V}{N}\right] \tag{4–23}$$

$$= -kT \ln\left[\left(\frac{2\pi mkT}{h^2}\right)^{3/2} \frac{kT}{p}\right]. \tag{4–24}$$

Equation (4–23) also follows from $F = N\mu = A + pV$ and Eqs. (4–14) and (4–15) for A and pV (Problem 4-3).

In thermodynamics one finds that, for an ideal gas,

$$\mu(p, T) = \mu^0(T) + kT \ln p, \tag{4–25}$$

where $\mu^0(T)$, often called a "standard chemical potential" or "standard free energy," is an integration constant. For a monatomic gas, Eq. (4–24) provides a statistical-mechanical evaluation of this integration constant (based on the zeros of energy and entropy already referred to):

$$\mu^0(T) = -kT \ln\left[\left(\frac{2\pi mkT}{h^2}\right)^{3/2} kT\right]. \tag{4–26}$$

It should be noted that the quantity in brackets here has dimensions of pressure, and must be in the same units (e.g., atm, mm Hg, erg·cc^{-1}, etc.) as p in Eq. (4–25). Clearly the numerical values of the separate terms on the right of Eq. (4–25) depend on the choice of the pressure unit, but their sum gives a $\mu(p, T)$ which is independent of this arbitrary choice. In several places in this book it will prove convenient to introduce standard chemical potentials $\mu^0(T)$. The above point about pressure units must always be kept in mind when a $\mu^0(T)$ is encountered. Of course, in computing any truly operational quantity, the combination in Eq. (4–25), or $pe^{\mu^0/kT}$, will occur, and the matter will then automatically take care of itself [e.g., there is no complication with Eq. (4–24) where the two terms in Eq. (4–25) are kept together].

This completes our summary of the basic thermodynamic functions of an ideal monatomic gas. Other functions (e.g., the heat content H) follow from those already given. From this first example, we can see the power of the statistical-mechanical method: it provides an explicit and detailed theoretical prediction of all the thermodynamic properties of the system. We have to reserve comparisons of equations such as (4–21) and (4–24) with experiment (entropy, crystal vapor pressure, chemical equilibrium constants, etc.) until we have discussed some other systems (see Chapters 5, 9, and 10).

For a binary mixture of dilute monatomic gases, according to Eq. (3–15),

$$Q = \frac{q_1^{N_1} q_2^{N_2}}{N_1! N_2!}, \tag{4–27}$$

where

$$q_1 = \frac{V}{\Lambda_1^3} \quad \text{and} \quad q_2 = \frac{V}{\Lambda_2^3}. \tag{4–28}$$

In (4–28), Λ_1 and Λ_2 differ only through m_1 and m_2. From

$$\begin{aligned}\ln Q = & -N_1 \ln N_1 + N_1 + N_1 \ln q_1(V, T) \\ & -N_2 \ln N_2 + N_2 + N_2 \ln q_2(V, T)\end{aligned} \tag{4-29}$$

and Eqs. (1-31) through (1-36), we find easily (Problem 4-4)

$$pV = (N_1 + N_2)kT, \tag{4-30}$$

$$E = \tfrac{3}{2}(N_1 + N_2)kT, \tag{4-31}$$

$$S = N_1 k \ln\left(\frac{Ve^{5/2}}{\Lambda_1^3 N_1}\right) + N_2 k \ln\left(\frac{Ve^{5/2}}{\Lambda_2^3 N_2}\right), \tag{4-32}$$

etc. The familiar thermodynamic formula for the entropy of mixing two ideal gases follows from Eq. (4-32) (Problem 4-5).

4-3 Grand ensemble and others. It is instructive to try out partition functions other than the canonical ensemble partition function on this simple system. We have in fact already derived the equation of state from the grand partition function in Eq. (3-12). We should note that Eq. (3-12) also gives

$$\frac{\overline{N}}{q} = \frac{\overline{N}\Lambda^3}{V} = \lambda = e^{\mu/kT}, \tag{4-33}$$

which is a slightly disguised form of Eq. (4-23) for μ.

As another example, let us use the partition function $\Delta(N, p, T)$ of Eqs. (1-87), (1-91), and (2-12):

$$-\frac{F}{kT} = \ln \Delta = \ln \int_0^\infty Q(N, V, T)e^{-pV/kT}\, d\left(\frac{pV}{kT}\right). \tag{4-34}$$

Here V is continuously variable, so we use an integration over V instead of a sum. The choice of the dimensionless variable of integration is somewhat arbitrary,* but pV/kT is particularly convenient for this purpose. If we let $x = pV/kT$,

$$\begin{aligned}\Delta &= \frac{1}{N!}\left(\frac{kT}{p\Lambda^3}\right)^N \int_0^\infty x^N e^{-x}\, dx \\ &= \left(\frac{kT}{p\Lambda^3}\right)^N.\end{aligned} \tag{4-35}$$

Then

$$F = N\mu = -kT \ln \Delta = -NkT \ln\left(\frac{kT}{p\Lambda^3}\right), \tag{4-36}$$

---

* See S. M., pp. 63 and 68.

which is the same as Eq. (4–24). Also,

$$S = -\left(\frac{\partial F}{\partial T}\right)_{p,N} = Nk \ln\left(\frac{kTe^{5/2}}{p\Lambda^3}\right), \tag{4–37}$$

in agreement with Eq. (4–21), and

$$V = \left(\frac{\partial F}{\partial p}\right)_{T,N} = \frac{NkT}{p}. \tag{4–38}$$

We shall not work out the details of the microcanonical ensemble *ab initio.** But we can easily find $\Omega(N, V, E)$ from the canonical ensemble equation (4–20) and the relations $S = k \ln \Omega$ and $kT = 2E/3N$:

$$\Omega(N, V, E) = \left[\left(\frac{4\pi mE}{3h^2}\right)^{3/2} \frac{Ve^{5/2}}{N^{5/2}}\right]^N. \tag{4–39}$$

This is a tremendously large number, of order e^N or $10^{10^{20}}$. As a partial check (Problem 4–6), we note that [using Eq. (1–79)]

$$\frac{1}{kT} = \left(\frac{\partial \ln \Omega}{\partial E}\right)_{V,N} = \frac{3N}{2E}. \tag{4–40}$$

4–4 Internal degrees of freedom. So far in this chapter we have treated monatomic molecules as mass points with three translational degrees of freedom, using the energy states of a particle of mass m in a cubical box of volume V. Of course, an atom is not a mass point but has electronic and nuclear substructure. Associated with this substructure are electronic and nuclear energy states, deducible in principle from quantum mechanics but in practice, in most cases, from spectroscopy. We wish to show briefly here how these "internal degrees of freedom" influence the thermodynamic functions of the system.

The first comment to make is that the translational Hamiltonian is rigorously separable from the electronic and nuclear Hamiltonians. Second, the electronic and nuclear Hamiltonians are also separable, to a very high accuracy. Hence the basic equation is (3–14):

$$Q = \frac{1}{N!}(q_t q_e q_n)^N. \tag{4–41}$$

We have already discussed q_t. We now turn to a consideration of the nuclear and electronic energy levels, and hence of q_n and q_e.

Because of the extremely large separation between the two lowest nuclear energy levels [$\Delta\epsilon_n = O(1 \text{ Mev})$; or $\Delta\epsilon_n = O(kT)$ when $T =$

* See Mayer and Mayer, pp. 109–117.

$O(10^{10}{}^\circ\text{K})$], at ordinary temperatures the atom is certain to be in the nuclear ground state. This follows from Eq. (3–32). We shall denote the degeneracy of the nuclear ground state by ω_{n1}. The situation is similar, but not so extreme, for the electronic energy levels. The separation $\Delta\epsilon_e$ between the two lowest levels is usually of the order of 1 ev; $\Delta\epsilon_e = O(kT)$ when $T = O(10{,}000^\circ\text{K})$. Therefore again, according to Eq. (3–32), at ordinary temperatures (say $T < 1000^\circ\text{K}$) the atom is almost certain to be in the electronic ground state. However, since there are exceptions to this, we mention a few particular cases to give some idea of the range in behavior.

In the inert gases, He, Ne, A, etc., $\Delta\epsilon_e$ is large, of order 10–20 ev. Hence only the ground state is important. The degeneracy ω_{e1} of the electronic ground state ($^1S_0$) is unity for these atoms.*

For the alkali metal atoms, $\Delta\epsilon_e = O(1.5\text{ ev})$. Here again excited electronic states can be ignored at ordinary temperatures. In this case $\omega_{e1} = 2$ ($^2S_{1/2}$) owing to the unpaired electron spin. Similarly, for the hydrogen atom, $\Delta\epsilon_e = 10.2$ ev and $\omega_{e1} = 2$.

In the case of the halogen atoms F, Cl, Br, and I, $\Delta\epsilon_e = 0.050$, 0.11, 0.46, and 0.94 ev, respectively. The degeneracies of the first two levels (the third level has a significantly higher energy and can be ignored here) are 4 ($^2P_{3/2}$) and 2 ($^2P_{1/2}$), respectively. Therefore the fraction of halogen atoms in the first excited level is (Eq. 3–32)

$$\eta_{e2} = \frac{\omega_{e2}e^{-\epsilon_{e2}/kT}}{\omega_{e1}e^{-\epsilon_{e1}/kT} + \omega_{e2}e^{-\epsilon_{e2}/kT}} = \frac{2e^{-\Delta\epsilon_e/kT}}{4 + 2e^{-\Delta\epsilon_e/kT}}, \qquad (4\text{–}42)$$

$$\Delta\epsilon_e = \epsilon_{e2} - \epsilon_{e1}\,.$$

At 1000°K, we find (Problem 4–7) $\eta_{e2} = 0.22$, 0.12, 0.0024, and 9×10^{-6} for F, Cl, Br, and I, respectively. Certainly for F and Cl, then, and possibly Br, we have to retain two terms in the electronic partition function.

We choose as zero of energy here an atom in its ground electronic and nuclear states and at the bottom (classically, at rest) of the translational potential box or well (Fig. 4–2). Then q_t (Eq. 4–10) is unchanged and

$$q_n = \omega_{n1}, \qquad \epsilon_{n1} = 0, \qquad (4\text{–}43)$$

$$q_e(T) = \omega_{e1} + \omega_{e2}e^{-\epsilon_{e2}/kT}, \qquad \epsilon_{e1} = 0, \qquad \Delta\epsilon_e = \epsilon_{e2}. \qquad (4\text{–}44)$$

The most common case (the inert gases) is $q_e = \omega_{e1} = 1$. All thermodynamic functions now follow in a routine way from Eqs. (4–41), (4–43), and (4–44).

---

* For spectroscopic notation and facts, see G. Herzberg, *Atomic Spectra and Atomic Structure*. New York: Dover, 1944.

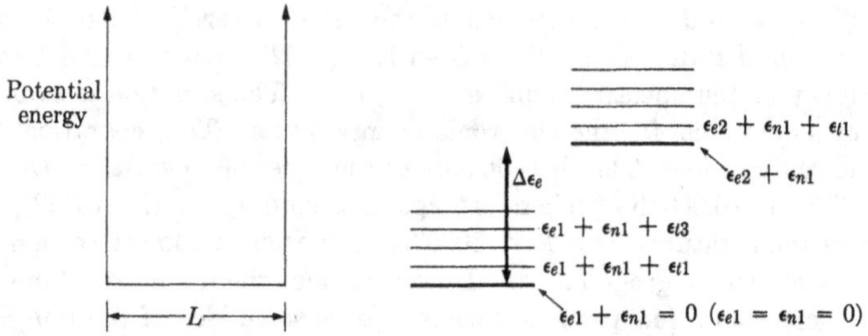

FIG. 4-2. Arbitrary location of energy zero.

For example, the equation of state ($pV = NkT$) is unchanged, but

$$E = NkT^2\left(\frac{\partial \ln q_t}{\partial T} + \frac{d \ln q_e}{dT}\right)$$

$$= \frac{3}{2}NkT + \frac{N\omega_{e2}\epsilon_{e2}e^{-\epsilon_{e2}/kT}}{q_e} \tag{4-45}$$

$$= N\bar{\epsilon}_t + N\bar{\epsilon}_e\,. \tag{4-46}$$

Similar equations can be derived for C_V, S, etc. (Problem 4-8.)

In the special case $q_n = \omega_{n1}$ and $q_e = \omega_{e1}$, the nuclear and electronic degrees of freedom make no contribution to p, E, C_V, etc., but they do contribute to all functions involving the entropy: S, A, F, μ, etc. Thus,

$$\ln Q = N \ln\left(\frac{Ve\omega_{e1}\omega_{n1}}{\Lambda^3 N}\right), \tag{4-47}$$

$$S = Nk \ln\left(\frac{Ve^{5/2}\omega_{e1}\omega_{n1}}{\Lambda^3 N}\right) = S_t + k \ln(\omega_{e1}\omega_{n1})^N, \tag{4-48}$$

$$\mu = -kT \ln\left(\frac{V\omega_{e1}\omega_{n1}}{\Lambda^3 N}\right). \tag{4-49}$$

However, in accordance with the convention established with Eq. (2-43), we shall usually *omit* (Chapter 22 is an exception) the nuclear degeneracy ω_{n1} from all thermodynamic functions. Because of cancellation, it does not ordinarily contribute to measurable thermodynamic quantities (entropy change with temperature, equilibrium constants, vapor pressure, etc.). We do not extend this convention to ω_{e1}, however, because when $\omega_{e1} > 1$ in the gas phase, the stable state at 0°K (e.g., Cl_2 crystal, H_2 crystal, sodium metal, etc.) is nondegenerate.* Hence in many cases,

* See Fowler and Guggenheim, pp. 199-205.

unlike ω_{n1}, ω_{e1} does *not* cancel. Also, as we have pointed out, excited electronic states occasionally have to be taken into account.

Problems

4–1. Calculate the temperature at which $\Lambda^3 N/V = 0.0001$ for He gas at 1 atm pressure (use the equation of state $pV = NkT$). (Page 76.)

4–2. Derive Eq. (4–10) for q by substituting Eq. (4–2) into Eq. (4–8), and replacing the sum by a triple integral over l_x, l_y, and l_z. (Page 77.)

4–3. Obtain Eq. (4–23) for μ from Eqs. (4–14) and (4–15) for A and pV, and $N\mu = A + pV$. (Page 80.)

4–4. Verify the fact that Eqs. (4–30) through (4–32) for a binary gas mixture follow from Eq. (4–29). (Page 81.)

4–5. Derive an equation for ΔS in the following process: initial state = (a) N_1 molecules of ideal gas 1 at T in a container of volume V_1, plus (b) N_2 molecules of ideal gas 2 at T in a separate container of volume $V_2 = N_2V_1/N_1$; and final state = N_1 molecules of 1 and N_2 of 2 in a single container of volume $V_1 + V_2$, at T. This is usually called the entropy of mixing, ΔS_{mix}. (Page 81.)

4–6. Confirm formulas already obtained for p and μ (ideal monatomic gas) by using the microcanonical ensemble equations (1–80), (1–81), and (4–39). (Page 82.)

4–7. Verify the values of η_{e2} calculated from Eq. (4–42) for F, Cl, Br, and I at 1000°K. (Page 83.)

4–8. Derive an equation for C_{Ve}, the electronic contribution to C_V, from Eq. (4–45). Verify that $C_{Ve} \to 0$ as $T \to 0$ or $T \to \infty$, and passes through a maximum in between. Derive an expression for S_e. Check that $S_e \to k \ln \omega_{e1}^N$ as $T \to 0$, and $S_e \to k \ln (\omega_{e1} + \omega_{e2})^N$ as $T \to \infty$. (Page 84.)

4–9. Use Problem 3–3 to show that the asymptotic behavior of C_{Ve} as $T \to \infty$ is $C_{Ve} = \text{constant}/T^2$.

4–10. For argon gas (assumed ideal) at 1 atm pressure and 25°C, calculate A, E, and μ in cal·mole^{-1} and C_V and S in cal·mole^{-1}·deg^{-1} (see Table 9–2). Take $\omega_{e1} = 1$ and omit ω_{n1}.

4–11. For a one-component ideal gas, show that the maximum term (a) in Ξ occurs at $N^* = q\lambda$, and (b) in Δ occurs at $V^* = NkT/p$.

4–12. Show that the result of Problems 3–1 and 3–6, $q = Vf(T)$, implies that the number $\Phi(\epsilon)$ of molecular quantum states with energy less than ϵ is proportional to V.

Supplementary Reading

Fowler and Guggenheim, Chapter 3.
Mayer and Mayer, Chapters 5 and 6.
Tolman, Chapter 14.

CHAPTER 5

MONATOMIC CRYSTALS

In this chapter we investigate the thermodynamic properties of monatomic crystals, especially the heat capacity C_V. The molecules in a crystal vibrate around equilibrium positions which are arranged in a regular lattice. Intermolecular forces, together with the pressure and temperature, determine the lattice structure and the spacing and the nature of the motion of the molecules in the neighborhood of their equilibrium positions. Offhand then, it would appear that a treatment of this system would belong more properly in Part III (which is concerned with systems of *dependent* molecules). But it turns out that despite the importance of intermolecular forces in a crystal, so long as the molecular vibrations are small, the system can be decomposed *mathematically* into *independent subsystems* (normal modes of vibration).

In the first section we treat a monatomic crystal in a very approximate manner, using the model of Einstein. This model has the advantage of conceptual and mathematical simplicity and leads to results that are qualitatively (but not quantitatively) correct. The remainder of the chapter is then devoted to more accurate approaches to the problem, with emphasis on the Debye approximation.

5–1 Einstein model of a monatomic crystal. We begin with some comments that establish the point of view adopted in the Einstein model. This discussion will also be useful later, since it pertains as well to certain approximate models ("cell theories") of the liquid state (Chapter 16).

The type of lattice structure is assumed given. Each molecule in the crystal is surrounded by a group of first (nearest) "neighbors," a more distant group of second neighbors, etc. The central molecule vibrates in the vicinity of its equilibrium or lattice point in a force field which is the sum of the separate forces exerted by all the neighbors on the central molecule. In the case of inert gas molecules, for example, these intermolecular forces are of the van der Waals (dispersion) type with a rather short range (see Appendix IV and Fig. 5–1), so that usually only the first and second neighbors make an appreciable contribution to the total force. The potential energy of a central molecule in this force field has a minimum at the equilibrium or lattice point, by definition. The potential energy increases in all directions as the central molecule departs from its equilibrium position and becomes very large, owing to van der Waals repulsion,

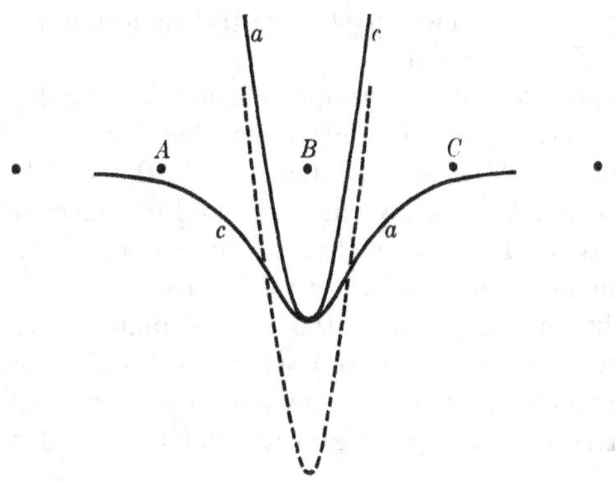

FIG. 5–1. One-dimensional crystal with nearest-neighbor distance r^*. Curves relate to potential energy of B, as a function of position, when other molecules $(A, C, \ldots)$ are fixed at equilibrium (lattice) points. Curve a is B-A potential energy, and curve c is B-C potential energy. The dashed curve is the sum of a and c, i.e., the total potential energy of B if second-neighbor interactions are neglected.

when a first neighbor is approached (Fig. 5–1). Thus each molecule is confined to a "cell" or "cage" bounded by its first neighbors.

The nearest-neighbor distance in the lattice (at low pressures) will correspond approximately to that distance (r^*) between a pair of molecules which leads to a minimum in the intermolecular potential energy (Figs. IV–1 and 5–1). The correspondence will not be exact, however, because of second-neighbor, and higher, effects and also because of thermal expansion when $T > 0°$K.

Each molecule has three degrees of freedom—translational degrees which have degenerated into vibrational degrees because of the cage of nearest neighbors. The vibrations about equilibrium positions can be treated as small vibrations if the temperature is not too high. We limit ourselves to this case. Of course, as the temperature is increased, the vibrations will become more violent, lattice imperfections and molecular migrations will become more frequent, and eventually the crystal will melt.

We assume for the sake of simplicity that each molecule vibrates in its own cell *independently* of the vibrations of its neighbors. Actually the motions of neighbors are coupled to each other (Section 5–2), but we ignore this complication here. To compute the potential field in which a given molecule moves, we can assume, say, that all other molecules in the crystal are frozen in their equilibrium positions. This potential will be not quite spherically symmetrical. A feature of the Einstein model is to insist on spherical symmetry as a further slight approximation. In other words, the vibrational motion of a central molecule is assumed *isotropic* about the equilibrium point.

Even though this approach is approximate, it provides a quite straightforward connection between intermolecular forces and the potential, referred to above, which plays a crucial role in the model. In the canonical ensemble, the exact geometry (numbers and distances of neighbors) of a cell or cage is fixed by the molecular volume V/N (or number density N/V) and the assumed type of crystal lattice.

Let $\varphi(r)$ be the potential of the central molecule (Fig. 5–1 shows a simplified one-dimensional version), where r is the distance from the center of the cell (equilibrium point). The zero of energy is infinitely separated molecules at rest. We expand $\varphi(r)$ about $r = 0$, and discard cubic and higher terms:

$$\varphi(r) = \varphi(0) + \tfrac{1}{2}fr^2 + \cdots$$

or

$$\varphi(x, y, z) - \varphi(0) = \tfrac{1}{2}f(x^2 + y^2 + z^2) + \cdots, \tag{5–1}$$

where x, y, and z are cartesian coordinates with the center of the cell as origin, and the force constant $f = (d^2\varphi/dr^2)_{r=0}$. The linear term is missing because $d\varphi/dr = 0$ at $r = 0$. The restriction to small vibrations enters here when we drop higher terms in the expansion of φ. The classical motion of the central molecule of mass m in the potential field (Eq. 5–1) is that of a three-dimensional isotropic harmonic oscillator with frequency

$$\nu = \frac{1}{2\pi}\sqrt{\frac{f}{m}}. \tag{5–2}$$

The curvature of φ at $r = 0$ (that is, f) is clearly a function of V/N; therefore ν is also a function of V/N. Of course the same is true of $\varphi(0)$.

In this model, by assumption, the total Hamiltonian for the system of N molecules is the sum of N independent and equivalent Hamiltonians, one for each molecule. Furthermore, for each molecule, we have by Eq. (5–1) that the x-, y-, and z-motions are independent and equivalent. Altogether, then, the system decomposes into an aggregation of $3N$ independent one-dimensional harmonic oscillators, all of classical frequency ν. These are the "subsystems" in the present model. The partition function for the system is

$$Q(N, V, T) = e^{-N\varphi(0;V/N)/2kT}\, q\,(V/N, T)^{3N}, \tag{5–3}$$

where q is the partition function of a one-dimensional harmonic oscillator of classical frequency ν, with zero of energy at the bottom of the parabolic potential well [as in Eq. (5–1)]. The factor $e^{-N\varphi(0)/2kT}$ in Eq. (5–3) would be the partition function of a hypothetical system with all molecules at rest at their lattice points; $N\varphi(0)/2$ is the potential energy (a negative quantity) of such a system relative to a zero at infinite separation of all molecules. The factor of two is necessary to avoid counting each intermolecular interaction twice. In the complete Q in Eq. (5–3) the zero of energy corresponds to all molecules infinitely separated and at rest. The electronic ground state of the atoms of the crystal is assumed to be nondegenerate in Eq. (5–3).

Next, we have to find q. The energy levels of a one-dimensional harmonic oscillator are nondegenerate, and given by

$$\epsilon_n = (n + \tfrac{1}{2})h\nu, \qquad n = 0, 1, 2, \cdots, \tag{5–4}$$

where ν is the classical frequency. Then

$$q = \sum_{n=0}^{\infty} e^{-\epsilon_n/kT} = e^{-h\nu/2kT} \sum_{n=0}^{\infty} (e^{-h\nu/kT})^n \tag{5–5}$$

$$= \frac{e^{-h\nu/2kT}}{1 - e^{-h\nu/kT}} = \frac{e^{h\nu/2kT}}{e^{h\nu/kT} - 1}$$

$$= \frac{e^{-\Theta/2T}}{1 - e^{-\Theta/T}}, \tag{5–6}$$

where $\Theta = h\nu/k$ and we have used the fact that $e^{-h\nu/kT} < 1$. The parameter Θ has dimensions of temperature and is called the "characteristic temperature." It is a function of V/N. In a typical case Θ is of the order of magnitude of 300°K and ν of order 6×10^{12} sec^{-1}. This frequency is about ten times smaller than the internal vibrational frequency in a typical diatomic molecule (Chapter 8), primarily because the forces between molecules in a monatomic crystal are in general weak compared with chemical bond forces.

At high temperatures, when $T \gg \Theta$,

$$q \to \frac{1 - (\Theta/2T) + \cdots}{1 - [1 - (\Theta/T) + \cdots]} \to \frac{T}{\Theta} = \frac{kT}{h\nu}. \tag{5–7}$$

The same result is obtained if we integrate instead of sum in Eq. (5–5). This is legitimate if $T \gg \Theta$. Then

$$q = e^{-h\nu/2kT} \int_0^\infty (e^{-h\nu/kT})^n \, dn$$

$$= \left(1 - \frac{h\nu}{2kT} + \cdots\right)\left(-\frac{1}{\ln e^{-h\nu/kT}}\right) = \frac{kT}{h\nu}.$$

At low temperatures, $q \to e^{-h\nu/2kT}$ (each oscillator is in its ground state).

We should emphasize that throughout this chapter, when we speak of the asymptotic behavior of various properties of a crystal at high temperatures we do not mean to imply that the temperature can be increased indefinitely. For in this case complications, not included in our model, would enter. For example, cubic and higher terms in the potential energy can no longer be neglected; crystal imperfections and the possibility of melting must be considered; etc.

We now derive expressions for A, E, C_V, and S from Q in Eq. (5–3). We have

$$A = -kT \ln Q = \frac{N\varphi(0)}{2} - 3NkT \ln \left(\frac{e^{-\Theta/2T}}{1 - e^{-\Theta/T}}\right) \tag{5–8}$$

and

$$E = kT^2 \left(\frac{\partial \ln Q}{\partial T}\right)_{N,V} = -\Theta k \left(\frac{\partial \ln Q}{\partial \Theta/T}\right)_{N,V}$$

$$= \frac{N\varphi(0)}{2} + \frac{3Nh\nu}{2} + 3NkT\left(\frac{\Theta/T}{e^{\Theta/T} - 1}\right). \tag{5–9}$$

The second term on the right in Eq. (5–9) is the zero-point vibrational energy, and the third term is the vibrational energy in excess of this minimum. At low temperatures,

$$E \rightarrow \frac{N\varphi(0)}{2} + \frac{3Nh\nu}{2}, \tag{5–10}$$

and at high temperatures,

$$E \rightarrow 3NkT. \tag{5–11}$$

In Eq. (5–10) E is the negative of the heat of sublimation of an Einstein crystal at 0°K (since E is zero for the infinitely dilute equilibrium vapor phase at 0°K).

Only the last term in Eq. (5–9) contributes to the heat capacity:

$$C_V = \left(\frac{\partial E}{\partial T}\right)_{N,V} = 3Nk\left(\frac{\Theta}{T}\right)^2 \frac{e^{\Theta/T}}{(e^{\Theta/T} - 1)^2}. \tag{5–12}$$

We see that C_V/Nk is a universal function of T/Θ, according to this model. This function is shown in Fig. 5–2. In general Θ is different for different crystals and also varies with V/N for the same crystal. But if the temperature scales are properly compressed or expanded, all experimental C_V/Nk curves, at least for monatomic crystals, should coincide. This is an example of a law of "corresponding states": for two crystals with Θ_1 and Θ_2, if the first crystal is at T_1, then the "corresponding temperature" T_2 for the second crystal is $T_2 = T_1\Theta_2/\Theta_1$; the two crystals

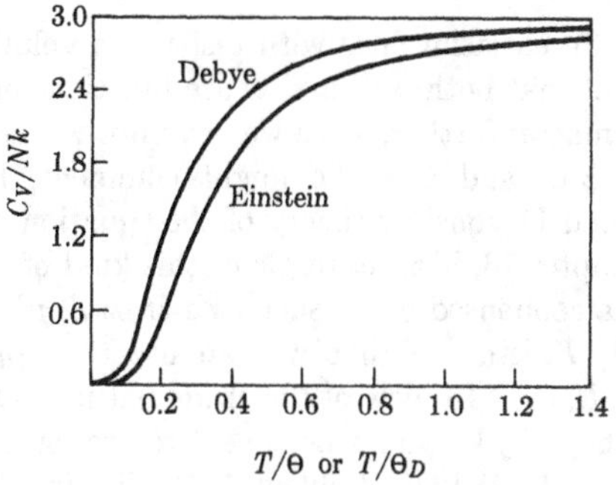

FIG. 5-2. Heat capacity of monatomic crystals according to the theories of Einstein and Debye.

have the same value of C_V/Nk at "corresponding temperatures" T_1 and T_2. Actually, it is found that experimental C_V/Nk curves for monatomic crystals can be made practically to coincide by adjusting the temperature scales in this way, but the common experimental heat capacity curve thus obtained does not fit the Einstein function exactly, though it does approximately. We shall return to this point in Section 5-3 in connection with Debye's theory.

The limiting forms of Eq. (5-12) are

$$C_V \to 3Nk \quad \text{as} \quad T \to \infty, \tag{5-13}$$

$$C_V \to 3Nk\left(\frac{\Theta}{T}\right)^2 e^{-\Theta/T} \quad \text{as} \quad T \to 0. \tag{5-14}$$

The high-temperature limit agrees with experiment ("law of Dulong and Petit"), but C_V approaches zero much more rapidly in Eq. (5-14) than is the case experimentally. Experimentally, $C_V \propto T^3$ as $T \to 0$ (see Section 5-3).

From Eqs. (5-8) and (5-9) for A and E we find

$$S = \frac{E}{T} - \frac{A}{T} = 3Nk\left[\frac{\Theta/T}{e^{\Theta/T} - 1} - \ln\left(1 - e^{-\Theta/T}\right)\right]. \tag{5-15}$$

As $T \to 0$, $S \to 0$. That is, $\Omega \to 1$. Here Ω refers to the total number of quantum states of a system of N independent and distinguishable three-dimensional isotropic harmonic oscillators (or $3N$ one-dimensional oscillators). As $T \to 0$, the system sinks into its lowest energy state (with $\Omega = 1$): each molecule is in its vibrational ground state [$n = 0$ in Eq. (5-4) for all $3N$ vibrational degrees of freedom].

We can easily set down a formal expression (Problem 5-1) for the pressure p by differentiating $\ln Q$ with respect to volume, as usual. Specifically, we recall that both $\varphi(0)$ and Θ are functions of V/N. But we do not pursue this matter further, since we have not written out here explicit expressions for $\varphi(0)$ and Θ. We might comment, however, that the Lennard-Jones and Devonshire theory of the equation of state of a liquid, presented in Chapter 16, is an example of this kind of calculation.

Actually, for a condensed phase such as a crystal, pV is in general small compared to A, E, etc., so that we can use the approximations $F = N\mu \cong A$, $H \cong E$, etc. In view of the crude nature of the Einstein model itself, the dropping of pV cannot be considered serious.

The chemical potential μ (Problem 5–1) also involves derivatives of $\varphi(0)$ and Θ with respect to V/N. However, an approximate μ follows immediately from $N\mu \cong A$ and Eq. (5–8):

$$\mu \cong \frac{A}{N} = \frac{\varphi(0)}{2} - 3kT \ln\left(\frac{e^{-\Theta/2T}}{1 - e^{-\Theta/T}}\right). \tag{5–16}$$

Finally, we make a few comments on the application of the grand partition function to an Einstein crystal. In the first place, the sum $\Xi = \sum_N Q(N, V, T)\lambda^N$, where Q is given by Eq. (5–3), cannot be carried out explicitly in general because of the dependence of $\varphi(0)$ and Θ on N (V constant). So there is no advantage in using Ξ instead of Q. Secondly, a simplified version of the Einstein model is often used in which the crystal is assumed to be incompressible. Here it would appear that Ξ might be useful, but there are complications which again lead to the conclusion that Q is the partition function of choice. In an incompressible crystal with a given type of lattice, the cell geometry (nearest-neighbor distance, etc.) is regarded as preassigned and fixed. Since $\varphi(0)$ and Θ should now be considered constants, the sum $\sum_N Q(N, V, T)\lambda^N$ becomes easy. However, V is no longer an independent variable: $V = Nv$, where v, the volume per molecule, is constant. That is, V is simply proportional to N. The basic thermodynamic equation for A becomes

$$dA = -S\,dT + \mu\,dN,$$

and $A = N\mu$. We still have

$$A(N, T) = -kT \ln Q(N, T),$$

but

$$\begin{aligned}\ln \sum_N Q(N, T)\lambda^N &= \ln \sum_N [e^{-\varphi(0)/2kT}\, q(T)^3\lambda]^N \\ &= \text{function of } \mu \text{ and } T = \ln [e^{-A(N^*,T)/kT} e^{N^*\mu/kT}] \\ &= 0,\end{aligned}$$

just as in Eq. (1–89) (where there is one additional independent thermodynamic variable). Thus, because of the loss of the volume as an independent variable, the grand partition function requires special instead of routine handling.* One finds without difficulty, by methods* which we shall not discuss here,

$$\frac{1}{\lambda} = e^{-\varphi(0)/2kT} q(T)^3. \tag{5–17}$$

This result also follows, but much more simply, from

$$N\mu = A = -kT \ln Q \tag{5–18}$$

or

$$\mu = -kT\left(\frac{\partial \ln Q}{\partial N}\right)_T. \tag{5–19}$$

Equation (5–17) is the same as Eq. (5–16). Thus dropping the pV term leads to the same μ as assuming incompressibility.

5–2 General treatment of molecular vibrations in a monatomic crystal. Let us begin this section by indicating how, in principle, one would handle the problem of small vibrations in a crystal exactly. The nature of the problem is clearly exhibited even by a one-dimensional crystal, so we consider this relatively simple case (Fig. 5–3). For concreteness, suppose only first- and second-neighbor interactions are significant. Let $u(r)$ be the intermolecular pair potential for any two atoms of the crystal (Figs. IV–1 and 5–1). Let there be N atoms (N is very large, so end effects are unimportant) in a length L, and $a = L/N$.

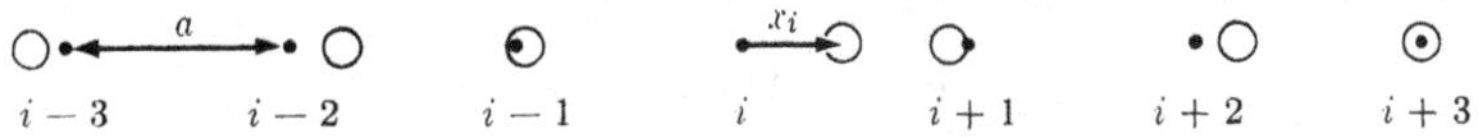

FIG. 5–3. One-dimensional crystal. Dots represent lattice or equilibrium points; circles represent atoms.

If all the atoms are at their lattice points, the nearest-neighbor separation is a. In general, let x_i be the position of atom i with its own lattice point chosen as origin. The distance between atom i and atom $i + 1$ is $a + x_{i+1} - x_i$, etc. The total potential energy of the crystal is then (neglecting end effects)

$$U(x) = \sum_{i=1}^{N} [u(a + x_{i+1} - x_i) + u(2a + x_{i+2} - x_i)]. \tag{5–20}$$

---

* S. M., Chapters 2 and 3.

With all atoms at lattice points, each $x_i = 0$, and

$$U(0) = N[u(a) + u(2a)]. \tag{5-21}$$

We might digress briefly to indicate the relation between Eq. (5–20) and the Einstein model of the previous section. First, $U(0)$ corresponds to $N\varphi(0)/2$. Second, the equation of motion of molecule j, according to Eq. (5–20), is

$$m\ddot{x}_j = -\frac{\partial U}{\partial x_j} = -u'(a + x_j - x_{j-1}) - u'(2a + x_j - x_{j-2})$$
$$+u'(a + x_{j+1} - x_j) + u'(2a + x_{j+2} - x_j).$$

To "uncouple" the motion of molecule j from that of molecules $j - 2$, $j - 1$, $j + 1$, and $j + 2$, as is required by the Einstein model (each molecule vibrates independently), we can assume as an approximation that these neighboring molecules are all fixed at their lattice points: $x_{j-2} = 0$, etc. Then the equation of motion becomes

$$\begin{aligned} m\ddot{x}_j &= -u'(a + x_j) - u'(2a + x_j) \\ &\quad +u'(a - x_j) + u'(2a - x_j) \\ &= -\frac{d\varphi(x_j)}{dx_j}, \end{aligned}$$

where

$$\varphi(x_j) = u(a + x_j) + u(2a + x_j) + u(a - x_j) + u(2a - x_j).$$

This potential for molecule j corresponds to $\varphi(r)$ in Eq. (5–1). Also,

$$\varphi(0) = 2[u(a) + u(2a)],$$

as already indicated above.

Returning now to Eq. (5–20), we can expand quantities of the form $u(a + \delta)$ and $u(2a + \eta)$, which appear in Eq. (5–20), in powers of δ and η:

$$u(a + \delta) = u(a) + u'(a)\,\delta + \tfrac{1}{2}u''(a)\,\delta^2 + \cdots,$$
$$u(2a + \eta) = u(2a) + u'(2a)\eta + \tfrac{1}{2}u''(2a)\eta^2 + \cdots.$$

If we substitute expansions of this kind (valid for small vibrations) for $u(a + x_i - x_{i-1})$, etc., in Eq. (5–20), we find that

$$U(x) = U(0) + \sum_i \text{(terms of type } x_i^2, x_i x_{i-1}, x_i x_{i-2}, x_i x_{i+1}, x_i x_{i+2})$$
$$+ \cdots. \tag{5-22}$$

Here $U(0)/N$ and the coefficients of the quadratic terms are functions of a only. The linear terms cancel because $U(x)$ has a minimum at $x_i = 0$ (all i).

The kinetic energy is simply $\sum_i m\dot{x}_i^2/2$, a sum of independent terms, one for each coordinate x_i. The total energy $H(x, \dot{x})$ is the sum of $U(x)$ and the kinetic energy. Unfortunately, because of the cross terms ($x_i x_{i-1}$, etc.) in Eq. (5–22), $H - U(0)$ is *not* separable, as in Eq. (3–1) [$U(0)$ itself is not important in the present connection—it has to do only with locating the zero of energy]. Hence, as matters stand, we do not have a system of independent subsystems as defined in Chapter 3.

The situation is saved, however, at least in principle, by a mathematical theorem which states that a linear transformation to a set of new coordinates,

$$\begin{aligned} \xi_1 &= \alpha_{11}x_1 + \alpha_{12}x_2 + \cdots + \alpha_{1N}x_N, \\ \xi_2 &= \alpha_{21}x_1 + \alpha_{22}x_2 + \cdots + \alpha_{2N}x_N, \\ &\vdots \\ \xi_N &= \alpha_{N1}x_1 + \alpha_{N2}x_2 + \cdots + \alpha_{NN}x_N, \end{aligned} \tag{5-23}$$

can always be found* such that the energy in the new coordinates, $H(\xi, \dot{\xi})$, retains the "diagonalized" form (i.e., no cross terms) of the kinetic energy and *also* has a diagonalized potential energy $U(\xi)$. That is, $H(\xi, \dot{\xi})$ can be written

$$H(\xi, \dot{\xi}) = U(0) + \frac{1}{2}\sum_i f_i\xi_i^2 + \frac{1}{2}\sum_i M_i\dot{\xi}_i^2, \tag{5-24}$$

where the f_i are "effective" force constants, functions of the coefficients of the quadratic terms in Eq. (5–22) [which in turn depend on $u''(a)$ and $u''(2a)$], and the M_i are "effective" masses, functions of m. The function $H - U(0)$ is now separable. For each of the new coordinates (called *normal coordinates*), ξ_i, there is an independent contribution,

$$H_i = \tfrac{1}{2}f_i\xi_i^2 + \tfrac{1}{2}M_i\dot{\xi}_i^2, \tag{5-25}$$

to $H - U(0)$. Hence, as in Section 3–1, the Schrödinger equation for the system is separable into N independent Schrödinger equations, each derived from an expression of the form (5–25). The Hamiltonian function H_i is in fact that of a harmonic oscillator and leads to nondegenerate energy levels

$$\epsilon_{in} = (n + \tfrac{1}{2})h\nu_i, \qquad n = 0, 1, 2, \ldots,$$

* See L. Pauling and E. B. Wilson, Jr., *Introduction to Quantum Mechanics*. New York: McGraw-Hill, 1935, pp. 282–290.

where

$$\nu_i = \frac{1}{2\pi}\sqrt{\frac{f_i}{M_i}}. \tag{5-26}$$

Each frequency ν_i is a function of the thermodynamic variable $a = L/N$.

Since this is a system of distinguishable molecules (each molecule restricted to the neighborhood of a labeled lattice point), no factor $N!$ appears in Q. We have, as in Eq. (3-5),

$$Q = e^{-U(0)/kT} \prod_{i=1}^{N} q(\Theta_i), \tag{5-27}$$

where $\Theta_i = h\nu_i/k$ (for each ν_i) and $q(\Theta_i)$ is the harmonic oscillator partition function, Eq. (5-6), for characteristic temperature Θ_i. With the factor $e^{-U(0)/kT}$ included in Q, the zero of energy corresponds to infinitely separated atoms at rest.

The conclusion we reach (and it is the same for three-dimensional as for one-dimensional crystals) is that, despite the importance of intermolecular forces in a monatomic crystal, small vibrations in the crystal can be decomposed rigorously into *independent* normal modes of vibration. Hence we are dealing here with a system of *independent subsystems*. This essentially solves the problem in principle [see Eq. (5-27)], but in practice we still have to find the N frequencies ν_i, a very difficult mathematical problem in most cases.

The normal coordinate problem also arises in studying the internal vibrations of polyatomic molecules (Chapter 9). In this case chemical bonds hold the atoms together, so the vibrational force constants are generally larger than in a crystal. Also, the number of atoms in a molecule is relatively small, so "edge effects" cannot be neglected, as in a macroscopic crystal. In principle, of course, there is no distinction between a crystal and a polyatomic molecule. That is, a crystal may be regarded as a giant molecule, usually with relatively weak bonds. If there are N atoms in a monatomic crystal, there are $3N$ degrees of freedom altogether, of which three are associated with the translational motion of the whole crystal and three more are concerned with the rotation of the crystal. There are then $3N - 6$ vibrational degrees of freedom. But with $N = O(10^{20})$, we can take this number of vibrational degrees to be $3N$, without noticeable error.

To illustrate the basic principles involved in normal coordinate analysis, in Appendix V we work out the problem in detail for an example of a hypothetical one-dimensional triatomic molecule.

Let us bypass the normal coordinate question temporarily and continue a little further with the formal analysis, assuming the normal frequencies

ν_i are known. In three dimensions, we have

$$Q = e^{-N\varphi(0)/2kT} \prod_{i=1}^{3N} q(\Theta_i), \tag{5-28}$$

where there are now $3N$ normal modes of vibration, $q(\Theta_i)$ is given by Eq. (5-6), and we have defined $\varphi(0)$ as $2U(0)/N$. The quantity $\varphi(0)$ has the same physical significance as in the preceding section: $\varphi(0)$ is the potential energy of interaction between one particular molecule and the other molecules in the crystal when all molecules are at their lattice points. It should be understood that Eq. (5-28) is not restricted to any specific model, such as Eq. (5-20), for example. Rather, a normal coordinate analysis of the small vibrations in the crystal is assumed to have been carried out whatever the nature of the forces between the molecules. From a thermodynamic point of view, $\varphi(0)$ and each Θ_i are functions of V/N.

Because of the very large number, $3N$, of frequencies ν_i, it is convenient and legitimate to introduce a continuous frequency distribution $g(\nu)$ such that $g(\nu)\,d\nu$ is the number of normal modes with frequencies between ν and $\nu + d\nu$. Then, from Eqs. (5-6) and (5-28),

$$-\ln Q = \frac{N\varphi(0)}{2kT} + \int_0^\infty \left[\ln\left(1 - e^{-h\nu/kT}\right) + \frac{h\nu}{2kT}\right] g(\nu)\,d\nu, \tag{5-29}$$

where

$$\int_0^\infty g(\nu)\,d\nu = 3N. \tag{5-30}$$

Thus we no longer need to know all the separate ν_i's; the function $g(\nu)$ is sufficient. In Eq. (5-29), $\varphi(0)$ and $g(\nu)$ are functions of V/N; a more explicit notation would be $g(\nu;\, V/N)$.

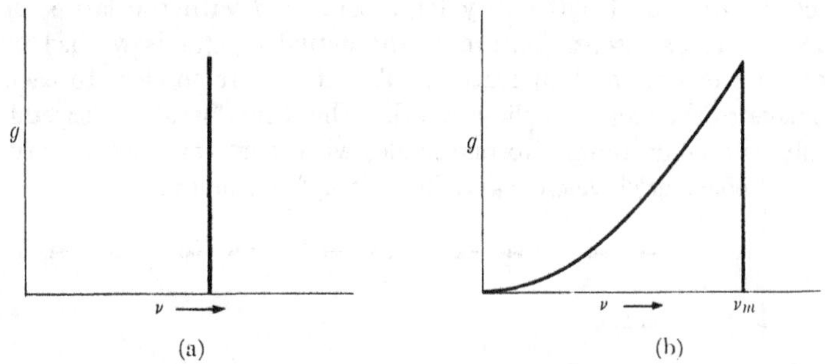

FIG. 5-4. Frequency distribution $g(\nu)$ for crystal. (a) Einstein approximation. (b) Debye approximation.

From Q above, one can now derive equations for thermodynamic functions such as E, C_V, p, μ, etc., which are formally exact but not very useful unless $g(\nu; V/N)$ is known, which is not usually the case. For example, we find for the heat capacity,

$$C_V = k\int_0^\infty \frac{(h\nu/kT)^2 e^{h\nu/kT} g(\nu)}{(e^{h\nu/kT}-1)^2}\, d\nu\,, \tag{5-31}$$

which is just the sum, as we should expect, of $3N$ terms of the Einstein type (Eq. 5-12).

In present notation, it is clear that Einstein's rather crude approximation (Section 5-1) amounts to taking $g(\nu)$ as a Dirac δ-function (Fig. 5-4a). The discrepancy between Einstein's $g(\nu)$ and more accurate frequency distributions will become apparent in Sections 5-3 and 5-4.

Attempts that have been made to calculate $g(\nu)$ exactly, from various models concerning the intermolecular forces, will be summarized briefly in Section 5-4. However, because of the mathematical difficulty of the problem, this rigorous approach is not very fruitful in practice. Therefore we shall first consider the very successful approximation of Debye, which is a compromise between Einstein's extremely simple model and an exact treatment.

5-3 The Debye approximation. In classical mechanics, when a single normal mode of vibration of a crystal is excited, the value of the associated normal coordinate varies periodically with time. As is pointed out in Appendix V, in such a normal mode each atom in the crystal vibrates about its equilibrium position with the same period (or frequency) and phase as that of the normal coordinate. Figure 5-5 shows two possible normal modes for a one-dimensional crystal, one with short "wavelength" ($\lambda = 2a$) and one with relatively long wavelength ($\lambda = 10a$). In normal modes with wavelengths very large compared with the lattice spacing, the atomic or discrete nature of the actual crystal is washed out and becomes an unimportant feature. The atoms are so close to each other relative to the length of the wave that the wave "sees" the crystal essentially as a continuum. Normal modes with long wavelengths may therefore be considered elastic waves in a virtual continuum.

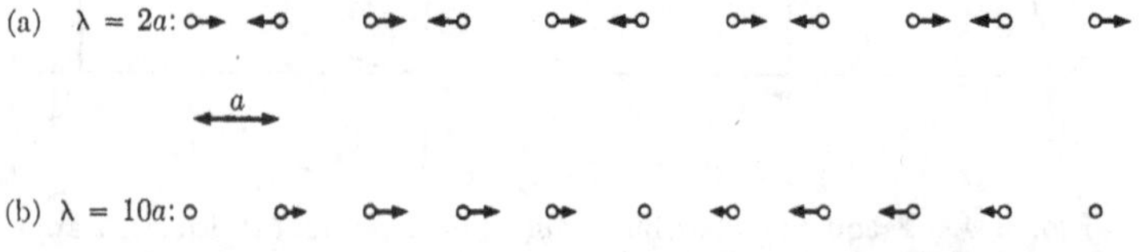

FIG. 5-5. Two normal vibrational modes for a one-dimensional crystal.

This is the origin of Debye's approximation. The above remarks about a continuum are exact for sufficiently long wavelengths (or low frequencies, see Appendix VI). Thus the correct asymptotic form of $g(\nu)$ for small ν can be deduced by treating the crystal as an elastic continuum. The approximation of Debye is to assume that this low-frequency form of $g(\nu)$ is correct for *all* frequencies.

It is shown in Appendix VI that in a three-dimensional continuum $g(\nu) = \alpha\nu^2$, where α is given by Eq. (VI-30). This is, then, the explicit low-frequency form of $g(\nu)$ that Debye adopts for all frequencies. The total number of normal modes is $3N$, so a cut-off at a maximum frequency ν_m is necessary (Fig. 5-4b). That is,

$$\int_0^{\nu_m} g(\nu)\,d\nu = 3N = \frac{\alpha\nu_m^3}{3},$$

$$\begin{aligned} g(\nu) &= \frac{9N\nu^2}{\nu_m^3} \qquad (0 \le \nu \le \nu_m) \\ &= 0 \qquad (\nu > \nu_m), \end{aligned} \tag{5-32}$$

as in Eq. (VI-34). From a thermodynamic point of view, ν_m in Eq. (5-32) is a function of V/N. With the complete frequency distribution assigned in Eq. (5-32), thermodynamic functions follow in a routine way from Eq. (5-29).

Before turning to the thermodynamic functions, let us digress briefly to make a comment on the range of validity of Debye's assumption $g(\nu) = \alpha\nu^2$, which we know to be accurate at sufficiently low frequencies. The question is, what do we mean by sufficiently low frequencies? Normal modes with wavelengths $\lambda \gg (V/N)^{1/3}$, where $(V/N)^{1/3}$ is of the order of the lattice spacing, are treated accurately by a continuum theory; say, rather arbitrarily, $\lambda \ge 10(V/N)^{1/3}$. If we use $\lambda = v_3/\nu$ and $\nu_m = O[(N/V)^{1/3}v_3]$ from Eqs. (VI-22) and (VI-33), we have the condition $\nu \le \nu_m/10$ for the range of validity of $g(\nu) = \alpha\nu^2$. See also Figs. 5-8 and 5-10, below, where the exact $g(\nu)$ can be compared with the Debye $g(\nu)$.

For the energy, according to Debye, we have

$$E = kT^2\left(\frac{\partial \ln Q}{\partial T}\right)_{N,V}$$

$$= \frac{N\varphi(0)}{2} + \frac{9NkT}{\nu_m^3}\int_0^{\nu_m}\left(\frac{h\nu}{2kT} + \frac{h\nu/kT}{e^{h\nu/kT} - 1}\right)\nu^2\,d\nu \tag{5-33}$$

$$= \frac{N\varphi(0)}{2} + \frac{9NkT}{u^3}\int_0^{u}\left(\frac{x}{2} + \frac{x}{e^x - 1}\right)x^2\,dx, \tag{5-34}$$

where

$$x = \frac{h\nu}{kT}, \qquad u = \frac{h\nu_m}{kT}. \tag{5-35}$$

The resemblance between Eq. (5–33) and Eq. (5–9) for the Einstein model should be noted. One of the integrals in Eq. (5–34) is easy, but the other has to be evaluated numerically:

$$E = \frac{N\varphi(0)}{2} + \frac{9Nh\nu_m}{8} + 3NkTD(u), \tag{5-36}$$

where

$$D(u) = D(h\nu_m/kT) = \frac{3}{u^3}\int_0^u \frac{x^3\,dx}{e^x - 1}. \tag{5-37}$$

In Eq. (5–36), $\varphi(0)$, ν_m, and u are functions of V/N. The asymptotic behavior of $D(u)$ is:

$$D(u) \to \frac{3}{u^3}\int_0^\infty \frac{x^3\,dx}{e^x - 1} = \frac{\pi^4}{5u^3} \quad \text{as} \quad T \to 0 \text{ and } u \to \infty, \tag{5-38}$$

$$D(u) \to \frac{3}{u^3}\int_0^u \frac{x^3\,dx}{(1 + x + \cdots) - 1} \to 1 \quad \text{as} \quad T \to \infty \text{ and } u \to 0. \tag{5-39}$$

Therefore

$$E \to \frac{N\varphi(0)}{2} + \frac{9Nh\nu_m}{8} + \frac{3N\pi^4 h\nu_m}{5}\left(\frac{kT}{h\nu_m}\right)^4 \quad \text{as} \quad T \to 0. \tag{5-40}$$

The second term on the right is the vibrational zero-point energy (Problem 5–2). Also,

$$E \to \frac{N\varphi(0)}{2} + 3NkT \quad \text{as} \quad T \to \infty. \tag{5-41}$$

This is the same as Eq. (5–11) (Einstein model), and is the energy predicted by classical statistical mechanics (Chapter 6).

For the heat capacity, we have (Problem 5–3)

$$C_V = \left(\frac{\partial E}{\partial T}\right)_{N,V} = 3Nk\frac{\partial}{\partial T}[TD(u)]$$

$$= 3Nk\left[4D(u) - \frac{3u}{e^u - 1}\right]. \tag{5-42}$$

At this point we introduce the notation

$$u = \frac{h\nu_m}{kT} = \frac{\Theta_D}{T}, \qquad \Theta_D = \frac{h\nu_m}{k}, \tag{5-43}$$

which is often used. We call Θ_D the "Debye temperature." It refers to the cut-off frequency ν_m. In the Einstein theory Θ refers to the *only* frequency ν. Θ_D is different for different crystals and is a function of V/N for any given crystal. According to Eq. (5-42), C_V/Nk is a universal function of T/Θ_D. This function is shown in Fig. 5-2. Thus the Debye approximation, like Einstein's, leads to a law of corresponding states. The asymptotic behavior of C_V is:

$$C_V \to \frac{12Nk\pi^4}{5}\left(\frac{T}{\Theta_D}\right)^3 \quad \text{as} \quad T \to 0, \tag{5-44}$$

$$C_V \to 3Nk\left(4 - \frac{3u}{1 + u + \cdots - 1}\right) \to 3Nk \quad \text{as} \quad T \to \infty. \tag{5-45}$$

That is, the Debye theory predicts $C_V \propto T^3$ at low temperatures (the criterion is approximately $T < \Theta_D/12$) and leads to the Dulong-Petit value of C_V at high temperatures.

Monatomic crystals do, in fact, follow a law of corresponding states rather closely. Figure 5-6 shows a superposition of heat-capacity points for Al, Cu, Pb, and C (diamond) after suitable adjustment of the temperature scale, taken from a paper by Lewis and Gibson. Many other sub-

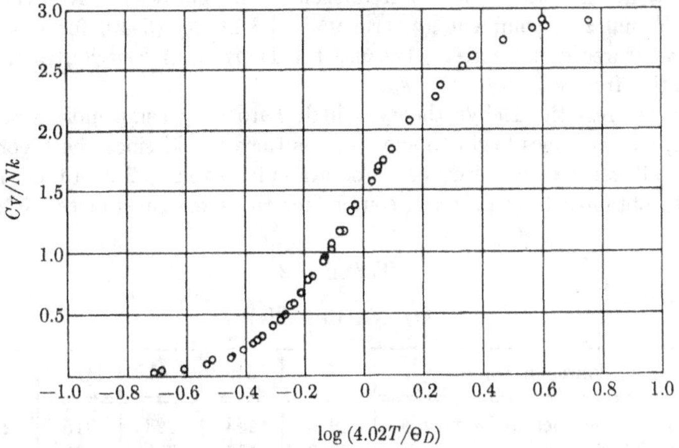

FIG. 5-6. Superimposed experimental heat capacity points for Al, Cu, Pb, and C (diamond).

TABLE 5-1

LOW-TEMPERATURE HEAT CAPACITIES

(C_V in cal·mole^{-1}·deg^{-1})

T, °K	Fe, $\frac{10^2 C_V^{1/3}}{T}$	T, °K	Al, $\frac{10^2 C_V^{1/3}}{T}$
32.0	1.67	19.1	2.12
33.1	1.70	23.6	2.03
35.2	1.77	27.2	2.01
38.1	1.73	32.4	1.95
42.0	1.64	33.5	2.00
46.9	1.71	35.1	1.97

stances, for example Ag, Hg, Tl, and Zn, also have curves that coincide with those included in the figure. Furthermore, the Debye function (5–42) fits the experimental curve, over the whole temperature range, in a very satisfactory manner.

The T^3 law for C_V at low temperatures is also followed quite accurately, as seen in two typical cases in Table 5–1. According to Eq. (5–44), $C_V^{1/3}/T$ should be a constant; we see from the table that it very nearly is. We should expect the Debye theory to be essentially exact for the heat capacity (and entropy) at low temperatures, because only the low-frequency modes are excited at low temperatures, and the Debye assumption $g(\nu) = \alpha\nu^2$ is correct for low frequencies (Problem 5–4). However, the vibrational zero-point energy term $9Nh\nu_m/8$ in Eq. (5–40) for E at low temperatures is *not* correct, because the Debye $g(\nu)$ contributes to this quantity from $\nu = 0$ to $\nu = \nu_m$.

If we press the Debye theory a little harder by using more sensitive tests, discrepancies begin to show up, as they should since the theory is, after all, an approximate one. For example, Table 5–2 contains values of Θ_D obtained from different sources for the same substances. Ideally,

TABLE 5-2

Θ_D VALUES (°K)

Source of Θ_D	C	Fe	Al	Cu	Ag
Entire experimental C_V curve	1860	453	398	315	215
T^3 part of C_V curve	2230	455	385	321	—
Experimental elastic constants and Eqs. (VI-15, 16, 33)	—	—	402	332	214

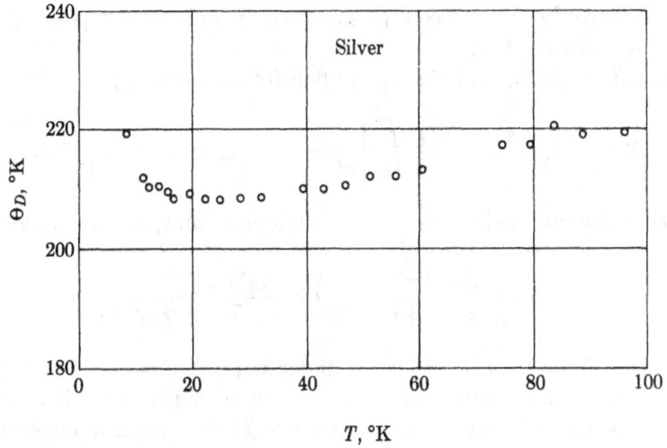

FIG. 5-7. Θ_D for silver as function of temperature.

values of Θ_D should be independent of source. A more rigorous test is shown in Fig. 5-7. For each experimental value of C_V for silver, Θ_D is calculated from Eq. (5-42), using the temperature of the C_V measurement. If the Debye theory were exact, points in Fig. 5-7 would lie near a horizontal line (Θ_D = constant, independent of T). Similar nonhorizontal $\Theta_D - T$ curves are found quite generally for other substances as well.

Table 5-3 (from Blackman*) presents average Θ_D values for a few more substances. From Fig. 5-2 we see that the classical Dulong-Petit value of C_V is practically reached at room temperature for many of these crystals. As mentioned in Section 5-1, a value of $\Theta_D = 300$°K corresponds to

TABLE 5-3

Θ_D VALUES (°K)

Element	Θ_D	Element	Θ_D	Element	Θ_D
Li	430	Cr	405	Hg	90
Na	160	Ca	230	Be	980
K	99	Mo	375	Mg	330
Au	185	Pt	225	Zn	240
Pb	86	W	315	Cd	165

* M. BLACKMAN, *Handbuch der Physik* (Springer, Berlin) **7.1**, 325 (1955).

$\nu_m = 6 \times 10^{12}\ \text{sec}^{-1}$, which is small compared with typical diatomic molecule vibration frequencies.

From Eqs. (5–29) and (5–34), we find for the entropy,

$$S = \frac{E - A}{T} = \frac{9Nk}{u^3} \int_0^u \left[\frac{x}{e^x - 1} - \ln (1 - e^{-x}) \right] x^2\, dx. \quad (5\text{–}46)$$

At low temperatures [put $u = \infty$ in the upper limit, as in Eq. (5–38)],

$$S \to \frac{9Nk}{u^3} \left(\frac{\pi^4}{15} + \frac{\pi^4}{45} \right) = \frac{4\pi^4 Nk}{5} \left(\frac{T}{\Theta_D} \right)^3. \quad (5\text{–}47)$$

Thus $S \to 0$ as $T \to 0$. In the nondegenerate ($\Omega = 1, S = 0$) ground state of the entire system (crystal), each vibrational mode is in its nondegenerate ground state. In writing Eq. (5–29), nuclear degeneracy is omitted as usual, and the electronic ground state has been assumed nondegenerate.

Finally, let us consider the vapor pressure $p_0(T)$ of a monatomic crystal at very low temperatures. As $T \to 0$, $p_0 \to 0$. Hence, for the crystal, we can drop the $p_0 V$ term in $\mu = (A + p_0 V)/N$, and we can assume that the gas phase is ideal. Then, for the crystal, from Eqs. (5–40) and (5–47),

$$\mu = \frac{A}{N} = \frac{E - TS}{N} = -\Lambda_0 - \frac{\pi^4 k T^4}{5\Theta_D^3}, \quad (5\text{–}48)$$

where

$$-\Lambda_0 = \frac{\varphi(0)}{2} + \frac{9h\nu_m}{8}. \quad (5\text{–}49)$$

The quantity Λ_0 is the heat of sublimation per molecule at 0°K. Equation (5–48) may be regarded as exact provided we use the experimental Λ_0, since the zero-point energy term in Eq. (5–49) is not correct, as already mentioned. We may calculate Θ_D (Section 5–4) in Eq. (5–48), or it may be obtained from experimental elastic-constant or heat-capacity measurements at low temperature and low pressure.

To obtain the vapor pressure p_0, we use the thermodynamic criterion for phase equilibrium: we set μ_{crystal} from Eq. (5–48) equal to μ_{gas} from Eq. (4–24). We find

$$\ln p_0 = \frac{5}{2} \ln T - \frac{\Lambda_0}{kT} - \frac{\pi^4}{5} \left(\frac{T}{\Theta_D} \right)^3 + \ln \left[\left(\frac{2\pi m k}{h^2} \right)^{3/2} k \right]. \quad (5\text{–}50)$$

We should emphasize here that the same zeros of energy (isolated gas molecule at rest) and entropy ($\Omega = 1$ for a system of point masses, i.e., the crystal, at $T = 0$) have been chosen for both phases. If a term in the ground-state nuclear degeneracy had been included in μ_{crystal} and in μ_{gas}, these terms would cancel in Eq. (5–50). The electronic ground state has been assumed nondegenerate in both crystal and gas.

In the equilibrium between a crystal and its vapor, as in all equilibria, competition and compromise between energy and entropy effects determine the equilibrium point. Here, the crystal is more stable from an energetic point of view (Λ_0); i.e., it has a lower energy than the gas because of intermolecular interactions in the crystal. But the gas is more stable insofar as the entropy is concerned; i.e., the gas molecules are in a much higher state of disorder and have a higher Ω and entropy. The molecules distribute themselves between the two phases in such a way as to equalize the chemical potentials in the two phases: the chemical potentials depend on both energy and entropy. An equivalent statement is that the molecules distribute themselves between the two phases so as to minimize the Helmholtz free energy A of the combined system *crystal* + *gas* if V, N, and T for the combined system are constant (or F is minimized if p, N, and T are constant) (see Sections 10–1 and 10–2).

The last term in Eq. (5–50) is a constant, call it i (the "vapor pressure constant"), the only term in the equation independent of temperature. Now i, which in thermodynamics appears as an integration constant on integrating the Clausius-Clapeyron equation for $d \ln p_0/dT$, can be evaluated experimentally by using data on the vapor pressure, heat of sublimation, and heat capacities of gas and crystal. Experimental values of i are found to agree* with theoretical values of i to within experimental error. Historically, this was a very significant check of the methods of quantum-statistical mechanics. Although i can be obtained as an *experimental* thermodynamic quantity, thermodynamics by itself is powerless, of course, to *predict* or *explain* i-values. Also, i cannot be deduced from purely classical-mechanical considerations (since h is included in i).

5–4 Exact treatments of the frequency distribution problem. A detailed discussion of attempts to compute $g(\nu)$ from first principles would be far outside the scope of this introductory book. However, we shall give here a brief summary, which the reader may find instructive, of exact, or practically exact, *results* obtained in certain special cases. Other sources must be consulted for details and derivations. We shall not discuss thermodynamic functions, since $g(\nu)$ is available for just a few idealized models (Problem 5–5).

* See Fowler and Guggenheim, pp. 199–202.

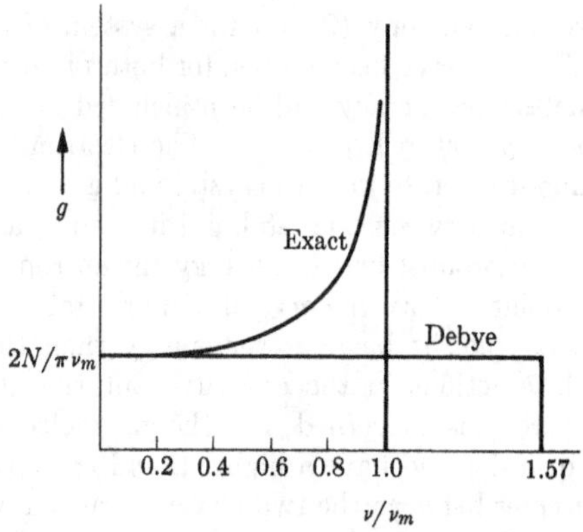

FIG. 5–8. Frequency distributions for one-dimensional crystal with nearest-neighbor interactions only.

The first case we consider is a *one-dimensional* lattice with lattice spacing a and nearest-neighbor interactions only. The Hamiltonian is

$$H = U(0) + \frac{1}{2}\sum_i m\dot{x}_i^2 + \frac{f}{2}\sum_i (x_{i+1} - x_i)^2. \quad (5\text{–}51)$$

This is a special case of the Hamiltonian associated with Eq. (5–22) (i.e., second-neighbor interactions are omitted). Specifically, the force constant f in Eq. (5–51) is just $u''(a)$ in the earlier notation (Problem 5–6). The frequency distribution problem here is rather easy and can be solved exactly. The result* is

$$g(\nu) = \frac{2N}{\pi}\frac{1}{(\nu_m^2 - \nu^2)^{1/2}}, \quad (5\text{–}52)$$

where

$$\nu_m = \frac{1}{\pi}\sqrt{\frac{f}{m}}. \quad (5\text{–}53)$$

The function $g(\nu)$ is shown in Fig. 5–8. At low frequencies,

$$g(\nu) \rightarrow \frac{2N}{\pi\nu_m}. \quad (5\text{–}54)$$

In the continuum theory (Appendix VI), which must agree with Eq. (5–54) at low frequencies, according to Eq. (VI–11) there are n

* Mayer and Mayer, pp. 246–248. See also Blackman, *loc. cit.*, pp. 330–331.

modes between $\nu = 0$ and $\nu = v_1 n/2Na$. That is,

$$\int_0^{v_1 n/2Na} g\, d\nu = n = \frac{g v_1 n}{2Na},$$

or

$$g = \frac{2Na}{v_1}. \tag{5–55}$$

Equating g in Eqs. (5–54) and (5–55), we find

$$v_1 = \pi a \nu_m = a\sqrt{f/m}. \tag{5–56}$$

This equation expresses the macroscopic quantity v_1 ("velocity of sound") in terms of the molecular parameters a, f, and m. In the Debye approximation for a one-dimensional crystal, g is a constant out to a cut-off frequency that we denote here, to avoid confusion, by ν_m^D. We have

$$g\nu_m^D = N = \frac{2Na\nu_m^D}{v_1},$$

or

$$\nu_m^D = \frac{v_1}{2a} = \frac{1}{2}\sqrt{\frac{f}{m}}. \tag{5–57}$$

This is larger than the true ν_m by a factor 1.57 (Fig. 5–8).

Next, suppose we have a linear crystal with lattic spacing a, force constant f (nearest-neighbor interactions only), but with two alternating

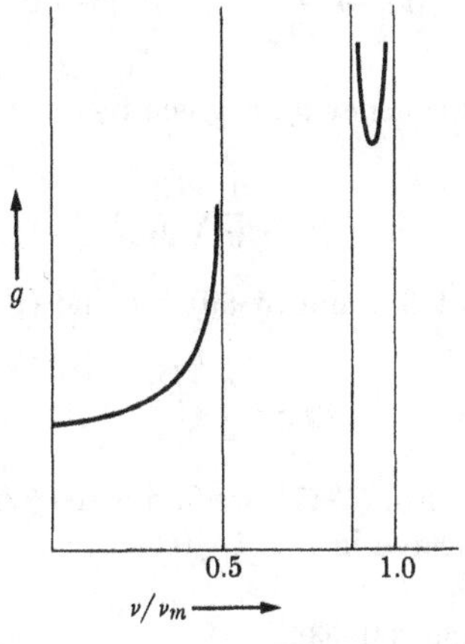

FIG. 5–9. Exact frequency distribution for one-dimensional crystal with $m_1/m_2 = 3$.

kinds of atoms with masses m_1 and m_2. This problem can also be handled exactly without difficulty.* We shall not give any details, but merely show in Fig. 5–9 a plot of $g(\nu)$ against ν for $m_1/m_2 = 3$. There are two branches in $g(\nu)$. The low-frequency branch is called the "acoustical" branch, and the high-frequency branch the "optical" branch. In normal modes belonging to the acoustical branch, neighboring atoms are displaced in the same direction (as in Fig. 5–5b). In the optical branch, neighboring atoms are displaced in opposite directions (as in Fig. 5–5a). A gap between branches, as in Fig. 5–9, is common in one-dimensional crystals with two kinds of atoms, but not invariable.

In a very elegant paper, Montroll† was able to find the exact $g(\nu)$ for a particular *two-dimensional* crystal. The lattice studied was simple square, with one kind of atom, and lattice spacing a. First- and second-neighbor interactions were taken into account. The parameters of the problem are a, the mass m, the nearest-neighbor force constant $f = u''(a)$ (called α by Montroll), and second-neighbor force constant $2\gamma = u''(2^{1/2}a)$. It is convenient to introduce

$$\tau = \frac{1}{1 + (f/2\gamma)}\cdot \tag{5–58}$$

When $\tau = 0$, second-neighbor interactions are absent. An exact analytical expression for $g(\nu)$ was found by Montroll in the special case $\tau = 1/3$, or $f/2\gamma = 2$. This value of τ corresponds to elastic isotropy. The function $g(\nu)$ is shown in Fig. 5–10. The low-frequency behavior of $g(\nu)$ agrees with continuum theory, $g \propto \nu$:

$$g(\nu) \to \frac{8N\nu}{\pi\nu_m^2} \qquad \text{as} \quad \nu \to 0, \tag{5–59}$$

where the maximum frequency ν_m is given by

$$\nu_m = \frac{1}{\pi}\sqrt{\frac{3f}{2m}}\cdot \tag{5–60}$$

On comparing Eqs. (VI–35) and (5–59), we find (Problem 5–7)

$$\bar{v}_2 = \frac{a}{2}\sqrt{\frac{3f}{m}}\cdot \tag{5–61}$$

If we calculate ν_m^D as in Eq. (5–57), we find here $\nu_m^D/\nu_m = (\pi/2)^{1/2} = 1.25$ (Problem 5–8), as indicated in Fig. 5–10.

---

* Blackman, *loc. cit.*, pp. 331–333.

† *J. Chem. Phys.* **15,** 575 (1947). See also M. SMOLLETT, *Proc. Phys. Soc. London* **65A,** 109 (1952).

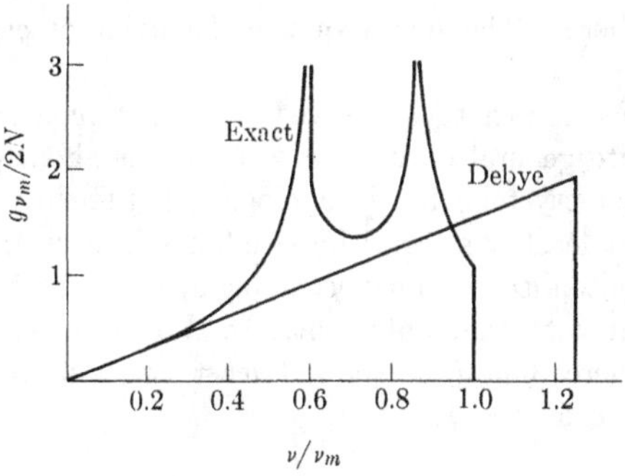

FIG. 5-10. Frequency distributions for special case ($\tau = \frac{1}{3}$) of two-dimensional square crystal.

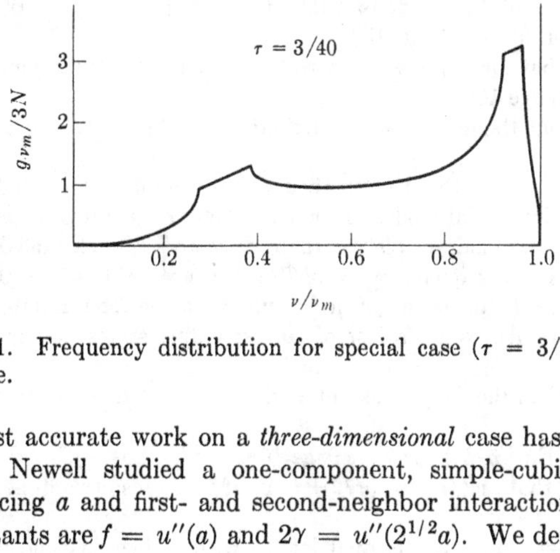

FIG. 5-11. Frequency distribution for special case ($\tau = 3/40$) of a simple cubic lattice.

The most accurate work on a *three-dimensional* case has been done by Newell.* Newell studied a one-component, simple-cubic lattice with lattice spacing a and first- and second-neighbor interactions. Again the force constants are $f = u''(a)$ and $2\gamma = u''(2^{1/2}a)$. We define

$$\tau = \frac{1}{2 + (f/2\gamma)}. \tag{5-62}$$

Although Newell did not obtain an exact $g(\nu)$ for any τ, his calculations of $g(\nu)$ are accurate to one or two percent for τ up to almost $1/10$. Furthermore, he was able to describe the exact location and type of singularities in $g(\nu)$ for $0 \leq \tau \leq 1/10$. Figure 5-11 shows $g(\nu)$ for $\tau = 3/40$. There are five singularities in the slope of $g(\nu)$, but $g(\nu)$ itself is finite and con-

* G. F. NEWELL, *J. Chem. Phys.* **21**, 1877 (1953).

tinuous everywhere. The low-frequency behavior of $g(\nu)$ is $g(\nu) \propto \nu^2$, as expected.

Finally, we mention that Van Hove* was able to come to the following definite and quite general qualitative conclusions about $g(\nu)$ for two- or three-dimensional crystals, by a topological argument. In two dimensions, $g(\nu)$ has at least one logarithmic infinity in each frequency branch and at least a finite discontinuity occurring at $\nu = \nu_m$. Montroll's results are in agreement with this statement. In three dimensions, $g(\nu)$ is continuous everywhere, but $dg/d\nu$ has at least two infinite discontinuities, and $dg/d\nu = -\infty$ at $\nu = \nu_m$.

PROBLEMS

5–1. Derive equations for p and μ for the Einstein model, considering $\varphi(0)$ and Θ functions of V/N. Show that $(\partial A/\partial N)_{V,T}$ and $(A + pV)/N$ give the same expression for μ. (Page 91.)

5–2. Show that $9Nh\nu_m/8$ is the zero-point energy for the frequency distribution (5–32). (Page 100.)

5–3. Work out the details of the derivation of Eq. (5–42) for the Debye C_V. (Page 100.)

5–4. Show that, in the Debye theory, the number of excited [i.e., $n > 0$ in Eq. (5–4)] vibrational modes in the frequency range ν to $\nu + d\nu$, at temperature T, is proportional to x^2e^{-x}, where $x = h\nu/kT$. The maximum in this function occurs at a frequency $\nu' = 2kT/h$; hence $\nu' \to 0$ as $T \to 0$. (Page 102.)

5–5. If an exact calculation of $g(\nu)$ could be made for a number of different monatomic crystals, would a law of corresponding states be expected for C_V? (Page 105.)

5–6. Show that the force constant f in Eq. (5–51) is equal to $u''(a)$. (Page 106.)

5–7. Verify Eq. (5–61) for $\bar{v}_2$. (Page 108.)

5–8. Show that $\nu_m^D/\nu_m = (\pi/2)^{1/2}$ in Montroll's two-dimensional case. (Page 108.)

5–9. Consider a one-dimensional lattice with lattice spacing r^*. All atoms are at equilibrium points except the "central" atom, which moves in the potential field of its two nearest neighbors. This is a one-dimensional Einstein model. Take Eq. (IV–1) for the intermolecular pair potential. Find $\varphi(0)$ and f in the one-dimensional form of Eq. (5–1) in terms of r^* and ϵ.

5–10. Consider a system of one-dimensional oscillators, all with characteristic temperature Θ [Eqs. (5–4) and (5–6)]. Derive an expression for the fraction P_n of these oscillators in the energy level ϵ_n. Calculate P_0, P_1, and P_2 for $T = 4\Theta$, $T = \Theta$, and $T = \Theta/4$.

---

* L. VAN HOVE, *Phys. Rev.* **89**, 1189 (1953).

5–11. Derive an equation for the vapor pressure $p_0(T)$ of an Einstein crystal, assuming the vapor is an ideal gas and using Eq. (5–16). What choice of V/N in $\varphi(0)$ and Θ is appropriate?

5–12. The heat capacity C_V of a monatomic solid at 300°K is $2R$ per mole. Use the Einstein theory to calculate the frequency ν.

5–13. Use thermodynamic connections between (a) E and C_V and (b) S and C_V to check the self-consistency of Eqs. (5–40), (5–44), and (5–47).

5–14. Calculate Θ_D for Fe from the value of C_V at 32.0°K given in Table 5–1.

5–15. Use Fig. 5–2 and the Θ_D values in Table 5–3 to estimate C_V in cal·mole^{-1}·deg^{-1} for Pb, Mg, and Be at 25°C.

Supplementary Reading

Blackman, M., *Handbuch der Physik* (Springer, Berlin) **7.1,** 325 (1955).
Fowler and Guggenheim, Chapter 4.
Mayer and Mayer, Chapter 11.
Slater, Chapters 13, 14, 15.
Tolman, Chapter 14.

CHAPTER 6

CLASSICAL STATISTICAL MECHANICS

It is well known that quantum-mechanical results for a given system go over asymptotically into classical mechanical equations in the limit of large quantum numbers. Also, in the quantum-mechanical canonical ensemble partition function (for one or many particles), terms corresponding to the higher quantum numbers make more and more important contributions to the sum as the temperature increases. We may anticipate, then, that at sufficiently high temperatures, the quantum partition function should approach asymptotically a partition function developed from a classical mechanical starting point.

We consider classical statistical mechanics in this chapter. No results are obtainable from classical statistics which cannot be found as limiting laws from quantum statistics, but often the classical method is easier to use. We shall, however, refrain from discussing the *principles* of classical statistical mechanics in any detail. Although this is a very elegant subject, we shall adopt the point of view here—for lack of space—that the quantum method provides the more general postulatory foundation and hence that classical statistics does not require separate development since it follows as a special case from the quantum postulates. The reader interested in the principles of classical statistical mechanics *per se* cannot do better than to read Tolman's masterly exposition.

The treatment of the transition from quantum to classical statistics in this chapter will be inductive, and proofs will be omitted for the more complicated cases. This subject will, however, be discussed in a more general way in Chapter 22.

6–1 Introductory examples. We consider two special cases in this section. The first is a one-dimensional harmonic oscillator, with mass m and classical frequency ν, restricted to the x-axis (equilibrium point $x = 0$). The particular question of interest is, what is the classical analog of the quantum-mechanical equation (5–5),

$$q = \sum_{n=0}^{\infty} e^{-\epsilon_n/kT}? \tag{6–1}$$

This is a sum of $e^{-(\text{energy})/kT}$ over all possible quantum states of the system. The corresponding classical expression is the sum (or actually integral, since the classical state can vary continuously) of $e^{-(\text{energy})/kT}$ over all possible *classical* states of the system. The classical energy is just the Hamiltonian function $H(q, p)$, which for this system is

$$H(q, p) = \tfrac{1}{2}fx^2 + \tfrac{1}{2}m\dot{x}^2$$
$$= 2\pi^2 m\nu^2 q^2 + \frac{p^2}{2m}, \tag{6-2}$$

where f is the force constant [see Eq. (5-2)], $q = x$, and $p = m\dot{x}$ (Appendix VII). We use q and p as independent variables for H because the equations of classical statistics which we shall use below turn out to be simpler with this choice, as one might anticipate from Eqs. (VII-7). The classical state (position and velocity) is specified by assigning values to q and p. Both variables range continuously from $-\infty$ to $+\infty$. Hence we have

$$q_{\text{class}} = c \iint_{-\infty}^{+\infty} e^{-H(q,p)/kT}\, dq\, dp, \tag{6-3}$$

where c is a constant. We have to choose c so that the classical equation (6-3) gives the same result as the high-temperature limit of the quantum equation (5-6), namely, $q \to kT/h\nu$. The fact that Planck's constant h appears in this limit shows that c could not possibly be deduced from purely classical considerations. What we are essentially doing here, in evaluating c, is establishing a statistical-mechanical correspondence principle for this special case.

We substitute Eq. (6-2) in Eq. (6-3) and carry out the two integrations. We find $q_{\text{class}} = c\,kT/\nu$. Therefore $c = 1/h$ in order to satisfy $q_{\text{class}} = q\,(T \to \infty)$.

The q, p space over which the integration is carried out in Eq. (6-3) is called the *phase space* of the system, and the integral itself is called the *phase integral*. Any point in phase space corresponds to a definite classical state. In quantum mechanics, such a precise (point) specification of state is not possible. According to the uncertainty principle, the state of the system cannot be located in phase space more closely than within an area $\Delta q\, \Delta p$ of order h. This is consistent with the value just found for c, because

$$h \sum_{\substack{\text{quantum}\\\text{states}}} e^{-(\text{energy})/kT} \to \iint_{\substack{\text{classical}\\\text{phase space}}} e^{-(\text{energy})/kT}\, dq\, dp \tag{6-4}$$

implies that the volume of classical phase space corresponding to one quantum state, in the limit of large quantum numbers, is h. The precise value h cannot, however, be deduced from the uncertainty principle itself, because this principle reads, to be exact, $\Delta q\, \Delta p \geq h/4\pi$ (where Δq and Δp are standard deviations).

Incidentally, the relation of Eq. (6–4) to the uncertainty principle confirms the convenience of the choice of independent variables q, p (instead of, say, $q, \dot{q}$ or some other choice).

An alternative deduction of the factor h in Eq. (6–4), or $1/h$ in Eq. (6–3), is the following. Let us draw paths of constant energy, $H(q, p) = \epsilon =$ constant, in classical q, p space. From Eq. (6–2), we see that these paths are ellipses

$$\frac{q^2}{\alpha^2} + \frac{p^2}{\beta^2} = 1, \tag{6–5}$$

where

$$\alpha^2 = \frac{\epsilon}{2\pi^2 m\nu^2}, \qquad \beta^2 = 2m\epsilon,$$

and the area of an ellipse is

$$\text{Area} = \pi\alpha\beta = \frac{\epsilon}{\nu}. \tag{6–6}$$

In particular, let ellipses be drawn, as in Fig. 6–1, for energies $\epsilon = \epsilon_{n-1}$, ϵ_n, and ϵ_{n+1}, where ϵ_n is the quantum-mechanical energy level (Eq. 5–4) for a large quantum number n. The area between two successive ellipses in Fig. 6–1 is then the area of classical phase space associated with one quantum state. We find from Eqs. (5–4) and (6–6),

$$\text{Area}_{n+1} - \text{Area}_n = (n + \tfrac{3}{2})h - (n + \tfrac{1}{2})h = h, \tag{6–7}$$

which confirms our previous result. To pursue the matter a little further, suppose we assign an area $h/2$ on either side of the $\epsilon = \epsilon_n$ ellipse to the quantum state n, as indicated by the shaded region in Fig. 6–1. For large

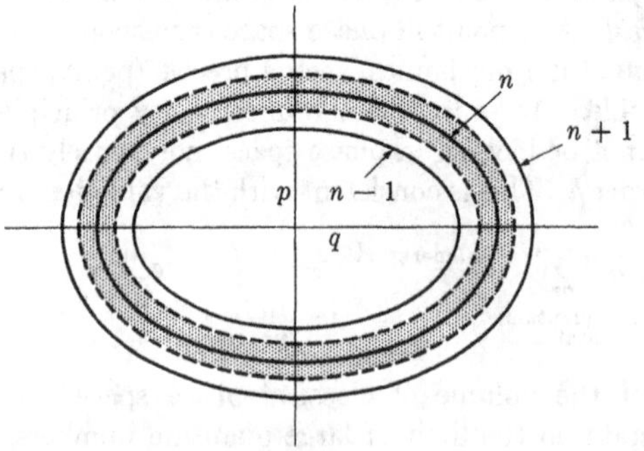

FIG. 6–1. Classical constant-energy paths in phase space for one-dimensional harmonic oscillator (schematic); n = quantum number.

n, the classical energy $H(q, p) = \epsilon$ is almost constant ($\epsilon \cong \epsilon_n$) throughout the shaded area (more precisely, the fractional variation in ϵ within the shaded area becomes smaller as n increases). We can therefore break up the classical integral over q, p space into discreet (as in the quantum sum over states) contributions from each "shaded area" (of magnitude h), because the integrand is essentially constant throughout each such area. Thus,

$$\iint e^{-H(q,p)/kT}\, dq\, dp \cong \sum_n e^{-\epsilon_n/kT}\, h,$$

as before. This argument becomes exact as $T \to \infty$ and higher quantum numbers dominate in the sum.

As a second example, consider a particle in a cubical box of volume V. At sufficiently high temperatures [Eqs. (4–8) and (4–10)],

$$q = \sum_{l_x, l_y, l_z = 1}^{\infty} e^{-\epsilon_{l_x l_y l_z}/kT} \to \left(\frac{2\pi mkT}{h^2}\right)^{3/2} V. \tag{6–8}$$

The classical analog of the above sum is

$$q_{\text{class}} = c' \iiint_{-\infty}^{+\infty} \iiint_V e^{-H(p_x, p_y, p_z)/kT}\, dx\, dy\, dz\, dp_x\, dp_y\, dp_z, \tag{6–9}$$

where

$$H(p_x, p_y, p_z) = \frac{1}{2m}\,(p_x^2 + p_y^2 + p_z^2). \tag{6–10}$$

The momentum $p_x = m\dot{x}$, etc. (Appendix VII). The potential energy is zero in the box and infinite outside the box. Therefore $e^{-H/kT}$ is zero outside, and the integral need be carried out only over the inside of volume V. We find

$$q_{\text{class}} = c'(2\pi mkT)^{3/2} V. \tag{6–11}$$

Comparison with Eq. (6–8) shows that $c' = 1/h^3$.

From these two examples we might surmise that, in general, if there are n degrees of freedom, the phase integral has to be divided by h^n to give the classical (high-temperature) partition function. This is consistent with the uncertainty principle (a factor h for each product $dq\, dp$). This conjecture turns out to be correct. It can be checked, as above, for each special case as it arises. Also, a general quantum-mechanical justification can be given (Chapter 22). We therefore adopt this correspondence principle as a general rule.

6–2 More general systems. We set down in this section a few basic equations in classical statistical mechanics that we shall need throughout the rest of the book. These pertain to more general systems than those considered in the preceding section.

Suppose that the Hamiltonian for one molecule, in a system of independent molecules, is separable in the form

$$H = H_{\text{class}} + H_{\text{quant}}, \tag{6–12}$$

where H_{class} refers to n degrees of freedom that can be treated classically (e.g., translation and often rotation; see Chapters 8 and 9), and H_{quant} refers to the remaining degrees of freedom that cannot be treated classically (e.g., electronic states and vibration). Then, by Eqs. (3–13), (3–14), and Section 6–1,

$$q = q_{\text{class}} q_{\text{quant}}, \tag{6–13}$$

where

$$q_{\text{class}} = \frac{1}{h^n} \int e^{-H_{\text{class}}(q,p)/kT}\, dq_1\, dp_1 \ldots dq_n\, dp_n. \tag{6–14}$$

Alternatively, of course, we can use in place of q_{class} the high-temperature limit of the quantum-mechanical q for the n degrees of freedom included in Eq. (6–14). But usually the direct classical calculation is easier.

As was mentioned at the beginning of this chapter, the qualitative criterion for the legitimate employment of classical statistics in a partition function q for a single molecule (or in the q belonging to some of the degrees of freedom of a single molecule) is that quantum states with large quantum numbers contribute heavily to q (Problem 6–1). This will be the case if $kT \gg \Delta\epsilon$, where $\Delta\epsilon$ is the magnitude of the energy separation between successive energy levels (Problem 6–2). This is the same criterion we have already used [see Eq. (4–8) and following Eq. (5–7)] for replacing the quantum sum for q by integral. The use of an integral in place of the quantum sum is therefore equivalent to a classical treatment.

We turn now to the consideration of a system of N indistinguishable monatomic molecules in a volume V with translational degrees of freedom only and no intermolecular forces. The classical phase integral in this case is [see Eq. (6–9) and Appendix VII]

$$I = \int e^{-H(p)/kT}\, dx_1\, dy_1\, dz_1\, dp_{x1}\, dp_{y1}\, dp_{z1} \ldots \\ \cdots dx_N\, dy_N\, dz_N\, dp_{xN}\, dp_{yN}\, dp_{zN}, \tag{6–15}$$

where

$$H(p) = \frac{1}{2m}(p_{x1}^2 + \cdots + p_{zN}^2). \tag{6–16}$$

All the integrations in Eq. (6–15) are easy, and we find

$$I = [(2\pi mkT)^{3/2} V]^N. \tag{6–17}$$

We cannot simply set Q_{class} equal to I/h^{3N} here because we are concerned with a system of N indistinguishable molecules, and not just a single molecule as in the previous examples in this chapter. In fact, we have already found in Chapter 4 that the high-temperature quantum-mechanical Q for this system is given by Eq. (4–12), and Q_{class} must agree with this result. That is, $Q_{\text{class}} = I/h^{3N}N!$ or

$$Q_{\text{class}} = \frac{1}{N!h^{3N}} \int e^{-H(p)/kT}\, dx_1 \ldots dp_{zN} = \frac{1}{N!} \left(\frac{V}{\Lambda^3}\right)^N. \tag{6–18}$$

We may understand the division by $N!$ in this equation as follows. In the integral I over all classical states (q, p) of the system, the molecules are treated as *distinguishable*. For example, let $N = 2$, and let $d\tau$ in Fig. 6–2 represent an element of volume $dx\, dy\, dz\, dp_x\, dp_y\, dp_z$ in the phase space of a single molecule and $d\tau'$ represent another such element of volume located at a different position. Then there are two separate contributions to the integral I counted (a) when molecule 1 is in $d\tau$ and 2 is in $d\tau'$, and (b) when 2 is in $d\tau$ and 1 is in $d\tau'$. In general, with N molecules, there would be $N!$ separate contributions of this type arising from all the possible permutations of the molecules. Actually (i.e., from the point of view of quantum mechanics), the molecules are indistinguishable, and the interchange of particles in the space of Fig. 6–2 does not lead to new states. Thus the classical phase integral overcounts the states by a factor $N!$, and this is corrected for by the division indicated in Eq. (6–18). It should be noted that in classical theory $d\tau$ and $d\tau'$ may be arbitrarily small so that we never have the complication [see Eq. (3–9)] that two molecules are in the same classical state.

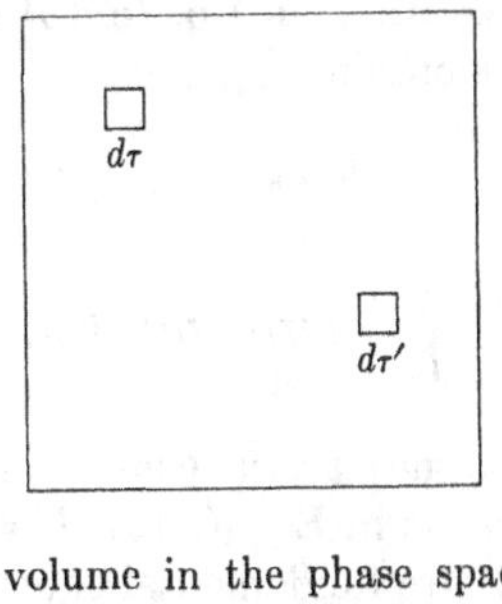

FIG. 6–2. Elements of volume in the phase space of a single monatomic molecule.

We have already seen in Chapter 4 that the condition (4–6) must be satisfied in order that the result given in Eq. (4–12) or (6–18) be correct, i.e., in order for classical statistics to be valid. We have also seen in the discussion following Eq. (4–8) that the condition (4–6) for absence of quantum effects is very much more stringent than the condition $\Delta E = O(\Delta\epsilon) \ll kT$; (4–6) is therefore not only a necessary but also a sufficient condition for the validity of classical statistics. By "classical statistics" here we mean use of (a) the factor $1/N!$, and (b) the classical phase integral.

Now suppose intermolecular forces are present in the system just discussed, so that there is a potential energy $U(x_1, \ldots, z_N)$ that depends on the location of the molecules in the volume V. The classical partition function here is

$$Q_{\text{class}} = \frac{1}{N!h^{3N}} \int e^{-H(q,p)/kT}\, dx_1 \ldots dp_{zN}, \tag{6–19}$$

where

$$H(q, p) = \frac{1}{2m}(p_{x1}^2 + \cdots + p_{zN}^2) + U(x_1, \ldots, z_N). \tag{6–20}$$

The explanation following Eq. (6–18) of the use of $N!$ in that equation is valid irrespective of the existence of a potential-energy contribution to H. Therefore a factor $1/N!$ is also inserted in Eq. (6–19), though in this case we do not have available the high-temperature limit of the quantum Q as a check. This check is provided, however, in Chapter 22. As to the criterion for the use of classical statistics in this case, we note that the paragraph preceding Eq. (4–8) gives a quite general interpretation of the condition (4–6). Hence we may consider (4–6) to be applicable regardless of intermolecular forces and whether the system is condensed or not. The values of $\Lambda^3 N/V$ in Table 4–1 are for liquids and become particularly pertinent at this point. We see, for example, that at its boiling point liquid neon, but not liquid argon, should show noticeable quantum effects. Hence Eq. (6–19) can be applied to liquid argon at its boiling point.

The momentum integrations in Eq. (6–19) can be carried out immediately, as before, and we obtain

$$Q_{\text{class}} = \frac{Z_N}{N!\Lambda^{3N}}, \tag{6–21}$$

where

$$Z_N = \int_V e^{-U(x_1,\ldots,z_N)/kT}\, dx_1 \ldots dz_N. \tag{6–22}$$

In Eq. 6–22, Z_N is called the classical configuration integral. In the absence of intermolecular forces, as in Eq. (6–18), $U = 0$ and $Z_N = V^N$. Equations (6–21) and (6–22) are fundamental equations in the study of monatomic, classical, imperfect gases and liquids (Part III).

Finally, consider a multicomponent system of molecules which may be polyatomic. Suppose that the Hamiltonian for the whole system is separable, as in Eq. (6–15). Then

$$H = H_{\text{class}} + H_{\text{quant}}, \tag{6–23}$$

$$Q = Q_{\text{class}}\, Q_{\text{quant}} \tag{6–24}$$

$$= \frac{Q_{\text{quant}}}{\prod_i (N_i! h^{n_i N_i})} \int e^{-H_{\text{class}}/kT}\, dq_{\text{class}}\, dp_{\text{class}}, \tag{6–25}$$

where n_i is the number of classical degrees of freedom for a molecule of component i. Usually (see Chapters 8 and 9) H_{class} includes translation and rotation, while H_{quant} includes internal vibration (and possibly an electronic contribution). Because of the relatively strong forces involved in the internal vibrations of polyatomic molecules, these vibrations are not perturbed much by intermolecular forces, and hence are approximately separable, as in Eq. (6–25).

6–3 Phase space and ensembles in classical statistics. Although we shall not use this material later in the book, for completeness we make a few very brief comments about the classical phase space of an isolated macroscopic system, of volume V, containing molecules of one or more species. These remarks would serve as an introduction to a discussion of the principles of classical statistical mechanics if we were going to give an independent development of this subject. However, as already mentioned, in this book we regard classical statistics as a limiting form of quantum statistics.

Let the total number of coordinates q_i required to locate the positions of all molecules be n. For example, if the system contains N monatomic molecules, $n = 3N$. Thus n is an extremely large number. For each coordinate q_i, we define a conjugate momentum p_i by Eq. (VII–6). The phase space of the system is therefore $2n$ dimensional: n coordinates and n momenta. The state of this classical system at any time t is completely specified if the position and velocity components of each molecule are specified, i.e., if all the q's and p's are assigned definite values. All this information is condensed into the location of a single point in the $2n$ dimensional phase space. Such a point is called a *phase point* or "representative point." That is, the point represents the complete classical state of the system.

With the forces of the system given, assignment of the position of a phase point in phase space at time t completely determines the future (and past) trajectory, or path, of the point as it moves through phase space in accordance with the laws of mechanics. The equations of motion of the phase point are, in fact, Hamilton's equations, (VII–7). In principle, this system of $2n$ first-order differential equations can be integrated to give $q_1(t), \ldots, q_n(t), p_1(t), \ldots, p_n(t)$. The $2n$ constants of integration would be fixed by knowing the location of the phase point at some time $t = t_0$. Of course, in practice, such an integration is quite hopeless.

Now suppose we make up a total of $\mathfrak{N}$ isolated systems, all with the same V and number of molecules of each species, and all replicas—as far as thermodynamic properties are concerned—of the single experimental system of interest. This is an ensemble, just as in Chapter 1. The detailed classical state of each system can be represented by a point in the same phase space. The points move independently, each along its own trajectory, since each system is isolated. The whole ensemble then appears in phase space as a "cloud" of moving representative points with an essentially continuous density ($\mathfrak{N} \to \infty$). A very important and interesting theorem, due to Liouville, which we shall not prove or use, is that in the immediate neighborhood of any particular phase point, as it moves along its trajectory, the density of phase points remains constant.

Just as in quantum mechanics, we find it necessary to replace the desired single-system time average of mechanical variables, such as pressure, by instantaneous ensemble averages (*first postulate*). But in order to calculate ensemble averages, we have to know the density of the cloud of representative points in the various portions of phase space. Corresponding to the quantum postulate (Section 1–2) of equal probability for each quantum state of an isolated system ($E, V, N_1, N_2, \ldots$ fixed), we have the analogous classical postulate (*second postulate*) of constant density of phase points throughout the region of phase space between the surfaces $E =$ constant and $E + \delta E =$ constant, where δE is arbitrarily small. As in Section 1–3, we can deduce from this a density of phase points proportional to $e^{-\beta H(q,p)}$ [compare Eq. (6–19) for example] for an ensemble representative of a closed system in contact with a heat bath, etc.

The two postulates above, combined, lead to the classical "ergodic hypothesis": the representative point of a single isolated system spends equal amounts of time, over a long period of time, in equal volumes of phase space between the surfaces $E =$ constant and $E + \delta E =$ constant, where δE is arbitrarily small.

6–4 Maxwell-Boltzmann velocity distribution. As an example of an application of Section 6–2, let us deduce here the important formulas for the classical translational velocity and kinetic-energy distributions. We can keep the discussion quite general: we are by no means restricted to ideal monatomic gases only. Specifically, let us consider any one-component system for which the internal vibrations are separable, and translation and rotation can be treated classically (e.g., liquid nitrogen).

This is a special case of Eq. (6-25):

$$Q = \frac{Q_{\text{vib}}}{N!h^{3N}h^{n_rN}} \int e^{-H/kT}\, dq_x\, dp_x\, dq_\theta\, dp_\theta, \tag{6-26}$$

where q_x refers to $3N$ translational coordinates, q_θ to n_rN rotational coordinates (n_r per molecule), and

$$H = K_x(p_x) + K_\theta(q_\theta, p_\theta) + U(q_x, q_\theta). \tag{6-27}$$

The first term on the right of Eq. (6-27) is the translational kinetic energy (Eq. 6-16); the second term is the rotational kinetic energy (in general a function of both rotational coordinates and momenta, see Chapters 8 and 9); and the third term is the intermolecular potential energy. Intermolecular forces in general depend on rotational orientations of molecules as well as on distances between centers of mass, so U is a function of q_θ and q_x.

By analogy with Eq. (1-28) and as a consequence of the fact that classical statistical mechanics is a limiting form of quantum-statistical mechanics, we can state that

$$\frac{e^{-H/kT}\, dq_x\, dp_x\, dq_\theta\, dp_\theta}{\int e^{-H/kT}\, dq_x\, dp_x\, dq_\theta\, dp_\theta} \tag{6-28}$$

is the probability that the system will be observed in the classical translation-rotation state $dq_x\, dp_x\, dq_\theta\, dp_\theta$, irrespective of the (independent) quantum-vibrational state. On integrating (6-28) over q_x, q_θ, and p_θ, we then have that

$$\frac{e^{-K_x/kT}\, dp_x \int e^{-(K_\theta+U)/kT}\, dq_x\, dq_\theta\, dp_\theta}{\int e^{-K_x/kT}\, dp_x \int e^{-(K_\theta+U)/kT}\, dq_x\, dq_\theta\, dp_\theta} = \frac{e^{-K_x/kT}\, dp_x}{\int e^{-K_x/kT}\, dp_x} \tag{6-29}$$

is the probability that the system is in the translational momentum state dp_x, irrespective of other conditions (q_x, q_θ, etc.). Finally, since K_x has independent and equivalent contributions (Eq. 6-16) from each of the N molecules, we can integrate (6-29) with respect to all translational momentum components except those belonging to any one molecule (say molecule 1, but we drop the subscript 1 for convenience), and find

$$\frac{e^{-(p_x^2+p_y^2+p_z^2)/2mkT}\, dp_x\, dp_y\, dp_z}{(2\pi mkT)^{3/2}} \tag{6-30}$$

for the probability that any one molecule is in the translational momentum state $dp_x\, dp_y\, dp_z$, or for the fraction of all molecules in this state, irrespective of other conditions. The integral over p_x, p_y, and p_z ($-\infty$ to $+\infty$)

has been carried out in the denominator of (6–30) with the result indicated. It should be emphasized that (6–30) applies to any system encompassed by Eq. (6–26). Actually, the restriction to a one-component system is not necessary. Each component in a multicomponent system has its own molecular mass and its own independent momentum distribution, as in (6–30).

If we change from cartesian momentum components p_x, p_y, p_z to the corresponding spherical coordinates, p, θ, φ defined by

$$p^2 = p_x^2 + p_y^2 + p_z^2, \qquad p_x = p \sin\theta \cos\varphi,$$

$$p_z = p\cos\theta, \qquad p_y = p\sin\theta\sin\varphi,$$

then

$$\frac{e^{-p^2/2mkT}\, p^2 \sin\theta \, dp \, d\theta \, d\varphi}{(2\pi mkT)^{3/2}} \tag{6–31}$$

is the fraction of molecules in the momentum state $dp\, d\theta\, d\varphi$. On integrating (6–31) over θ and φ, we have

$$\frac{4\pi e^{-p^2/2mkT} p^2\, dp}{(2\pi mkT)^{3/2}} \tag{6–32}$$

for the fraction of molecules with momentum between p and $p + dp$. Finally, if we put $p = mv$ and multiply by N, (6–32) becomes

$$4\pi N \left(\frac{m}{2\pi kT}\right)^{3/2} e^{-mv^2/2kT} v^2 \, dv \tag{6–33}$$

for the *number* of molecules with velocities between v and $v + dv$. This is the well-known Maxwell-Boltzmann velocity distribution. If we write $\epsilon = mv^2/2$ for the translational kinetic energy of a molecule, (6–33) transforms into

$$\frac{2\pi N}{(\pi kT)^{3/2}} e^{-\epsilon/kT} \epsilon^{1/2} \, d\epsilon \tag{6–34}$$

for the number of molecules with values of ϵ between ϵ and $\epsilon + d\epsilon$. This agrees (Problem 6–3) with Eq. (4–11), deduced from quantum statistics for a special case (ideal monatomic gas).

If Eq. (6–30) is integrated over, say, p_y and p_z, we obtain

$$\frac{e^{-p_x^2/2mkT}\, dp_x}{(2\pi mkT)^{1/2}} = \left(\frac{m}{2\pi kT}\right)^{1/2} e^{-mv_x^2/2kT} \, dv_x \tag{6–35}$$

for the fraction of molecules with x-component of momentum or velocity in dp_x or dv_x.

Applications of the above results are included in the problems.

Problems

6–1. Show that in a system of independent harmonic oscillators, all with frequency ν, the quantum number of the state with energy equal to the average energy per oscillator is $n = kT/h\nu$, in the classical limit ($kT/h\nu \gg 1$). (Page 116.)

6–2. Show that at high temperatures, q for a simple harmonic oscillator (Eq. 5–6) can be expanded in the form

$$q = x^{-1}\left(1 - \frac{x^2}{24} + \cdots\right), \qquad x = \frac{h\nu}{kT}.$$

For what value of $kT/h\nu$ is the approximation $q = kT/h\nu$ accurate to 0.01%? Show that $\Delta\epsilon/kT = h\nu/kT$ for a harmonic oscillator, where $\Delta\epsilon$ is the energy difference between successive energy levels. (Page 116.)

6–3. Verify that the Maxwell-Boltzmann energy distribution from Eq. (4–11) (quantum statistics) is the same as (6–33) (classical statistics). (Page 122.)

6–4. Give an argument analogous to Eqs. (6–5) through (6–7), showing that the area of phase space per quantum state is h for a particle in a one-dimensional box.

6–5. Prove that the surfaces in phase space $H(q, p) = E$ (constant) and $H(q, p) = E'$, where $E \neq E'$, can never cross.

6–6. Show that in the classical Hamiltonian, any independent term, potential or kinetic, of the form $a\xi^2$ ($\xi = q$ or p; $a =$ constant) will lead to a contribution $kT/2$ (per term) to the energy E and $k/2$ to C_V. Examples are Eqs. (5–25), (6–2), and (6–20).

6–7. Show that according to (6–33): (a) the most probable velocity is $v^* = (2kT/m)^{1/2}$; (b) the mean velocity is $\bar{v} = (8kT/\pi m)^{1/2}$; (c) the root mean-square velocity is $(\overline{v^2})^{1/2} = (3kT/m)^{1/2}$; and (d) the mean translational kinetic energy per molecule is $3kT/2$.

6–8. Calculate $\bar{v}$ in cm/sec for H_2 and O_2 at 0°C.

6–9. Show that according to (6–35): (a) the most probable velocity component v_x is zero; (b) $\bar{v}_x = 0$; (c) $(\overline{v_x^2})^{1/2} = (kT/m)^{1/2}$; and (d) verify that $\overline{v^2} = \overline{v_x^2} + \overline{v_y^2} + \overline{v_z^2} = 3\overline{v_x^2}$.

Supplementary Reading

Fowler and Guggenheim, Chapter 3.
Mayer and Mayer, Chapters 1 and 2.
Rushbrooke, Chapter 4.
S. M., Chapter 1.
ter Haar, Chapters 1–3.
ter Haar, D., *Rev. Mod. Phys.* **27**, 289 (1955).
Tolman, Chapters 2–4.

CHAPTER 7

INTRODUCTION TO LATTICE STATISTICS: ADSORPTION, BINDING, AND TITRATION PROBLEMS

In the first two sections of this chapter we consider problems in which molecules of one species can be "bound" on one-, two-, or three-dimensional arrays of sites presented by regular arrangements of molecules or molecular subunits of a different species. A bound molecule executes three-dimensional vibrational motion in the neighborhood of its site. The most familiar example is the binding or adsorption of gas molecules on the two-dimensional lattice of sites presented by the surface of a crystal. But one-dimensional cases (e.g., binding of ions from solution on a linear polyelectrolyte molecule) and three-dimensional cases (e.g., absorption of hydrogen atoms by palladium) are also common. Irrespective of the dimensionality, the system of molecules attached to a set of sites is sometimes referred to as a lattice gas.

We restrict the discussion in this chapter to models in which the binding on any one site (or, possibly, small group of sites) is *independent* of the binding on the remaining sites. Chapter 14 is devoted to the more complicated problems in lattice statistics in which this independence is lacking because of interactions between molecules bound on neighboring sites.

Section 7–3 is concerned with the similar problem of binding of molecules on (indistinguishable) groups of sites which are free to move (e.g., binding of small molecules or ions on protein molecules in solution, or the hydrogen ion equilibrium of acid molecules in solution, etc.). Interactions between different groups of sites (e.g., between two protein molecules) are not included here but will be treated in Chapter 19.

In Section 7–4 we discuss the elasticity of a linear polymer made up of units each of which can be in a "short state" or a "long state." This system, we shall find, is formally equivalent to the lattice gas of Section 7–1. Adsorption of another molecular species on such a polymer is also studied in Section 7–4.

7–1 Ideal lattice gas (Langmuir adsorption theory). An ideal lattice gas is a system of N molecules bound not more than one per site to a set of M equivalent, distinguishable, and independent sites, and without interactions between bound molecules. The arrangement of the M sites in space is immaterial. In the Langmuir adsorption model, the sites are arranged in a regular two-dimensional array on the surface of a crystal, and the bound molecules come from a gas phase which is in equilibrium

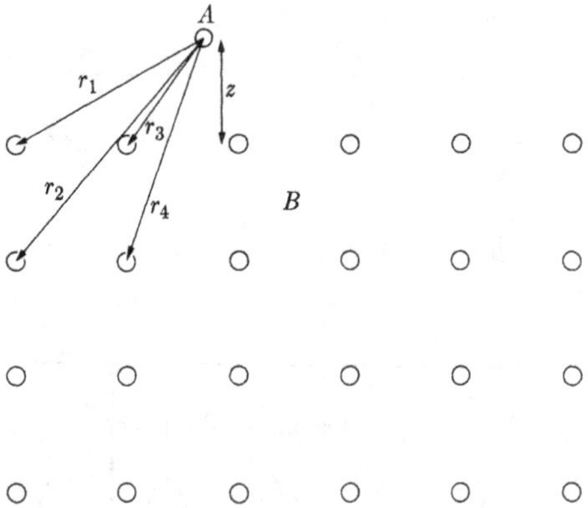

FIG. 7-1. Adsorbed molecule A near surface of crystal B.

with the lattice gas (adsorbed phase). For concreteness and because of its importance, we shall discuss this particular (adsorption) problem explicitly here, but the treatment is not restricted to it, as will be evident from Section 7-4 and Chapter 14.

As an example, suppose the monatomic gas A is adsorbed on the surface 100-plane of a simple cubic lattice of solid B. We assume for simplicity that the forces holding the solid together are much stronger than the adsorption forces, so that the solid is essentially unperturbed by the presence of gas molecules on its surface. Thus the solid merely plays the role of providing a potential field for the adsorbed molecules. The thermodynamic system we consider consists, then, of gas molecules "bound" in this potential field.

Suppose that the potential energy of interaction $u(r)$ between a molecule of A and a molecule of B is of the Lennard-Jones type (Appendix IV). The adsorbing force holding an A molecule to the surface of B is then the sum of a number of such interactions. In Fig. 7-1 is shown an adsorbed molecule A at a distance z from the surface of B (taken as the xy-plane), and a few of the distances r_i which should be substituted in $u(r)$ to give the total potential energy of interaction of A with all molecules of the solid, for this particular location of A:

$$U = \sum_i u(r_i). \tag{7-1}$$

In Eq. (7-1), U is a function of x, y, and z, the coordinates of A. If we hold x and y fixed and consider the dependence of U on z (i.e., along a line perpendicular to the surface), $U(z)$ will have a qualitative appearance similar to

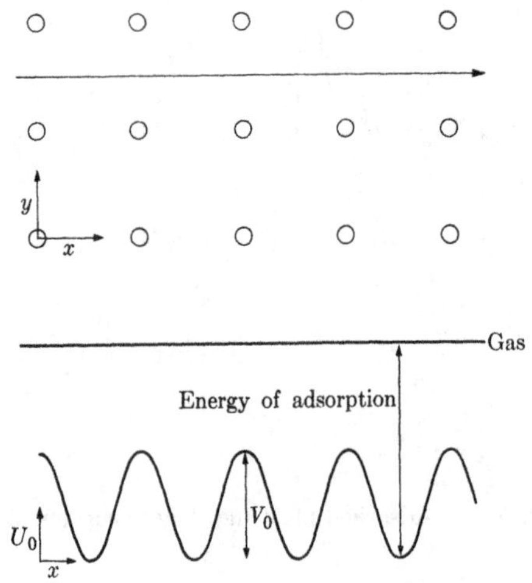

FIG. 7–2. Periodic variation over surface of depth U_0 of the potential well $U(z)$ for an adsorbed molecule.

$u(r)$ in Fig. IV–1 and the approximate mathematical form (Problem 7–1)

$$U(z) = \frac{\text{const}}{z^9} - \frac{\text{const}'}{z^3}. \tag{7–2}$$

Thus the free translational motion of a gas molecule in the z-direction is replaced on adsorption by vibration in the potential well associated with Eq. (7–2). In the harmonic-oscillator approximation, the frequency of vibration ν_z is determined by the curvature of $U(z)$ at its minimum. Usually ν_z is of order 10^{12} sec^{-1} (Problem 7–2).

For different choices of the fixed values of x and y, $U(z)$ will be different (in other words, U is a function of x and y also). Thus $U(z)$ directly above a B molecule will be different from $U(z)$ above the center of a square of B molecules. In particular, the depth U_0 of the potential well in $U(z)$ will vary periodically in both x- and y-directions. Hence U_0 is a function of x and y. Figure 7–2 illustrates this variation, where U_0 is plotted against x along one particular line (y constant) in the surface, as indicated in the figure. It is clear from Fig. 7–2 that motion parallel to the surface involves passing over potential barriers V_0. With van der Waals forces, these barriers are of order 0.3 to 1 kcal·mole^{-1}. At temperatures sufficiently low that the thermal energy kT of the adsorbed molecules is small compared with the height of the barrier V_0, the molecules will be trapped ("localized") in the neighborhood of potential minima in $U(x, y, z)$, except for occasional passages over the barrier or evaporation and recondensation (necessary for equilibrium with the gas phase). When kT is large compared with V_0, the periodic variation in U_0 becomes unimportant, and the adsorbent surface, in effect, becomes a continuum.

The Langmuir model corresponds to localized adsorption ($kT \ll V_0$). At the end of this section we shall also discuss briefly the case of a dilute "mobile" ($kT \gg V_0$) adsorbed phase. The transition from localized to mobile adsorption, as T is increased, is an interesting problem which will be dealt with in Chapter 9.

In localized adsorption, the adsorbed molecule has three vibrational degrees of freedom (replacing three translational degrees in the gas). We have already mentioned the vibration in the z-direction. In addition, there is vibration in x- and y-directions (if these are the normal coordinates for the two-dimensional motion) around the minima of $U_0(x, y)$, with frequencies ν_x and ν_y. These frequencies are usually a little less than 10^{12} sec^{-1}.

The partition function for a single adsorbed molecule, in the harmonic-oscillator approximation, is then

$$q(T) = q_x q_y q_z e^{-U_{00}/kT}, \tag{7-3}$$

where q_x, q_y, and q_z are one-dimensional harmonic-oscillator partition functions (Eq. 5-6) with frequencies ν_x, ν_y, and ν_z, respectively, and with zero of energy in each case at the bottom of the potential well $U(x, y, z)$. As explained above, the crystal surface is assumed to present an external and fixed potential field, so that q is a function of T only. Since we want to investigate the equilibrium between adsorbed and gas molecules, we must choose the same zero of energy for the two phases. We take an isolated gas molecule [i.e., $z = \infty$ in Eq. (7-2)] at rest as the zero. With this zero, we have to insert the Boltzmann factor in U_{00} (a negative quantity) in Eq. (7-3), where U_{00} is the potential energy at the minima in $U(x, y, z)$ [or in $U_0(x, y)$]. Then the heat of adsorption per molecule at 0°K is $U_{00} + h(\nu_x + \nu_y + \nu_z)/2$.

If we had a system of N sites and N molecules, we would write $Q = q^N$, just as in the Einstein model for a crystal. But here the number of sites (equivalent but distinguishable) is $M \geq N$. There is therefore a configurational degeneracy not present in an Einstein crystal (it would be present if the crystal had lattice vacancies), which must be taken into account. For each quantum state of an Einstein crystal there are here $M!/N!(M-N)!$ quantum states—this being the number of ways N indistinguishable molecules can be distributed among M labeled sites. Hence

$$Q(N, M, T) = \frac{M!q(T)^N}{N!(M-N)!}, \tag{7-4}$$

and

$$\ln Q = M \ln M - N \ln N - (M - N) \ln (M - N) + N \ln q. \qquad (7\text{–}5)$$

The basic thermodynamic equation for the lattice gas (or adsorbed phase) is

$$dE = T\,dS - \Phi\,dM + \mu\,dN, \qquad (7\text{–}6)$$

where M is proportional to the volume, area, or length, depending on the dimensionality of the lattice. That is, M is the "external variable" in this case, and Φ is essentially a pressure, though it has dimensions of energy here. Thus, in two dimensions, if we write $\mathcal{A} = M\alpha$ where $\mathcal{A}$ = area and α is a constant, then Φ/α is the so-called surface or spreading pressure (dimensions of energy per unit area) φ.

From Eqs. (7–5) and (7–6) we have

$$\frac{\Phi}{kT} = \left(\frac{\partial \ln Q}{\partial M}\right)_{N,T} = -\ln (1 - \theta) = \theta + \tfrac{1}{2}\theta^2 + \tfrac{1}{3}\theta^3 + \cdots, \qquad (7\text{–}7)$$

$$\frac{\mu}{kT} = -\left(\frac{\partial \ln Q}{\partial N}\right)_{M,T} = \ln \frac{\theta}{(1 - \theta)q}, \qquad (7\text{–}8)$$

where $\theta = N/M$, the fraction of sites occupied. Equation (7–7) is the equation of state. Note that in the limit as $\theta \to 0$, $\Phi \to \theta kT$ (which is the same, in three dimensions, as $p \to \rho kT$, where $\rho = N/V$). If the adsorbed phase is in equilibrium with a gas phase at pressure p [assumed ideal as in Eq. (4–25)],

$$\frac{\mu}{kT} = \ln \frac{\theta}{(1 - \theta)q} = \frac{\mu_{\text{gas}}}{kT} = \frac{\mu^0(T)}{kT} + \ln p, \qquad (7\text{–}9)$$

or

$$\theta(p, T) = \frac{\chi(T)p}{1 + \chi(T)p}, \qquad \chi(T) = q(T)e^{\mu^0(T)/kT}. \qquad (7\text{–}10)$$

Equation (7–10) is the Langmuir "adsorption isotherm," giving the amount of gas adsorbed as a function of gas pressure at a fixed temperature. When $p \to 0$, $\theta \to \chi p$, and when $p \to \infty$, $\theta \to 1$. Many adsorption systems follow this equation approximately. Statistical mechanics provides all the necessary molecular details to calculate $\chi(T)$ explicitly (Problem 7–3).

The entropy of the lattice gas (Eq. 1–33) can be put in the form (Problem 7–4)

$$S = S_{\text{config}} + S_{\text{vib}}, \qquad (7\text{–}11)$$

where

$$S_{\text{config}} = k \ln \frac{M!}{N!(M - N)!}, \qquad (7\text{–}12)$$

$$S_{\text{vib}} = Nk\left(\ln q + T\frac{d\ln q}{dT}\right). \tag{7-13}$$

We note that S_{config} is independent of temperature, whereas $S_{\text{vib}} \to 0$ as $T \to 0$ (just as for an Einstein crystal). If equilibrium could be maintained to $T = 0$, we would have $S > 0$ at $T = 0$. This is not an exception to the third law of thermodynamics (Section 2-4), since the combined system *sites* + *lattice gas* is not a pure crystal, but a binary mixture. That is, a lattice gas can exist only in the presence of an additional molecular matrix (e.g., a crystal) that provides the sites.

For future reference, let us also apply the grand partition function to this system. We have

$$\begin{aligned}\Xi = e^{\Phi M/kT} &= \sum_{N=0}^{M} \frac{M!(q\lambda)^N}{N!(M-N)!} \\ &= (1+q\lambda)^M,\end{aligned} \tag{7-14}$$

or

$$\frac{\Phi}{kT} = \ln(1+q\lambda), \tag{7-15}$$

where $\lambda = e^{\mu/kT}$. This agrees with the result obtained by eliminating θ between Eqs. (7-7) and (7-8). Also, from Eq. (1-67),

$$\overline{N} = \lambda\left(\frac{\partial \ln \Xi}{\partial \lambda}\right)_{M,T} = \frac{Mq\lambda}{1+q\lambda}, \tag{7-16}$$

which is essentially just Eq. (7-10).

For contrast we consider a dilute, *mobile* adsorbed phase (two-dimensional ideal gas). In this case the adsorbed molecules still vibrate in the z-direction with partition function q_z, but have free translational motion in the xy-plane. The xy partition function for one molecule is [see Eq. (6-9)]

$$q_{xy} = \frac{2\pi mkT}{h^2}\,\mathcal{A}, \tag{7-17}$$

where $\mathcal{A}$ = area. Then

$$Q = \frac{1}{N!}q^N, \qquad q = q_{xy}q_z e^{-U_{00}/kT}. \tag{7-18}$$

For present purposes, we replace $\Phi\, dM$ in Eq. (7-6) by $\varphi\, d\mathcal{A}$. Then we find

$$\frac{\varphi}{kT} = \left(\frac{\partial \ln Q}{\partial \mathcal{A}}\right)_{N,T} = \frac{N}{\mathcal{A}}, \tag{7-19}$$

$$\frac{\mu}{kT} = -\left(\frac{\partial \ln Q}{\partial N}\right)_{\alpha,T} = \ln \frac{N}{q} = \frac{\mu_{\text{gas}}}{kT} = \frac{\mu^0(T)}{kT} + \ln p. \tag{7-20}$$

The adsorption isotherm is then

$$\frac{N}{\alpha} = \chi'(T)p, \qquad \chi'(T) = \left(\frac{2\pi mkT}{h^2}\right) q_z e^{-U_{00}/kT} e^{\mu^0/kT}. \tag{7-21}$$

It is interesting to compare the predicted relative amounts of localized and mobile adsorption in the limit as $p \to 0$ where Eq. (7–21) is valid and $\theta \to \chi p$ in Eq. (7–10), assuming all conditions the same except for the xy-motion. We find from Eqs. (7–10) and (7–21),

$$\frac{N_{\text{mobile}}}{N_{\text{localized}}} = \frac{(2\pi mkT/h^2)\alpha}{q_x q_y}, \tag{7-22}$$

where $\alpha = \alpha/M$ = area per site. The numerator on the right is the partition function for a molecule moving freely in a two-dimensional box of area α. The denominator is the partition function of a two-dimensional oscillator, which would have a numerical value less than the numerator because, in classical language, there is a Boltzmann factor $\exp[-(f_x x^2 + f_y y^2)/2kT]$ (Eq. 5–1) in the integrand of the configuration integral for an oscillator, whereas the integrand is unity [Eq. (6–22), $N = 1$] for a particle in a two-dimensional box. Therefore, we find $N_{\text{mobile}} > N_{\text{localized}}$.

7–2 Grand partition function for a single independent site or subsystem. In Section 3–4 we discussed a canonical ensemble of subsystems. Problem 3–5 extends the argument to a grand ensemble of subsystems. Because of its importance, we discuss here the same question as in Problem 3–5, but use a different approach: we start with the ordinary grand partition function and show that it reduces to a product of subsystem grand partition functions ξ.

Consider a macroscopic system of M equivalent, independent, and distinguishable sites on each of which any number s, from zero to a maximum m, of molecules can be "bound" (the nature of the binding or association with the site is immaterial here). Let $q(s) = \sum_j e^{-\epsilon_j(s)/kT}$ be the site partition function when s molecules are bound to the site. If there is a total of N molecules bound on the M sites, and if the number of sites having s molecules bound is a_s, then the canonical ensemble partition function for the system of M sites is

$$Q(N, M, T) = \sum_{\mathbf{a}} \frac{M! q(0)^{a_0} q(1)^{a_1} \cdots q(m)^{a_m}}{a_0! a_1! \cdots a_m!}, \tag{7-23}$$

where the sum is over all sets* $\mathbf{a} = a_0, a_1, \ldots, a_m$ satisfying the restrictions

$$\sum_{s=0}^{m} a_s = M, \qquad \sum_{s=0}^{m} s a_s = N. \tag{7-24}$$

This is a straightforward extension of Eq. (7-4) which, in the present notation, would be written

$$Q = \sum_{\mathbf{a}} \frac{M! q(0)^{a_0} q(1)^{a_1}}{a_0! a_1!},$$

where

$$a_0 + a_1 = M, \qquad a_1 = N,$$

$$q(0) = 1, \qquad q(1) = q.$$

That is, in the Langmuir problem, $s = 0$ (site empty) or 1 (site occupied).

The grand partition function is

$$\Xi = \sum_{N=0}^{mM} Q(N, M, T)\lambda^N$$

$$= \sum_{\mathbf{a}} \frac{M! q(0)^{a_0}[q(1)\lambda]^{a_1}[q(2)\lambda^2]^{a_2} \cdots [q(m)\lambda^m]^{a_m}}{a_0! \cdots a_m!}, \tag{7-25}$$

where we have used the second of Eqs. (7-24) in rewriting λ^N in the last step, and the only restriction on the sets $\mathbf{a}$ is now the first of Eqs. (7-24), since we want to sum over all possible values of N for given M. By the multinomial theorem [this is just a generalization of the binomial theorem used in obtaining Eq. (7-14)],

$$\Xi(\lambda, M, T) = \xi(\lambda, T)^M, \tag{7-26}$$

where

$$\xi = q(0) + q(1)\lambda + \cdots + q(m)\lambda^m$$

$$= \sum_{s=0}^{m} q(s)\lambda^s. \tag{7-27}$$

This is a good example of a summation having been made easier by passing

* Here and throughout the rest of the book we use a boldface letter to represent a set of numbers with subscripts 1, 2, ... For example, we use $\mathbf{N}$ for $N_1, N_2, \ldots$ (in analogy with the components of a vector). We did not introduce this notation in Chapter 1 for $N_1, N_2, \ldots$ and $\mu_1, \mu_2, \ldots$ because there we wanted to emphasize the one-component case for simplicity. Hence we used N instead of $\mathbf{N}$, etc. However, $\mathbf{N}$ is generally preferable because it is more explicit.

from Q to Ξ. The sum ξ has exactly the form of a grand partition function, but it pertains to only a single site instead of a macroscopic system. It is rather obvious that if the sites were all different, but still independent of each other, $\Xi = \xi_1\xi_2 \ldots \xi_M$, where ξ_i is a sum, as in Eq. (7–27), for site i only. The sites must be independent of each other for this simplification in Ξ to occur. Equations (3–5) and (3–6) are analogs in the canonical ensemble.

If, when all sites are equivalent, we consider each site to be an open subsystem, then the macroscopic system of M sites may be regarded as a grand ensemble of subsystems (see Problem 3–5). The chemical potential and temperature of a single open subsystem are well-defined thermodynamic properties, for they are determined by the chemical potential and temperature of the macroscopic reservoir with which the subsystem is in contact. Incidentally, in adsorption problems, this reservoir is real and not imaginary: it is the gas or solution, in equilibrium with the adsorbed phase, providing the molecules or ions being adsorbed.

The average number of molecules in the macroscopic system is

$$\overline{N} = \lambda\left(\frac{\partial \ln \Xi}{\partial \lambda}\right)_{M,T} = M\lambda\left(\frac{\partial \ln \xi}{\partial \lambda}\right)_T, \tag{7–28}$$

or

$$\bar{s} = \frac{\overline{N}}{M} = \lambda\left(\frac{\partial \ln \xi}{\partial \lambda}\right)_T = \frac{\sum_{s=0}^{m} sq(s)\lambda^s}{\sum_{s=0}^{m} q(s)\lambda^s}, \tag{7–29}$$

where $\bar{s}$ is the average number of molecules per site. Thus the equation relating $\bar{s}$ and ξ (subsystem) is formally the same as that relating $\overline{N}$ and Ξ (macroscopic system). In the differentiation in Eq. (7–29), m is held constant, which is analogous to holding M constant in Eq. (7–28).

The "pressure" Φ is related to the site or subsystem grand partition function ξ by

$$e^{\Phi M/kT} = \Xi = \xi^M,$$

$$\Phi = kT \ln \xi(\lambda, T). \tag{7–30}$$

Thus, whenever we are concerned with a system of independent, distinguishable, and open sites (subsystems), we can go directly to Eq. (7–29) to determine the mean population $\bar{s}$ and also the population distribution [since $q(s)\lambda^s/\xi$ is clearly the probability of a population s] of each site. This is a simpler procedure than using the full grand partition function, but of course leads to the same results. The mean population $\bar{s}$ as a function of λ is essentially the adsorption isotherm in an adsorption system.

The situation with respect to fluctuations in s in a subsystem is completely analogous to that of fluctuations in the energy of independent molecules in Eq. (3-24). We find from Eqs. (7-29) and (2-7),

$$\overline{s^2} - (\bar{s})^2 = kT\left(\frac{\partial \bar{s}}{\partial \mu}\right)_T = \frac{kT}{M}\left(\frac{\partial \overline{N}}{\partial \mu}\right)_{M,T}$$

$$= \frac{\overline{N^2} - (\overline{N})^2}{M},$$

or

$$\frac{\sigma_s}{\bar{s}} = M^{1/2}\frac{\sigma_N}{\overline{N}},$$

just as in Problem 3-8. Since $\sigma_N/\overline{N}$ is of order $M^{-1/2}$, σ_s is of the same order of magnitude as $\bar{s}$. That is, the probability distribution in s is broad, as should be expected, and not sharp.

We now consider some applications of Eq. (7-29). First, an ideal lattice gas. Here, as was pointed out following Eq. (7-24), $s = 0$ or 1, $q(0) = 1$, and $q(1) = q$. Therefore,

$$\xi = 1 + q\lambda,$$

$$\bar{s} = \theta = \lambda\left(\frac{\partial \ln \xi}{\partial \lambda}\right)_T = \frac{q\lambda}{1 + q\lambda}, \tag{7-31}$$

which is the same as Eqs. (7-10) and (7-16).

Next, suppose we have a system of M independent pairs of sites:

$$\begin{matrix} & \bullet 1 & \bullet 1 & \bullet 1 & \\ \cdots & & & & \cdots \\ & \times 2 & \times 2 & \times 2 & \end{matrix}$$

Such a system of sites might occur on a linear polymer, for example. The two sites in a pair are different: q_1 is the partition function [as in Eq. (7-3)] for a molecule bound on a site of type 1, and q_2 for a site of type 2. Also, when both sites of a pair are occupied, let there be a potential energy of interaction w between the two bound molecules. So long as there is no interaction between molecules on different pairs, the introduction of w does not affect the independence of the pairs. Thus, a *pair* of sites is the independent subsystem here, and $s = 0, 1$, or 2. If we consider all possible states of a pair of sites with s molecules bound, we see that

$$q(0) = 1, \qquad q(1) = q_1 + q_2, \qquad q(2) = q_1 q_2 e^{-w/kT}.$$

Therefore

$$\xi = 1 + (q_1 + q_2)\lambda + q_1 q_2 e^{-w/kT}\lambda^2, \tag{7-32}$$

and

$$\bar{s} = \frac{\overline{N}}{M} = \frac{(q_1 + q_2)\lambda + 2q_1 q_2 e^{-w/kT}\lambda^2}{1 + (q_1 + q_2)\lambda + q_1 q_2 e^{-w/kT}\lambda^2}. \tag{7-33}$$

This is the adsorption isotherm. Except for a factor of two, this reduces to Eq. (7–31) for an ideal lattice gas if $w = 0$ and $q_1 = q_2 = q$. The factor of two has its origin in the fact that the total number of sites is $2M$ in Eq. (7–33) and M in Eq. (7–31).

As a third example, we derive the well-known B.E.T. (Brunauer-Emmett-Teller) equation for multimolecular adsorption. The model, which is physically unrealistic, is the following. A surface has M independent, distinguishable, and equivalent sites, on each of which an indefinite number of molecules can be adsorbed in a vertical pile, so that $s = 0, 1, 2, \ldots$. Let q_1 be the partition function for the bottom molecule in a pile ("first layer"), q_2 for the next molecule (second layer), etc. The positions in a pile are distinguishable in the same sense as in a one-dimensional Einstein crystal. Then

$$\xi = 1 + q_1\lambda + q_1q_2\lambda^2 + q_1q_2q_3\lambda^3 + \cdots, \tag{7–34}$$

and

$$\bar{s} = \frac{\overline{N}}{M} = \frac{q_1\lambda + 2q_1q_2\lambda^2 + 3q_1q_2q_3\lambda^3 + \cdots}{\xi}. \tag{7–35}$$

This is the adsorption isotherm. Now consider the special case $q_2 = q_3 = q_4 = \ldots$. That is, molecules in the first layer (next to the surface) have a partition function q_1, and all others q_2. (Second and higher layers are supposed to be "liquid-like," though a one-dimensional pile is a very poor model for a liquid.) With this simplification,

$$\begin{aligned}\bar{s} &= \frac{q_1\lambda(1 + 2q_2\lambda + 3q_2^2\lambda^2 + \cdots)}{1 + q_1\lambda(1 + q_2\lambda + q_2^2\lambda^2 + \cdots)} \\ &= \frac{q_1\lambda}{(1 - q_2\lambda + q_1\lambda)(1 - q_2\lambda)} \\ &= \frac{cx}{(1 - x + cx)(1 - x)},\end{aligned} \tag{7–36}$$

where

$$\begin{aligned}c &= q_1/q_2, \\ x &= q_2\lambda = q_2\lambda_{\text{gas}}, \\ &= q_2e^{\mu^0/kT}p.\end{aligned} \tag{7–37}$$

Equation (7–36) is the B.E.T. adsorption isotherm.

A typical isotherm ($c = 157$) is shown in Fig. 7–3. Many experimental isotherms with this type of qualitative behavior are known. When $c \gg 1$, adsorption in the first layer is strongly favored relative to higher layers, so the first layer is almost completely filled before higher layers begin. This accounts for the "knee" in the isotherm (Fig. 7–3) near $\bar{s} = 1$. Up to

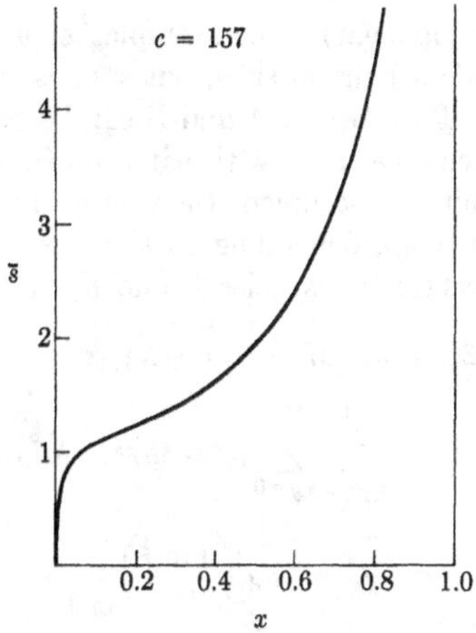

FIG. 7-3. B.E.T. adsorption isotherm for $c = 157$.

almost $\bar{s} = 1$, $\bar{s} \cong cx/(1 + cx)$ (Langmuir equation) when $c \gg 1$. On thermodynamic grounds we know that $\bar{s} \to \infty$ when $p \to p_0$ (the vapor pressure of liquid adsorbate at the temperature of the experiment; we assume $T < T_c$, the critical temperature). That is, when p approaches p_0, bulk liquid will begin to condense on the adsorbing surface. On the other hand, $\bar{s} \to \infty$ when $x \to 1$, according to Eq. (7-36). Therefore, from Eq. (7-37), we have

$$q_2 e^{\mu^0/kT} = \frac{1}{p_0}, \tag{7-38}$$

which is equivalent to the vapor pressure equation for an Einstein crystal (Problem 5-11).

In the B.E.T. theory, the surface pressure Φ/α is determined by

$$e^{\Phi M/kT} = \Xi = \xi^M,$$

or

$$\frac{\Phi}{kT} = \ln \xi = \ln \left(\frac{1 - q_2\lambda + q_1\lambda}{1 - q_2\lambda} \right)$$

$$= \ln \left(\frac{1 - x + cx}{1 - x} \right). \tag{7-39}$$

We note that $\Phi \to \infty$ as $x \to 1$. But, thermodynamically, Φ/α is necessarily finite at $x = 1$. This is a serious fault of the B.E.T. theory.

The general method of this section can easily be extended to cases in which more than one species of molecule can be bound to a site (i.e., the subsystem is multicomponent). For example, let us return to the system above of M independent pairs of sites, but suppose now the two sites in a pair are equivalent. Two species A and B can be bound on the sites (not more than one molecule per site) with partition functions q_A and q_B, and if both sites in a pair are occupied, there is an intermolecular potential energy w_{AA}, w_{AB}, or w_{BB}, depending on the species making up the pair. The general equations for two species A and B are (Problem 7–5)

$$\Xi(\lambda_A, \lambda_B, M, T) = \xi(\lambda_A, \lambda_B, T)^M, \tag{7–40}$$

$$\xi = \sum_{s_A=0}^{m_A} \sum_{s_B=0}^{m_B} q(s_A, s_B)\lambda_A^{s_A}\lambda_B^{s_B}, \tag{7–41}$$

$$\bar{s}_A = \frac{\overline{N}_A}{M} = \lambda_A \left(\frac{\partial \ln \xi}{\partial \lambda_A}\right)_{\lambda_B, T}, \tag{7–42}$$

and a similar equation for $\bar{s}_B$. Equation (7–41) is a two-component grand partition function for a subsystem. In the special case under consideration,

$$\begin{aligned} \xi = q(0, 0) + q(1, 0)\lambda_A + q(0, 1)\lambda_B + q(2, 0)\lambda_A^2 \\ + q(1, 1)\lambda_A\lambda_B + q(0, 2)\lambda_B^2, \end{aligned} \tag{7–43}$$

where

$$\begin{aligned} q(0, 0) = 1, \qquad q(1, 0) &= 2q_A, \qquad q(0, 1) = 2q_B, \\ q(2, 0) &= q_A^2 e^{-w_{AA}/kT}, \\ q(1, 1) &= 2q_A q_B e^{-w_{AB}/kT}, \\ q(0, 2) &= q_B^2 e^{-w_{BB}/kT}. \end{aligned} \tag{7–44}$$

Then

$$\bar{s}_A = \frac{\overline{N}_A}{M} = \frac{q(1, 0)\lambda_A + 2q(2, 0)\lambda_A^2 + q(1, 1)\lambda_A\lambda_B}{\xi}, \tag{7–45}$$

with a similar equation for $\bar{s}_B$. This is equivalent to Problem 7–6 if $w_{AA} = w_{AB} = w_{BB} = 0$.

7–3 Systems composed of independent and indistinguishable subsystems. Here we extend the treatment of the previous section to systems in which binding or adsorption takes place on sites which are attached to freely moving adsorbent molecules. Examples are: adsorption of ions from solution onto protein or other macromolecules; titration curves (i.e., binding of H^+) of polybasic acids or polyacidic bases; adsorption of a gas on small dust particles; etc. The subsystem here consists of one adsorbent molecule (protein, etc.) and s bound (H^+, etc.) molecules, where $0 \leq$

$s \leq m$ as before. We still assume, for simplicity, that the adsorbent molecules are unperturbed in their other properties by the presence of adsorbed molecules. We shall use a language appropriate to a gaseous system, since we have not yet discussed solutions. But we shall see in Chapter 19 that, under suitable conditions, if the adsorbent molecules are in a solvent rather than a vacuum, we merely have to replace the adsorbent partial pressure p by the osmotic pressure Π. The gas is assumed ideal—that is, sufficiently dilute so that intermolecular interactions are unimportant. However, interactions between adsorbed molecules on the same adsorbent molecule are included. Interactions between adsorbent molecules will be discussed in Chapter 19.

The system we consider is shown schematically in Fig. 7-4. There are M adsorbent molecules in a volume V, to each of which is attached from zero to m adsorbed molecules, with a total of N adsorbed molecules. Free adsorbate molecules are not included in the system for convenience and also because this omission leads to equations which are completely analogous to the osmotic equations of Chapter 19. Since the gas is assumed ideal (each component behaves independently), no error is introduced by this procedure, and the pressure p of the system as we have defined it is actually the partial pressure of the adsorbent molecules in the adsorbent-adsorbate gas mixture.

This is a two-component system. The basic thermodynamic equations are

$$\begin{aligned} dE &= T\,dS - p\,dV + \mu\,dN + \mu'\,dM, \\ A &= E - TS = -pV + \mu N + \mu' M, \end{aligned} \tag{7-46}$$

where μ' is the adsorbent chemical potential.

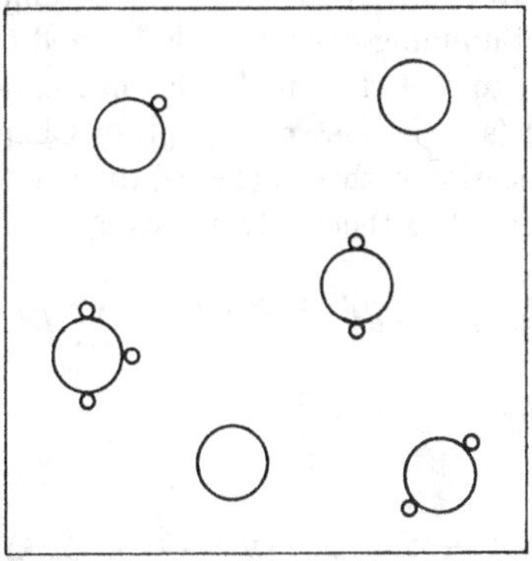

FIG. 7-4. Adsorbent molecules (large) with adsorbed molecules (small) attached.

We let $q(s)$ be the partition function of an adsorbent molecule with s adsorbed molecules attached. This will include, as factors, the translational [hence each $q(s)$ is proportional to V], rotational, etc., partition functions of an adsorbent ($s = 0$) molecule. Also, for the adsorbed molecules, $q(s)$ will include a factor q^s [as in Eq. (7-3)], possibly a configurational factor or factors (for distributing s molecules on m sites), and Boltzmann factors in the potential energy of interaction between the s adsorbed molecules, if such interactions are present. The quotient $q(s)/q(0)$, which we denote by $q_0(s)$, refers to the adsorbed molecules only and corresponds to $q(s)$ in the preceding section [in the preceding section, $q(0) = 1$].

We now treat this system as an ideal gas mixture of $m + 1$ different molecular species, corresponding to the different numbers of adsorbed molecules, $s = 0, 1, \ldots, m$. If there are a_s molecules of species s in the system, then the canonical ensemble partition function is [according to Eq. (3-15)]

$$\frac{q(0)^{a_0}}{a_0!} \cdot \frac{q(1)^{a_1}}{a_1!} \cdots \frac{q(m)^{a_m}}{a_m!}.$$

However, as a partition function we want a sum over all possible states of the system, not for fixed $a_0, \ldots, a_m$, but rather for all possible $a_0, \ldots, a_m$ consistent with fixed N and M. Therefore

$$Q(N, M, V, T) = \sum_{\mathbf{a}} \frac{q(0)^{a_0} \ldots q(m)^{a_m}}{a_0! \ldots a_m!}, \tag{7-47}$$

subject to the restrictions (7-24). Equation (7-47) also follows from Eq. (7-23) on dividing by $M!$, as should be expected.

We wish next to multiply Q by $\lambda = e^{\mu/kT}$ and sum, as in Eq. (7-25). But first we have to digress briefly to consider what kind of a partition function this will give us. Summing over N with M fixed corresponds to a system open with respect to adsorbed molecules but closed with respect to adsorbent molecules (see the last paragraph of Chapter 1). From the rule mentioned in connection with Eq. (1-92), the partition function, which we denote by Γ, is related to thermodynamics by

$$\Gamma(M, V, \mu, T) = e^{(pV - \mu' M)/kT} = \sum_{N=0}^{mM} Q(N, M, V, T)\lambda^N \tag{7-48}$$

$$= \frac{q(0)^M \xi^M}{M!}, \tag{7-49}$$

where

$$\xi = 1 + q_0(1)\lambda + \cdots + q_0(m)\lambda^m. \tag{7-50}$$

In Eq. (7-49), $q(0)$ is a function of the thermodynamic variables V and T, and ξ is a function of μ and T.

The average number of adsorbed molecules is

$$\overline{N} = \frac{\sum_N NQ\lambda^N}{\Gamma} = \lambda\left(\frac{\partial \ln \Gamma}{\partial \lambda}\right)_{M,V,T} = M\lambda\left(\frac{\partial \ln \xi}{\partial \lambda}\right)_T. \qquad (7\text{–}51)$$

Thus Eq. (7–29), either as it stands or with $q(s)$ replaced by $q_0(s)$, is also valid here: the number of adsorbed molecules per subsystem depends on the nature of the group of m sites, but not on whether they are moving around or not—just what one would expect intuitively.

It is instructive to consider the grand partition function as well (system open with respect to both components):

$$e^{pV/kT} = \Xi(V, \mu, \mu', T) = \sum_{N,M} Q(N, M, V, T)\lambda^N \lambda'^M$$

$$= \sum_{M=0}^{\infty} \frac{q(0)^M \xi^M \lambda'^M}{M!} = e^{q(0)\xi\lambda'}. \qquad (7\text{–}52)$$

Now

$$\overline{M} = \lambda'\left(\frac{\partial \ln \Xi}{\partial \lambda'}\right)_{V,\mu,T} = q(0)\xi\lambda' = \frac{pV}{kT}. \qquad (7\text{–}53)$$

This is the expected equation of state. Also,

$$\overline{N} = \lambda\left(\frac{\partial \ln \Xi}{\partial \lambda}\right)_{V,\mu',T} = q(0)\lambda'\lambda\left(\frac{\partial \xi}{\partial \lambda}\right)_T$$

$$= \frac{\overline{M}}{\xi}\lambda\left(\frac{\partial \xi}{\partial \lambda}\right)_T, \qquad (7\text{–}54)$$

which is equivalent to Eq. (7–51).

The simplest application of Eq. (7–51) is to a system of adsorbent molecules each of which has only one site for binding. The hydrogen ion equilibrium of a dilute monobasic acid would be an example. Then Eqs. (7–51) and (7–31) are applicable, and we get just the Langmuir adsorption isotherm. This is also the familiar equation of a simple titration curve (Problem 7–7) if we plot θ as a function of $\ln \lambda$ rather than λ (λ is proportional to the concentration of adsorbate in a dilute solution).

If each adsorbent molecule has two sites and there is an interaction energy w between two adsorbed molecules on the same adsorbent, we obtain Eq. (7–33) again. An example would be the binding of hydrogen ions by, say, a diamine with two different sites (—R—NH_2 and —R′—NH_2). In this case w is positive (repulsion between two hydrogen ions). The relation between the statistical mechanical equation (7–33) and the equivalent thermodynamic expression involving successive dissociation constants is explored in Problem 7–8.

If each adsorbent molecule has m sites and there are no interactions, then we again get the Langmuir equation (Problem 7–9).

7–4 Elasticity of and adsorption on a linear polymer chain. The first problem we examine here is superficially unrelated to the preceding sections of the chapter. Actually, though, there is an extremely close analogy, as will be seen below. We consider a single linear polymer chain composed of units each of which can be in a short state α of length l_α or a long state β of length l_β ($l_\beta > l_\alpha$). The units are interconvertible (as in the chemical reactions of Chapter 10). Thus if a pulling force τ is applied to the chain, some α units will be converted into longer β units and the chain will lengthen. This model is crudely representative of some real systems: the $\alpha - \beta$ transition in fibrous proteins; the helix-random coil transition in solutions of proteins and nucleic acids ($\tau = 0$); and possibly the elasticity of muscle and some textiles.

Neighboring units in the chain are assumed here to be independent of each other. Interactions between units will be considered in Chapter 14.

Let M be the total number of units, with M_α and M_β of the two types. We choose M_α to be independent; then $M_\beta = M - M_\alpha$. The fundamental thermodynamic equation is

$$dE = T\,dS + \tau\,dl + \mu'\,dM, \tag{7–55}$$

where l is the length,

$$l = l_\alpha M_\alpha + l_\beta(M - M_\alpha), \tag{7–56}$$

and μ' refers to a single unit (either α or β). If we replace l and M as independent variables in Eq. (7–55) by M_α and M, we have from Eq. (7–56),

$$dE = T\,dS - \tau(l_\beta - l_\alpha)\,dM_\alpha + (\mu' + \tau l_\beta)\,dM. \tag{7–57}$$

We let $j_\alpha(T)$ and $j_\beta(T)$ represent the partition functions of one α and one β unit respectively. The explicit forms used for j_α and j_β would depend on the particular system of interest. The ratio of j_α to j_β reflects the relative intrinsic (i.e., unbiased by a force; $\tau = 0$) stability of the two kinds of units.

The canonical ensemble partition function is

$$Q(M_\alpha, M, T) = \frac{M!\,j_\alpha^{M_\alpha} j_\beta^{M-M_\alpha}}{M_\alpha!(M - M_\alpha)!}, \tag{7–58}$$

where the configurational factor is the number of ways of distributing M_α α units among a total of M possible positions in the chain. We then

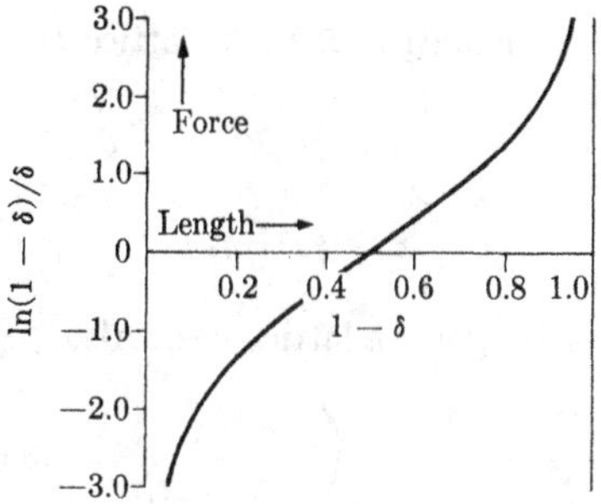

FIG. 7-5. Length-force relation for linear polymer chain of α and β units.

find from Eqs. (7-57) and (7-58),

$$\frac{\tau(l_\beta - l_\alpha)}{kT} = -\frac{1}{kT}\left(\frac{\partial A}{\partial M_\alpha}\right)_{M,T} = \left(\frac{\partial \ln Q}{\partial M_\alpha}\right)_{M,T}$$

$$= \ln\left(\frac{1-\delta}{\delta}\cdot\frac{j_\alpha}{j_\beta}\right), \tag{7-59}$$

where $\delta = M_\alpha/M$. Equation (7-59) is essentially the length-force relation (Fig. 7-5), since l and δ are related, from Eq. (7-56), by

$$1 - \delta = \frac{l - Ml_\alpha}{M(l_\beta - l_\alpha)}\cdot \tag{7-60}$$

At zero force, $\tau = 0$,

$$\frac{\delta}{1-\delta} = \frac{M_\alpha}{M_\beta} = \frac{j_\alpha}{j_\beta}\cdot \tag{7-61}$$

This is the intrinsic stability ratio (or "equilibrium constant"; see Chapter 10) referred to above. When $\tau \to \infty$, $\delta \to 0$, and $l \to Ml_\beta$ (all "long" units).

The similarity between Eqs. (7-8) and (7-59) will be obvious. The chemical potential of a lattice gas and the force on a polymer chain play similar roles in altering the equilibrium ratio of the two possible states in each case (empty and occupied sites; α and β units).

Let us consider also the partition function (Problem 7-10)

$$Y = \sum_{M_\alpha=0}^{M} Q(M_\alpha, M, T)\eta^{M_\alpha}, \tag{7-62}$$

where

$$\eta = e^{-\tau(l_\beta - l_\alpha)/kT}. \tag{7-63}$$

In Eq. (7–62), Y is the analog of Ξ for a lattice gas. Using Eq. (7–58), we obtain

$$Y = \xi^M, \tag{7-64}$$

where

$$\xi = j_\beta + j_\alpha \eta. \tag{7-65}$$

This corresponds to $1 + q\lambda$ for a lattice gas. Also,

$$\overline{M}_\alpha = \frac{\sum_{M_\alpha} M_\alpha Q \eta^{M_\alpha}}{Y} = \eta\left(\frac{\partial \ln Y}{\partial \eta}\right)_{M,T} = M\eta\left(\frac{\partial \ln \xi}{\partial \eta}\right)_T,$$

or

$$\delta = \frac{\overline{M}_\alpha}{M} = \eta\left(\frac{\partial \ln \xi}{\partial \eta}\right)_T = \frac{j_\alpha \eta}{j_\beta + j_\alpha \eta}, \tag{7-66}$$

which agrees with Eq. (7–59).

We turn next to a more complicated situation. Suppose that there are x equivalent and independent sites on each unit of the polymer chain for adsorption of another molecular species. An example would be the binding of hydrogen ions. In this case the length-force curve of the polymer would depend on the hydrogen ion concentration. Let q_α (q_β) be the partition function of one adsorbed molecule on a site of an α (β) unit. In general, $q_\alpha \neq q_\beta$, so now we have the possibility of, say, altering the length of the polymer chain at constant force as a consequence of a change in the concentration (or pressure, if a gas) of the adsorbate. For example, if $q_\alpha > q_\beta$, an increase in adsorbate concentration will cause some β units to go over into α units, thus shortening the chain. It is conceivable that an effect of this kind could be involved in muscle contraction. This problem is very simple to treat by generalizing ξ above (Eq. 7–65), but it is rather cumbersome using Q (Problem 7–11).

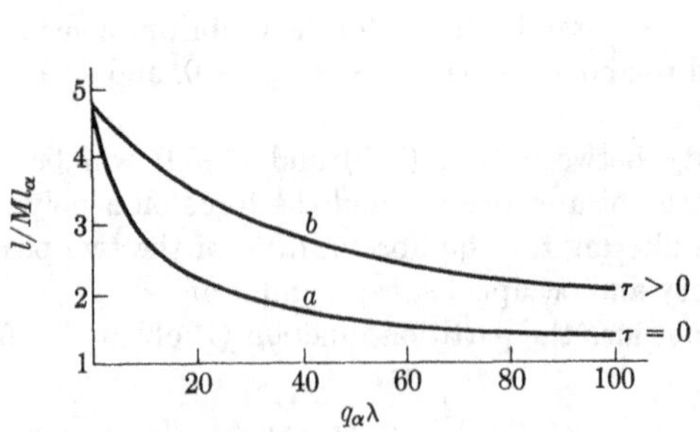

FIG. 7–6. Shortening of chain with increasing adsorbate concentration at constant τ in the arbitrary special case $x = 1$, $l_\beta/l_\alpha = 5$, $j_\alpha/j_\beta = 1/10$ (β units longer and more stable than α), and $q_\beta = 0$ (no adsorption on β units). Curve a: $\tau = 0$ (rest length as function of λ). Curve b: $\exp[\tau(l_\beta - l_\alpha)/kT] = 3.5$.

The thermodynamic equation here is

$$dE = T\,dS - \tau(l_\beta - l_\alpha)\,dM_\alpha + (\mu' + \tau l_\beta)\,dM + \mu\,dN, \qquad (7\text{–}67)$$

where μ and N refer to adsorbed molecules. We set up a function $\xi(\eta, \lambda, T)$ which is analogous to the two-component ξ in Eq. (7–41). We need a term in ξ for each possible state of a single unit. Thus, the unit can be of type β with $0, 1, \ldots, x$ molecules adsorbed (q_β), or of type α with $0, 1, \ldots, x$ molecules adsorbed (q_α). The appropriate combination of Eq. (7–65) and Problem 7–9 is

$$\xi = j_\beta(1 + q_\beta\lambda)^x + j_\alpha\eta(1 + q_\alpha\lambda)^x. \qquad (7\text{–}68)$$

Then

$$\delta = \eta\left(\frac{\partial \ln \xi}{\partial \eta}\right)_{\lambda,T} = \frac{j_\alpha\eta(1 + q_\alpha\lambda)^x}{\xi}, \qquad (7\text{–}69)$$

$$\bar{s} = \lambda\left(\frac{\partial \ln \xi}{\partial \lambda}\right)_{\eta,T} = x[\theta_\alpha\delta + \theta_\beta(1 - \delta)], \qquad (7\text{–}70)$$

where

$$\theta_\alpha = \frac{q_\alpha\lambda}{1 + q_\alpha\lambda}, \qquad \theta_\beta = \frac{q_\beta\lambda}{1 + q_\beta\lambda}. \qquad (7\text{–}71)$$

Equations (7–71) give the fraction of sites of the two types which are occupied, as in Eq. (7–16). In the limit as $\lambda \to \infty$, $\bar{s} \to x$ (all sites occupied) and

$$\delta \to \frac{j_\alpha\eta q_\alpha^x}{j_\beta q_\beta^x + j_\alpha\eta q_\alpha^x}.$$

If we solve Eq. (7–69) for η, we find for the length-force relation

$$-\ln \eta = \frac{\tau(l_\beta - l_\alpha)}{kT} = \ln\left[\frac{1-\delta}{\delta}\cdot\frac{j_\alpha}{j_\beta}\cdot\left(\frac{1 + q_\alpha\lambda}{1 + q_\beta\lambda}\right)^x\right]. \qquad (7\text{–}72)$$

This shows the effect of adsorbate concentration (proportional to λ). If $q_\alpha > q_\beta$, the force τ must be greater to achieve the same length (or δ) when $\lambda > 0$ as compared with $\lambda = 0$. This is because the presence of adsorbate molecules favors short units in this case. Figure 7–6 illustrates an equivalent effect: the shortening of the chain with increase in λ at constant τ when $q_\alpha > q_\beta$.

Problems

7–1. Assume that the solid is a continuum with number density $\rho = N/V$ and occupies the semi-infinite region $z \leq 0$. Let $u(r)$ in Eq. (IV–1) be the interaction potential between a gas molecule and one molecule of the solid. Show that the total interaction between a gas molecule at $z > 0$ and the entire solid is

$$U(z) = \frac{\epsilon r^{*12}\pi\rho}{45z^9} - \frac{\epsilon r^{*6}\pi\rho}{3z^3}. \tag{7–73}$$

(Page 126.)

7–2. Show that the frequency of simple harmonic motion at the bottom of the potential well (7–73) is

$$\nu_z = 1.351\left(\frac{\epsilon r^*\rho}{m}\right)^{1/2}.$$

(Page 126.)

7–3. Consider argon gas adsorbed on a solid at 200°K, according to Langmuir's equation, (7–10). In Eq. (7–3), take U_{00} as 1500 cal·mole^{-1} and $\nu_x = \nu_y = \nu_z = 5 \times 10^{12}$ sec^{-1}. Calculate θ when $p = 1$ atm. (Page 128.)

7–4. Verify Eq. (7–11) for the entropy of an ideal lattice gas. Make a rough plot of S/Mk as a function of θ over the entire range $0 \leq \theta \leq 1$. (Page 128.)

7–5. Extend the derivation of Eq. (7–26) to include Eq. (7–40) for adsorption from a binary mixture. (Page 136.)

7–6. Apply the ξ-method to a two-component (A, B) ideal lattice gas (independent sites; each site can be empty or can be occupied by an A or a B). Show that

$$\frac{\overline{N}_A}{M} = \frac{q_A\lambda_A}{1 + q_A\lambda_A + q_B\lambda_B}.$$

(Page 136.)

7–7. Put Eq. (7–31) in the form

$$\ln\frac{\lambda}{\lambda_{1/2}} = \ln\frac{\theta}{1-\theta},$$

where $\lambda_{1/2}$ is the value of λ when $\theta = \frac{1}{2}$. Make a rough plot of θ against $\ln(\lambda/\lambda_{1/2})$ ("titration curve"). Compare with Fig. 7–5. (Page 139.)

7–8. Consider a system of adsorbent molecules, each with two adsorption sites (q_1 and q_2). Let w be the interaction energy when both sites of a pair are occupied. Let σ_0, σ_1, and σ_2 be the fractions of adsorbent molecules with zero, one, and two adsorbed molecules. Let ρ be the concentration of adsorbate molecules in equilibrium with adsorbed molecules. For a dilute solution, from thermodynamics,

$$\mu = \mu^* + kT\ln\rho, \tag{7–74}$$

where μ^* is independent of ρ. We define successive thermodynamic dissociation constants by

$$K_{(1)} = \frac{\sigma_1\rho}{\sigma_2}, \qquad K_{(2)} = \frac{\sigma_0\rho}{\sigma_1}.$$

Show that the average number $\bar{s}$ of adsorbed molecules per pair is

$$\bar{s} = \frac{[\rho/K_{(2)}] + [2\rho^2/K_{(1)}K_{(2)}]}{1 + [\rho/K_{(2)}] + [\rho^2/K_{(1)}K_{(2)}]}. \tag{7-75}$$

This is a purely thermodynamic result. Next, we define an "intrinsic" dissociation constant K_1 for sites of type 1 by

$$K_1 = \frac{\rho(1 - \theta_1^0)}{\theta_1^0}, \qquad \theta_1^0 = \frac{q_1\lambda}{1 + q_1\lambda}. \tag{7-76}$$

These equations refer to adsorption on *isolated* sites of type 1. Show that Eqs. (7-74) and (7-76) give

$$K_1 = \frac{e^{-\mu^*/kT}}{q_1}, \qquad K_2 = \frac{e^{-\mu^*/kT}}{q_2}. \tag{7-77}$$

Using this notation, show that the statistical-mechanical equation (7-33) becomes

$$\bar{s} = \frac{[(1/K_1) + (1/K_2)]\rho + (2e^{-w/kT}\rho^2/K_1K_2)}{1 + [(1/K_1) + (1/K_2)]\rho + (e^{-w/kT}\rho^2/K_1K_2)}. \tag{7-78}$$

Finally, compare Eqs. (7-75) and (7-78) to deduce the connections

$$K_{(1)} = (K_1 + K_2)e^{w/kT}, \qquad \frac{1}{K_{(2)}} = \frac{1}{K_1} + \frac{1}{K_2}. \tag{7-79}$$

Give a physical interpretation of Eqs. (7-79). (Page 139.)

7-9. For a system of adsorbent molecules each of which has m equivalent and independent sites for adsorption, show that $\xi = (1 + q\lambda)^m$ and that the adsorption isotherm is the same as the Langmuir equation, (7-31). (Page 140.)

7-10. Use the method of Eq. (1-92) to find the characteristic thermodynamic function for the partition function Y. Use this result and thermodynamics to show that

$$\overline{M}_\alpha = \eta\left(\frac{\partial \ln Y}{\partial \eta}\right)_{M,T}.$$

(Page 141.)

7-11. Derive Eq. (7-72) for the length-force relation of an adsorbing polymer chain from the canonical partition function $Q(M_\alpha, M, N, T)$. (Page 142.)

7-12. Derive the properties of an ideal lattice gas using the N, Φ, T partition function, Eq. (1-87).

7-13. Deduce, from Eq. (7-6), the thermodynamic equation for a lattice gas (or adsorbed phase)

$$\Phi(p) = kT\int_0^p \theta(p')\,d\ln p' \qquad (T \text{ constant}),$$

assuming that the equilibrium gas phase (pressure p) is ideal. Use this equation to check the self-consistency of Eqs. (7-10) and (7-15).

7–14. If adsorption in the B.E.T. theory is limited to n layers (by pore structure in the adsorbent, for example), show that

$$\bar{s} = \frac{cx[1 - (n+1)x^n + nx^{n+1}]}{(1-x)(1-x+cx-cx^{n+1})}.$$

7–15. Derive the B.E.T. equation, (7–36), from the complete grand partition function Ξ. For a given N, let N_1 be the number of molecules in the first layer and N' be the number in higher layers, where $N_1 + N' = N$. Then $Q(N, M, T)$ will involve a sum over either N_1 or N'.

7–16. Show that $\overline{N} = \lambda(\partial \ln \Gamma/\partial\lambda)$ in Eq. (7–51) follows from $kT \ln \Gamma = pV - \mu' M$ and Eq. (7–46).

7–17. Discuss the equilibrium between gas and adsorbed phase (Langmuir model) in terms of an energy-entropy competition (see Section 5–3).

Supplementary Reading

BAND, Chapter 5.

FOWLER and GUGGENHEIM, Chapter 10.

HILL, T. L., in *Advances in Catalysis*, Vol. 4. New York: Academic Press, 1952.

S. M., Chapter 7 and Appendixes 4 and 5.

CHAPTER 8

IDEAL DIATOMIC GAS

In Chapter 4 we studied an ideal monatomic gas. We extend the discussion of ideal gases here to those composed of diatomic molecules, and then in Chapter 9 we consider polyatomic molecules. An important application of the results obtained in these two chapters is to chemical equilibrium and chemical kinetics in ideal gas mixtures (Chapters 10 and 11).

8–1 Independence of degrees of freedom. We shall present in this chapter and the following one the simplest possible treatments, based on the rather good approximation of independence of degrees of freedom. Brief mention of this approximation has already been made in Chapter 3 [see Eqs. (3–13) and (3–14)], but we discuss it a little more fully here.

Consider a diatomic molecule consisting of two nuclei (masses m_1 and m_2) and several electrons. In an exact quantum-mechanical treatment of this system, all these particles and their (coulombic) interactions have to be included in a single Schrödinger equation. However, because the electrons are very light compared with the nuclei and hence move much faster on the average, it is an excellent approximation (called the Born-Oppenheimer approximation) to consider the nuclei fixed while studying the electronic motion. The procedure, then, is the following. For nuclei fixed at some distance apart r, we write and solve the Schrödinger equation in the coordinates of the electrons. The potential energy in this equation is made up of electron-electron, electron-nucleus, and nucleus-nucleus coulombic interactions. The last of these interactions is of course constant for constant r. A set of (electronic) energy levels is obtained, which will depend on the value of r. In particular, let the dependence of the ground state electronic energy on internuclear distance be denoted by $u_e(r)$. This function can be calculated accurately only for very simple molecules (one and two electrons). In more complicated cases it must be approximated or deduced semiempirically from experimental information. The typical form of $u_e(r)$ is shown in Fig. 8–1. It is qualitatively similar in appearance to the van der Waals interaction curve, Fig. IV–1, but is much deeper (of the order of 100 kcal·mole^{-1}), and the minimum occurs at smaller values of r (often 1 or 2 A). The depth of the $u_e(r)$ well is of course essentially the energy of the chemical bond between the two atoms. The dashed curve in Fig. 8–1 is a schematic plot of the first excited electronic energy level as a function of r.

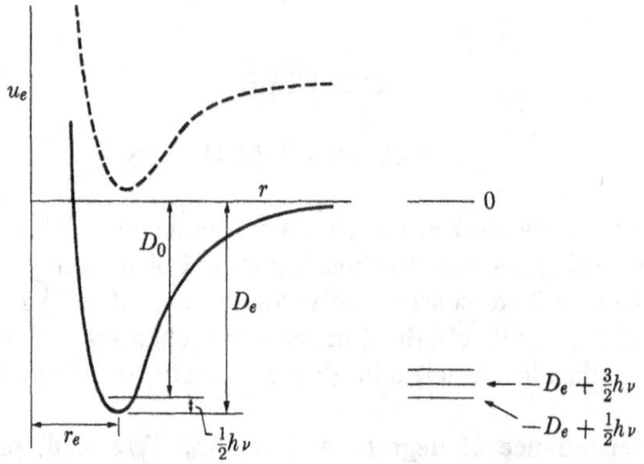

FIG. 8–1. Ground (and first excited) state electronic energy u_e as a function of internuclear separation r. The first few vibrational levels are shown.

We can now investigate the (relatively slow) motion of the two nuclei. The work required to change the value of r when the molecule is in its ground electronic state is determined by $u_e(r)$. That is, $u_e(r)$ serves as the potential function for the nuclear motion. A second Schrödinger equation is then written, this time in the six coordinates (x_1, y_1, z_1, x_2, y_2, z_2) of the two nuclei with $u_e(r)$ as the potential function. Since we want to consider the molecule to be in a box of volume V, we must add that the potential becomes infinite when the center of mass of the molecule is outside this box. The next step is to change to the coordinates x, y, z, r, θ, φ, where x, y, and z refer to the position of the center of mass and r, θ, and φ are indicated in Fig. 8–2. With these new coordinates, the equation separates into two equations, one in x, y, and z (translation) and one in r (vibration), θ and φ (rotation). The x, y, z equation is just that of a particle of mass $m_1 + m_2$ in a box of volume V. This problem has already been discussed in Section 4–1. We can take over the results obtained there simply by replacing m by $m_1 + m_2$. The energy levels from the r, θ, φ equation are difficult to calculate [for one thing, $u_e(r)$ is not a simple function] and are not simply additive (vibration + rotation). Therefore we make two further approximations. First, we replace $u_e(r)$ by a parabola which fits $u_e(r)$ in the neighborhood of its minimum. Second, we separate the vibrational and rotational motions from each other by assuming that, as far as rotation (θ, φ) is concerned, the molecule has a fixed internuclear distance r_e, where r_e is the ("equilibrium") value of r at the minimum in $u_e(r)$. The rotational motion is therefore approximated as that of a rigid, linear, free rotator and the vibrational motion as that of a simple harmonic oscillator (a problem already discussed in Section 5–1).

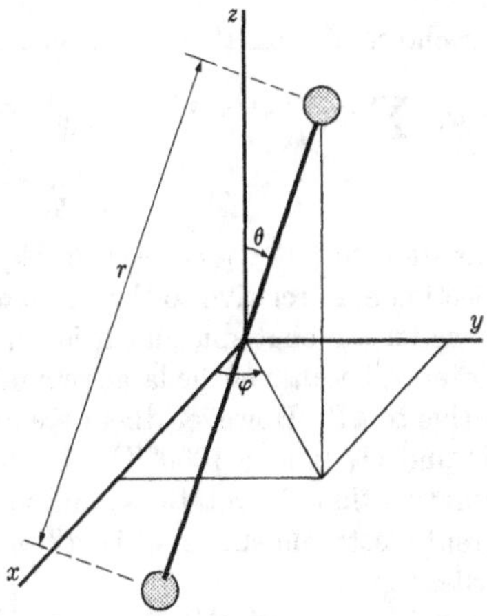

FIG. 8–2. Vibrational (r) and rotational (θ, φ) coordinates of a diatomic molecule. The center of mass is at the origin.

We have then the basic equations [see (3–13) and (3–14)] for an ideal diatomic gas,

$$H = H_t + H_r + H_v + H_e, \tag{8–1}$$

$$\epsilon = \epsilon_t + \epsilon_r + \epsilon_v + \epsilon_e, \tag{8–2}$$

$$Q = \frac{1}{N!} q^N, \qquad q(V, T) = q_t(V, T)q_r(T)q_v(T)q_e(T). \tag{8–3}$$

The rotational, vibrational, and electronic states are obviously independent of the volume, so the corresponding q's are functions of T only. We shall discuss q_v and q_r in Sections 8–2 and 8–3, respectively. For q_t, as already mentioned, we have from Eq. (4–10),

$$q_t = \left[\frac{2\pi(m_1 + m_2)kT}{h^2}\right]^{3/2} V. \tag{8–4}$$

In the notation of Eq. (4–42), q_e is simply

$$q_e = \omega_{e1} e^{-\epsilon_{e1}/kT}. \tag{8–5}$$

We shall usually choose the zero of energy as the separated atoms ($r \to \infty$ in Fig. 8–1) at rest. If the vibrational energy levels are denoted by ϵ_n, relative to the customary choice of the bottom of the potential energy well $u_e(r)$ as zero, and if the depth of the well (Fig. 8–1) is $D_e(>0)$, then

$q_e q_v$ relative to our choice of separated atoms as zero is

$$q_e q_v = \omega_{e1} \sum_n e^{-(-D_e+\epsilon_n)/kT} = \underbrace{\omega_{e1} e^{D_e/kT}}_{q_e} \underbrace{\sum_n e^{-\epsilon_n/kT}}_{q_v}.$$

Thus it is self-consistent to put $\epsilon_{e1} = -D_e$ in Eq. (8–5), as we shall do, and write q_v (in Section 8–2) relative to the bottom of the well as zero.

Ordinarily, just as for a monatomic gas, it is only the ground electronic state which is of interest because of the large separation between electronic energy levels, relative to kT. However, there are exceptions; for example, NO (above 15°K) and O_2 (above 1000°K). In these cases, assuming as a further approximation that the rotational and vibrational states are the same in the different electronic states, q_e is still a separate factor in Eq. (8–3), and we write

$$q_e = \omega_{e1} e^{-\epsilon_{e1}/kT} + \omega_{e2} e^{-\epsilon_{e2}/kT}. \tag{8–6}$$

As a consequence of Eq. (8–3), many of the thermodynamic functions receive additive contributions from translation, rotation, etc. For example,

$$A = -kT \ln Q = A_t + A_r + A_v + A_e, \tag{8–7}$$

where

$$A_t = -NkT \ln \left(\frac{q_t e}{N}\right), \tag{8–8}$$

$$A_r = -NkT \ln q_r, \qquad A_v = -NkT \ln q_v, \qquad \text{etc.} \tag{8–9}$$

Equation (8–8) is the same as Eq. (4–14), except for the mass. Also,

$$E = kT^2 \left(\frac{\partial \ln Q}{\partial T}\right)_{N,V} = E_t + E_r + E_v + E_e, \tag{8–10}$$

where

$$E_t = \tfrac{3}{2}NkT, \tag{8–11}$$

$$E_r = NkT^2 \frac{d \ln q_r}{dT}, \qquad \text{etc.} \tag{8–12}$$

Similar equations can be written for C_V, S, etc.

The equation of state,

$$p = kT \left(\frac{\partial \ln Q}{\partial V}\right)_{N,T} = NkT \left(\frac{\partial \ln q_t}{\partial V}\right)_T = \frac{NkT}{V}, \tag{8–13}$$

does not depend on $q_r(T)$, $q_v(T)$, etc. [see also Eq. (3–12) and Problems 3–1 and 3–6].

To continue the discussion, we need further details concerning q_v and q_r.

8-2 Vibration. The vibrational motion of a diatomic molecule involves only one coordinate, the internuclear distance r. Thus, there is only one vibrational degree of freedom. For the potential energy that determines the vibrational motion, we take

$$u_e(r) = -D_e + \tfrac{1}{2}f(r - r_e)^2, \tag{8-14}$$

as already explained, where f is the force constant,

$$f = \left(\frac{d^2u_e}{dr^2}\right)_{r=r_e} \tag{8-15}$$

(Problem 8-1). Since we wish to use the bottom of the potential well as the zero of vibrational energy, we drop D_e in Eq. (8-14) and include it instead in Eq. (8-5) for q_e, as mentioned earlier.

Let the line between centers of the two nuclei be the ζ-axis, with arbitrary origin and, say, ζ_2 (location of nucleus of mass m_2) $> \zeta_1$. Then $r = \zeta_2 - \zeta_1$, and

$$u_e = \tfrac{1}{2}f(\zeta_2 - \zeta_1 - r_e)^2.$$

The classical equations of motion are

$$m_1\ddot{\zeta}_1 = -\frac{\partial u_e}{\partial \zeta_1} = f(\zeta_2 - \zeta_1 - r_e),$$

$$m_2\ddot{\zeta}_2 = -\frac{\partial u_e}{\partial \zeta_2} = -f(\zeta_2 - \zeta_1 - r_e).$$

We multiply the top equation by $-m_2$, the bottom by m_1, add the two equations, and divide by $m_1 + m_2$. Then

$$\mu\ddot{x} = -fx, \tag{8-16}$$

where

$$x = r - r_e, \qquad \mu = \frac{m_1m_2}{m_1 + m_2}. \tag{8-17}$$

This is the equation of motion of a one-dimensional harmonic oscillator with mass μ (called the "reduced mass"). This is the standard symbol used, and is of course not to be confused with the chemical potential. The frequency of the classical motion is

$$\nu = \frac{1}{2\pi}\sqrt{\frac{f}{\mu}}. \tag{8-18}$$

In quantum mechanics, the energy states are

$$\epsilon_n = (n + \tfrac{1}{2})h\nu, \qquad n = 0, 1, 2, \ldots, \tag{8-19}$$

where ν is the classical frequency, (8-18).

The vibrational partition function is therefore (Eq. 5–6)

$$q_v = \frac{e^{-\Theta_v/2T}}{1 - e^{-\Theta_v/T}}, \tag{8–20}$$

where $\Theta_v = h\nu/k$. The vibrational contributions E_v, C_{Vv}, S_v, etc., to the thermodynamic functions of an ideal diatomic gas are just those of N one-dimensional harmonic oscillators all of the same frequency ν. These are the same functions as for an Einstein crystal ($3N$ one-dimensional oscillators), except for a factor of three. For example,

$$E_v = NkT^2 \frac{d \ln q_v}{dT} = Nk\left(\frac{\Theta_v}{2} + \frac{\Theta_v}{e^{\Theta_v/T} - 1}\right), \tag{8–21}$$

and

$$C_{Vv} = \left(\frac{\partial E_v}{\partial T}\right)_N = Nk\left(\frac{\Theta_v}{T}\right)^2 \frac{e^{\Theta_v/T}}{(e^{\Theta_v/T} - 1)^2}. \tag{8–22}$$

A law of "corresponding states" obviously applies, since C_{Vv}/Nk is a function of Θ_v/T only.

Although the Einstein equations are applicable, the values of ν and therefore Θ_v are in general higher here by a factor of about ten because of the fact that chemical bonds are relatively strong (D_e) and stiff (f) relative to van der Waals and other types of "bonds" in most crystals. In principle, $u_e(r)$ and therefore f, ν, and Θ_v can be calculated from quantum mechanics. But in practice, these quantities are usually deduced from the experimental vibrational spectrum. Table 8–1 gives a few values of Θ_v and Θ_r (defined in Section 8–3).

To take two specific numerical examples, the value $\Theta_v = 3340°K$ for N_2 corresponds to $\nu = 6.96 \times 10^{13}\ \text{sec}^{-1}$. At 300°K, $\Theta_v/T = h\nu/kT = 11.1$. Hence the fraction of molecules in excited vibrational states ($n > 0$) at this temperature is

$$1 - \frac{e^{-\epsilon_0/kT}}{q_v} = e^{-\Theta_v/T} = e^{-11.1}.$$

This order of magnitude is typical for most diatomic molecules, as is clear from Table 8–1. That is, at room temperature, the vibrational degree of freedom is practically "unexcited" and makes only a very small contribution to C_V, etc. (Problem 8–2). On the other hand, a few molecules have relatively loose vibrations and small values of Θ_v. For example, $\Theta_v/T = 1.57$ for Br_2 at 300°K, and the fraction of Br_2 molecules in excited vibrational states at this temperature is 0.21.

The vibrational contribution to the heat capacity C_V follows the Einstein curve in Fig. 5–2 and ranges from zero at low temperatures to Nk

TABLE 8-1

PARAMETERS FOR DIATOMIC MOLECULES

	Θ_v, °K	Θ_r, °K	r_e, A	D_0, ev
H_2	6210	85.4	0.740	4.454
N_2	3340	2.86	1.095	9.76
O_2	2230	2.07	1.204	5.08
CO	3070	2.77	1.128	9.14
NO	2690	2.42	1.150	5.29
HCl	4140	15.2	1.275	4.43
HBr	3700	12.1	1.414	3.60
HI	3200	9.0	1.604	2.75
Cl_2	810	0.346	1.989	2.48
Br_2	470	0.116	2.284	1.97
I_2	310	0.054	2.667	1.54

(i.e., R per mole) at high temperatures. The upper limit is predicted by classical statistical mechanics (see Problem 6-6). The value $0.1Nk$ for C_{Vv} is reached at $T = \Theta_v/5.8$, $0.5Nk$ at $T = \Theta_v/3.0$, and $0.9Nk$ at $T = \Theta_v/1.1$. In the Br_2 example above (at 300°K), $C_{Vv} = 0.82Nk$.

In Table 8-1 and Fig. 8-1, D_0 is the dissociation energy of the diatomic molecule at 0°K. The relation to D_e is

$$D_0 = D_e - \tfrac{1}{2}h\nu.$$

Although D_e cannot be measured by a direct experiment, it may be calculated from experimental values of D_0 and ν.

8-3 Rotation. The two ends of a symmetrical diatomic molecule are indistinguishable, just as two monatomic molecules of the same species in a gas are indistinguishable. This feature leads to symmetry complications and the involvement of nuclear spin in a quantum-statistical treatment of the rotation of these molecules. We postpone such a treatment until Chapter 22. Fortunately, as we shall see below, a quantum treatment is really necessary only for very light molecules (e.g., hydrogen) at low temperatures. Hence, the postponement referred to is not a serious limitation.

These complications do not arise with unsymmetrical diatomic molecules. Hence we can begin with a discussion of the quantum-mechanical q_r for such molecules. In quantum mechanics one finds that the energy levels of a rigid linear rotator are

$$\epsilon_j = \frac{j(j+1)h^2}{8\pi^2 I}, \qquad j = 0, 1, 2, \ldots,$$

with degeneracy $\omega_j = 2j + 1$, and where I is the moment of inertia about the center of mass. For a diatomic molecule, $I = \mu r_e^2$ (Problem 8–3). Therefore,

$$q_r(T) = \sum_j \omega_j e^{-\epsilon_j/kT} = \sum_{j=0}^{\infty} (2j + 1)e^{-j(j+1)\Theta_r/T}, \tag{8–23}$$

where

$$\Theta_r = \frac{h^2}{8\pi^2 Ik}. \tag{8–24}$$

In Eq. 8–24, Θ_r is the characteristic temperature for rotation. The separation $\Delta\epsilon$ between successive rotational levels, relative to kT, is Θ_r/T multiplied by a simple function of j. Hence the sum in Eq. (8–23) may be replaced by an integral when $T \gg \Theta_r$. This will lead to the high-temperature or classical limit of q_r. Table 8–1 includes values of Θ_r for a number of diatomic gases. It will be seen from the table that for ordinary gases at ordinary temperatures or even fairly low temperatures, T is indeed much larger than Θ_r. In principle again, r_e, and hence I, may be calculated from quantum mechanics (Fig. 8–1), but in practice Θ_r, I, and r_e are deduced from the experimental vibrational-rotational or pure rotational spectrum.

The sum in Eq. (8–23) cannot be put in closed form. However, it is easy to use numerically, as it stands, at low temperatures (Problem 8–4). At high temperatures,

$$\begin{aligned} q_r &\to \int_0^{\infty} (2j + 1)e^{-j(j+1)\Theta_r/T}\, dj \\ &= \int_0^{\infty} e^{-j(j+1)\Theta_r/T}\, d[j(j + 1)] \\ &= \frac{T}{\Theta_r} = \frac{8\pi^2 IkT}{h^2}. \end{aligned} \tag{8–25}$$

When T is not quite high enough* for use of Eq. (8–25),

$$q_r = \frac{T}{\Theta_r}\left(1 + \frac{\Theta_r}{3T} + \cdots\right). \tag{8–26}$$

The corresponding equation for q_v, incidentally, is given in Problem 6–2.

As a further check on the validity of Eq. (6–14) for q_{class}, we have

$$q_{\text{class}} = \frac{1}{h^2} \iint_{-\infty}^{+\infty} \int_0^{2\pi} \int_0^{\pi} e^{-H_r/kT}\, d\theta\, d\varphi\, dp_\theta\, dp_\varphi, \tag{8–27}$$

* Mayer and Mayer, pp. 151–154.

where (Problem 8-5)

$$H_r = \frac{1}{2I}\left(p_\theta^2 + \frac{p_\varphi^2}{\sin^2\theta}\right). \tag{8-28}$$

On carrying out the integrations (Problem 8-6), we get Eq. (8-25) again.

The expression in Eq. (8-25) is also the classical partition function of a particle of mass μ moving freely on the surface of a sphere of radius r_e [see Eq. (7-17)]:

$$q_r = \frac{2\pi\mu kT}{h^2}(4\pi r_e^2) = \frac{8\pi^2 IkT}{h^2}.$$

The above discussion pertains to unsymmetrical diatomic molecules. The principal conclusion we reach, in view of Table 8-1, is that we can generally regard rotation as classical—or fully "excited," as it is sometimes put. Table 8-1 also contains Θ_r values for symmetrical molecules. Except for hydrogen (and deuterium), these are of the same order of magnitude as for unsymmetrical molecules. Hence the same conclusion: generally rotation of symmetrical molecules can be treated classically. We may therefore apply Eq. (8-27) to symmetrical molecules also, but with one correction. For exactly the same reason that division by $N!$ is included in Eq. (6-18), we have to divide here by two if the diatomic molecule is symmetrical. If we do not do this, the two indistinguishable configurations in Fig. 8-3 will both be counted in the integration over θ and φ in Eq. (8-27) [see the discussion following Eq. (6-18)].

Thus we can write for the high-temperature (classical) q_r,

$$q_r = \frac{T}{\sigma\Theta_r} = \frac{8\pi^2 IkT}{\sigma h^2}, \tag{8-29}$$

where $\sigma = 1$ for unsymmetrical molecules and $\sigma = 2$ for symmetrical ones. The constant σ is called the symmetry number.

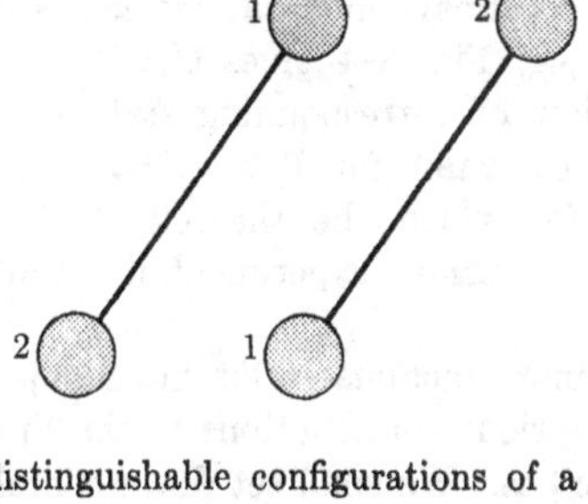

FIG. 8-3. Two indistinguishable configurations of a symmetrical diatomic molecule.

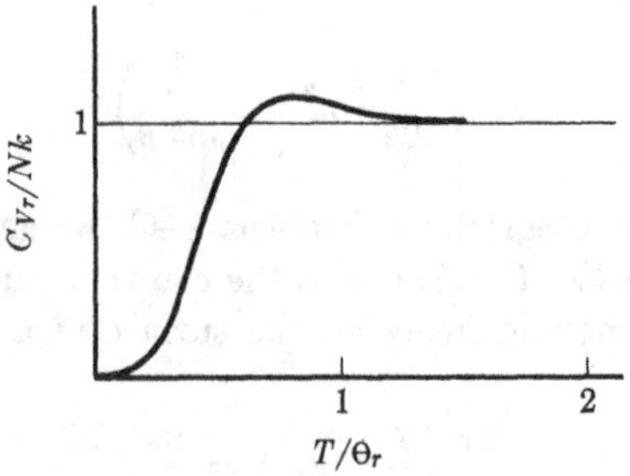

FIG. 8-4. Rotational heat capacity of an unsymmetrical diatomic molecule.

Equation (8–29) leads to very simple rotational contributions to thermodynamic functions. For example,

$$A_r = -NkT \ln \left(\frac{T}{\sigma\Theta_r}\right), \tag{8–30}$$

$$E_r = NkT^2 \frac{d \ln q_r}{dT} = NkT, \tag{8–31}$$

$$S_r = \frac{E_r - A_r}{T} = Nk \ln \left(\frac{Te}{\sigma\Theta_r}\right), \tag{8–32}$$

$$C_{Vr} = \left(\frac{\partial E_r}{\partial T}\right)_N = Nk. \tag{8–33}$$

Equations (8–31) and (8–33) could have been predicted immediately from Eq. (8–28) for H_r and from Problem 6–6. There are two rotational degrees of freedom (θ, φ), and E_r represents kinetic energy only (unlike E_v, which has equal kinetic and potential contributions). Incidentally, the rotation of a diatomic molecule in an imperfect gas, liquid or solid, does involve a potential energy and is therefore not "free" rotation, as in the present situation (ideal gas).

We can also derive low-temperature rotational contributions to thermodynamic functions for unsymmetrical molecules directly from the sum (8–23). For example, Fig. 8–4 gives C_{Vr}/Nk as a function of T/Θ_r. Again we have a law of corresponding states. The heat capacity has practically its classical value for $T > 1.5\Theta_r$. An important example is HD ($\Theta_r = 64°K$), for which the theoretical C_{Vr} curve, including the maximum, has been confirmed experimentally (Problem 8–7).

8–4 Thermodynamic functions. Our main object in this section is to bring together the various contributions to the thermodynamic functions of an ideal diatomic gas. We shall set down complete equations only for the most important special case: translation and rotation treated classically; vibration treated quantum-mechanically; and electronic degrees of freedom unexcited (i.e., only the ground electronic state included). Before writing the equations, we emphasize again that (a) the present analysis, in which the various contributions are additive, is somewhat approximate; and (b) the question of classical (fully excited) *vs.* quantum *vs.* unexcited (i.e., ground state only) for a particular type of degree of freedom is determined by the magnitude of the appropriate energy-level separation relative to kT. In typical cases (Problem 8–8):

$$\Delta\epsilon\ (\text{translation}) = O(10^{-18}\ \text{ev}), \qquad \Delta\epsilon\ (\text{vibration}) = O(0.3\ \text{ev}),$$

$$\Delta\epsilon\ (\text{rotation}) = O(5 \times 10^{-4}\ \text{ev}), \qquad \Delta\epsilon\ (\text{electronic}) = O(5\ \text{ev}),$$

$$kT(300^\circ\text{K}) = O(0.03\ \text{ev}).$$

The criteria are:

Classical:	$\Delta\epsilon \ll kT,$	$\Theta \ll T.$
Quantum:	$\Delta\epsilon = O(kT),$	$\Theta = O(T).$
Unexcited:	$\Delta\epsilon \gg kT,$	$\Theta \gg T.$

In the special case referred to in the above paragraph, we find from Eq. (8–7), etc.,

$$-\frac{A}{NkT} = \ln\left[\frac{2\pi(m_1+m_2)kT}{h^2}\right]^{3/2}\frac{Ve}{N} + \ln\frac{8\pi^2 IkT}{\sigma h^2} - \frac{h\nu}{2kT} - \ln\left(1 - e^{-h\nu/kT}\right) + \frac{D_e}{kT} + \ln\omega_{e1}; \tag{8–34}$$

$$\frac{E}{NkT} = \frac{3}{2} + \frac{2}{2} + \frac{h\nu}{2kT} + \frac{h\nu/kT}{e^{h\nu/kT}-1} - \frac{D_e}{kT}; \tag{8–35}$$

$$\frac{C_V}{Nk} = \frac{3}{2} + \frac{2}{2} + \left(\frac{h\nu}{kT}\right)^2 \frac{e^{h\nu/kT}}{(e^{h\nu/kT}-1)^2}; \tag{8–36}$$

$$\frac{S}{Nk} = \ln\left[\frac{2\pi(m_1+m_2)kT}{h^2}\right]^{3/2}\frac{Ve^{5/2}}{N} + \ln\frac{8\pi^2 IkTe}{\sigma h^2} + \frac{h\nu/kT}{e^{h\nu/kT}-1} - \ln\left(1 - e^{-h\nu/kT}\right) + \ln\omega_{e1}; \tag{8–37}$$

$$pV = NkT. \tag{8–38}$$

Ordinarily, $\omega_{e1} = 1$. The molecules O_2 and NO are exceptions (see below). The order of terms in these equations is: translation; rotation; vibration; and electronic. The heat capacity C_V is thus $5Nk/2$ (translation, rotation) for $T \ll \Theta_v = h\nu/k$ and rises to $7Nk/2$ (vibration classical) for $T \gg \Theta_v$. The form of the C_V curve between these two limits (Einstein curve in Fig. 5–2) has been verified experimentally in a number of cases. At very low temperatures, rotation is no longer classical, and C_V falls from $5Nk/2$ to $3Nk/2$ as the temperature decreases. It is not practical to observe this effect experimentally, however, except for very light gases (e.g., HD). Figure 8–4 (unsymmetrical molecules) gives C_{V_r} for HD if we put $\Theta_r = 64°K$. For HD, Θ_v is 6100°K, so that the two observable transitions (owing to exciting rotation and vibration) in C_V from $3Nk/2$ to $5Nk/2$ to $7Nk/2$ are very widely separated on the temperature scale.

For the chemical potential, we obtain [see the remarks in connection with Eq. (4–26)]

$$\frac{F}{NkT} = \frac{A}{NkT} + \frac{pV}{NkT} = \frac{\mu}{kT} = \frac{\mu^0(T)}{kT} + \ln p, \tag{8–39}$$

where

$$\frac{\mu^0(T)}{kT} = -\ln\left[\frac{2\pi(m_1+m_2)kT}{h^2}\right]^{3/2} kT - \ln\frac{8\pi^2 IkT}{\sigma h^2} + \frac{h\nu}{2kT}$$
$$+ \ln(1 - e^{-h\nu/kT}) - \frac{D_e}{kT} - \ln\omega_{e1}. \tag{8–40}$$

In the above equations the zero of energy has been chosen as the separated atoms at rest, and the zero of entropy corresponds to a single ($\Omega = 1$) translation-rotation-vibration-electronic quantum state for the macroscopic system of N diatomic molecules, i.e., the pure crystal at 0°K.

We have already referred briefly to the agreement found between experimental C_V measurements and the above theory. Experimental checks of the entropy equation, (8–37), will be mentioned in the next chapter. In addition, Eq. (8–39) for the chemical potential has been verified by measurements of crystal vapor pressures* [compare Eq. (5–50)] and experimental gaseous equilibrium constants (Chapter 10).

For NO and O_2 we must use† Eq. (8–6) for q_e:

$$q_e = e^{D_e/kT}[\omega_{e1} + \omega_{e2}e^{-(\epsilon_{e2}-\epsilon_{e1})/kT}], \tag{8–41}$$

$$\text{NO}: \omega_{e1} = 2, \qquad \omega_{e2} = 2, \qquad (\epsilon_{e2} - \epsilon_{e1})/k = 178°\text{K},$$

$$\text{O}_2: \omega_{e1} = 3, \qquad \omega_{e2} = 2, \qquad (\epsilon_{e2} - \epsilon_{e1})/k = 11{,}300°\text{K}.$$

Equation (8–41) leads to a significant electronic contribution to C_V for NO, with a maximum predicted (Problem 8–9) at 74°K. For a con-

* See Fowler and Guggenheim, pp. 202–205.
† See Fowler and Guggenheim, pp. 102–106, for further details.

siderable temperature range (see Table 8–1) on both sides of 74°K, $C_{Vt} + C_{Vr} = 5Nk/2$ and $C_{Vv} = 0$. The experimental variation of C_V with T in this region is therefore due to C_{Ve}. Theory and experiment have been checked between 125°K and 180°K.

Finally, it should be mentioned that instead of deriving thermodynamic functions from the slightly idealized energy levels of a simple harmonic oscillator and rigid rotator, one can use instead actual energy levels deduced from spectroscopy. This is obviously a more accurate procedure, but one which we shall not discuss. The approximation we have used in this chapter takes care very adequately of all the first-order effects.

Problems

8–1. The empirical Morse function for $u_e(r)$ is

$$u_e(r) = D_e\{[1 - e^{-a(r-r_e)}]^2 - 1\}.$$

Find the force constant f in terms of D_e and a. (Page 151.)

8–2. Calculate C_{Vv} in cal·mole^{-1}·deg^{-1} for N_2 and Cl_2 at 300°K, 900°K, and 2700°K. (Page 152.)

8–3. From the definitions of center of mass and moment of inertia, show that $I = \mu r_e^2$ for a diatomic molecule. (Page 154.)

8–4. For HD gas at 96°K, use Eq. (8–23) to calculate q_r, E_r (cal·mole^{-1}), and C_{Vr} (cal·mole^{-1}·deg^{-1}). Take $\Theta_r = 64$°K. Compare this value of q_r with that obtained from the classical q_r. What are the values of E in cal·mole^{-1} and C_p in cal·mole^{-1}·deg^{-1} for HD at 96°K? Also, find the fraction of HD molecules in each of the first four rotational energy levels at 32°K, 96°K, and 256°K. (Page 154.)

8–5. In the free rotation of a diatomic molecule with center of mass fixed at the origin, each nucleus moves over the surface of a sphere. Express the kinetic energy of the two nuclei in cartesian coordinates, change to spherical coordinates, and hence show that the kinetic energy is

$$\tfrac{1}{2}I(\dot{\theta}^2 + \dot{\varphi}^2 \sin^2 \theta).$$

From this, derive Eq. (8–28). Show that a single particle of mass μ moving on a sphere of radius r_e has this same kinetic energy. (Page 155.)

8–6. Derive the result $q_r = T/\Theta_r$ from the classical phase integral, Eq. (8–27). (Page 155.)

8–7. Deduce the value of Θ_r for HD from $\Theta_r = 85.4$°K for H_2. (Page 156.)

8–8. Verify the orders of magnitude listed near the beginning of Section 8–4 for $\Delta\epsilon$ (translation, rotation, etc.) What wavelengths of light in A and wave numbers in cm^{-1} do these energies correspond to ($\Delta\epsilon = h\nu = hc/\lambda$)? (Page 157.)

8–9. With the aid of Problem 4–8, verify that C_{Ve} for NO has a maximum at about 74°K. (Page 158.)

8–10. From spectroscopy it is found that $\Theta_r = 9.0$°K for HI. Calculate, (a) the moment of inertia of HI in cgs units, and (b) the equilibrium internuclear separation r_e of HI in A.

8–11. Calculate the fraction of Cl_2 molecules in the first four vibrational states at 200°K, 800°K, and 3000°K.

8–12. From Table 8–1 calculate the vibrational force constant f in dyne·cm^{-1}, the frequency ν, and wave number in cm^{-1} for HCl, Cl_2, and I_2.

8–13. From Table 8–1 calculate D_e in cal·mole^{-1} for H_2.

8–14. Calculate S and C_p in cal·mole^{-1}·deg^{-1} and μ in cal·mole^{-1} for N_2 and HBr at 25°C and 1 atm pressure (see Table 9–2). The value of ω_{e1} is unity for both gases.

8–15. Calculate the work (in cal·mole^{-1}) necessary to stretch the chemical bond in HCl from $r = r_e$ to $r = r_e + 0.1$ A.

Supplementary Reading

Fowler and Guggenheim, Chapter 3.

Herzberg, G., *Molecular Spectra and Molecular Structure.* 2nd ed. New York: Prentice-Hall, 1950.

Mayer and Mayer, Chapters 6 and 7.

Rushbrooke, Chapters 6 and 8.

CHAPTER 9

IDEAL POLYATOMIC GAS

In this chapter we make a direct extension to polyatomic molecules of the methods and approximations used in the previous chapter for diatomic molecules. One new feature arises: hindered internal rotation (in molecules such as ethane), which is considered in Section 9–5. We also include in this chapter a section (9–6) on the transition from localized to mobile adsorption (these terms were introduced in Section 7–1). Offhand this appears to be a completely unrelated subject, but actually this transition can be described by the same equations as those developed for the torsional oscillation→free internal rotation transition in ethane, etc.

9–1 Potential energy surface. A polyatomic molecule consists of three or more nuclei and, usually, many electrons. To deduce the thermodynamic properties of an ideal (i.e., very dilute) polyatomic gas, we first need the quantum-mechanical energy levels for a single polyatomic molecule in a box of volume V. This problem is extremely, if not hopelessly, complicated unless we make the same simplifications as in the previous chapter for diatomic molecules. Actually, there are several refinements that could have been introduced without much complication in our treatment of diatomic molecules, but such refinements are not practical here because of the greater complexity of the problem.

For a given fixed configuration of the nuclei of one molecule, we first study the quantum-mechanical problem of the motion of the electrons. This leads in principle to a set of electronic energy levels of which only the ground state is of much interest to us. If we vary the location of the nuclei relative to each other, the ground state electronic energy u_e varies. The (relative) nuclear configuration of a diatomic molecule can be represented by a single variable r, the internuclear distance. One can then use a curve (Fig. 8–1) to show how u_e depends on relative nuclear configuration. But for a polyatomic molecule, u_e is a function of $3n - 5$ variables for a linear molecule, or $3n - 6$ variables for a nonlinear molecule, where n is the number of nuclei in the molecule. This follows because, of the total of $3n$ coordinates needed to locate the n nuclei in space, three are used up on the position of the center of mass of the molecule and two (linear) or three (nonlinear) on the rotational orientation of the whole molecule in space. The remainder are the coordinates (vibrational degrees of freedom) having to do with the position of the nuclei *relative* to each other and are the only ones on which u_e depends. Thus if

$n \geq 3$, u_e is a function of at least three variables. We have to imagine then that u_e is represented by a surface in a space of at least four dimensions. If the molecule is a stable molecule, this u_e surface necessarily has a minimum (in some cases, several equivalent minima). The nuclear configuration corresponding to the minimum in u_e is the stable configuration of the molecule ($r = r_e$ in Fig. 8–1).

The ground state electronic energy u_e is used as the potential function for the nuclear motion (Born-Oppenheimer approximation). Hence the u_e surface is often called a potential surface. An exact calculation of the potential surface from quantum mechanics is too difficult mathematically, so an approximation, or more often, experimental information, is used to construct it (near the minimum).

The translational motion (3 degrees of freedom) of the center of mass of the molecule is separable and again leads to (Eq. 8–4)

$$q_t = \left[\frac{2\pi(\sum_i m_i)kT}{h^2}\right]^{3/2} V, \tag{9–1}$$

where the sum is over all atoms in the molecule.

The rotational motion (two or three degrees of freedom) is made separable by assuming that, as far as rotation is concerned, the molecule has the *rigid* structure corresponding to the minimum in the potential surface.

An analysis of the vibrational motion ($3n - 5$ or $3n - 6$ degrees of freedom) is rendered practical by use of a quadratic approximation [see for example, Eq. (V–1)] to the true potential surface at the minimum. Even with this approximation, a normal coordinate analysis of the vibrational motion is still required.

Because of the large separation in electronic energy levels, we shall be concerned with only the ground electronic state. Hence

$$q_e = \omega_{e1} e^{-\epsilon_{e1}/kT}. \tag{9–2}$$

For ordinary chemically saturated molecules, $\omega_{e1} = 1$. Probably the most convenient choice of the zero of energy is completely separated atoms at rest. If the depth of the potential minimum is D_e (>0) relative to this reference point, we put $\epsilon_{e1} = -D_e$, and choose the bottom of the potential surface as the zero of vibrational energy.

Having carried over the approximations of Section 8–1 to polyatomic molecules, we can apply Eqs. (8–1) through (8–3) and (8–7) through (8–13) here without modification.

Potential energy surfaces are also discussed in Section 11–1.

9–2 Vibration. Let us assign cartesian coordinates to each nucleus in the molecule, $x_1, y_1, z_1, \ldots, x_n, y_n, z_n$, using as respective origins the positions of the nuclei when the molecule is in its stable or equilibrium (bottom of u_e) configuration. For small vibrations about the equilibrium configuration, the Hamiltonian will have the form of Eq. (V-1), but extended to three dimensions. The potential energy u_e (relative to the minimum) will be a quadratic function of relative coordinates such as $x_2 - x_1$, $x_3 - x_1$, $y_2 - y_1$, etc. To make further progress we must perform a normal coordinate analysis as in Appendix V. This can be carried out for fairly complicated molecules. Symmetry in the molecular structure and simplifications in the assumed form of u_e help. Of the $3n$ normal coordinates, five (linear) or six (nonlinear) will turn out to be associated with motion of zero frequency, that is, with translation or rotation. The remaining $3n - 5$ or $3n - 6$ normal coordinates are connected with vibration (except for internal rotation—a complication discussed in Section 9-5 and not present in the molecules under consideration up to that section).

Corresponding to the diagrams (V-10) and (V-11) of normal modes in one dimension, the following are simple schematic examples in three dimensions:

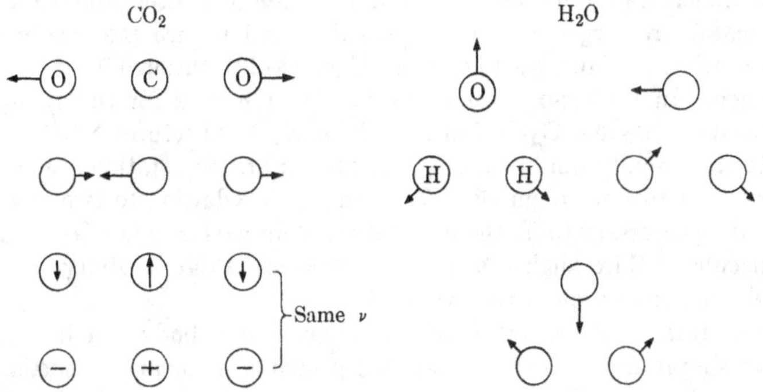

The + and − signs refer to displacements perpendicular to the paper.

Each normal mode of vibration makes an independent contribution to thermodynamic functions such as E, C_V, S, etc., as we have already seen in Section 5-2 in discussing crystals. Incidentally, we remind the reader that the vibration problems in molecules and in crystals do not differ at all in principle. In practice, (a) the force constants in molecules are usually larger, and (b) the number of atoms in crystals is so large that we can ignore (i) translational and rotational degrees of freedom and (ii) edge effects.

If the normal frequencies $\nu_1, \nu_2, \ldots, \nu_{n'}$, where $n' = 3n - 5$ or $3n - 6$, are known, we have immediately (see Sections 5-2 and 8-2) equations such as

$$q_v = \prod_{i=1}^{n'} \frac{e^{-\Theta_i/2T}}{1 - e^{-\Theta_i/T}}, \tag{9-3}$$

$$E_v = Nk \sum_{i=1}^{n'} \left(\frac{\Theta_i}{2} + \frac{\Theta_i}{e^{\Theta_i/T} - 1} \right), \tag{9-4}$$

$$C_{V_v} = Nk \sum_{i=1}^{n'} \left[\left(\frac{\Theta_i}{T} \right)^2 \frac{e^{\Theta_i/T}}{(e^{\Theta_i/T} - 1)^2} \right], \tag{9-5}$$

where $\Theta_i = h\nu_i/k$.

In principle, u_e, and hence the ν_i, can be calculated by quantum mechanics. In practice, the ν_i for polyatomic molecules must always be deduced empirically from vibrational (infrared and Raman) spectra.

Numerical examples will be included in Section 9–4.

9–3 Rotation. If the equilibrium configuration (minimum in u_e) of a polyatomic molecule is linear, as for example in CO_2, C_2H_2, N_2O, etc., the rotational problem is exactly that already discussed in Section 8–3. The molecules are heavier here, the moments of inertia (about the center of mass) are larger, and the classical equations are practically always appropriate. Thus we take over Eqs. (8–29) through (8–33) without change. In the examples mentioned above, $\sigma = 2$ for the symmetrical molecules CO_2 and C_2H_2, but $\sigma = 1$ for N_2O (structure: NNO).

If the equilibrium structure is nonlinear, three rotational coordinates instead of two are required. For example, in addition to two angles, say θ and φ, necessary to fix the orientation of an axis in space (as for a linear molecule), a third angle is also needed because of the possibility of rotation of the molecule around the axis itself.

The rotational properties of a nonlinear rigid body can be expressed most simply in terms of the so-called principal moments of inertia, which we now define. First, we locate the center of mass of the equilibrium configuration of the molecule. To do this, we use the following definition: if the center of mass is chosen as origin of cartesian coordinates for all nuclei in the molecule, then (Problem 9–1)

$$\sum_i m_i x_i = \sum_i m_i y_i = \sum_i m_i z_i = 0. \tag{9-6}$$

Now consider any straight line passing through the center of mass. A moment of inertia can be calculated about this line: $I = \sum_i m_i d_i^2$, where d_i is the perpendicular distance of the mass m_i from the line. On this line, mark off a distance on both sides of the center of mass proportional to $I^{-1/2}$ calculated about the line. Now choose other lines through the center of mass, and on each mark off distances proportional to (same proportionality constant) $I^{-1/2}$ about the line. The locus of the marks is an ellipsoid,* called the ellipsoid of inertia. The longest marked-off line is one of the principal axes of the ellipsoid, and the moment of inertia about this line (that is, the minimum I) is called a principal moment of inertia. The shortest line is also a principal axis, and the associated I (the maximum I) is a principal moment. The third principal moment is that about the line perpendicular to both the lines just referred to. Let us call the three principal moments of inertia I_A, I_B, and I_C. If all three moments are equal, the molecule is called a spherical top (e.g., CH_4); if only two are equal, the molecule is a symmetrical top (e.g., C_6H_6 and $CHCl_3$); if all three moments are different, the molecule is an unsymmetrical top (e.g., H_2O).

The rotational energy levels for the most general case (an unsymmetrical top) are very complicated. Hence a general quantum-statistical treatment is awkward on this account, as well as on account of quantum symmetry considerations (as with H_2) for some molecules (see Chapter 22). Fortunately, a classical treatment is legitimate except in the case of very light (hydrogen-containing) gases at low temperatures (e.g., CH_4 below 80°K). We consider the classical case only. Now even the classical Hamiltonian† for a rigid unsymmetrical rotator is a little too complicated to derive here (since this is not a book on mechanics). We confine ourselves, therefore, to the statement that insertion of the classical Hamiltonian into the classical phase integral leads, after a rather straightforward integration,† to (Problem 9-2)

$$q_r = \frac{\pi^{1/2}}{\sigma}\left(\frac{8\pi^2 I_A kT}{h^2}\right)^{1/2}\left(\frac{8\pi^2 I_B kT}{h^2}\right)^{1/2}\left(\frac{8\pi^2 I_C kT}{h^2}\right)^{1/2}. \tag{9-7}$$

As before, the symmetry number σ has been introduced in Eq. (9-7) to correct for repeated counting of indistinguishable configurations in the classical phase integral. The symmetry number is the number of different ways in which the molecule can achieve, by rotation, the same (i.e., counting like atoms as indistinguishable) orientation in space. For example, we can see that $\sigma = 12$ for CH_4 (tetrahedral symmetry) as follows. If we imagine, say, that the tetrahedron is sitting on a table with one hydrogen up and three down, then there are *three* ways of arranging the "down" hydrogens (rotations of 0°, 120°, 240°) while keeping the same hydrogen up. But there are *four* possibilities for the "up" hydrogen.

---

* For proof, see Chapter 5 of H. Goldstein, *Classical Mechanics*. Reading, Mass.: Addison-Wesley, 1950.

† See Mayer and Mayer, pp. 191–194.

Therefore, $\sigma = 3 \times 4$. Similarly, $\sigma = 2$ for H_2O, $\sigma = 3$ for NH_3, $\sigma = 4$ for C_2H_4, and $\sigma = 12$ for C_6H_6.

Equation (9–7) can also be written, in obvious notation,

$$q_r = \frac{\pi^{1/2}}{\sigma}\left(\frac{T^3}{\Theta_A\Theta_B\Theta_C}\right)^{1/2}. \tag{9-8}$$

The rotational contributions to some of the thermodynamic functions of nonlinear molecules are then:

$$A_r = -NkT \ln \frac{\pi^{1/2}}{\sigma}\left(\frac{T^3}{\Theta_A\Theta_B\Theta_C}\right)^{1/2}, \tag{9-9}$$

$$E_r = \tfrac{3}{2}NkT, \tag{9-10}$$

$$C_{Vr} = \tfrac{3}{2}Nk, \tag{9-11}$$

$$S_r = Nk \ln \frac{\pi^{1/2}}{\sigma}\left(\frac{T^3e^3}{\Theta_A\Theta_B\Theta_C}\right)^{1/2}. \tag{9-12}$$

Here the rotational (kinetic) energy is $3NkT/2$ instead of $2NkT/2$, as for linear molecules, because of three instead of two rotational degrees of freedom. We do not need to know I_A, I_B, and I_C to evaluate E_r, C_{Vr}, etc., but these quantities are required for the entropy and any thermodynamic function which involves the entropy (e.g., μ, A, etc.). The moments of inertia can be computed if the molecular structure is known. This structure (i.e., the nuclear configuration at the minimum in u_e) is too difficult to deduce by a quantum-mechanical calculation, so recourse must be taken to experimental information (rotational spectrum, electron and x-ray diffraction, etc.).

9–4 Thermodynamic functions. We first bring together the various contributions to the main thermodynamic functions. For linear polyatomic molecules, Eqs. (8–34) through (8–38) apply, except that $m_1 + m_2$ must be replaced by $\sum_i m_i$, and each term involving the vibrational frequency ν must be replaced by a sum of similar terms over $3n - 5$ vibrational frequencies. For nonlinear molecules,

$$-\frac{A}{NkT} = \ln\left[\frac{2\pi(\sum_i m_i)kT}{h^2}\right]^{3/2}\frac{Ve}{N} + \ln\frac{\pi^{1/2}}{\sigma}\left(\frac{T^3}{\Theta_A\Theta_B\Theta_C}\right)^{1/2} - \sum_{i=1}^{3n-6}\left[\frac{h\nu_i}{2kT} + \ln\left(1 - e^{-h\nu_i/kT}\right)\right] + \frac{D_e}{kT} + \ln \omega_{e1}; \tag{9-13}$$

$$\frac{E}{NkT} = \frac{3}{2} + \frac{3}{2} + \sum_{i=1}^{3n-6} \left(\frac{h\nu_i}{2kT} + \frac{h\nu_i/kT}{e^{h\nu_i/kT} - 1} \right) - \frac{D_e}{kT}; \tag{9-14}$$

$$\frac{C_V}{Nk} = \frac{3}{2} + \frac{3}{2} + \sum_{i=1}^{3n-6} \left(\frac{h\nu_i}{kT} \right)^2 \frac{e^{h\nu_i/kT}}{(e^{h\nu_i/kT} - 1)^2}; \tag{9-15}$$

$$\frac{S}{Nk} = \ln \left[\frac{2\pi(\sum_i m_i)kT}{h^2} \right]^{3/2} \frac{Ve^{5/2}}{N} + \ln \frac{\pi^{1/2}e^{3/2}}{\sigma} \left(\frac{T^3}{\Theta_A \Theta_B \Theta_C} \right)^{1/2}$$

$$+ \sum_{i=1}^{3n-6} \left[\frac{h\nu_i/kT}{e^{h\nu_i/kT} - 1} - \ln (1 - e^{-h\nu_i/kT}) \right] + \ln \omega_{e1}; \tag{9-16}$$

$$pV = NkT. \tag{9-17}$$

The chemical potential follows immediately from $\mu = (A + pV)/N$ (Problem 9-3).

We now turn to examples of applications of the equations for C_V/Nk. Aside from the constants 5/2 (linear) and 6/2 (nonlinear), the only contributions come from the sum over frequencies in Eq. (9-15). We use frequencies obtained from infrared and Raman spectra.

(1) For CO_2, the four values of $\Theta_i = h\nu_i/k$ are 1890, 3360, 954, and 954°K. The respective vibrational terms in Eq. (9-15) are then calculated to be 0.086, 0.000, 0.483, and 0.483 at 312°K, or a total of 1.05. Hence C_V/Nk at 312°K is $2.50 + 1.05 = 3.55$. The experimental value is 3.53.

(2) For BF_3, the Θ_i are 1270, 995, 2070, 2070, 631, and 631°K. The respective terms in Eq. (9-15), at 278°K, are 0.221, 0.378, 0.033 (×2), and 0.662 (×2). The total is 1.99. Hence C_V/Nk is 4.99. The experimental value is 4.89. At 189°K, we calculate from the Θ_i that $C_V/Nk = 5.05$, whereas the experimental value is 5.04.

TABLE 9-1

HEAT CAPACITY OF N_2O

T, °K	C_V/Nk (calc.)	C_V/Nk (expt.)
203	3.06	3.10
293	3.61	3.61
390	4.08	4.05
467	4.49	4.46
625	4.89	4.90
733	5.15	5.15

(3) For N_2O, the Θ_i are 850, 850, 1840, and 3200°K. Table 9–1 compares with experiment, over a considerable temperature range, values of C_V calculated from these vibrational frequencies.

The agreement between theory and experiment in the above examples is very good. Furthermore, this kind of agreement is found quite generally. However, the above treatment of the vibrational degrees of freedom is *not* adequate for molecules with internal rotation, such as ethane. We return to this subject in the next section.

The entropy provides a somewhat more severe test of the theory than the heat capacity, since the rotational and translational contributions to the entropy are not just constants. Table 9–2 compares values of the molar entropy at 1 atmosphere pressure for a number of gases, obtained in two ways: (a) the "spectroscopic entropy" calculated from Eqs. (4–21) (monatomic), (8–37) (diatomic or linear polyatomic), or (9–16) (nonlinear polyatomic), using spectroscopic data as the source of the moments of inertia, vibrational frequencies, and ground state electronic degeneracies; and (b) the "calorimetric entropy" calculated from experimental heat capacities and heats of phase transitions. That is, we are using in (a) the statistical-mechanical relation $S = k \ln \Omega$, and in (b) the thermodynamic equation $S = g(T)$ (Eq. 2–35). Both these expressions relate to the same zero of entropy, as explained at the end of Section 2–4, and should lead to results which agree with each other. This is seen to be the case in Table 9–2. Three comments about this table should be made: (1) If the real gas is not effectively ideal at 1 atm pressure and the temperature indicated, the calorimetric value of

TABLE 9–2

ENTROPY OF GASES AT 1 ATMOSPHERE PRESSURE

Gas	T, °K	S_{spect}, cal·deg^{-1}·mole^{-1}	S_{cal}, cal·deg^{-1}·mole^{-1}
A	298.1	37.0	36.4
Cd	298.1	40.1	40.0
Zn	298.1	38.5	38.4
Hg	298.1	41.8	41.3
N_2	298.1	45.8	45.9
O_2	298.1	49.0	49.1
HCl	298.1	44.6	44.5
HBr	298.1	47.5	47.6
NH_3	239.7	44.1	44.1
CO_2	194.7	47.5	47.6
CH_3Br	276.7	58.0	57.9

S has been corrected for gas imperfection by use of equation-of-state data (the correction is usually less than 0.3 cal·deg^{-1}·mole^{-1}). (2) The spectroscopic values of S for the diatomic molecules are slightly more accurate than would follow from Eq. (8–37), since the actual spectroscopic vibration-rotation energy levels have been used (see the last paragraph of Section 8–4). (3) The spectroscopic method is capable of greater accuracy than the calorimetric method, e.g., the deviations in Table 9–2 for A and Hg must be attributed to experimental error in the calorimetric work.

There are, however, a few molecules for which a discrepancy between S_{spect} and S_{cal} is noted. Some of these are included in Table 9–3. In every case, $S_{\text{cal}} < S_{\text{spect}}$. The explanation for CO, for example, is generally accepted to be the following. S_{spect} is calculated from the number of quantum states Ω of the gas and should therefore be considered the true entropy of the gas (relative to $\Omega = 1$ as zero). Solid CO, in its true equilibrium state at 0°K, would have each CO molecule in some definite orientation (corresponding to the lowest possible energy) and $\Omega = 1$. However, the CO molecule is effectively very symmetrical. It has a very small dipole moment and is isoelectronic with N_2. Hence, in actually preparing the crystal at low temperatures, a random mixture of the two orientations CO and OC, rather than one particular orientation, is frozen into the crystal at each molecular position. Thus the crystal is in a metastable state, and the rate of the process metastable state (mixed orientation) $\rightarrow$ stable state (definite orientation) is negligibly small at low temperatures. The metastable crystal has $\Omega = 2^N$ at 0°K, instead of $\Omega = 1$. Since the calorimetric measurements are made on the metastable crystal, we have to write [see Eqs. (2–42) and (2–43)]

$$S(T) = S(0) + g(T) = S(0) + S_{\text{cal}},$$

where $S(0) = k \ln 2^N$ instead of $S(0) = k \ln 1 = 0$, as usual. If $R \ln 2$ is added to S_{cal} for CO in Table 9–3, the agreement with S_{spect} becomes satisfactory.

TABLE 9–3

ENTROPY OF GASES AT 1 ATMOSPHERE PRESSURE

Gas	T, °K	S_{spect}, cal·deg^{-1}·mole^{-1}	S_{cal}, cal·deg^{-1}·mole^{-1}	$S_{\text{spect}} - S_{\text{cal}}$	
CO	298.1	47.3	46.2	1.1	$R \ln 2 = 1.4$
NNO	184.6	48.5	47.4	1.1	$R \ln 2 = 1.4$
CH_3D	99.7	39.5	36.7	2.8	$R \ln 4 = 2.7$
H_2O	298.1	45.1	44.3	0.8	$R \ln (3/2) = 0.8$
D_2O	298.1	46.7	45.9	0.8	$R \ln (3/2) = 0.8$

The explanation for NNO is presumed to be the same as for CO. In the case of CH_3D, there is an obvious four-fold effective symmetry of orientation, so that the discrepancy is $R \ln 4$. The argument for H_2O and D_2O is rather more involved,* and we omit it. The conclusion is that the discrepancy should be $R \ln (3/2)$. In all these cases, satisfactory agreement with experiment is achieved after the correction is made.

The next chapter, on chemical equilibrium in ideal gases, will provide tests of our theoretical expressions for the chemical potential of ideal gases (monatomic, diatomic, and polyatomic).

9-5 Hindered internal rotation in ethane. There are many molecules, especially hydrocarbons, with internal rotation about bonds. Ethane is a relatively simple prototype, and we confine our rather qualitative discussion to this case. One of the 18 normal modes of vibration in ethane corresponds to a torsional oscillation of one methyl group relative to the other. This mode is neither Raman nor infrared active, so direct spectroscopic information about the frequency is unobtainable. However, if we measure the heat capacity of ethane and subtract the contributions to the heat capacity of the translational, (rigid) rotational and 17 spectroscopically available vibrational degrees of freedom, the remainder must be the contribution C of the torsional mode. This calculation of C can be carried out at several temperatures. Values of C obtained in this way follow the solid part of curve 1, Fig. 9-1, rather closely. The one-dimensional Einstein heat-capacity curve which comes nearest to fitting these data over the whole temperature range is curve 2 in Fig. 9-1 ($h\nu/k = 350°K$), but the fit is not at all good. We have to conclude,

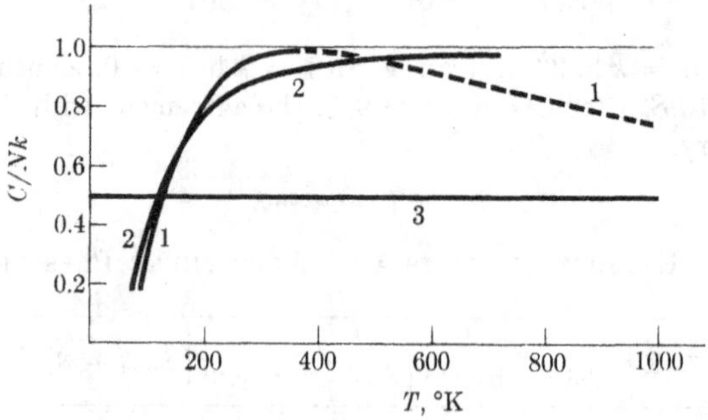

FIG. 9-1. Contribution of torsional mode to heat capacity of ethane. Curve 1: restricted rotator with $V_0 = 3100 \text{ cal}\cdot\text{mole}^{-1}$. Curve 2: simple harmonic oscillator with $h\nu/k = 350°K$. Curve 3: free rotator.

* See Fowler and Guggenheim, pp. 214–215.

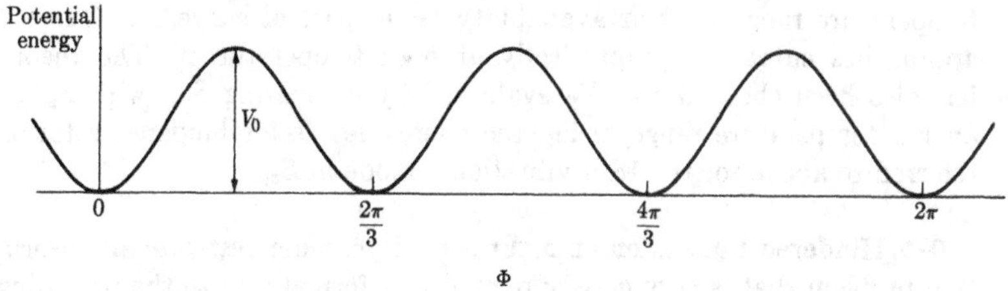

FIG. 9-2. Potential energy hindering rotation in ethane.

then, that this degree of freedom is not a simple harmonic (torsional) oscillation.

A second possibility is that the degree of freedom in question can be considered to be free rotation, about the C—C bond, of one methyl group with respect to the other. However, one free rotational degree of freedom at the temperatures in question would give the classical result $C/Nk = 1/2 =$ constant. This is shown as curve 3 in Fig. 9-1. Clearly this possibility must also be discarded.

The correct explanation has turned out to be intermediate between these two extremes. If Φ is the angle between one methyl group and the other, looking along the C—C bond and taking $\Phi = 0$ when the two methyl groups are "staggered" (not lined up or "eclipsed"), then the potential energy (ground-state electronic energy) of the molecule varies periodically with Φ. This variation can be represented approximately by

$$\tfrac{1}{2}V_0(1 - \cos 3\Phi), \tag{9-18}$$

as shown in Fig. 9-2. That is, the rotation of one methyl group relative to the other is not free, but is hindered by a periodic potential, with barrier V_0. However, when $kT \gg V_0$, the rotation is effectively free and C/Nk approaches the value 1/2. On the other hand, when $kT \ll V_0$, the system is trapped in one of the three minima in the potential, and the motion is simple harmonic (torsional) oscillation. In this limit, C/Nk approaches the low-temperature behavior of a one-dimensional Einstein heat-capacity curve (Eq. 5-14).

The following steps are necessary to take care of an arbitrary temperature [e.g., $kT = O(V_0)$]: (a) the one-dimensional Schrödinger equation in the coordinate Φ must be solved, using the potential (9-18), to give the energy levels; (b) the energy levels are inserted in the partition function $q_\Phi = \sum_j e^{-\epsilon_j/kT}$; and (c) C/Nk is computed from temperature derivatives of q_Φ, as usual. This work can be carried out numerically. Of course, the function $C(T)/Nk$ will be different for different choices of V_0. Curve 1 in Fig. 9-1 is actually the theoretical curve for $V_0 = 3100$ cal·mole^{-1}. This curve fits the experimental points over the temperature range of their availability (solid part of curve 1). Curve 1 approaches curve 3 asymptotically at high temperatures. The theory has also been checked and V_0 evaluated by comparing S_{cal} with S_{spect}, over a temperature range, using the energy levels for hindered rotation referred to above for the 18th vibrational mode in S_{spect}.

9–6 Hindered translation on a surface. Here we investigate an adsorption problem that is very closely related in a formal way to the preceding section. In Section 7–1 we discussed the two extreme kinds of monolayer adsorption: localized and mobile. Here we consider the transition from one kind to the other. Suppose that we have a simple square lattice of adsorption sites on the surface of a crystal, with nearest-neighbor distance a. The surface area is $\mathcal{a} = L^2 = Ma^2$, where M is the number of sites. We restrict the discussion to the case of a very dilute monolayer: $N \ll M$, where N is the number of adsorbed molecules. The xy-motion of an adsorbed molecule is assumed independent of the z-motion. The z-motion is taken care of by q_z, a one-dimensional harmonic oscillator partition function, as in Eqs. (7–3) and (7–18). We represent the potential energy function for the xy-motion (see Section 7–1 and Fig. 7–2), approximately, by

$$U_0(x, y) = U_{00} + \frac{1}{2} V_0\left(1 - \cos\frac{2\pi x}{a}\right) + \frac{1}{2} V_0\left(1 - \cos\frac{2\pi y}{a}\right). \quad (9\text{–}19)$$

The resemblance to (9–18) for hindered rotation in ethane is obvious. At low temperatures, the molecules vibrate about the minima in $U_0(x, y)$ with frequency (Problem 9–4)

$$\nu_x = \nu_y = \left(\frac{V_0}{2ma^2}\right)^{1/2}. \quad (9\text{–}20)$$

Equation (7–3) is applicable in this limit. At high temperatures, the motion is free translation and Eq. (7–17) is appropriate. At intermediate temperatures, Eq. (9–19) must be used as it stands. We have then to find the quantum-mechanical energy levels and partition function for the potential $U_0(x, y)$. This problem becomes the same as that in the preceding section, after a suitable correlation between the notations.

The exact treatment of this problem, which is complicated, was omitted in Section 9–5 and will be omitted here. We present instead a very accurate (over the entire temperature range) approximate solution which was introduced and tested against exact results by Pitzer and Gwinn for the hindered rotation problem. The approximation is to write the xy-partition function as

$$q_{xy} = q_{\text{class}} \times \frac{q_{\text{har osc—quant}}}{q_{\text{har osc—class}}}, \quad (9\text{–}21)$$

where q_{class} is the classical q_{xy} using the potential $U_0 - U_{00}$ in Eq. (9–19), $q_{\text{ho-q}}$ is the quantum harmonic oscillator partition function for motion about the minima in (9–19), and $q_{\text{ho-c}}$ is the classical limit of $q_{\text{ho-q}}$. It is easy to see that Eq. (9–21) has the right asymptotic properties:

$$T \to \infty, \qquad q_{\text{ho-q}} \to q_{\text{ho-c}}, \qquad q_{xy} \to q_{\text{class}},$$

$$T \to 0, \qquad q_{\text{class}} \to q_{\text{ho-c}}, \qquad q_{xy} \to q_{\text{ho-q}}.$$

The complete partition function for the system is

$$Q = \frac{1}{N!}\, q^N, \tag{9–22}$$

where

$$q = q_{xy} q_z e^{-U_{00}/kT}, \tag{9–23}$$

with q_{xy} given by Eq. (9–21).

Next we find q_{class}. We have

$$q_{\text{class}} = \frac{1}{h^2} \int_{-\infty}^{+\infty}\!\!\int\!\!\int_0^L\!\!\int e^{-H/kT}\, dx\, dy\, dp_x\, dp_y, \tag{9–24}$$

where

$$H = \frac{p_x^2}{2m} + \frac{p_y^2}{2m} + \frac{1}{2} V_0 \left(2 - \cos\frac{2\pi x}{a} - \cos\frac{2\pi y}{a}\right).$$

Then

$$q_{\text{class}} = \frac{2\pi mkT}{h^2}\, e^{-V_0/kT} \left[\int_0^L \exp\left(\frac{V_0}{2kT}\cos\frac{2\pi x}{a}\right) dx\right]^2.$$

If we put $u = V_0/2kT$ and $\theta = 2\pi x/a$,

$$q_{\text{class}} = \frac{2\pi mkT}{h^2}\, e^{-2u} \left[\frac{L}{2\pi}\int_0^{2\pi} e^{u\cos\theta}\, d\theta\right]^2.$$

The integral is equal to $2\pi I_0(u)$, where $I_0(u)$ is a modified Bessel function of the first kind. Hence

$$q_{\text{class}} = \frac{2\pi mkT}{h^2}\, \mathcal{a} e^{-2u} I_0^2(u). \tag{9–25}$$

We can obtain $q_{\text{ho-c}}$ easily from q_{class}, using the limit $T \to 0$ (i.e., $u \to \infty$):

$$q_{\text{ho-c}} = q_{\text{class}}(u \to \infty) = M \left(\frac{kT}{h\nu_x}\right)^2, \tag{9–26}$$

where ν_x is given by Eq. (9–20) and we have used the property

$$\lim_{u\to\infty} I_0(u) = \frac{e^u}{(2\pi u)^{1/2}}.$$

The factor M in Eq. (9–26) is to be expected, since there are M sites in the area $\mathcal{A}$ over which the phase integral in Eq. (9–24) is extended. Finally, in view of Eqs. (9–26), (5–6), and (5–7), we must have

$$q_{\text{ho-q}} = M\left(\frac{e^{-h\nu_x/2kT}}{1 - e^{-h\nu_x/kT}}\right)^2. \tag{9–27}$$

We can now substitute Eqs. (9–25) through (9–27) into Eq. (9–21). We obtain

$$q_{xy} = \frac{2\pi M u e^{-2u} e^{-2Ku} I_0^2(u)}{(1 - e^{-Ku})^2}, \tag{9–28}$$

where

$$K = \left(\frac{2h^2}{ma^2 V_0}\right)^{1/2}.$$

The adsorption isotherm is linear here (since the adsorbed phase is very dilute), but the proportionality constant between N and p (Eq. 7–20) is a function of temperature:

$$N = (q_{xy} q_z e^{-U_{00}/kT} e^{\mu^0/kT})p. \tag{9–29}$$

Other thermodynamic properties can be deduced from Eq. (9–22) in

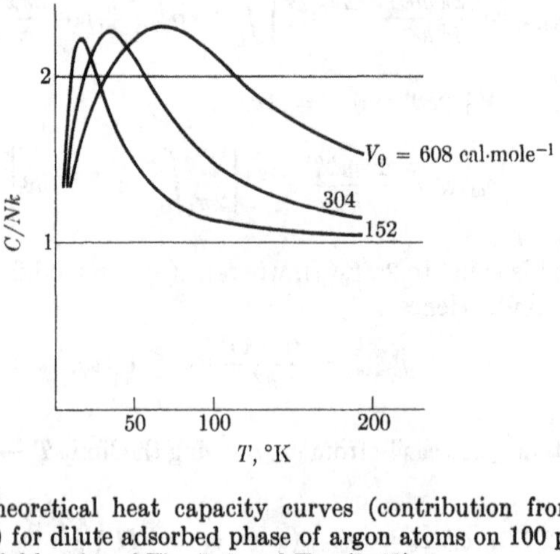

FIG. 9–3. Theoretical heat capacity curves (contribution from motion in plane of surface) for dilute adsorbed phase of argon atoms on 100 plane of KCl. V_0 is the potential barrier of Fig. 7–2 and Eq. (9–19).

the usual way. For example, one finds that the xy-contribution to the heat capacity (constant N and $\mathcal{A}$) is

$$\frac{C}{Nk} = -1 + 2u\left[u - \frac{I_1(u)}{I_0(u)} - u\frac{I_1^2(u)}{I_0^2(u)}\right] + \frac{2K^2u^2e^{-Ku}}{(1 - e^{-Ku})^2}. \quad (9\text{–}30)$$

This equation is illustrated in Fig. 9–3 for argon atoms on a KCl 100-plane. V_0 is probably about 200 cal·mole^{-1}. Experimental curves of this type have not been observed as yet. In general, the maximum in the heat-capacity curve is predicted to occur at about $kT = V_0/5$. These curves belong to the same family as curve 1 in Fig. 9–1.

Problems

9–1. Locate the center of mass of the linear molecule HCN, given that the equilibrium internuclear distances are 1.157 A for CN and 1.059 A for HC. Calculate the moment of inertia about the center of mass. Carry out the same calculations for DCN. (Page 164.)

9–2. For a spherical top, the rotational energy levels are

$$\epsilon_j = \frac{j(j+1)h^2}{8\pi^2 I_A}, \qquad j = 0, 1, 2, \ldots,$$

with degeneracy $\omega_j = (2j+1)^2$. Ignore molecular symmetry, replace the sum in q_r by an integral, and verify Eq. (9–7) for the case $I_A = I_B = I_C$. (Page 165.)

9–3. Derive expressions for μ and $\mu^0(T)$ for nonlinear polyatomic molecules. (Page 167.)

9–4. Derive Eq. (9–20) for ν_x from the potential (9–19). (Page 172.)

9–5. H_2O has an OH distance 0.958 A and an HOH bond angle of 104°27′. Find the center of mass and calculate I_A, I_B, and I_C (see Problem 9–11).

9–6. CH_4 has a tetrahedral structure with a CH distance 1.094 A. Calculate $I_A = I_B = I_C$.

9–7. Show that in localized adsorption, in the limit as $N/M \to 0$,

$$Q = \frac{M!q(T)^N}{N!(M-N)!} \to \frac{1}{N!}(Mq)^N.$$

Equation (9–22) with (9–26) is an example. The equation of state is $\varphi\mathcal{A} = NkT$, in view of Problems 3–1 and 3–6. This is verified by Eq. (7–7).

9–8. Calculate ν_x from Eq. (9–20), using $V_0 = 200$ cal·mole^{-1}, $a = 4$ A and m for argon.

9–9. From Eq. (9–28), derive Eq. (9–30) for C/Nk.

9–10. In CO_2, the CO distance is 1.161 A. The vibrational characteristic temperatures are $\Theta_i = 1890, 3360, 954$, and 954°K. Calculate the entropy of CO_2 per mole at 1 atm pressure and 194.7°K (see Table 9–2).

9–11. The normal vibrational frequencies for H_2O are 3650, 1590, and 3760 in cm^{-1}. The moments of inertia are 1.024, 1.921, and 2.947×10^{-40} gm·cm^2. Calculate S and μ per mole at 1 atm pressure and 25°C (see Table 9–3).

9–12. Use the data given in the text to calculate C_V for N_2O at 733°K (see Table 9–1).

Supplementary Reading

FOWLER and GUGGENHEIM, Chapters 3 and 5.

HERZBERG, G., *Infrared and Raman Spectra of Polyatomic Molecules.* New York: Van Nostrand, 1945.

MAYER and MAYER, Chapter 8.

RUSHBROOKE, Chapter 9.

CHAPTER 10

CHEMICAL EQUILIBRIUM IN IDEAL GAS MIXTURES

In previous chapters we have shown how the thermodynamic properties of ideal monatomic, diatomic, and polyatomic gases can be calculated by statistical-mechanical methods. An important application of this work is to chemical equilibria occurring in ideal gas mixtures. The object is to deduce equilibrium constants for such reactions, using spectroscopic information about the individual molecules. Equilibrium constants obtained in this way are often more accurate than those found by direct measurement. Chemical equilibria in imperfect gases will be considered in Chapter 15.

10–1 General relations. The statistical-mechanical basis of the second law of thermodynamics was examined in Section 2–3. In particular, we considered the second law in the form $S \rightarrow$ maximum (isolated system), $A \rightarrow$ minimum (N, V, T constant), $F \rightarrow$ minimum (N, p, T constant), etc. Since from these statements of the second law, by thermodynamic arguments, one can deduce results such as $\Delta\mu = 0$ for phase or chemical equilibria, we may take the position that the equilibrium criterion $\Delta\mu = 0$ has been given a statistical foundation. We have already adopted this attitude, for example, in studying the vapor pressure of a Debye crystal (Section 5–3) or of an adsorbed phase (Chapter 7). However, because of the importance of chemical equilibria, and for variety, we deviate from this policy here on two counts: (a) In this section we give the thermodynamic derivation of $\Delta\mu = 0$ for a homogeneous (i.e., one-phase) chemical reaction from the equilibrium criterion $F \rightarrow$ minimum (N, p, T constant). (b) In the next section we verify the thermodynamic result in (a) for a special case, using a purely statistical argument.

We now turn to the derivation of $\Delta\mu = 0$ for a chemical reaction. Consider a closed one-phase system, at equilibrium, at pressure p, temperature T, and containing the numbers of molecules N_A, N_B, N_C, N_1, N_2, ... The system may be a gas, liquid, or solid. The equilibrium of the system includes equilibrium with respect to a chemical reaction, say

$$\nu_A A + \nu_B B \rightleftarrows \nu_C C, \qquad (10\text{–}1)$$

where ν_A, ν_B, and ν_C are small integers. The species 1, 2, ... are present but do not participate in the reaction. The Gibbs free energy of this multicomponent system is

$$F = N_A\mu_A + N_B\mu_B + N_C\mu_C + N_1\mu_1 + N_2\mu_2 + \cdots .$$

Now consider the infinitesimal process of converting, say, $\nu_A\, d\xi$ molecules of A and $\nu_B\, d\xi$ molecules of B into $\nu_C\, d\xi$ molecules of C. Since the process is an infinitesimal one, none of the intensive properties of the system will change appreciably and hence all the μ's are constant during the process. The change in F in the process is therefore

$$\begin{aligned} dF &= \mu_A\, dN_A + \mu_B\, dN_B + \mu_C\, dN_C \\ &= (-\nu_A\mu_A - \nu_B\mu_B + \nu_C\mu_C)\, d\xi \equiv \Delta\mu\, d\xi. \end{aligned}$$

No terms in μ_1, μ_2, etc., appear here, since N_1 = constant, etc. Since the process occurs in a closed system (molecules of A, B, and C are interconverted but no molecules go in or out of the system) at constant p and T, and the system is at equilibrium, we must have, according to the second law, $dF = 0$. We therefore conclude, since $d\xi$ is arbitrary, that $\Delta\mu = 0$ or

$$\nu_A\mu_A + \nu_B\mu_B = \nu_C\mu_C. \tag{10–2}$$

The chemical potentials here must be assigned the values they have after chemical equilibrium has been achieved. This result is general; it is *not* restricted to an ideal gas mixture, or even to gaseous reactions.

Now let us apply Eq. (10–2) to the special case in which we are interested in this chapter—a chemical reaction occurring in an ideal gas mixture. First we note that the expression for the chemical potential of each species i, in terms of N_i, V, and T, is the same as if only that species existed in the system (this is true, of course, only for an ideal gas mixture). To see this, consider, say, a binary system. Then from Eq. (3–15) [see also Eq. (4–27)],

$$Q(N_1, N_2, V, T) = \frac{q_1(V, T)^{N_1}}{N_1!} \cdot \frac{q_2(V, T)^{N_2}}{N_2!}, \tag{10–3}$$

where q_1 and q_2 may refer to monatomic, diatomic, or polyatomic molecules. We find easily from

$$\left(\frac{\partial \ln Q}{\partial N_1}\right)_{N_2, V, T} = -\frac{\mu_1}{kT}$$

that

$$-\frac{\mu_1}{kT} = \ln \frac{q_1(V, T)}{N_1}, \tag{10–4}$$

just as for a one-component gas.

If the reaction (10–1) occurs between molecules A, B, and C in an ideal gas mixture, the equilibrium condition is (10–2), where each chemical potential is given by an equation of the form (10–4). Substituting Eq.

(10–4) in (10–2), we find

$$\frac{N_C^{\nu_C}}{N_A^{\nu_A} N_B^{\nu_B}} = \frac{q_C^{\nu_C}}{q_A^{\nu_A} q_B^{\nu_B}}. \tag{10–5}$$

The N's in this equation refer, of course, to the numbers of molecules present after the chemical equilibrium has been established. Now, each q is equal to V multiplied by a function of temperature only [see Problems 3–1 and 3–6 and Eqs. (4–10), (8–4) and (9–1)], so that q/V is a function of temperature only. Therefore, the quotient of q's in Eq. (10–5) is in general a function of both V and T, but we have

$$\frac{\rho_C^{\nu_C}}{\rho_A^{\nu_A} \rho_B^{\nu_B}} = \frac{(q_C/V)^{\nu_C}}{(q_A/V)^{\nu_A}(q_B/V)^{\nu_B}} = K(T), \tag{10–6}$$

where $\rho_C = N_C/V$, etc., and $K(T)$ is a *function of temperature only*, called the equilibrium constant.

As the partial pressures are $p_i = \rho_i kT$, we also have

$$\frac{p_C^{\nu_C}}{p_A^{\nu_A} p_B^{\nu_B}} = \frac{(q_C kT/V)^{\nu_C}}{(q_A kT/V)^{\nu_A}(q_B kT/V)^{\nu_B}}$$

$$= (kT)^{\nu_C - \nu_A - \nu_B} K(T) = K_p(T), \tag{10–7}$$

which is, in general, a different equilibrium constant, often used.

In thermodynamics, one substitutes

$$\mu_i = \mu_i^0(T) + kT \ln p_i$$

for each component in Eq. (10–2) and obtains

$$\frac{p_C^{\nu_C}}{p_A^{\nu_A} p_B^{\nu_B}} = e^{-\Delta F^0/kT} = K_p(T), \tag{10–8}$$

where

$$\Delta F^0(T) \equiv \nu_C \mu_C^0(T) - \nu_A \mu_A^0(T) - \nu_B \mu_B^0(T). \tag{10–9}$$

This is called the "standard free energy change" for the reaction. Equations (10–7) and (10–8) are, of course, equivalent and are interrelated by

$$\frac{\mu_i^0(T)}{kT} = -\ln \frac{q_i kT}{V} \tag{10–10}$$

for each participant in the reaction [see, for example, Eqs. (4–26) and (8–40)].

In Eqs. (10–6) and (10–7), or indeed in *any* application of $\Delta\mu = 0$, it is always understood that the zeros of energy chosen for the different molecular species (or states, in a phase equilibrium) involved must be self-consistent. For example, one can choose as zero for each molecule the separated atoms of which the molecule is composed, at rest. Since the same atoms occur on both sides of the chemical equation (10–1), this choice necessarily leads to the same zero for reactants and products. A number of specific examples will be found in Section 10–4.

10–2 Statistical derivation in a special case. In this section we derive Eq. (10–5) for a simple reaction, using a statistical argument which bypasses the thermodynamic equation (10–2) (but does use the second law of thermodynamics). We could use the more complicated reaction (10–1), but there would be no particular advantage in this since we merely want to illustrate the equivalence of the two approaches.

Consider an ideal gas mixture made up of N_A molecules of type A and N_B molecules of type B in a closed container with fixed V and T. We assume that in the absence of a catalyst, no chemical reaction takes place between A and B molecules. The canonical ensemble partition function for this binary system is

$$Q(N_A, N_B, V, T) = \frac{q_A^{N_A} q_B^{N_B}}{N_A! N_B!}, \tag{10–11}$$

and the Helmholtz free energy is $A = -kT \ln Q$. Suppose a catalyst is added, making possible the chemical reaction $A \rightleftarrows 2B$, which will take place spontaneously until equilibrium is established with respect to the reaction. Since the system is closed and has constant V and T, we know from the second law of thermodynamics that as the reaction proceeds the Helmholtz free energy will decrease, and that at equilibrium this free energy will have its minimum possible value consistent with V, T, and with the fixed amount of material in the container. (We can specify this amount by stating that each A molecule contains two B units, and the total number of B units is $2N_A + N_B = N =$ constant.) Translating this into statistical-mechanical language, we can say that the equilibrium point of the reaction for given V and T will correspond to the maximum possible value of Q (since $A = -kT \ln Q$) consistent with the restraint $2N_A + N_B = N =$ constant.

We need, then, to maximize Q in Eq. (10–11) subject to $2N_A + N_B = N$. We could use an undetermined multiplier, but this is unnecessary here because the restraint can be used instead to eliminate, say, N_B from Eq. (10–11) (see Appendix III). We have

$$\begin{aligned}\ln Q = N_A \ln q_A + (N - 2N_A) \ln q_B - N_A \ln N_A \\ - (N - 2N_A) \ln (N - 2N_A) + N - N_A,\end{aligned}$$

and hence

$$\left(\frac{\partial \ln Q}{\partial N_A}\right)_{N,V,T} = 0 = \ln \frac{q_A(N - 2N_A^*)^2}{q_B^2 N_A^*},$$

where N_A^* is the value of N_A making Q a maximum—that is, the value of N_A at equilibrium. Therefore,

$$\frac{N_B^{*2}}{N_A^*} = \frac{q_B^2}{q_A}, \tag{10–12}$$

which is the same as Eq. (10–5) for this special case.

10–3 Fluctuations in a simple chemical equilibrium. Here we consider a situation very similar to that in the preceding section: we have a binary gas mixture with numbers of molecules N_A and N_B, in a volume V at temperature T. The reaction $A \rightleftarrows B$ is then made possible by introduction of a catalyst. The system proceeds to equilibrium with V, T, and $N_A + N_B = N$ held constant.

We adopt a slightly more general point of view here than in the preceding section. That is, we use the partition function $Q(N, V, T)$ for the system described above, including *all* possible states of the system for given N, V, T; that is, including all possible values of N_A and N_B consistent with any assigned value of N. Thermodynamically, this is a one-component system (N). Then

$$Q(N, V, T) = \sum_{\substack{N_A, N_B \\ (N_A + N_B = N)}} \frac{q_A^{N_A} q_B^{N_B}}{N_A! N_B!} \tag{10–13}$$

$$= \frac{(q_A + q_B)^N}{N!}. \tag{10–14}$$

The treatment in the preceding section corresponds to the use of only the maximum term in the sum (10–13). With the retention of the full sum, we can easily investigate the extent to which we can expect to find fluctuations about the equilibrium concentrations predicted by equations such as (10–5) and (10–12).

The average value of N_A is clearly

$$\overline{N}_A = \frac{1}{Q} \sum_{\substack{N_A, N_B \\ (N_A + N_B = N)}} \frac{N_A q_A^{N_A} q_B^{N_B}}{N_A! N_B!}. \tag{10–15}$$

If Q in Eq. (10–13) is regarded in a purely formal way as a function of N,

q_A, and q_B, we note from Eq. (10–15) that

$$\overline{N}_A = q_A \left(\frac{\partial \ln Q}{\partial q_A} \right)_{q_B, N} . \tag{10–16}$$

Then, from Eqs. (10–14) and (10–16),

$$\overline{N}_A = \frac{N q_A}{q_A + q_B}, \tag{10–17}$$

and therefore, since $\overline{N}_A + \overline{N}_B = N$,

$$\frac{\overline{N}_B}{\overline{N}_A} = \frac{q_B}{q_A}, \tag{10–18}$$

in agreement with Eq. (10–5). Now if we differentiate

$$Q\overline{N}_A = \sum_{\substack{N_A, N_B \\ (N_A + N_B = N)}} \frac{N_A q_A^{N_A} q_B^{N_B}}{N_A! N_B!}$$

with respect to q_A, we obtain

$$\overline{N_A^2} - (\overline{N}_A)^2 = q_A \left(\frac{\partial \overline{N}_A}{\partial q_A} \right)_{q_B, N} . \tag{10–19}$$

Finally, from Eqs. (10–17) and (10–19), we have

$$\overline{N_A^2} - (\overline{N}_A)^2 = \frac{\overline{N}_A \overline{N}_B}{N} = \overline{N_B^2} - (\overline{N}_B)^2. \tag{10–20}$$

Fluctuations about the equilibrium composition are therefore very small, as we might expect: $\sigma_{N_A}/\overline{N}_A = O(N^{-1/2})$, the usual result (see Section 2–1).

10–4 Examples of chemical equilibria. In this section we discuss five particular illustrations of the above equations.

(a) *Isomeric equilibrium.* This is the case $A \rightleftarrows B$ already considered as a fluctuation problem in the preceding section. We wish to examine the equilibrium relation (10–18) here. As an example, suppose that some of the energy levels of A are lower than those of B (the translational levels would be the same in the two cases, because A and B have the same mass —so these are not involved), but that the levels of B are closer together, as shown schematically in Fig. 10–1. This would arise, for example, if (a) the chemical bonds in A are more stable than in B (i.e., the minimum in the ground electronic energy surface is lower for A), (b) the vibrational force constants are in general larger in A than in B (i.e., the chemical

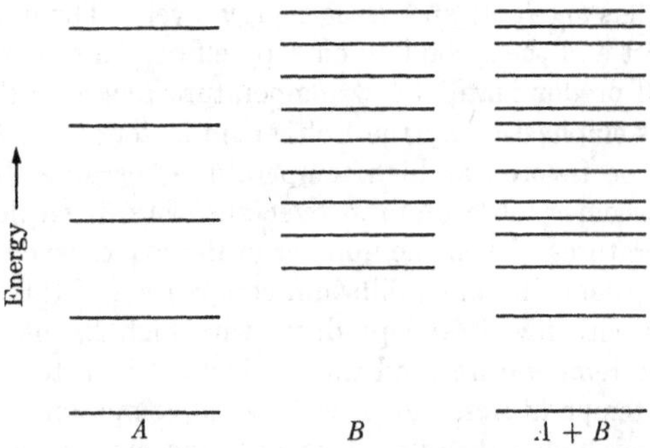

FIG. 10–1. Energy states for two isomers, A and B (schematic).

bonds are "stiffer" in A), and (c) the moments of inertia are larger in B than in A (i.e., the atoms are more spread out spatially in B).

We can take the point of view that a given molecule has accessible to it the full set of energy states indicated by $A + B$ in Fig. 10–1, with partition function q, but that these states can be classified into two subgroups called A and B, with partition functions q_A and q_B:

$$q_A = \sum_i e^{-\epsilon_i^A/kT}, \qquad q_B = \sum_j e^{-\epsilon_j^B/kT},$$

$$q = \sum_i e^{-\epsilon_i^A/kT} + \sum_j e^{-\epsilon_j^B/kT} = q_A + q_B.$$

There is a single Boltzmann distribution of molecules among all the levels $A + B$. The fraction of molecules in any level, say ϵ, is $e^{-\epsilon/kT}/q$. Therefore the fraction of molecules in all levels belonging to subgroup A is

$$\frac{N_A}{N_A + N_B} = \sum_i \left(\frac{e^{-\epsilon_i^A/kT}}{q}\right) = \frac{q_A}{q},$$

in agreement with Eq. (10–17).

The expression

$$\frac{N_B}{N_A} = \frac{q_B}{q_A} = \frac{\sum_j e^{-\epsilon_j^B/kT}}{\sum_i e^{-\epsilon_i^A/kT}}$$

shows how the density and energy of the two subgroups of quantum states determine the equilibrium composition N_B/N_A. The first few terms in the sum q_A are the largest of all, because the energies in the exponents are lowest. But the *number* of terms in q_B is larger, so that $q_B > q_A$ is still possible. In general, a substance is favored in an equilibrium (i.e., its q

is larger) if it has *low-lying* and *dense* energy levels. The first is primarily an energy effect and the second an entropy effect. In the example in Fig. 10–1, A would predominate at low temperatures, because the occupation of low levels is emphasized by the Boltzmann factor at low temperatures. But B would be favored at high temperatures, because the Boltzmann distribution becomes more uniform over all levels, regardless of energy, at high temperatures. Then the number or density of levels counts most. This is quite general: in an equilibrium competition of this sort between, say, reactants with low E and products with high S, the reactants will win out at low temperatures and the products at high temperatures. At intermediate temperatures, there will be a compromise between low energy (A) and high entropy (B) to give the system (as always) the minimum possible free energy $E - TS$ (N, V, T constant).

(b) *Ionization of hydrogen atoms.* The reaction we consider here is

$$\mathrm{H} \rightleftarrows \mathrm{H}^+ + e^-. \tag{10–21}$$

The concentrations of all species and the temperature are such that classical statistics and ideal gas behavior can be assumed. The zero of energy is chosen as separated proton and electron at rest. The ground state (the only state we need consider) of the hydrogen atom relative to this zero then has an energy $\epsilon_{e1} = -13.53$ ev. Because of electron spin, $\omega_{e1} = 2$ for the hydrogen atom. There is the same degeneracy $\omega = 2$ for the electron. The proton spin degeneracy can be ignored as usual (for nuclei), since it appears in both H and H^+ and cancels in the equilibrium constant. Then we have:

$$q_{\mathrm{H}^+} = q_{t(\mathrm{H}^+)} = \left(\frac{2\pi m_{\mathrm{H}^+} kT}{h^2}\right)^{3/2} V,$$

$$q_{e^-} = q_{t(e^-)} = 2\left(\frac{2\pi m_{e^-} kT}{h^2}\right)^{3/2} V,$$

$$q_{\mathrm{H}} = q_{t(\mathrm{H})} q_{e(\mathrm{H})} = 2\left(\frac{2\pi m_{\mathrm{H}} kT}{h^2}\right)^{3/2} V e^{-\epsilon_{e1}/kT}.$$

Hence, from Eq. (10–6), the equilibrium constant for the reaction (10–21) is

$$K(T) = \frac{\rho_{\mathrm{H}^+}\rho_{e^-}}{\rho_{\mathrm{H}}} = \left(\frac{2\pi m_{e^-} kT}{h^2}\right)^{3/2} e^{\epsilon_{e1}/kT}, \tag{10–22}$$

where we have neglected the difference in mass between H^+ and H. If the ionization is small, ρ_{H} is practically equal to ρ_{H}^0, the total concentration of H atoms placed in the volume V. Also, necessarily $\rho_{\mathrm{H}^+} = \rho_{e^-}$. In this case the fraction α of H atoms which have ionized at equilibrium is

$$\alpha = \frac{\rho_{H^+}}{\rho_H^0} = \left[\frac{1}{\rho_H^0}\left(\frac{2\pi m_{e^-}kT}{h^2}\right)^{3/2} e^{\epsilon_{e1}/kT}\right]^{1/2}. \tag{10–23}$$

As a numerical example, let us take $T = 10{,}000°K$ and $\rho_H^0 = 0.01$ mole·liter^{-1}. Then we find from Eq. (10–23), $\alpha = 7.8 \times 10^{-3}$ (Problem 10–1). Equation (10–22) has been verified experimentally for $Cs \rightleftarrows Cs^+ + e^-$.

In the equilibrium (10–21), the hydrogen atoms are favored by a much lower energy, but the dissociated state is favored by the entropy. The numerical calculation above shows that very high temperatures are needed for the entropy to overcome the energy in this particular competition.

(c) *Dissociation of hydrogen molecules.* We assume that the temperature is such that the rotation in H_2 is classical and that the hydrogen atoms themselves ionize [as in (b)] to a negligible extent. The ground-state degeneracy of H_2 is unity. The zero of energy is chosen as separated atoms at rest. On this basis, the ground state energy of H_2 is (Section 8–1) $\epsilon_{e1} = -D_e = -4.722$ ev. The partitions functions are

$$q_H = 2\left(\frac{2\pi m_H kT}{h^2}\right)^{3/2} V,$$
$$q_{H_2} = q_t q_r q_v q_e = \left(\frac{4\pi m_H kT}{h^2}\right)^{3/2} V \cdot \frac{T}{2\Theta_r} \cdot \frac{e^{-\Theta_v/2T}}{1 - e^{-\Theta_v/T}} \cdot e^{-\epsilon_{e1}/kT}. \tag{10–24}$$

Then, for the reaction

$$H_2 \rightleftarrows 2\,H, \tag{10–25}$$

we have, from Eq. (10–6),

$$K(T) = \frac{\rho_H^2}{\rho_{H_2}} = \frac{(q_H/V)^2}{(q_{H_2}/V)}.$$

If the fraction α of dissociated molecules is small, $\rho_{H_2} \cong \rho_{H_2}^0$ (total concentration of H_2 placed in V), and

$$2\alpha = \frac{\rho_H}{\rho_{H_2}^0} = \left[\frac{1}{\rho_{H_2}^0}\left(\frac{4\pi m_H kT}{h^2}\right)^{3/2} \frac{\Theta_r}{T} \cdot \frac{1 - e^{-\Theta_v/T}}{e^{-\Theta_v/2T}} \cdot e^{\epsilon_{e1}/kT}\right]^{1/2}. \tag{10–26}$$

Thus, aside from $\rho_{H_2}^0$ and T, α depends on the mass m_H, the equilibrium internuclear separation r_e, the vibrational force constant f, and the ground-state electronic energy ϵ_{e1}. If we use the values of Θ_r and Θ_v for H_2 in Table 8–1, we find at 1000°K and $\rho_{H_2}^0 = 0.01$ mole·liter^{-1}, $\alpha = 4.6 \times 10^{-9}$ (Problem 10–2). Again, as in (b) above, hydrogen molecules are favored in the equilibrium by the energy, but hydrogen atoms are favored by the entropy. This will in fact always be the case in a dissociative equilibrium.

(d) *Isotopic exchange reaction.* Consider the reaction

$$H_2 + D_2 \rightleftarrows 2\,HD \tag{10–27}$$

at temperatures such that rotation is classical and vibration is unexcited. The Born-Oppenheimer approximation has interesting consequences in a reaction of this type. These follow from the fact that in this approximation the ground-state electronic energy $u_e(r)$ must be the same function for all three species: H_2, D_2, and HD. Therefore r_e, f, and ϵ_{e1} [see (c) above] must all be the same for the three species. Hence, in the quotient

$$K(T) = \frac{\rho_{HD}^2}{\rho_{H_2}\rho_{D_2}} = \frac{p_{HD}^2}{p_{H_2}p_{D_2}} = \frac{q_{HD}^2}{q_{H_2}q_{D_2}}, \tag{10–28}$$

the translational factors cancel except for the masses; the rotational factors cancel except for the reduced masses ($I = \mu r_e^2$) and the symmetry numbers; and the electronic factors cancel completely. Only the zero-point vibrational terms $e^{-\Theta_v/2T}$ [see Eq. (10–24)] are left in the (unexcited) vibrational partition functions, and in these the frequencies are the same except for the proportionality to $\mu^{-1/2}$ (Eq. 8–18). Specifically, the separate quotients in Eq. (10–28) are easily seen to be:

$$\text{Translation:}\quad \frac{(m_H + m_D)^3}{(4m_Hm_D)^{3/2}},$$

$$\text{Rotation:}\quad \frac{16m_Hm_D}{(m_H + m_D)^2},$$

$$\text{Vibration:}\quad \exp\left\{-\frac{h\nu_{HD}}{kT}\left[1 - \frac{m_D^{1/2} + m_H^{1/2}}{2^{1/2}(m_H + m_D)^{1/2}}\right]\right\}.$$

Putting these together we find

$$K(T) = 2\frac{(m_H + m_D)}{m_H^{1/2}m_D^{1/2}}\exp\left\{-\frac{h\nu_{HD}}{kT}\left[1 - \frac{m_D^{1/2} + m_H^{1/2}}{2^{1/2}(m_H + m_D)^{1/2}}\right]\right\}. \tag{10–29}$$

Using $\nu_{HD} = 3770\ \text{cm}^{-1}$, we calculate $K = 3.46$, 3.67, 3.77, and 3.81 for $T = 383$, 543, 670, and 741°K. These are in good agreement with the respective experimental values 3.50, 3.85, 3.8, and 3.70. The theory must, in fact, be considered more accurate than the experimental work.

Isotope effects are not usually this large, of course. Note that if we put $m_H = m_D$ in Eq. (10–29), $K = 4$ (owing to the symmetry numbers). For the chlorine isotopes 35 and 37, $K = 4(1 - \delta)$, with $\delta = O(10^{-4})$.

(e) *"Water-gas" reaction.* As a final example, we calculate the equilibrium constant K for the so-called "water-gas" reaction,

$$CO_2 + H_2 \rightleftarrows CO + H_2O. \tag{10-30}$$

We restrict ourselves to temperatures high enough so that rotation is classical in H_2. Then

$$\begin{aligned} K(T) &= \frac{\rho_{CO}\rho_{H_2O}}{\rho_{CO_2}\rho_{H_2}} = \frac{p_{CO}p_{H_2O}}{p_{CO_2}p_{H_2}} = \frac{q_{CO}q_{H_2O}}{q_{CO_2}q_{H_2}} \\ &= \frac{q'_{CO}q'_{H_2O}}{q'_{CO_2}q'_{H_2}} e^{-\Delta\epsilon_{e1}/kT}, \end{aligned} \tag{10-31}$$

where

$$\Delta\epsilon_{e1} = \epsilon_{e1}(CO) + \epsilon_{e1}(H_2O) - \epsilon_{e1}(CO_2) - \epsilon_{e1}(H_2), \tag{10-32}$$

and q' means the partition function q with the electronic part q_e omitted: $q' = q_t q_r q_v$. For all these molecules, $\omega_{e1} = 1$. The zero of energy is the separated atoms (C, O, H) at rest. Actually, we do not need to know each ϵ_{e1} separately relative to this zero, but only the combination $\Delta\epsilon_{e1}$. For this reaction (from combined thermal and spectroscopic information), $\Delta\epsilon_{e1} = 7000$ cal·mole^{-1}. Incidentally, the energy change for the reaction at 0°K involves the zero-point vibrational energies as well, and is given by

$$\begin{aligned} \Delta E(0) &= \Delta\epsilon_{e1} + (h/2)\Delta\left(\sum_i \nu_i\right) \\ &= 7000 + 2640 = 9640 \text{ cal·mole}^{-1}, \end{aligned}$$

where the vibrational frequencies can be found in Table 8-1 and Problems 9-10 and 9-11. The rotational data are included in the same sources ($I_{CO_2} = 71.9 \times 10^{-40}$ gm·cm^2). With this information ($\Delta\epsilon_{e1}$, ν's, I's), K can be calculated by use of expressions for q_t, q_r, and q_v which we shall not write down again. One finds $K = 0.45$ at 900°K and 1.41 at 1200°K (actually using spectroscopic energy levels instead of the slightly approximate equations in Chapters 8 and 9), compared with the respective experimental values 0.46 and 1.37.

Problems

10–1. Calculate the fraction of hydrogen atoms ionized at 10,000°K when $\rho_{\text{H}}^0 = 0.01$ mole·liter^{-1}. (Page 185.)

10–2. Calculate the fraction of hydrogen molecules dissociated into atoms when $\rho_{\text{H}_2}^0 = 0.01$ mole·liter^{-1} and (a) $T = 1000$°K, (b) $T = 5000$°K. (Page 185.)

10–3. Repeat the argument of Section 10–2, but maximize the partition function $\Delta(N_A, N_B, p, T)$ [see Eq. (4–35)] at constant N, p, T instead of Q at constant N, V, T.

10–4. In the example of Section 10–3, show that

$$\overline{N_A N_B} - \overline{N}_A \overline{N}_B = -\frac{\overline{N}_A \overline{N}_B}{N}.$$

Why the negative sign?

10–5. Use the data in Table 8–1 to calculate K at 700°K for the reaction $2\text{HI} \rightleftarrows \text{H}_2 + \text{I}_2$.

10–6. Calculate K at 1000°K for the reaction $\text{I}_2 \rightleftarrows 2\text{I}$ (see Section 4–4 and Table 8–1).

10–7. Calculate Θ_v for H_2, given that $\nu = 3770$ cm^{-1} for HD.

10–8. Calculate Θ_v for H_2, given that, for H_2, $D_e = 4.722$ ev and $D_0 = 4.454$ ev.

10–9. Calculate K for the "water-gas" reaction at 1200°K using data in the text.

Supplementary Reading

Fowler and Guggenheim, Chapter 5.
Mayer and Mayer, Chapter 9.
Rushbrooke, Chapter 11.

CHAPTER 11

THE RATE OF CHEMICAL REACTIONS IN IDEAL GAS MIXTURES

Strictly speaking, the subject indicated in the chapter title is outside the scope of this book, which is devoted to equilibrium statistical mechanics. However, Eyring's approximate absolute reaction rate theory has a quasi-equilibrium foundation: it is based on an application of the chemical equilibrium theory of the preceding chapter. For this reason, and because of its importance, it seems appropriate to include an account of the Eyring theory in the present work. But the treatment we give will be very brief, and we shall not consider any detailed special cases. The reader should consult the book by Glasstone, Laidler, and Eyring (see the Supplementary Reading list) for further details.

More exact approaches to this problem will not be discussed here, since they cannot be put in quasi-equilibrium form.

An example of a nonchemical application of Eyring's theory is presented in Section 11–3: the surface diffusion of a dilute, localized monolayer.

11–1 Potential surfaces. There are two distinct stages in the Eyring theory. The first is the purely quantum-mechanical one of calculating the ground-state electronic energy surface (potential surface, for short) for the reaction, and the second is the statistical-mechanical calculation of the reaction rate. This division is the same as that which we have encountered in calculating the thermodynamic functions of, say, an ideal polyatomic gas (Chapter 9). In this latter problem, we first have to find the potential surface of the molecule by quantum mechanics (or obtain equivalent information empirically from spectroscopy). This surface (see Section 9–1) determines the equilibrium structure of the molecule, the moments of inertia, the vibrational force constants and normal coordinates, and the depth of the potential well in the surface relative, say, to separated atoms as zero. With this information, we can then turn to the statistical-mechanical problem of deducing the thermodynamic functions.

We discuss the potential-surface part of the rate problem in this section, and the statistical-mechanical part in the next section.

For ease of visualization, let us consider a hypothetical one-dimensional reaction

$$A + BC \rightarrow AB + C. \qquad (11\text{–}1)$$

Three atoms and (in one dimension) three nuclear coordinates are in-

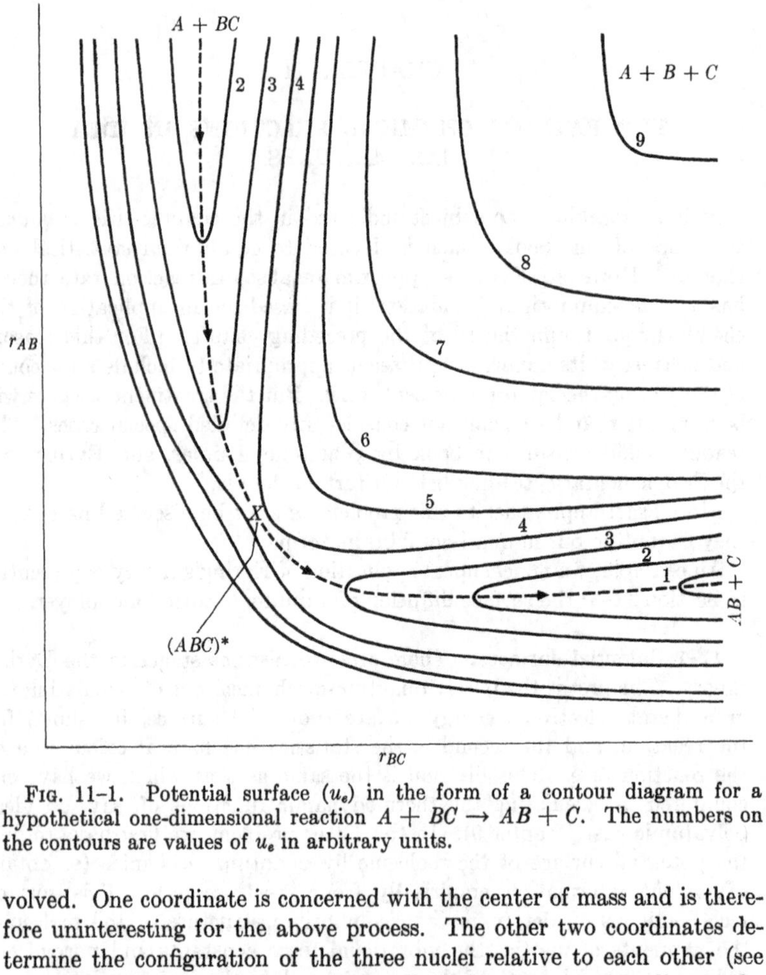

FIG. 11–1. Potential surface (u_e) in the form of a contour diagram for a hypothetical one-dimensional reaction $A + BC \rightarrow AB + C$. The numbers on the contours are values of u_e in arbitrary units.

volved. One coordinate is concerned with the center of mass and is therefore uninteresting for the above process. The other two coordinates determine the configuration of the three nuclei relative to each other (see Appendix V). For example, we might choose for these two coordinates the internuclear distances r_{AB} and r_{BC}. For given values of r_{AB} and r_{BC} the ground-state electronic energy $u_e(r_{AB}, r_{BC})$ is calculated. From a large number of such values of r_{AB} and r_{BC}, one can construct a potential surface in the form of a contour diagram, which in a typical case might appear as in Fig. 11–1. The valley at the upper left corresponds to an A atom and a diatomic BC molecule. (The curvature of the surface at the bottom of and perpendicular to the valley determines the vibrational frequency in BC; the depth of the valley is a measure of the energy of the bond BC.) The valley at the lower right corresponds to the state $AB + C$.

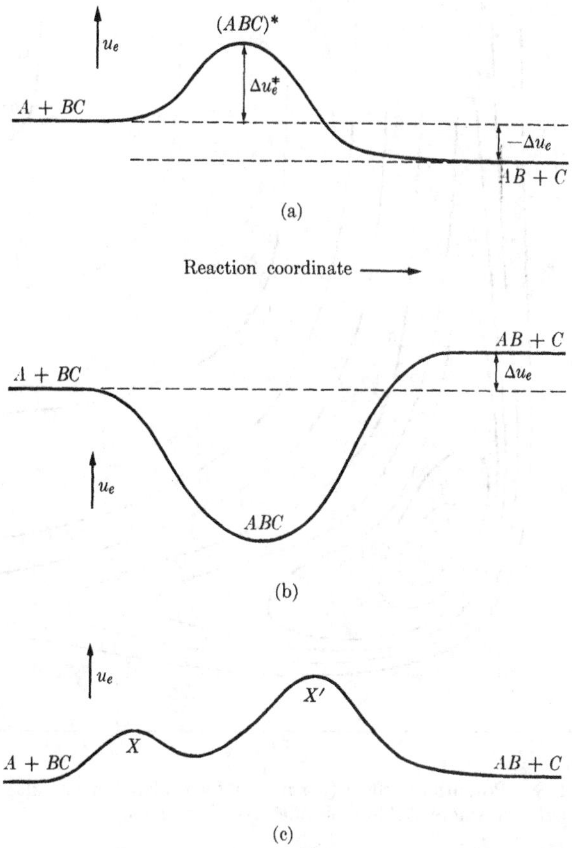

FIG. 11–2. (a) Potential energy (u_e) along the reaction path in Fig. 11–1. Note the potential barrier. (b) Corresponding curve from Fig. 11–3. Note the minimum in u_e (stable molecule, ABC). (c) Corresponding curve from Fig. 11–4.

The high plateau is $A + B + C$. When reaction (11–1) occurs, the lowest possible path from reactants ($A + BC$) to products ($AB + C$) is the dashed line in Fig. 11–1. The highest point on this path is marked X in the figure. This is the "activated state," and the triatomic system A, B, C at this point is referred to as an "activated complex," denoted by $(ABC)^*$. If one plots the potential energy u_e along the dashed path of Fig. 11–1 as a function of the distance along the path (called the "reaction coordinate"), one obtains a curve as in Fig. 11–2(a). The height Δu_e^* of the potential energy barrier which must be overcome is called the "activation energy."

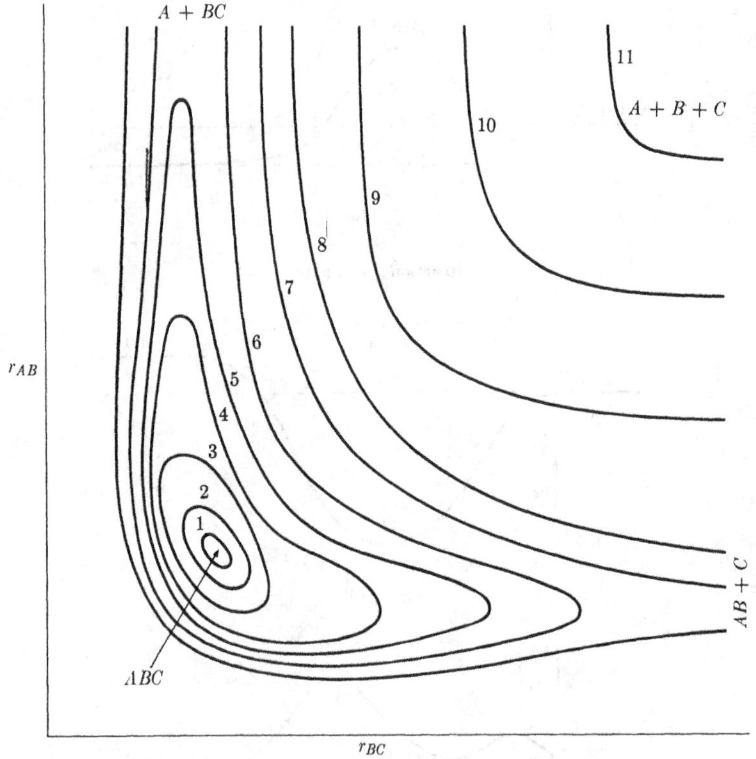

FIG. 11–3. Potential surface (u_e) in the form of a contour diagram in one-dimensional case where stable molecule ABC is formed.

For the reverse reaction,

$$AB + C \rightarrow A + BC, \tag{11–2}$$

the reaction path in Fig. 11–1 must be reversed in direction. In this example the activated state and complex (X in Fig. 11–1) are the same for forward (11–1) and reverse (11–2) reactions. Let

$$\Delta u_e = u_e(AB + C) - u_e(A + BC). \tag{11–3}$$

This is Δu_e for the reaction as written in (11–1) and has the same meaning as $\Delta\epsilon_{e1}$ in Eq. (10–32). In Fig. 11–1 it is determined by the difference in levels of the two valleys. In Figs. 11–1 and 11–2(a), Δu_e is negative. If $\Delta u_e^{\ddagger}$ is the activation energy for the forward reaction, $\Delta u_e^{\ddagger} - \Delta u_e$ is the activation energy for the reverse reaction.

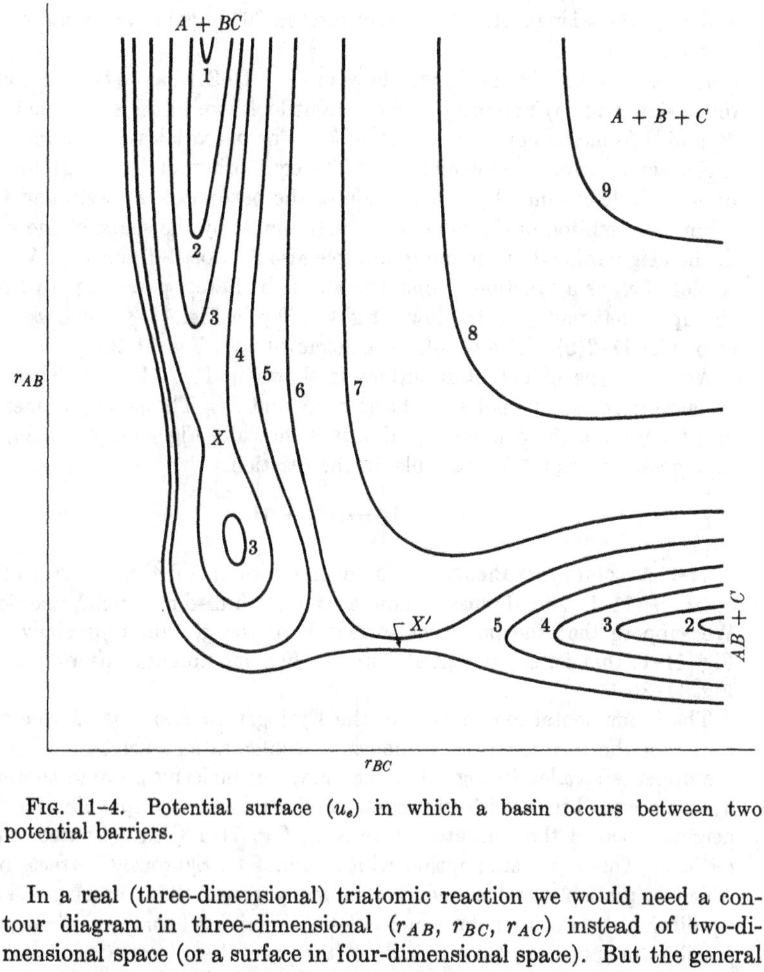

FIG. 11-4. Potential surface (u_e) in which a basin occurs between two potential barriers.

In a real (three-dimensional) triatomic reaction we would need a contour diagram in three-dimensional (r_{AB}, r_{BC}, r_{AC}) instead of two-dimensional space (or a surface in four-dimensional space). But the general concepts introduced above for a hypothetical one-dimensional reaction remain the same. Incidentally, it should be noticed that linear configurations of A, B, C are included in the potential surface (in fact, activated complexes in triatomic reactions are usually linear), and we recall that linear molecules have four, not three, vibrational coordinates. However, this is not contradictory to the statement that u_e is a function of the three variables r_{AB}, r_{BC}, and r_{AC} only, for two of the normal modes in a linear molecule are degenerate (i.e., of the same frequency). For example, the positions of the atoms in the two degenerate modes of CO_2 [shown preceding Eq. (9-3)] are expressible by the same sets of values of r_{AB}, r_{BC}, and r_{AC} (the triangle ABC is merely rotated 90° in going from one mode to the other).

In contrast to Fig. 11–1, we show in Fig. 11–3 what a typical (one-dimensional again) potential surface might look like in the event that A, B, and C formed a stable molecule ABC. The potential well in Fig. 11–3 is characteristic of a stable molecule. The equilibrium molecular geometry of ABC is determined by the location of the bottom of the well, and the vibrational motion of the molecule is determined by the shape of the well in the neighborhood of the minimum [see also Section 9–1 and Eq. (V–1)]. A plot of u_e as a function of distance along the lowest possible path from the upper left valley to the lower right valley in Fig. 11–3 would appear as in Fig. 11–2(b). This should be contrasted with Fig. 11–2(a).

Another type of potential surface is shown in Fig. 11–4. A basin is situated between two potential barriers, X and X'. The potential energy as a function of the reaction coordinate is shown in Fig. 11–2(c). A basin may possibly occur,* for example, in the reaction

$$H + H_2 \rightarrow H_2 + H.$$

11–2 Absolute rate theory. As a concrete example, let us return to the reaction (11–1) and discuss it now as a real three-dimensional reaction. We suppose that the potential surface is of the general type shown in Fig. 11–1; that is, the potential "profile" has one potential barrier, as in Fig. 11–2(a).

The fundamental assumption of the Eyring theory is that, during the course of the reaction (11–1), molecular configurations corresponding to the upper left valley in Fig. 11–1 (i.e., reactant molecules) are in thermodynamic equilibrium with molecular configurations corresponding to the neighborhood of the activated state X in Fig. 11–1 (i.e., activated complexes). This is an assumption which cannot be rigorously correct, but which is probably rather accurate in many cases. This assumption of equilibrium between reactant molecules and activated complexes makes it possible for us to use the methods of the preceding chapter on chemical equilibria to deduce the concentration (defined below) of activated complexes. From this knowledge, as we shall see, we can then calculate the number of reactants passing over the barrier, from upper left to lower right, per unit time and per unit volume of the system. This is the desired reaction rate.

An activated complex is very much like an ordinary stable molecule. It has a definite mass ($m_A + m_B + m_C$ in this example) and a definite

* See Glasstone, Laidler, and Eyring (Supplementary Reading list), p. 108, and R. E. Weston, Jr., *J. Chem. Phys.* **31**, 892 (1959).

nuclear configuration (corresponding to the position of the X in Fig. 11-1 at the top of the potential barrier). This configuration determines the moments of inertia and the symmetry number. Thus we can immediately write (assuming the potential surface is available) the translational and rotational partition functions for an activated complex, just as we did in Chapter 9 for a stable molecule.

Furthermore, we can carry out a normal-coordinate analysis for the vibrational frequencies, based on the shape of the potential surface in the neighborhood of the activated state. If the activated complex is linear, there will be $3n - 5$ normal vibrational modes, otherwise there will be $3n - 6$. The activated state is located at a saddle point in the potential surface. That is, although the activated state is a maximum point along the reaction coordinate, it is a minimum in other directions (e.g., in Fig. 11-1, in the direction perpendicular to the reaction coordinate). This feature will appear automatically in the normal-coordinate analysis when the potential energy u_e is expressed in terms of the normal coordinates ξ_i [see Eq. (V-13)]. Necessarily (i.e., by definition of a normal coordinate), u_e will be a sum of squared terms in the ξ_i, and the coefficients will be positive as usual (Eq. V-13) *except* for one coordinate, call it ξ, which will have a negative coefficient. This particular normal coordinate is the rigorous equivalent of what we have hitherto been loosely calling the reaction coordinate. The coefficient of ξ^2 in u_e is negative because the potential surface *falls off* on both sides of the activated state along this direction (and this direction only).

From the normal-coordinate analysis we thus obtain $3n - 6$ (linear) or $3n - 7$ (nonlinear) ordinary vibrational frequencies ν_i. The frequency associated with ξ is imaginary because the force constant (twice the coefficient of ξ^2) is negative. We can therefore construct, in the usual way, from the ν_i, a vibrational partition function for the activated complex—except we omit the factor belonging to the ξ-motion. Thus of the $3n$ nuclear degrees of freedom of an activated complex, $3n - 1$ can be handled just as with stable molecules. Only the reaction normal coordinate ξ requires special treatment.

Let q_v^* represent the vibrational partition function of the activated complex, omitting the ξ-factor. Also, let $q^* = q_t^* q_r^* q_v^* q_e^*$ be the complete partition function of the activated complex, just as for any polyatomic molecule—except, again, omitting the ξ degree of freedom.

We wish next to calculate the number of activated complexes per unit volume of the system which are in an infinitesimal range $d\xi$ of the reaction coordinate at the activated state. We shall call this number $\rho' \, d\xi$, so that ρ' is the number of activated complexes per unit volume and also per unit length along the reaction coordinate ξ at the activated state. We want the number $\rho'\,d\xi$ to include all activated complexes in $d\xi$ irrespective of the value of p_ξ, the momentum conjugate to ξ. For the element of phase space $d\xi\,dp_\xi$ in the coordinate ξ, the (classical) partition function is

$$\frac{1}{h}e^{-p_\xi^2/2m^*kT}\,d\xi\,dp_\xi, \tag{11-4}$$

where m^* is defined by $p_\xi = m^*\dot{\xi}$. It should be noted, incidentally, that there is always some arbitrariness in defining normal coordinates [see the constants C_i in Eqs. (V-12) and (V-14)], and so there is arbitrariness in m^* and ξ. But combinations of these quantities with physical significance (e.g., $p_\xi^2/2m^*$ or $d\xi\,dp_\xi$ above) are not arbitrary. Integration of (11-4) over p_ξ gives the ξ partition function for an activated complex in $d\xi$. The *complete* partition function for such an activated complex is then

$$q^* \left(\frac{2\pi m^*kT}{h^2}\right)^{1/2} d\xi. \tag{11-5}$$

With the assumption, already mentioned, of equilibrium between reactants (say A and BC, for concreteness) and activated complexes,

$$A + BC \rightleftarrows (ABC)^*, \tag{11-6}$$

we have, as in Eqs. (10-6) and (10-31),

$$\begin{aligned}\frac{\rho'\,d\xi}{\rho_A\rho_{BC}} &= \frac{(q^*/V)(2\pi m^*kT/h^2)^{1/2}\,d\xi}{(q_A/V)(q_{BC}/V)} \\ &= \frac{(q^{*\prime}/V)(2\pi m^*kT/h^2)^{1/2}\,d\xi}{(q_A'/V)(q_{BC}'/V)}e^{-\Delta u_e^*/kT},\end{aligned} \tag{11-7}$$

where q' means, as before, that q_e is omitted from q. The quantity Δu_e^* is defined for the process (11-6) by analogy with (11-2) and (11-3). If the ground electronic states are degenerate, a factor $\omega_{e1}^*/\omega_{e1}(A)\omega_{e1}(BC)$ must be included in Eq. (11-7) (ω_{e1}^* being the degeneracy of the activated complex).

Equation (11-7) provides us with an explicit equation for ρ', the number of activated complexes per unit volume of the system and per unit length along ξ at the activated state. Our next task is to calculate the number of activated complexes per unit volume which cross the potential barrier X per unit time in the direction of reaction (left to right in Fig. 11-1). Assuming all of these complexes become products, this is the rate of the reaction. Consider those activated complexes with values of $\dot{\xi}$ between $\dot{\xi}$ and $\dot{\xi} + d\dot{\xi}$. The fraction of all activated complexes which are in this class is, from (11-4),

$$f(\xi)\,d\xi \equiv \frac{e^{-m^*\dot{\xi}^2/2kT}\,d\dot{\xi}}{\int_{-\infty}^{+\infty} e^{-m^*\dot{\xi}^2/2kT}\,d\dot{\xi}} = \left(\frac{m^*}{2\pi kT}\right)^{1/2} e^{-m^*\dot{\xi}^2/2kT}\,d\dot{\xi}.$$

Suppose ξ increases as the activated state is approached from the reactant (left) side in Fig. 11-1. Then an activated complex with $\dot{\xi} > 0$ is proceeding along the reaction coordinate in the direction required for the reaction to take place. We note that activated complexes with a given value of $\dot{\xi} > 0$ will cross the potential barrier in unit time if they start at a distance from X not greater than a length of magnitude $\dot{\xi}$. The number of activated complexes per unit volume and per unit length along ξ having values of $\dot{\xi}$ in the interval $d\dot{\xi}$ is $\rho' f(\dot{\xi})\,d\dot{\xi}$. The number of activated complexes per unit volume in the length $\dot{\xi}$ along ξ having $\dot{\xi}$ in the interval $d\dot{\xi}$ is then $\dot{\xi}\rho' f(\dot{\xi})\,d\dot{\xi}$. This is the number of complexes per unit volume with $\dot{\xi}$ in $d\dot{\xi}$ which cross the barrier per unit time. To get the total number of complexes per unit volume crossing the barrier per unit time, we have to integrate $\dot{\xi}$ over the range $0 \leq \dot{\xi} \leq +\infty$:

$$\rho' \int_0^\infty \dot{\xi} f(\dot{\xi})\,d\dot{\xi} = \rho' \left(\frac{m^*}{2\pi kT}\right)^{1/2} \left(\frac{kT}{m^*}\right)$$

$$\equiv \mathbf{k}\rho_A\rho_{BC}, \tag{11-8}$$

which is the defining equation for $\mathbf{k}$. The quantity $\mathbf{k}\rho_A\rho_{BC}$ is the rate of the reaction, and $\mathbf{k}(T)$ is called, conventionally, the rate constant. If ρ' is obtained from Eq. (11-7), we find

$$\mathbf{k} = \frac{kT}{h} \frac{(q^{\ddagger\prime}/V)e^{-\Delta u_e^\ddagger/kT}}{(q'_A/V)(q'_{BC}/V)}. \tag{11-9}$$

If only a fraction κ (called the transmission coefficient) of complexes passing the potential barrier in the right direction actually proceed to products, then κ must be inserted as a factor on the right-hand side of Eq. (11-9). This situation arises, for example, in cases such as Fig. 11-4, where the system passes over a barrier (X) but then finds itself in a basin. The system may leave the basin (via X') to form products or return (via X) to reactants. In the reaction $H + H_2 \rightarrow H_2 + H$, the basin is symmetrical (if it exists), and it is usually assumed that $\kappa = 1/2$.

Equation (11-9) furnishes a straightforward statistical-mechanical recipe for calculating the rate constant $\mathbf{k}$. The potential surface must be available, however, and this is a very serious practical obstacle. Because of this, it is difficult to test the theory in a really satisfactory way. Of course, one does not expect exact agreement between theory and experiment, as the argument used to deduce Eq. (11–9) is not rigorous—the assumption of equilibrium between reactants and activated complexes being especially questionable. Also, if a transmission coefficient must be used, this introduces a somewhat nebulous feature into the theory since κ is in general difficult to evaluate.

For the reverse reaction (11–2), the rate constant is

$$\mathbf{k}' = \frac{kT}{h} \frac{(q^{\ddagger\prime}/V)e^{-(\Delta u_e^{\ddagger} - \Delta u_e)/kT}}{(q'_{AB}/V)(q'_C/V)}. \tag{11–10}$$

The equilibrium constant K for the reaction (11–1) is then

$$K = \frac{\mathbf{k}}{\mathbf{k}'} = \frac{(q'_{AB}/V)(q'_C/V)}{(q'_A/V)(q'_{BC}/V)} e^{-\Delta u_e/kT}, \tag{11–11}$$

in agreement with Eqs. (10–6) and (10–31).

11–3 A nonchemical application of the Eyring theory. The fundamental ideas in Eyring's theory of the rate of chemical reactions can be and have been applied to many physical rate processes as well. Merely to illustrate the possibilities, we consider here a particularly straightforward example, namely, the rate at which monatomic molecules adsorbed at localized sites on a surface jump from one site to another. This rate is of course closely related to the coefficient of surface diffusion. The model we consider is essentially that already discussed in Sections 7–1 and 9–6. We have a lattice of equivalent surface sites for adsorption, but we need not specify the lattice type. The number of adsorbed molecules is small, so that each one behaves independently. The potential in which a molecule moves is $U_0(x, y)$ [see, for example, Eq. (9–19)]. The potential wells in $U_0(x, y)$ are the sites for adsorption. The partition function for an adsorbed molecule at a site is given by Eq. (7–3).

To move from a given site to a nearest-neighbor site, a molecule must pass over a potential barrier of height V_0. The top of the barrier is the activated state for this process:

$$A \text{ (site)} \rightarrow A^* \text{ (top of barrier)} \rightarrow A \text{ (neighboring site)}.$$

Let ξ and η be the normal coordinates at the activated state, which is a saddle point in the surface U_0. We take ξ as the "reaction coordinate." That is, the coefficient of ξ^2 in the expansion of U_0 in powers of ξ and η about the activated state is negative, while the coefficient of η^2 is positive. Thus a molecule at the top of a potential barrier vibrates in the usual way in the z-direction (perpendicular to the surface) and also in the η-direction (perpendicular to the direction of passage from one site to the other, i.e., perpendicular to ξ), but not in the ξ-direction. We denote the z and η vibrational partition functions at the activated state by $q_z^{\ddagger}$ and $q_\eta^{\ddagger}$.

The partition function for a molecule in the element of length $d\xi$ at the top of a barrier is [see Eq. (11-5)]

$$q_z^{\ddagger} q_\eta^{\ddagger} \left(\frac{2\pi mkT}{h^2}\right)^{1/2} d\xi \, e^{-(U_{00}+V_0)/kT}. \tag{11-12}$$

Let $N'\, d\xi$ be the equilibrium number of molecules in a length $d\xi$ at the top of barriers, and let N be the equilibrium number of molecules in sites. Then, from Eq. (10-5), the ratio of these two numbers is

$$\frac{N'\, d\xi}{N} = \frac{q_z^{\ddagger} q_\eta^{\ddagger} (2\pi mkT/h^2)^{1/2}\, d\xi \, e^{-V_0/kT}}{q_x q_y q_z} \times \frac{M^{\ddagger}}{M}, \tag{11-13}$$

where $M^{\ddagger}/M$ is the ratio of the number of activated states to the number of sites (this ratio is two for a square lattice). By the same argument as in the preceding section (Eq. 11-8),

$$N' \left(\frac{m}{2\pi kT}\right)^{1/2} \left(\frac{kT}{m}\right) \tag{11-14}$$

is the number of molecules crossing a barrier (or the number of jumps being made from one site to another) per unit time, where N' is given by Eq. (11-13). We are assuming here that there are no "rebounds": $\kappa = 1$. If τ is the mean time a molecule spends at a site between jumps, then a second expression for the number of jumps occurring in unit time is N/τ. If we set N/τ equal to (11-14), we find

$$\frac{1}{\tau} = \frac{kT}{h} \cdot \frac{M^{\ddagger} q_z^{\ddagger} q_\eta^{\ddagger} e^{-V_0/kT}}{M q_x q_y q_z}. \tag{11-15}$$

This is the analog of Eq. (11-9). That is, $1/\tau$ is the rate constant for this process.

To obtain an estimate of the order of magnitude of τ in Eq. (11-15), we set $M^{\ddagger}/M = 2$, $q_\eta^{\ddagger} = q_y$ [as would be the case with Eq. (9-19)], $q_z^{\ddagger} = q_z$, and $q_x = kT/h\nu_x$ (classical). Then

$$\frac{1}{\tau} = 2\nu_x e^{-V_0/kT}. \tag{11-16}$$

This equation has the following approximate interpretation: $2\nu_x$ is the number of "attempts" per second the molecule makes to leave its site; $e^{-V_0/kT}$ is the probability that any particular attempt will be successful; and hence $1/\tau$ is the actual number of jumps a molecule makes from one site to another per second. If we take $\nu_x = 3 \times 10^{11}$ sec^{-1}, $V_0 = 500$ cal·mole^{-1}, and $T = 80°$K, then $e^{-V_0/kT} = 0.043$ and $\tau = 3.9 \times 10^{-11}$ sec.

Equation (11–15) also provides a theoretical equation for the coefficient of surface diffusion D, since D is related to τ (from the theory of random walks) by $D = Ca^2/\tau$, where a is the distance between nearest-neighbor sites and C is a constant of order unity which depends on the lattice type.

Problems

11–1. Write out the explicit forms for the partition functions in Eq. (11–9), assuming the triatomic complex $(ABC)^*$ is linear. Insert typical orders of magnitudes for the masses, frequencies, bond distances, etc., to estimate a magnitude for $\mathbf{k}$.

11–2. Derive an equation for $d \ln \mathbf{k}/dT$ from $\mathbf{k}$ in Problem 11–1.

11–3. Discuss the rate of diffusion of impurity atoms in a monatomic crystal from the point of view of Eyring's theory.

11–4. Consider the rate of evaporation of a dilute localized monatomic monolayer into the gas phase. (a) Use Eyring's method to derive an equation for $1/\tau_s$, where τ_s is the mean time a molecule spends on the surface before evaporating. (b) Derive the same expression for $1/\tau_s$ by equating the number of molecules condensing on the surface per unit area and per unit time with the number evaporating, at equilibrium:

$$\frac{p}{(2\pi mkT)^{1/2}} = \frac{(N/\mathcal{A})}{\tau_s},$$

where p is the equilibrium gas pressure and $N/\mathcal{A}$ is given by Eq. (7–10) in the limit as $\theta \to 0$.

Supplementary Reading

Fowler and Guggenheim, Chapter 12.

Frenkel, J., *Kinetic Theory of Liquids.* New York: Dover, 1955.

Glasstone, S., Laidler, K. J., and Eyring, H., *The Theory of Rate Processes.* New York: McGraw-Hill, 1941.

CHAPTER 12

IDEAL GAS IN AN ELECTRIC FIELD

Our principal object in this chapter is to deduce the thermodynamic properties of a very dilute gas in an electric field. In Chapter 15, this study will be extended briefly to slightly imperfect gases.

In Section 12–1 we give some necessary thermodynamic background, and in Section 12–2 we develop general statistical-mechanical equations (canonical ensemble). The material in the first two sections is quite general and would apply to any fluid (or isotropic) dielectric. The dilute-gas special case is then considered in Section 12–3. Finally, in Section 12–4, we discuss a somewhat related problem: a lattice of noninteracting magnetic dipoles in a magnetic field. This problem turns out to be formally the same as that of the ideal lattice gas in Chapter 7. The interacting magnetic dipole case (the Ising model for ferromagnetism) is included in Chapter 14.

12–1 Thermodynamic background. A number of alternative and equivalent thermodynamic formulations can be devised for a dielectric fluid in an electric field. Koenig* has given a very full discussion of this subject. The corresponding treatment for magnetic systems is contained in a paper by Guggenheim.† We confine ourselves here to the one particular formulation that is most convenient in the statistical mechanics of gases in an electric field. For condensed systems, there are some advantages to other choices.

Consider the parallel plate condenser in Fig. 12–1. The plate surface charge densities are $+\sigma$ and $-\sigma$, as indicated. The condenser is assumed to have a large enough plate area so that edge effects can be ignored. The volume V contains the dielectric fluid whose properties we are interested in. For simplicity, we take the fluid as one component with N molecules, but it could as well be multicomponent. The same is true in Section 12–2. One wall of the fluid container, parallel to the condenser plates, serves as a piston to vary the volume V. The equilibrium pressure on the piston is p. The regions between the fluid container and the condenser plates are evacuated. As a consequence of polarization of the dielectric in the field of the condenser plates, there are induced surface-charge densities $-\sigma'$

---

* F. O. Koenig, *J. Phys. Chem.* **41**, 597 (1937).
† E. A. Guggenheim, *Proc. Roy. Soc.* **155A**, 49, 70 (1936).

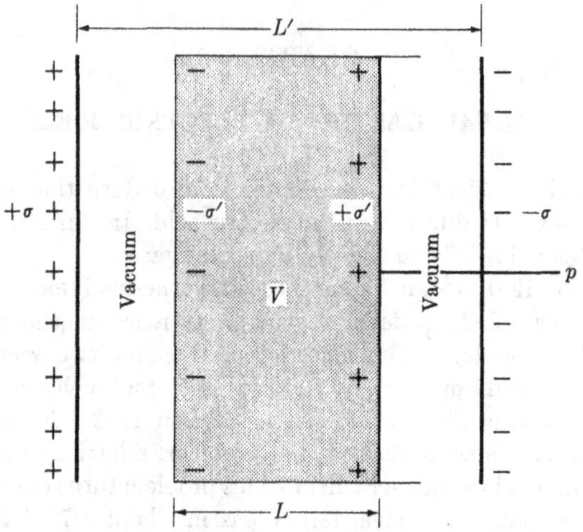

FIG. 12–1. Dielectric fluid in volume V between two condenser plates. V can be varied by a piston with pressure p.

and $+\sigma'$ on the inside surfaces of the indicated walls (Fig. 12–1) of the fluid container.

Let us consider the electric field acting at any point in the fluid. The field due to the condenser surface charge $+\sigma$ is $2\pi\sigma$, as can easily be deduced from Coulomb's law (Problem 12–1). The field arising from the surface charge $-\sigma$ is also $2\pi\sigma$ (acting in the same direction). The total field from the charge on the condenser plates is therefore $4\pi\sigma$. This field is called the dielectric displacement D; $D = 4\pi\sigma$. We regard D as a variable intensive parameter on which the thermodynamic properties of the fluid depends. D is an external field; that is, it can be controlled from outside the thermodynamic system (fluid) itself. At a point in V, D has the same value whether or not a dielectric fills V; that is, it is determined solely by σ. In addition to the field D due to external charges acting at the point in V, there is clearly also a field $-4\pi\sigma'$ due to charges induced on the surfaces of the dielectric. An equivalent point of view is that this second field arises from molecular dipoles throughout V oriented somewhat in the external field D. It is customary to use the symbol P (polarization) in place of σ'. Then the total electric field acting at a point in V is $D - 4\pi P$. This field is denoted by $\mathcal{E}$ (electric field strength):

$$\mathcal{E} = D - 4\pi P, \qquad D = \mathcal{E} + 4\pi P. \tag{12–1}$$

$\mathcal{E}$, D, and P all have the same sign. The dielectric constant ϵ is then defined by $D = \epsilon\mathcal{E}$; necessarily $\epsilon \geq 1$. In this chapter we shall avoid the use of the symbol ϵ as a molecular energy level.

We now consider the thermodynamic functions of the fluid molecules. In these functions we include only contributions from the molecules themselves, and exclude contributions that would be associated with the space V and field D if no molecules were present in V. This is a natural choice in a molecular theory. Thus, the energy E includes the molecular kinetic energy, the potential energy of interaction of the molecules with the external field D, and the potential energy of interaction between the molecules themselves.

In the expression

$$dE = DQ^* - DW + \mu\, dN, \tag{12–2}$$

there are two contributions to DW. One is the usual $p\,dV$ term associated with a volume change (N and D constant, $DQ^* = 0$), and the other is related to changes in D (N and V constant, $DQ^* = 0$). Consider the work that must be done on the entire system shown in Fig. 12–1 if σ is to be changed by $d\sigma$. Let $\mathcal{A}$ be the cross-sectional area of a condenser plate. Then we have to transport an amount of positive charge $\mathcal{A}\, d\sigma$ from the negative condenser plate to the positive plate. This charge has to be moved against a field $\mathcal{E}$ through the distance L (fluid) and against a field D through the distance $L' - L$ (vacuum). The total work done on the system is thus

$$\mathcal{E}L\mathcal{A}\, d\sigma + D(L' - L)\mathcal{A}\, d\sigma. \tag{12–3}$$

But to get the quantity we want, the work done *on the fluid*, we have to subtract from (12–3) the work which would have to be done to increase σ by $d\sigma$ when the volume V is evacuated. This work is

$$DL'\mathcal{A}\, d\sigma. \tag{12–4}$$

On subtracting (12–4) from (12–3), we get

$$\mathcal{E}L\mathcal{A}\, d\sigma - DL\mathcal{A}\, d\sigma = -PV\, dD. \tag{12–5}$$

This is work done *on* the fluid. The desired contribution to DW is $+PV\, dD$, since DW is work done *by* the fluid.

We note that $PV = (\sigma'\mathcal{A})L$, which is just the total dipole moment (charge × separation distance) of the fluid in V. We call the total moment M and replace PV by M. Finally, then, Eq. (12–2) becomes

$$dE = T\, dS - p\, dV - M\, dD + \mu\, dN. \tag{12–6}$$

Koenig has shown that the pressure p in this equation has the operational significance implied by the particular kind of volume change indicated in

Fig. 12-1. On integrating Eq. (12-6), holding intensive properties constant, we find

$$\begin{aligned} E &= TS - pV + \mu N, \\ F &= \mu N = A + pV. \end{aligned} \tag{12-7}$$

Alternative useful forms of Eq. (12-6) are

$$dA = -S\,dT - p\,dV - M\,dD + \mu\,dN, \tag{12-8}$$

$$d(pV) = S\,dT + p\,dV + M\,dD + N\,d\mu. \tag{12-9}$$

12-2 Statistical-mechanical background. We are concerned with a fluid system of N molecules in a volume V. The system is placed in the external electric field D (Fig. 12-1). We regard D as a parameter. The possible energy states of the system, when D has a particular value, must be found from quantum mechanics and are denoted by $E_j(N, V, D)$. The argument up to Eq. (1-13) in Chapter 1 is unchanged, but in Eq. (1-13) we now write (N constant)

$$dE_j = \left(\frac{\partial E_j}{\partial V}\right)_{N,D} dV + \left(\frac{\partial E_j}{\partial D}\right)_{N,V} dD.$$

Just as we defined p_j as the pressure in state E_j (Eq. 1-7), here we define

$$M_j = -\left(\frac{\partial E_j}{\partial D}\right)_{N,V}$$

as the total moment of the system in state E_j. That is, $-M_j\,dD$ is the total work that has to be done on the system, when in the state E_j, in order to increase D by dD. Then Eqs. (1-15) and (1-16) become

$$T\,dS = d\overline{E} + \overline{p}\,dV + \overline{M}\,dD, \tag{12-10}$$

where $\overline{M}$ is the ensemble average of the mechanical variable M_j:

$$\overline{M} = \sum_j P_j M_j. \tag{12-11}$$

Equations (1-29) through (1-31) are still obtained, for any assigned value of D, and we have

$$Q(N, V, T, D) = \sum_j e^{-E_j(N,V,D)/kT}, \tag{12-12}$$

$$A(N, V, T, D) = -kT \ln Q(N, V, T, D). \tag{12-13}$$

Thus the connection between thermodynamics and statistical mechanics is established. In particular, from Eqs. (12–8), (12–12), and (12–13), we find

$$\overline{M} = kT\left(\frac{\partial \ln Q}{\partial D}\right)_{N,V,T} = \sum_j P_j M_j. \tag{12–14}$$

In an open system (independent variables V, T, μ, D), the grand partition function is defined as

$$\Xi(V, T, \mu, D) = \sum_N Q_N(V, T, D)\lambda^N, \tag{12–15}$$

where $\lambda = e^{\mu/kT}$ and $Q_N(V, T, D) \equiv Q(N, V, T, D)$. We can easily verify from the rule (1–92) and Eq. (12–7) that $\Xi = e^{pV/kT}$ here, as in the absence of a field D. Equations (12–9) and (12–15) then provide us with all the necessary interconnections between Ξ and thermodynamics. For example,

$$\overline{M} = kT\left(\frac{\partial \ln \Xi}{\partial D}\right)_{T,V,\mu} \tag{12–16}$$

$$= \frac{kT}{\Xi}\sum_N \left(\frac{\partial Q_N}{\partial D}\right)_{V,T}\lambda^N = \frac{\sum_N M_N Q_N \lambda^N}{\sum_N Q_N \lambda^N}, \tag{12–17}$$

where $M_N(V, T, D)$ is new notation for the canonical ensemble average $\overline{M}(N, V, T, D)$ in Eq. (12–14). The average indicated in Eq. (12–16) is over N; each M_N in Eq. (12–17) is already averaged over j.

We shall return to the above equations for an open system when we discuss an imperfect gas in an electric field in Section 15–5 (see also Problem 12–2).

12–3 Dilute gas in an electric field. We consider a one-component gas which is dilute enough so that each molecule behaves independently and classical statistics may be used. Then in the canonical ensemble (Eq. 3–10),

$$Q_N(V, T, D) = \frac{Q_1(V, T, D)^N}{N!}, \tag{12–18}$$

where Q_1 here replaces the symbol q used earlier (for example, in Chapters 3, 4, 8, and 9). Equations (12–8), (12–13), and (12–18) then give for the total moment of the system,

$$M_N(V, T, D) = kT\left(\frac{\partial \ln Q_N}{\partial D}\right)_{N,V,T} = NkT\left(\frac{\partial \ln Q_1}{\partial D}\right)_{V,T}$$

$$= NM_1(D, T), \tag{12–19}$$

where M_1 is the canonical ensemble average dipole moment of a single molecule at temperature T and in the external field D. The chemical potential is

$$\mu = -kT\left(\frac{\partial \ln Q_N}{\partial N}\right)_{D,V,T} = -kT \ln \frac{Q_1}{N}; \qquad \lambda = \frac{N}{Q_1}. \tag{12-20}$$

The equation of state will be deduced below.

The problem is reduced by the above equations to one of finding the partition function $Q_1(V, T, D)$ of one molecule in a volume V, external field D, and temperature T. We consider, therefore, a single diatomic or polyatomic (without internal rotation) molecule, in a box of volume V, and in an electric field D. Suppose the molecule has a permanent dipole moment μ_0. Also, we assume that the field induces in the molecule a further, additive, moment αD (in the direction of the field), where α is the polarizability. The polarizability is a purely quantum-mechanical quantity, independent of T. The work necessary to create this induced moment is [compare Eq. (12-6)]

$$-\int_0^D \alpha D\, dD = -\frac{\alpha D^2}{2}.$$

Let θ be the angle between the direction of the field and the axis of the permanent moment μ_0 (Fig. 12-2a). In the presence of the field, the rotation of the molecule is not free, because of the dependence on θ of the potential energy of the permanent dipole (Fig. 12-2b):

$$U = -\mu_0 D \cos\theta.$$

[This problem should be compared with those associated with Eqs. (9-18) and (9-19).] The total electrostatic potential energy U_1 of a molecule in the field D is then

$$U_1 = -\tfrac{1}{2}\alpha D^2 - \mu_0 D \cos\theta. \tag{12-21}$$

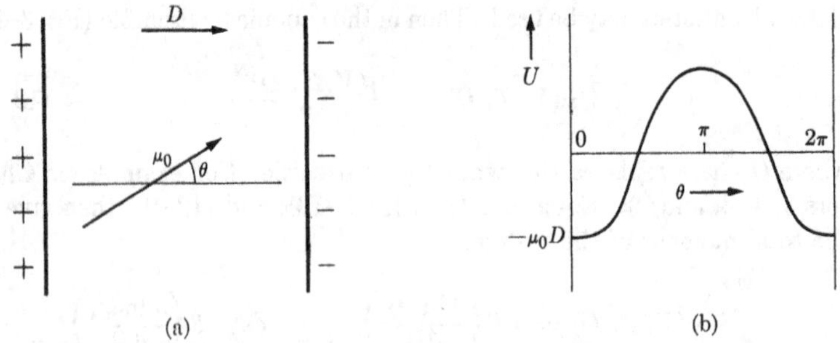

FIG. 12-2. (a) Permanent dipole μ_0 oriented at an angle θ with respect to the field D. (b) Potential energy of dipole as a function of θ.

We assume that U_1 simply makes an additive contribution to the Hamiltonian of the molecule. Let us call U_1 the rotational potential energy. The rotational Hamiltonians of Chapters 8 and 9 contain kinetic energy terms only. We treat rotation classically, as in these earlier chapters. Let φ be an azimuthal angle around the direction of the field D in Fig. 12–2(a). In the absence of a field, all orientations (θ, φ) of the axis of the moment μ_0 are equally probable. Therefore the probability of finding an orientation in the range $d\theta\, d\varphi$ is proportional to $\sin\theta\, d\theta\, d\varphi$. In the presence of a field, some orientations are favored over others because of Eq. (12–21). The partition function $Q_1(D)$ in the presence of a field is altered from that in the absence of a field, $Q_1(0)$, only through the rotational potential energy U_1. Thus we can easily take care of the effect of the field by extracting from $Q_1(0)$ the result, 4π, of integrating over θ and φ with $U_1 = 0$, and then include an integration over θ and φ in $Q_1(D)$:

$$Q_1(D) = \frac{Q_1(0)}{4\pi} \int_0^{2\pi} \int_0^{\pi} e^{-U_1/kT} \sin\theta\, d\theta\, d\varphi. \tag{12–22}$$

In the notation of Chapters 8 and 9,

$$Q_1(0) = q_t q_r q_v q_e, \tag{12–23}$$

where q_t, q_r, etc., are unchanged from our earlier treatment.

We digress to note at this point that the equation of state of the gas is

$$p = kT\left(\frac{\partial \ln Q_N}{\partial V}\right)_{N,D,T} = NkT\left(\frac{\partial \ln Q_1(0)}{\partial V}\right)_T = \frac{NkT}{V}. \tag{12–24}$$

We substitute Eq. (12–21) in Eq. (12–22) and find easily

$$Q_1(D) = Q_1(0) e^{\alpha D^2/2kT} \frac{\sinh y}{y}, \tag{12–25}$$

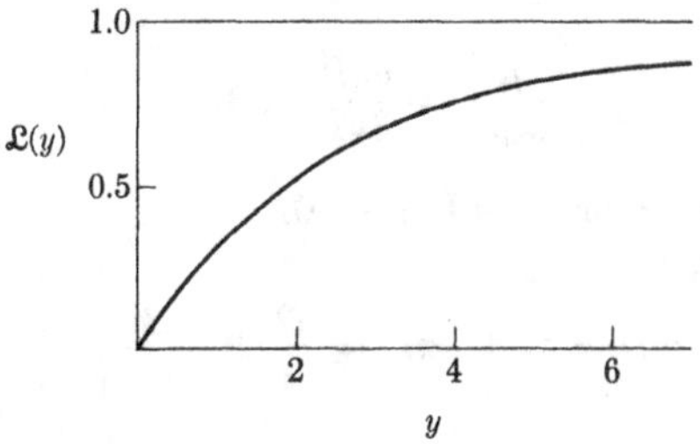

FIG. 12–3. Langevin function, $\mathcal{L}(y) = \coth y - y^{-1}$.

where $y = \mu_0 D/kT$. Then, from Eq. (12–19),

$$M_1 = \alpha D + \mu_0 \mathcal{L}(y), \tag{12–26}$$

where $\mathcal{L}(y)$ is the Langevin function (Fig. 12–3):

$$\mathcal{L}(y) = \coth y - \frac{1}{y} \cdot \tag{12–27}$$

In practice, $y \ll 1$. For example, if $\mu_0 = 1$ Debye, $D = 1000$ volts·cm^{-1}, and $T = 300°$K, $y = O(10^{-4})$. The expansion of $\mathcal{L}(y)$ about $y = 0$ is (Problem 12–3)

$$\mathcal{L}(y) = \frac{y}{3} - \frac{y^3}{45} + \cdots . \tag{12–28}$$

Ordinarily only the first term is required. Similarly,

$$\frac{\sinh y}{y} = 1 + \frac{y^2}{6} + \cdots .$$

Equations (12–25) and (12–26) become then

$$Q_1(D) = Q_1(0) e^{\alpha D^2/2kT}\left[1 + \frac{1}{6}\left(\frac{\mu_0 D}{kT}\right)^2\right], \tag{12–29}$$

$$M_1 = D\left(\alpha + \frac{\mu_0^2}{3kT}\right) \cdot \tag{12–30}$$

Here M_1 is the average dipole moment of one molecule; αD is the induced moment (independent of T), and $\mu_0^2 D/3kT \equiv \bar{\mu}(T)$ is the statistically averaged component (in the direction of the field) of the rotating permanent moment μ_0. If the permanent moment were completely oriented by the field [$D \to \infty$ or $T \to 0$; $\mathcal{L}(y) \to 1$], $\bar{\mu}$ would approach μ_0 itself. That is, $\bar{\mu}_{\max} = \mu_0$. It is clear that at ordinary field strengths and temperatures the rotation is almost free and the orientation of μ_0 in the field is very slight, for

$$\frac{\bar{\mu}}{\bar{\mu}_{\max}} = \frac{\mu_0 D}{3kT} \ll 1.$$

In the limit as $T \to \infty$ or $D \to 0$, $\bar{\mu} \to 0$.

From Eq. (12–1),

$$D = \mathcal{E} + 4\pi P = \frac{D}{\epsilon} + \frac{4\pi M}{V},$$

or

$$\frac{\epsilon - 1}{\epsilon} = \frac{4\pi M}{DV} \cdot \tag{12–31}$$

In the limit as $\rho = N/V \to 0$, we then have the following equation for the dielectric constant ϵ of the gas:

$$\epsilon - 1 = \frac{4\pi M_1 N}{DV} = 4\pi\rho\left(\alpha + \frac{\mu_0^2}{3kT}\right). \tag{12-32}$$

A plot of the experimental quantity $(\epsilon - 1)/4\pi\rho$ against $1/T$ will give α as the intercept and $\mu_0^2/3k$ as the slope. This well-known method of determining permanent dipole moments is treated in physical chemistry texts and will not be considered further here.

Finally, let us deduce some of the thermodynamic functions for a dilute gas in an electric field. First, from Eqs. (12-20) and (12-29),

$$\frac{\mu}{kT} = \frac{\mu(0)}{kT} - \frac{D^2}{2kT}\left(\alpha + \frac{\mu_0^2}{3kT}\right), \tag{12-33}$$

where $\mu(0)$ is the value of μ when $D = 0$ (see Chapters 8 and 9). Next, from Eqs. (12-8), (12-18), and (12-29), we find for the entropy

$$\frac{S}{Nk} = \frac{S(0)}{Nk} - \frac{1}{6}\left(\frac{\mu_0 D}{kT}\right)^2. \tag{12-34}$$

The entropy is reduced in the presence of a field because rotation is slightly hindered. Also, for the energy,

$$\frac{E}{NkT} = \frac{S}{Nk} - \frac{1}{N}\ln Q = \frac{E(0)}{NkT} - \frac{D^2}{kT}\left(\frac{\alpha}{2} + \frac{1}{3}\frac{\mu_0^2}{kT}\right). \tag{12-35}$$

The added terms in Eq. (12-35), with the field on, are clearly just the average value of U_1/kT (Problem 12-4).

The terms in D^2 in Eqs. (12-33) through (12-35) are very small, say of order 10^{-6} to 10^{-8} in typical cases. The leading term in each of these equations is of order unity.

12-4 Lattice of noninteracting magnetic dipoles. In this section we discuss an idealized magnetic problem. Suppose we have a lattice of M equivalent magnetic dipoles (associated, say, with electron or nuclear spins), each of which can exist in only two orientations or states: $\uparrow$, in the direction of the magnetic field H; or $\downarrow$, against the field. The potential energy of a dipole or spin is $-mH$ if oriented with the field ($\uparrow$), and $+mH$ if oriented against the field ($\downarrow$), where m is the magnetic moment. The dipoles are assumed not to interact with each other (or, more accurately, only "weak" interaction is allowed—see Chapter 3); each dipole behaves independently of the rest. This is a good model for nuclear spin systems but not for electron spin systems (e.g., ferromagnetism). In our discussion of ferromagnetism in Chapter 14 we shall introduce nearest-neighbor interactions.

Let N be the number of ↓ states and $M - N$ the number of ↑ states. For a given value of N, the total potential energy of the dipoles in the field is

$$mHN - mH(M - N) = (2N - M)mH.$$

For given M, H, and T, let $\overline{N}$ be the average value of N (N can range from 0 to M). Then the work necessary to increase H by dH is $-I\,dH$, where

$$I = (M - 2\overline{N})m. \tag{12-36}$$

In Eq. (12–36), I is the intensity of magnetization; it is the average excess number of ↑ states over ↓ states, multiplied by m. In the absence of a magnetic field, the two states have the same potential energy (zero), and therefore occur in equal numbers; hence $I = 0$ when $H = 0$.

The basic thermodynamic equations we use are*

$$dE = T\,dS - I\,dH + \mu\,dM, \tag{12-37}$$

$$dA = -S\,dT - I\,dH + \mu\,dM. \tag{12-38}$$

On integrating (H is intensive, I extensive),

$$A = E - TS = \mu M. \tag{12-39}$$

For a given value of N, there are $M!/N!(M - N)!$ possible arrangements of the two states ↑ and ↓ among the M positions. Also, for given N, the energy of the dipoles in the field H is $(2N - M)mH$, as already mentioned. Therefore, the canonical ensemble partition function for a system of M dipoles at T and H is

$$Q(M, T, H) = q^M \sum_{N=0}^{M} \frac{M!e^{(M-2N)mH/kT}}{N!(M - N)!}, \tag{12-40}$$

where $q(T)$ is the "internal" partition function of a dipole—a quantity we need not specify further. The summation in Eq. (12–40) is easy to carry out. We find

$$\begin{aligned} Q(M, T, H) &= q^M e^{MmH/kT}(1 + e^{-2mH/kT})^M \\ &= \left(2q \cosh \frac{mH}{kT}\right)^M. \end{aligned} \tag{12-41}$$

* In S. M., pp. 288–289, the quantity E_m is equivalent to $E + HI$ here, and the partition function Δ is equivalent to Q here.

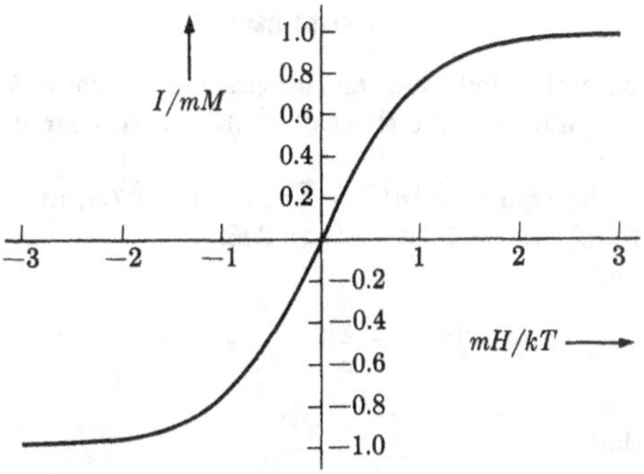

FIG. 12–4. Magnetization (I/mM) of lattice of noninteracting magnetic dipoles as a function of magnetic field (mH/kT).

From $A = -kT \ln Q$, and Eqs. (12–38) and (12–41), we find for the magnetization I,

$$I = (M - 2\overline{N})m = kT\left(\frac{\partial \ln Q}{\partial H}\right)_{M,T} = mM \tanh \frac{mH}{kT}. \quad (12\text{–}42)$$

This function is plotted in Fig. 12–4. When $H = 0$, $I = 0$ (i.e., $\overline{N} = M/2$). When $H \to +\infty$, all dipoles are oriented $\uparrow$, and $I = mM$ (i.e., $\overline{N} = 0$). When $H \to -\infty$, $I = -mH$ and $\overline{N} = M$.

The formal resemblance between this model and Langmuir adsorption (or an ideal lattice gas) is rather obvious. See also Section 7–4 on the elasticity of a polymer chain. The two states $\uparrow$ and $\downarrow$ correspond to an empty site and an occupied site, respectively. If we denote the fraction of dipoles in the state $\downarrow$ by θ, then

$$\frac{I}{mM} = 1 - 2\theta = \tanh \frac{mH}{kT}. \quad (12\text{–}43)$$

The Langmuir adsorption isotherm, $\theta = q\lambda/(1 + q\lambda)$ (Eq. 7–16), can be put in the same form:

$$1 - 2\theta = \frac{1 - q\lambda}{1 + q\lambda} = \tanh x, \quad (12\text{–}44)$$

where

$$x = -\ln (q\lambda)^{1/2}.$$

In the magnetic, adsorption, and polymer elasticity problems, the magnetic field, chemical potential, and force, respectively, play corresponding roles in shifting the relative population of the two states involved in each case.

Problems

12–1. Use Coulomb's law and an integration to show that the (normal) electric field of an infinite plane sheet of charge, with charge density σ, is $2\pi\sigma$. (Page 202.)

12–2. Deduce the equations $pV = \overline{N}kT$, $\lambda = \overline{N}/Q_1$, and $M = \overline{N}M_1$ from Eqs. (12–9), (12–15), and (12–18). (Page 205.)

12–3. Show that

$$\mathfrak{L}(y) = \frac{y}{3} - \frac{y^3}{45} + \cdots .$$

(Page 208.)

12–4. Show that

$$\overline{U}_1 = \frac{\int_0^\pi U_1 e^{-U_1/kT} \sin\theta\, d\theta}{\int_0^\pi e^{-U_1/kT} \sin\theta\, d\theta} = -D^2\left(\frac{\alpha}{2} + \frac{1}{3}\frac{\mu_0^2}{kT}\right) + \cdots .$$

(Page 209.)

12–5. Show that

$$\bar{\mu} = \frac{\int_0^\pi \mu_0 \cos\theta\, e^{\mu_0 D\cos\theta/kT} \sin\theta\, d\theta}{\int_0^\pi e^{\mu_0 D\cos\theta/kT} \sin\theta\, d\theta} = \frac{\mu_0^2 D}{3kT} + \cdots .$$

12–6. If $\mu_0 D \gg kT$, a gas molecule will oscillate about $\theta = 0$ instead of rotating. In this case it is appropriate to express $\bar{\mu}$ as a power series in T instead of T^{-1}. Find the first two terms in the series.

12–7. Obtain the basic thermodynamic functions for a sample of an ideal gas at a height h in a gravitational field. Take the gravitational potential energy as zero at $h = 0$. Find the density of the gas as a function of h, by use of the equilibrium condition $\mu(h) = \mu(0)$, where μ = chemical potential.

12–8. Calculate the value of $y = \mu_0 D/kT$ for $\mu_0 = 1$ Debye, $D = 1000$ volts·cm^{-1}, and $T = 300$°K.

12–9. Calculate $\epsilon - 1$ from Eq. (12–32) for H_2O vapor at 100°C and 1 atm pressure. Take $\alpha = 1.68 \times 10^{-24}$ cm^3 and $\mu_0 = 1.84$ Debye.

12–10. Derive an expression for the equilibrium ratio of the number of $\uparrow$ dipoles to the number of $\downarrow$ dipoles, for given H and T. Show that if suddenly the field is switched from H to $-H$, and if the distribution of dipoles between $\uparrow$ and $\downarrow$ remains unchanged an appreciable time before adjusting to the new direction of the field, then during the time before adjustment the system has, in effect, a negative temperature. This is a useful point of view in connection with some nuclear spin systems.

12–11. Derive equations for S/Mk and E/MmH as functions of kT/mH (take $q \equiv 1$).

12–12. Investigate fluctuations in N and I in the magnetic system of Section 12–4.

Supplementary Reading

Fowler and Guggenheim, Chapter 14.
Kittel, Section 18.
Mayer and Mayer, Chapter 15.
Rushbrooke, Chapter 10.
Wilson, Chapter 10.

CHAPTER 13

CONFIGURATION OF POLYMER MOLECULES AND RUBBER ELASTICITY

In this chapter we present a brief introduction to polymer configuration problems and to the theory of rubber elasticity. The treatment is based to a considerable extent on the work of James and Guth (Supplementary Reading list). A number of other topics having to do primarily with polymer molecules in solution will be discussed in Chapter 21.

The basic prototype for polymer chains can be represented as

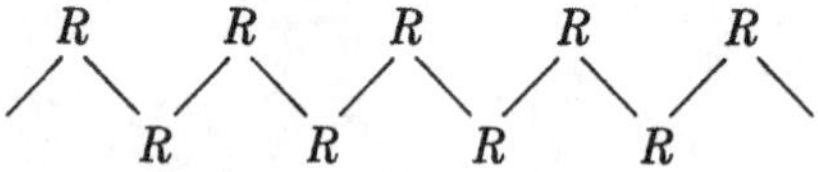

where R is the monomer unit, and the chain continues indefinitely at both ends. The configuration shown above is the fully extended configuration. Actually, because of rotation of the attached groups of R's around each R—R bond, a great many configurations are possible, of which the extended configuration is only one. One of the fundamental problems in polymer statistics is to deduce the relative number of configurations of a long polymer chain consistent with a specified end-to-end distance (Fig. 13–1). This problem is closely related to problems in brownian motion, random walks, diffusion, etc.

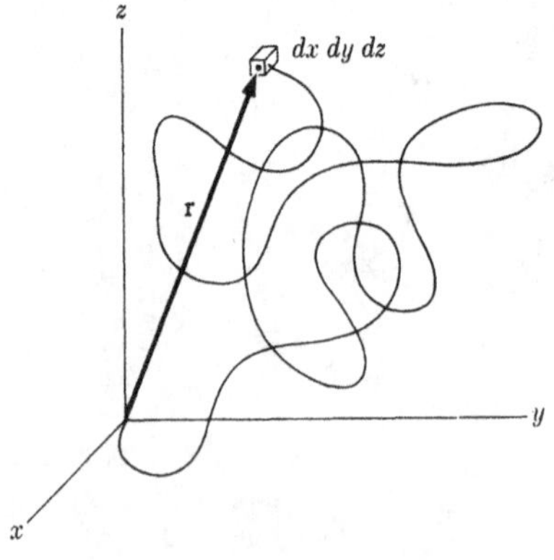

FIG. 13–1. Long polymer chain with end-to-end vector **r**.

One of the simplest polymer chains, polymethylene, has $R = CH_2$. Other well-known possibilities for R are:

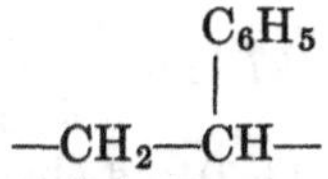

$$
\begin{array}{cc}
C_6H_5 & CH_3 \\
| & | \\
-CH_2-CH- & -CH_2-C=CH-CH_2- \\
\text{Polystyrene} & \text{Rubber}
\end{array}
$$

In typical cases the number of monomers in a chain might be from 100 to 10,000.

13-1 Freely jointed chain. To handle this problem we first have to generalize the formalism of Section 7–4. Consider a linear polymer chain made up of M units, where M is large enough so that one chain can be considered a thermodynamic system. Each unit can exist in the states $i = 1, 2, \ldots, n$ with partition functions $j_i(T)$ and lengths l_i. The total length of the chain is l. The system (chain) is characterized thermodynamically by l, M, T. The canonical ensemble partition function is then

$$Q(l, M, T) = \sum_{\mathbf{M}} M! \prod_{i=1}^{n} \frac{j_i^{M_i}}{M_i!}, \tag{13-1}$$

where M_i is the number of units with length l_i, and the sum is over all sets $\mathbf{M} = M_1, M_2, \ldots, M_n$ consistent with the restrictions

$$\sum_{i=1}^{n} M_i = M, \tag{13-2}$$

$$\sum_{i=1}^{n} l_i M_i = l. \tag{13-3}$$

Equation (13–1) is a rather obvious generalization of Eq. (7–58). Here, for purposes of symmetry, we choose l as independent variable instead of one of the M_i [M_α was used in Eq. (7–58)]. The appropriate thermodynamic equation is

$$dA = -S\,dT + \tau\,dl + \mu\,dM, \tag{13-4}$$

with

$$A = -kT \ln Q \tag{13-5}$$

and $\tau =$ force pulling on the chain.

The restriction (13–3) is troublesome; to avoid it we change to another partition function. We use the partition function

$$\Delta(\tau, M, T) = \sum_{l} Q(l, M, T) e^{\tau l / kT} \tag{13-6}$$

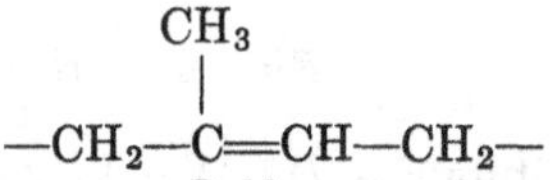

This is the analog of Eq. (1–87). The connection with thermodynamics is

$$dF = -S\,dT - l\,d\tau + \mu\,dM, \tag{13-7}$$

$$F = A - \tau l = \mu M = -kT \ln \Delta. \tag{13-8}$$

We substitute Eqs. (13–1) and (13–3) in Eq. (13–6) and obtain

$$\Delta(\tau, M, T) = \sum_{\mathbf{M}} M! \prod_{i=1}^{n} \frac{(j_i e^{\tau l_i/kT})^{M_i}}{M_i!},$$

where now the only restriction on sets **M** is (13–2). The sum can be carried out immediately, and we have

$$\Delta(\tau, M, T) = \left(\sum_{i=1}^{n} j_i e^{\tau l_i/kT}\right)^M \equiv \xi(\tau, T)^M, \tag{13-9}$$

This gives, for example, for the average length $\bar{l}$ of the chain at a given force τ,

$$\bar{l} = -\left(\frac{\partial F}{\partial \tau}\right)_{M,T} = kT\left(\frac{\partial \ln \Delta}{\partial \tau}\right)_{M,T} = MkT\left(\frac{\partial \ln \xi}{\partial \tau}\right)_T$$

$$= \frac{\sum_l lQ(l, M, T)e^{\tau l/kT}}{\sum_l Q(l, M, T)e^{\tau l/kT}} = M \frac{\sum_i l_i j_i e^{\tau l_i/kT}}{\sum_i j_i e^{\tau l_i/kT}}. \tag{13-10}$$

Equations (13–1) through (13–10) are formally the same as Eqs. (7–23) through (7–29). Therefore the notation $\Delta = \xi^M$ in Eq. (13–9) is appropriate. The partition function ξ for one unit has the same form as Δ in Eq. (13–6) for the entire chain [just as ξ in Eq. (7–27) resembles Ξ in Eq. (7–25)].

We now consider a special case, a chain of M units, each of length a, with "free" joints between units. That is, if we choose one end of any unit as origin, the other end of the unit moves freely (in the absence of a force on the chain) over the surface of a sphere with radius a (Fig. 13–2). The ends of the chain are a distance l apart and are on the x-axis. If the left end of the chain is considered fixed, we want to calculate, among other things, the equilibrium force τ along the x-axis necessary to hold the chain extended a distance l (Fig. 13–2). Real polymer chains do not have free joints between monomers (R units), but an approximate connection can be established between real chains and this idealized model (see Section 13–2).

The contribution of any one unit to l can range from $-a$ to $+a$. Thus l_i in Eq. (13–9) can vary continuously between these limits. We use x for this continuous variable. It is clear from Eq. (13–10) that j_i is proportional to the probability of a length l_i being observed when there is no force on

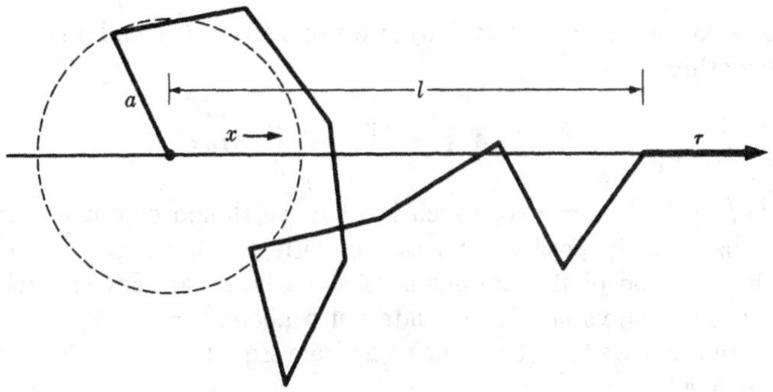

FIG. 13-2. Two-dimensional version of freely jointed chain. Each unit is of length a. The ends are on the x-axis and are a distance l apart.

the chain ($\tau = 0$). When there is a force, $j_i e^{\tau l_i/kT}$ is proportional to this probability. In the present problem, then, we let $j(x, T)\,dx$ be proportional to the probability that the end of a freely moving unit (Fig. 13-2) will have an x-component in dx. It is easy to see (Problem 13-1) that this probability is in fact independent of x (in the range $-a \leq x \leq +a$). Therefore we have

$$\xi = \int_{-a}^{+a} j(T) e^{\tau x/kT}\, dx$$

$$= \frac{2ja}{t} \sinh t, \tag{13-11}$$

where $t = \tau a/kT$. From Eq. (13-10) we find for the length-force relation

$$\bar{l} = MkT\left(\frac{\partial \ln \xi}{\partial \tau}\right)_T = Ma\mathfrak{L}(t), \tag{13-12}$$

or

$$t = \mathfrak{L}^{-1}\left(\frac{\bar{l}}{Ma}\right), \tag{13-13}$$

where $\mathfrak{L}$ is the Langevin function defined in Eq. (12-30) and $\mathfrak{L}^{-1}$ is the inverse Langevin function. The maximum extension is Ma; to achieve this we need $t \to \infty$. Figure 12-3 provides us with a plot of $\bar{l}/Ma$ (ordinate) against t (abscissa).

The occurrence of the same (Langevin) function here and in Section 12-3 is not surprising. In Section 12-3 we were dealing with freely rotating dipoles perturbed in their rotation by an electric field. Here we have freely rotating units of a chain perturbed in their rotation by a force pulling on the chain.

At small extensions ($\bar{l} \ll Ma$), we use $\mathcal{L}(t) = t/3$ and find the linear relationship

$$t = \frac{3\bar{l}}{Ma} \quad \text{or} \quad \tau = \frac{3kT\bar{l}}{Ma^2}. \tag{13–14}$$

Thus $\bar{l} = 0$ when $\tau = 0$, which is what we should expect on symmetry grounds (l can be positive or negative; with $\tau > 0$ we have $\bar{l} > 0$).

The question of fluctuations is of some interest. For example, when $\tau = 0$ does the value of l fluctuate much about $\bar{l} = 0$? By the methods of Section 2–1 we find (Problem 13–2) from Eq. (13–10) for the fluctuation in l at fixed τ,

$$\overline{l^2} - (\bar{l})^2 = kT\left(\frac{\partial \bar{l}}{\partial \tau}\right)_{M,T}. \tag{13–15}$$

From Fig. 12–3, we see that $\partial\bar{l}/\partial\tau$ is largest at small extensions. In the limit as $\tau \to 0$ (Eq. 13–14),

$$\frac{\overline{l_0^2}}{(Ma)^2} = \frac{\sigma_l^2}{(Ma)^2} = \frac{1}{3M}, \tag{13–16}$$

where we compare the fluctuation $\overline{l_0^2}$ with the maximum extension Ma instead of $\bar{l}$, since $\bar{l} \to 0$ as $\tau \to 0$. The subscript on $\overline{l_0^2}$ refers to $\tau = 0$. Thus, if $M = 3000$, $\sigma_l/Ma = 10^{-2}$, a rather significant fluctuation. The conclusion we draw from this is that chains of this size are not quite macroscopic in the thermodynamic sense. Hence, although quantities such as τ, $\bar{l}$, $\overline{l^2}$, T, etc., are well defined (see Sections 3–4 and 7–2) and equations of the type (13–10), (13–13), and (13–15) are valid, functions such as E, F, S, A, etc., for a single chain are slightly fuzzy in their thermodynamic significance. These latter functions become sharply defined, of course, for a system consisting of a *large number* of chains (as is always the case in practice).

From Eq. (13–16) we have that $\overline{l_0^2} = Ma^2/3$. Then Eq. (13–14) can be written

$$\tau = \frac{kT\bar{l}}{\overline{l_0^2}}. \tag{13–17}$$

Actually, the length-force equation in this form is applicable to small extensions of *any kind* of chain with $\bar{l}_0 = 0$. This relation follows directly from Eq. (13–15) and the necessary (by symmetry) linear dependence of τ on $\bar{l}$ at small $\bar{l}$.

The generalization of Eq. (13–17) to $\bar{l}_0 \neq 0$ (as, for example, in Section 7–4) and extensions that are not small can be deduced by a method which has very wide applicability in statistical mechanics. Define Δ_0 from Eq. (13-6) by

$$\Delta_0(M, T) \equiv \Delta(0, M, T) = \sum_l Q(l, M, T),$$

and rewrite Eq. (13-10) as

$$\bar{l} = \frac{1}{\Delta/\Delta_0}\left(\frac{\partial \Delta/\Delta_0}{\partial \tau/kT}\right)_{M,T}. \tag{13-18}$$

Now if we expand the exponential in Eq. (13-6), we find

$$\frac{\Delta}{\Delta_0} = \frac{\sum_{i=0}^{\infty} (1/i!) \sum_l Q(l, M, T)(\tau l/kT)^i}{\sum_l Q(l, M, T)} = \sum_{i=0}^{\infty} \frac{1}{i!}\left(\frac{\tau}{kT}\right)^i \overline{l_0^i},$$

where $\overline{l_0^i}$, a function of M and T in general, is the average value of l^i at *zero force*. This follows from Eq. (13-10), which states that the probability of the chain having a length l is proportional to $Qe^{\tau l/kT}$ when the force is τ and to Q when $\tau = 0$ [see also Eq. (2-13)]. The occurrence of "unperturbed" ($\tau = 0$) averages is the essential point here, since these are not so difficult to calculate. Equation (13-18) becomes, then,

$$\bar{l} = \frac{\sum_{i=1}^{\infty} [1/(i-1)!](\tau/kT)^{i-1}\overline{l_0^i}}{\sum_{i=0}^{\infty} (1/i!)\,(\tau/kT)^i\,\overline{l_0^i}}. \tag{13-19}$$

This gives $\bar{l} - \bar{l}_0$ as a power series in τ/kT, or vice versa. We shall not pursue this method further here (but see Problem 13-3).

As just indicated, $Q(l, M, T)$ is proportional to the probability that the free ($\tau = 0$) chain has a length l (for given M and T). An equivalent statement is that $Q(l, M, T)$ is proportional to the number of configurations (a configurational degeneracy) the chain can assume consistent with a length l, for given M and T. The dependence of this probability on l is perhaps the most important single property of a polymer chain. We are now in a position to deduce this dependence for a freely jointed chain (and more general chains) from Q. The same results can be obtained from the theory of random walks, without use of any of our statistical thermodynamical formalism.

The general method we employ is to integrate the length-force relation to obtain A and hence Q, using Eqs. (13-4) and (13-5). It should be noted that at this point we make use of macroscopic thermodynamics; hence in the following we are dealing implicitly with the limit of very long chains ($M \to \infty$). For the freely jointed chain,

$$dA = \tau\, dl = \frac{kT}{a}\,\mathcal{L}^{-1}\left(\frac{l}{Ma}\right) dl \qquad (T, M \text{ constant}),$$

and therefore

$$\frac{Q(l, M, T)}{Q(0, M, T)} = \exp\left[-\frac{1}{a}\int \mathcal{L}^{-1}\left(\frac{l}{Ma}\right) dl\right]. \tag{13-20}$$

This is the probability of a free chain having a length l relative to the probability of a length $l = 0$. It is also the ratio of the number of configurations of the chain with length l to the number with length zero. If we use the expansion (Problem 13-4)

$$\mathcal{L}^{-1}(x) = 3x + \tfrac{9}{5}x^3 + \cdots, \tag{13-21}$$

which can be deduced from Eq. (12-31), Eq. (13-20) becomes

$$\frac{Q(l, M, T)}{Q(0, M, T)} = \exp\left\{-M\left[\frac{3}{2}\left(\frac{l}{Ma}\right)^2 + \frac{9}{20}\left(\frac{l}{Ma}\right)^4 + \cdots\right]\right\}. \tag{13-22}$$

When $l \ll Ma$, we keep just the first term in this expansion and obtain the gaussian probability distribution ordinarily used,

$$\frac{Q(l, M, T)}{Q(0, M, T)} = e^{-3l^2/2Ma^2}. \tag{13-23}$$

This equation also follows directly on integrating the linear length-force equation, (13-14). Thus the gaussian probability distribution for the length of a free chain and the linear length-force relation for a chain under an extending force have the same limits of validity (Problem 13-5). To go beyond the linear length-force range, configurations of the chain with values of l outside the gaussian region become involved. That the "gaussian region" is in fact quite extensive can be seen as follows. The ratio of the correction term in Eq. (13-22) to the gaussian term is $(3/10)(l/Ma)^2$. Even for a very large extension, this quantity is small compared with unity. For example, take $M = 1000$ and an extension l' of ten times the root mean-square extension $(Ma^2/3)^{1/2}$. Then

$$\frac{3}{10}\left(\frac{l'}{Ma}\right)^2 = \frac{10}{M} = \frac{1}{100}.$$

We have been emphasizing the probability significance of Q and $Qe^{\tau l/kT}$ for the length of a chain with *fixed force*. But one must also keep in mind that Q has the usual connections with the thermodynamic properties of a chain with *fixed length* (fluctuating force). An example is the deduction of the linear $l - \tau$ relation from $Q(l)$ in Eq. (13-23) (Problem 13-6). Another example is the derivation of an equation for the dependence of the entropy S of a chain on its length l. In the present model, $Q(l, M, T)$ has the functional form (Eq. 13–20)

$$Q(l, M, T) = Q(0, M, T)f(l, M). \tag{13–24}$$

This relation, combined with Eq. (1–33) for S, leads immediately to

$$S(l, M, T) - S(0, M, T) = k \ln f = k \ln [Q/Q(0)]$$

$$= -\frac{k}{a}\int \mathcal{L}^{-1}\left(\frac{l}{Ma}\right) dl \tag{13–25}$$

$$= -\frac{3kl^2}{2Ma^2} = -\frac{3}{2}Mk\left(\frac{l}{Ma}\right)^2 \qquad (l \ll Ma). \tag{13–26}$$

The entropy is a maximum (largest number of configurations) at $l = 0$ and decreases with increasing l. The right side of Eq. (13–25) approaches $-\infty$ when $l \to Ma$. This, however, is pushing the model too far: a real polymer molecule, when fully extended, will not be rigid, but will have internal vibrational motion. The analog of this situation for an ideal gas is letting $V \to 0$ in Eq. (4–20).

It is possible to write Eq. (13–20) in an alternative and more explicit form. Thus, from Eqs. (13–11) and (13–13),

$$Q = e^{-A/kT} = \Delta\, e^{-\tau l/kT}$$

$$= \left[\frac{2ja \sinh \mathcal{L}^{-1}(l/Ma)}{\mathcal{L}^{-1}(l/Ma)}\right]^M \exp\left[-\frac{l}{a}\mathcal{L}^{-1}\left(\frac{l}{Ma}\right)\right],$$

or

$$\frac{Q(l, M, T)}{Q(0, M, T)} = \left[\frac{\sinh \mathcal{L}^{-1}(l/Ma)}{\mathcal{L}^{-1}(l/Ma)}\right]^M \exp\left[-\frac{l}{a}\mathcal{L}^{-1}\left(\frac{l}{Ma}\right)\right]. \tag{13–27}$$

Here again we should note that the limit $M \to \infty$ is implicit, since we have made use of the thermodynamic equivalence of the partition functions Q and Δ. Equation (13–22) may also be obtained from Eq. (13–27) (Problem 13–7).

From Eq. (13–19) we can derive a more general version of Eq. (13–23) [or Eq. (13–22)] for any polymer chain (see also Problem 13–3). We integrate $dA = \tau\, dl$, where

$$\tau = \frac{kT(l - l_0)}{\overline{l_0^2} - (\bar{l}_0)^2}, \tag{13–28}$$

and obtain

$$\frac{Q(l, M, T)}{Q(l_0, M, T)} = \exp\left[-\frac{1}{2}\frac{(l - l_0)^2}{\overline{l_0^2} - (\bar{l}_0)^2}\right], \tag{13–29}$$

where M is a number proportional to the mass of the polymer molecule. Thus a gaussian probability distribution about $l = l_0 \equiv \bar{l}_0$, for small extensions, is *always* found. Since in general $\bar{l}_0$ and $\overline{l_0^2}$ are functions of temperature, Eq. (13–26) is somewhat more complicated here. For example, if $\bar{l}_0(T) \equiv 0$, we find (Problem 13–8)

$$S(l, M, T) - S(0, M, T) = -\frac{Mk}{2}\frac{l^2}{M\overline{l_0^2}}\left[1 - \left(\frac{\partial \ln \overline{l_0^2}}{\partial \ln T}\right)_M\right]. \tag{13–30}$$

We should expect $\overline{l_0^2}$ to increase with temperature for a real molecule owing to increased freedom of rotation about chemical bonds in the chain.

As a final topic in this section, we consider briefly the one-dimensional version of a freely jointed chain. The chain has M units, each of length a. Each unit must now always lie on the x-axis so that the possible contributions of a unit to l are the two values $-a$ or $+a$. Thus the chain resembles a folding ruler. In random-walk language, this is a random walk along a line with each step of length $+a$ or $-a$. In Eq. (13–9), we take $l_1 = +a$, $l_2 = -a$, and $j_1 = j_2 = j$. Then

$$\Delta = \left(2j \cosh \frac{\tau a}{kT}\right)^M \tag{13–31}$$

and, from Eq. (13–10),

$$\bar{l} = Ma \tanh t, \qquad t = \tau a/kT, \tag{13–32}$$

$$t = \tanh^{-1}\left(\frac{\bar{l}}{Ma}\right) = \frac{\bar{l}}{Ma} + \frac{1}{3}\left(\frac{\bar{l}}{Ma}\right)^3 + \cdots. \tag{13–33}$$

Just as the three-dimensional freely jointed chain under a pulling force resembles a gas of dipolar molecules oriented by an electric field (Section 12–3), the one-dimensional freely jointed chain under a force resembles a system of magnetic dipoles in a magnetic field (Section 12–4). In particular, Eqs. (12–45) and (13–32) should be compared. Figure 12–4 is also a plot of $\bar{l}/Ma$ (ordinate) against t (abscissa).

The present one-dimensional problem is a special case of the model discussed at the beginning of Section 7–4 using different independent variables. The connection in notation is $l_\alpha = -a$, $l_\beta = +a$, and $j_\alpha = j_\beta = j$.

By the same methods as for the three-dimensional case, we find (Problem 13-9)

$$\frac{Q(l, M, T)}{Q(0, M, T)} = \exp\left[-\frac{1}{a}\int \tanh^{-1}\left(\frac{l}{Ma}\right) dl\right] \tag{13-34}$$

$$= \cosh^M\left(\tanh^{-1}\frac{l}{Ma}\right)\exp\left(-\frac{l}{a}\tanh^{-1}\frac{l}{Ma}\right) \tag{13-35}$$

$$= \exp\left\{-M\left[\frac{1}{2}\left(\frac{l}{Ma}\right)^2 + \frac{1}{12}\left(\frac{l}{Ma}\right)^4 + \cdots\right]\right\}, \tag{13-36}$$

and $\overline{l_0^2} = Ma^2$.

13-2 Gaussian probability distribution for free polymer molecules. In this section we discuss further the gaussian probability distribution for free ($\tau = 0$) polymer molecules with the usual property $\overline{l_0} = 0$. Since the whole section is concerned with free chains, we drop the subscript zero on $\overline{l_0}$, $\overline{l_0^2}$, etc.

We saw in the preceding section that if one end of a long polymer molecule is chosen as origin and the other end is forced to lie on a preassigned line passing through the origin, say the x-axis, then according to Eq. (13-29), the probability that the ends of the molecule will be separated by a distance l is proportional, for l not too large, to $\exp(-l^2/2\overline{l^2})$. Since the direction of the preassigned line is arbitrary, we can make the equivalent alternative statement that if one end of a polymer molecule is chosen as origin, the probability that the other end will lie in a specified volume element $dx\,dy\,dz$, a distance r from the origin (Fig. 13-1), is proportional to

$$e^{-r^2/2\overline{l^2}}\,dx\,dy\,dz.$$

As a next step, we can conclude that if one end of a polymer molecule is chosen as origin, the probability that the other end is at a distance between r and $r + dr$, *irrespective of direction*, is

$$P(r)\,dr = \frac{e^{-r^2/2\overline{l^2}}4\pi r^2\,dr}{\int_0^\infty e^{-r^2/2\overline{l^2}}4\pi r^2\,dr} = (2\pi\overline{l^2})^{-3/2}e^{-r^2/2\overline{l^2}}4\pi r^2\,dr. \tag{13-37}$$

This probability is normalized to unity. The average values of r^2 and r are

$$\overline{r^2} = \int_0^\infty r^2 P(r)\,dr = 3\overline{l^2}, \tag{13-38}$$

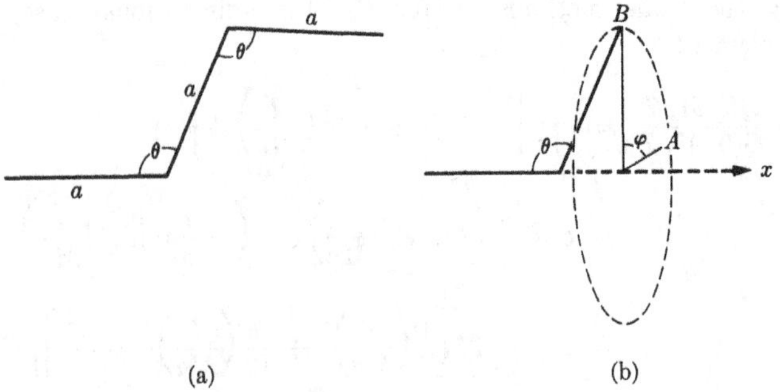

FIG. 13–3. Chain with fixed angle θ between units. In part (b), if the left unit is on the x-axis, the end of the right unit (point B) can be anywhere on the dotted circle. The angle φ is measured from some fixed point A.

$$\bar{r} = \int_0^\infty rP(r)\,dr = \left(\frac{8\overline{l^2}}{\pi}\right)^{1/2} = \left(\frac{8\overline{r^2}}{3\pi}\right)^{1/2}. \qquad (13\text{–}39)$$

Thus $\bar{l} = 0$ but $\bar{r} > 0$ (l can be positive or negative, but r is always positive). Using Eq. (13–38), Eq. (13–37) takes the more appropriate form

$$P(r)\,dr = \left(\frac{3}{2\pi\overline{r^2}}\right)^{3/2} e^{-3r^2/2\overline{r^2}} 4\pi r^2\,dr. \qquad (13\text{–}40)$$

Equation (13–40), it will be recalled, follows from the very general equation (13–29) and is therefore not restricted to any particular model. In various special cases, an explicit expression can be given for $\overline{r^2}$. For example, for the freely jointed chain of Section 13–1, $\overline{l^2} = Ma^2/3$, and hence $\overline{r^2} = Ma^2$. We now list, without proof,* some further results for idealized models of polymer molecules, which, however, are considerably more realistic than the freely jointed chain.

(1) If the chain has M units or bonds of length a, and θ is the fixed bond angle between successive bonds (Fig. 13–3a), and if rotation about bonds (see the angle φ in Fig. 13–3b) is free, then for large M,

$$\overline{r^2} = Ma^2\,\frac{(1 - \cos\theta)}{(1 + \cos\theta)}. \qquad (13\text{–}41)$$

The tetrahedral angle $\theta = 109.5°$ is the case of most interest: $\overline{r^2} = 2Ma^2$. If $\theta = 90°$, $\overline{r^2} = Ma^2$, as for a freely jointed chain.

* See Flory, pp. 414–422, for more details.

(2) If fixed bond angles θ_1 and θ_2 alternate (e.g., O—Si—O and Si—O—Si in the silicone chain), then

$$\overline{r^2} = Ma^2 \frac{(1 - \cos\theta_1)(1 - \cos\theta_2)}{(1 - \cos\theta_1 \cos\theta_2)} \tag{13–42}$$

for large M.

(3) Here we have the same situation as in (1) except that rotation about φ is hindered (see Section 9–5). For a hindering potential $V(\varphi)$ which is symmetrical about $\varphi = 0$,

$$\overline{r^2} = Ma^2 \frac{(1 - \cos\theta)(1 + \overline{\cos\varphi})}{(1 + \cos\theta)(1 - \overline{\cos\varphi})} \tag{13–43}$$

for large M and $\overline{\cos\varphi}$ not too near unity, where

$$\overline{\cos\varphi} = \frac{\int_0^\pi \cos\varphi\, e^{-V(\varphi)/kT}\, d\varphi}{\int_0^\pi e^{-V(\varphi)/kT}\, d\varphi}. \tag{13–44}$$

If $V(\varphi) \equiv 0$ or if $V(\varphi) = V(\varphi + 2\pi m^{-1})$, where $m \geq 2$, as in Eq. (9–18), then $\overline{\cos\varphi} = 0$ and Eq. (13–43) reduces to Eq. (13–41). However, actual polymer chains will not have this symmetry, and the $\overline{\cos\varphi}$ correction will be significant.

Although the models leading to Eqs. (13–41) through (13–43) are much more realistic than a freely jointed chain, they still cannot be taken too seriously. For example, bending and stretching of bonds have not been taken into account. Much more important, van der Waals (or other) attractions and, especially, repulsions between different units of the chain have been ignored. The neglect of van der Waals repulsions enters all the above models with the implicit assumption that the chain has a length but no thickness. Because of this complication alone, the polymer configuration problem differs significantly from ordinary random-walk problems: in a given polymer configuration, two parts of the chain cannot cross each other (occupy the same space), but there is no such restriction on random-walk (or diffusion) paths. In polymer language, this is called the excluded volume problem, and much recent theoretical work has been done on it.*

For the above reasons, detailed theories providing expressions for $\overline{r^2}$ in terms of a model are not very practical. Instead, one can regard $\overline{r^2}$

---

* See, for example, F. T. Wall and J. J. Erpenbeck, *J. Chem. Phys.* **30**, 634, 637 (1959). These authors find that $\overline{r^2} \propto M^b$, where $b = 1.18$ for a tetrahedral lattice.

in Eq. (13–40) as an empirical quantity to be determined by some physical property of the polymer molecules that can be related to $\overline{r^2}$.

An *approximate* semiempirical device that may be used to relate a real chain of *unknown* $\overline{r^2}$ to the simplest model above, the freely jointed chain, is the following. Bond angle restrictions exist between one monomer and the next in a real polymer molecule. But if we call, say, five or ten (depending on the stiffness of the chain) monomers one "statistical unit," then the (end-to-end) direction of one statistical unit is essentially independent of the direction of neighboring statistical units in the chain. In fact, enough monomers are included in a statistical unit to ensure this independence. Thus we can replace the actual restricted chain of monomers by an equivalent chain of freely jointed statistical units. If M is the number of monomers in the chain and n the number in a statistical unit, then the number of statistical units is $M' = M/n$. The length of a statistical unit, a', is estimated as the root mean-square end-to-end distance of a statistical unit (i.e., a chain of n monomers). Then, finally, in Eq. (13–40) we put

$$\overline{r^2} = M'a'^2, \tag{13–45}$$

as for a freely jointed chain. The excluded volume problem is ignored here.

If $\overline{r^2}$ is *known*, then M' and a' can be chosen in a unique way so that not only does the product $M'a'^2$ equal $\overline{r^2}$ but also so that the fully extended length of the effective freely jointed chain, $M'a'$, is equal to the fully extended length of the real chain, $l_{\max}$. That is, from the equations

$$\overline{r^2} = M'a'^2 \quad \text{and} \quad l_{\max} = M'a',$$

we deduce

$$M' = \frac{l_{\max}^2}{\overline{r^2}} \quad \text{and} \quad a' = \frac{\overline{r^2}}{l_{\max}}. \tag{13–46}$$

Again the excluded volume problem is ignored.

13–3 Rubber elasticity. Rubber consists of an isotropic network of long polymer chains. The space-filling property of the chains, referred to in Section 13–2 in connection with the excluded volume problem, is important here, for rubber is a condensed phase with some liquidlike properties. A rather good analogy to a sample of rubber is a large tightly packed collection of very long actively wiggling worms, with each end of each worm attached to one end of each of three other worms (to form a network). The junctions joining the ends of four chains (worms) together are called cross-links. A real network will of course have imperfections (chains with free ends, etc.) just as a real crystal has imperfections.

It is commonplace that rubber has rather unique elastic behavior. This behavior is a consequence of the special configurational properties of polymer molecules considered in the preceding sections. We shall give here only a very brief and semiphenomenological discussion of rubber elasticity. An adequate treatment of the details of polymer network theory would take us far beyond the scope of this book. The reader interested in this subject should see the papers of James and Guth (Supplementary Reading list). The alternative, simpler, but less satisfactory theory of rubber elasticity, due to Wall, will be presented in Chapter 21. This latter theory provides the starting point for the only existing theories of polymer and polyelectrolyte gels, etc.

Let us begin by summarizing the observed thermodynamic behavior of rubber for extensions up to the order of 300%. First, rubber is approximately incompressible (as are typical liquids); when rubber is stretched, the volume stays almost constant. We can therefore use the following rather accurate thermodynamic equations for a sample of rubber of definite mass (L = length of sample):

$$dE = T\,dS + \tau\,dL, \tag{13-47}$$

$$dA = -S\,dT + \tau\,dL, \tag{13-48}$$

$$\tau = \left(\frac{\partial A}{\partial L}\right)_T = \tau_E + \tau_S, \tag{13-49}$$

$$\tau_E = \left(\frac{\partial E}{\partial L}\right)_T, \tag{13-50}$$

$$\tau_S = -T\left(\frac{\partial S}{\partial L}\right)_T = T\left(\frac{\partial \tau}{\partial T}\right)_L, \tag{13-51}$$

where τ_E and τ_S are the energy and entropy contributions to the force τ. By measuring τ as a function of both L and T, τ_S can be calculated from $(\partial\tau/\partial T)_L$, and hence τ_E can be obtained from Eq. (13-49). It is found in this way that τ_E is approximately zero: the elasticity of rubber is an entropy effect. Thus E depends on T but not on L. The implication of this is that when rubber is extended, the intermolecular potential energy remains constant, which is not surprising for a condensed phase of constant volume, and also that the extension is made possible by sufficient uncoiling of the polymer chains but does not involve any bending or stretching of chemical bonds. This behavior is equivalent to that of an ideal gas: E is a function of T but not V; and in the equation analogous to Eq. (13-49), $p = p_E + p_S$, $p_E = 0$.

An alternative and equivalent experimental observation is that the force τ is directly proportional to T at constant L. From the relation

$$\tau \cong \tau_S = -T\left(\frac{\partial S}{\partial L}\right)_T, \tag{13-52}$$

we conclude, then, that $(\partial S/\partial L)_T$ is a function of L only. This is consistent with a split of the entropy into two parts:

$$S = S_1(T) + S_2(L), \tag{13-53}$$

where $S_2(L)$ is the entropy associated with the configurational degeneracy of the polymer chains of the network. Again there is an analogy with an ideal gas: $(\partial S/\partial V)_T$ is a function of V only; $S = S_1(T) + S_2(V)$. For an ideal gas, replace, in Eq. (13-52), τ by p, τ_S by p_S, and $-(\partial S/\partial L)_T$ by $(\partial S/\partial V)_T = Nk/V$.

On the basis of the above discussion, we postulate that the essential molecular mechanism determining the elasticity of rubber is the elasticity of the individual chains making up the network, and this in turn is determined by the configurational properties of the chains (Section 13-2). We have to superimpose on this postulate the facts that the volume is constant on stretching and that a hydrostatic pressure exists in the rubber, just as in any liquid.

Consider an isotropic cube of rubber, with edge L_0, when under no force. The volume is $V = L_0^3$. Now let a force τ extend the rubber in the x-direction so that $L \equiv L_x > L_0$. Then

$$L_y = L_z, \qquad V = L_0^3 = LL_yL_z = LL_y^2. \tag{13-54}$$

Let us examine the mechanical equilibrium at a surface of the stretched rubber perpendicular to the z-axis. There is an outward force pLL_y owing to the hydrostatic pressure, but this is just balanced by the inward force of the molecular chains. We cannot write a satisfactory and completely explicit expression for this inward force without a detailed study of the properties of the network. However, for the small extensions we are interested in, we can deduce from Eq. (13-17) for a single chain that the inward force exerted by a network of N chains will have the form $CNkTL_z$, since L_z will be proportional to l_z for a single chain. Here, C is a constant which depends on the structure of the network. On equating the inward and outward forces, and putting $L_z = L_y$, we find

$$p = \frac{CNkT}{L}. \tag{13-55}$$

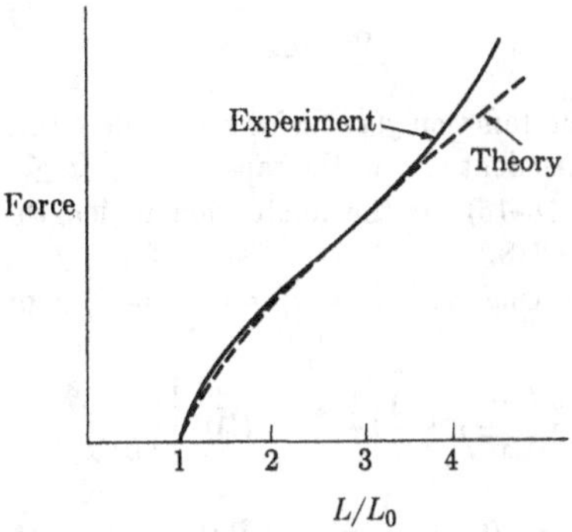

FIG. 13-4. Comparison of experimental and theoretical length-force relations for rubber in a typical case. The vertical scale has been adjusted to give best fit.

Next, consider the mechanical equilibrium at the surface (perpendicular to the x-axis) which is being pulled by an external force τ. Here $\tau + pL_y^2$ (outward force) is balanced by $CNkTL$ (inward force). Then

$$\tau = CNkTL - pL_y^2. \tag{13-56}$$

From Eqs. (13-54) and (13-55) this becomes

$$\tau = CNkTL_0\left(\alpha - \frac{1}{\alpha^2}\right), \tag{13-57}$$

where $\alpha = L/L_0$. This is the desired length-force equation, valid for small (up to about $\alpha = 3$) extensions. Of course $\alpha = 1$ when $\tau = 0$. The initial slope, $(d\tau/d\alpha)_{\alpha=1}$, is $3CNkTL_0$. Experiment and theory are compared in Fig. 13-4, where the vertical scale has been adjusted to give the best fit. The experimental "knee" is well reproduced by the theory. Deviations occur, as expected, at high extensions.

In the Wall theory of Chapter 21, $C = L_0^{-2}$.

PROBLEMS

13–1. Show that the function $j(x, T)$ for a unit in a three-dimensional freely jointed chain is independent of x in the range $-a \leq x \leq +a$. (Page 217.)

13–2. Derive Eq. (13–15) for the fluctuation in length of a chain under a constant force. (Page 218.)

13–3. For any molecule with $\overline{l_0} = \overline{l_0^3} = \cdots = 0$, show from Eq. (13–19) that

$$\bar{l} = \overline{l_0^2}\left(\frac{\tau}{kT}\right) + \frac{1}{2}\left[\frac{1}{3}\overline{l_0^4} - (\overline{l_0^2})^2\right]\left(\frac{\tau}{kT}\right)^3 + \cdots. \qquad (13\text{–}58)$$

Invert this series to get τ/kT in powers of $\bar{l}$, then integrate $dA = \tau\, dl$ to find

$$\frac{Q(l, T)}{Q(0, T)} = \exp -\left\{\frac{l^2}{2\overline{l_0^2}} + \frac{1}{8}\left[\frac{1}{(\overline{l_0^2})^2} - \frac{\overline{l_0^4}}{3(\overline{l_0^2})^4}\right]l^4 + \cdots\right\}. \qquad (13\text{–}59)$$

Equation (13–59) is exact only for $M \rightarrow \infty$, but Eq. (13–58) is exact in general. By comparing Eqs. (13–12) and (13–58), show that, for a freely jointed chain,

$$\overline{l_0^2} = \frac{Ma^2}{3} \quad \text{and} \quad \overline{l_0^4} = \frac{M^2a^4}{3}\left(1 - \frac{2}{5M}\right). \qquad (13\text{–}60)$$

(Page 219.)

13–4. Deduce the expansion of $\mathcal{L}^{-1}(x)$ from that of $\mathcal{L}(y)$ in Eq. (12–31). (Page 220.)

13–5. Use the gaussian form for $Q(l)$ (Eq. 13–29) to deduce Δ from Eq. (13–6) and $\bar{l}$ from Eq. (13–10). The result should agree with Eq. (13–28), of course. (Page 220.)

13–6. Derive the length-force equation, (13–14), from the canonical ensemble equations (13–4), (13–5), and (13–23). (Page 220.)

13–7. Deduce the probability expansion (13–22) from Eq. (13–27). (Page 221.)

13–8. Deduce the entropy equation (13–30) from Eqs. (1–33) and (13–29). (Page 222.)

13–9. Derive Eqs. (13–34) through (13–36) for a one-dimensional freely jointed chain. (Page 223.)

13–10. Obtain the equivalent of Eqs. (13–60) in Problem 13–3 for the one-dimensional freely jointed chain.

13–11. Discuss the problem of a two-dimensional freely jointed chain.

13–12. Derive Eq. (13–33) as a special case of Eq. (7–59).

13–13. Discuss the problem of a three-dimensional freely jointed chain in which each unit can have two lengths, a_α and a_β, with partition functions $j_\alpha(T)$ and $j_\beta(T)$ [in the notation of Eq. (13–11)]. Consider also the problem in which each unit can have any length between $a = 0$ and $a = a_m$ with equal probability. Incidentally, in an equivalent chain of statistical units, a gaussian distribution in a for the length of a statistical unit would be an appropriate approximation (in the text, we use a single length a').

13–14. Calculate $\overline{l_0^2}$ and $\overline{l_0^4}$ from the gaussian function (13–23). Compare with Problem 13–3.

13–15. Show the identity of Eqs. (13–34) and (13–35).

Supplementary Reading

Chandrasekhar, S., *Revs. Mod. Phys.* **15**, 1 (1943).

Flory, Chapters 10 and 11.

James, H. M., and E. Guth, *J. Chem. Phys.* **11**, 455 (1943); *J. Polymer Sci.* **4**, 153 (1949).

Treloar, L. R. G., *The Physics of Rubber Elasticity*. Oxford: 1958. Chapters 1–6.

Wilson, Chapter 14.

Part III

Systems of Interacting Molecules

CHAPTER 14

LATTICE STATISTICS

In Part III we discuss systems of molecules that exert intermolecular forces on each other. First we consider, in the present chapter, molecules confined to sites in a lattice (the "Ising problem"), and then turn in later chapters to mathematically more complicated systems (imperfect gases, liquids, etc.) in which the molecules are not restricted in this way. A feature of particular interest in all this work is the possibility of phase transitions (e.g., gas-liquid) when the temperature is low enough.

We have already considered, in Chapter 7 and in Sections 12–4 and 13–1, a number of problems involving lattice statistics. In all these cases, however, the individual sites or subsystems were independent of each other. In this chapter we extend the discussion to more complicated problems in which interactions between nearest-neighbor sites or subsystems exist. Second-neighbor, and higher, interactions are important in some cases but, for simplicity, we shall confine ourselves to models with nearest-neighbor interactions only.

Further applications of the results obtained in this chapter will be found in Chapters 16 (hole theory of liquids) and 20 (lattice solution theory).

A more detailed and advanced review of this subject is available in S. M., Chapter 7.

14–1 One-dimensional lattice gas (adsorption). We consider here the following model: we have a linear array of M equivalent sites ($M \to \infty$); each site may be empty or occupied by one molecule; the partition function of an isolated molecule on a site is $q(T)$; and when two nearest-neighbor sites are both occupied by molecules, there is a potential energy of interaction w between the molecules (in some cases w is a free energy, a function of T, and not a potential energy; also, the zero of energy for w is infinite separation). This is a realistic model for the adsorption of molecules or ions onto sites on a linear polymer molecule if the forces between the adsorbed molecules or ions are of sufficiently short range that only nearest-neighbor interactions need be taken into account. This might well be the case even for ions, in the presence of added electrolyte (see Chapter 18). The model is the same as that of Eqs. (7–3) through (7–16) except for the added complication here of the nearest-neighbor interaction energy w.

This one-dimensional problem has a special added importance in statistical mechanics by virtue of the fact that it can be solved easily and exactly.

An exact treatment has been achieved with great difficulty for the corresponding two-dimensional problem only in the special case that half the sites are occupied; it has not been achieved at all for the three-dimensional problem (except for various series expansions).

We shall use the canonical ensemble and the maximum-term method. A much more elegant matrix method is available (see S. M., pp. 312–314 and 323–324), but we avoid this in deference to readers not sufficiently equipped mathematically. The system is characterized thermodynamically [see Eq. (7–6)] by M sites, of which N are occupied, and the temperature T. Equation (7–4) gives the canonical ensemble partition function $Q(N, M, T)$ when $w = 0$. Our task is to generalize this Q for $w \neq 0$.

When the N molecules are distributed among the M sites in a particular configuration or arrangement with N_{11} nearest-neighbor pairs of sites both occupied, the interaction potential energy is $N_{11}w$. Actually, it proves convenient later to use here, instead of N_{11}, the variable N_{01}—the number of nearest-neighbor pairs of sites with one site empty (subscript 0) and one site filled (subscript 1). The relation between N_{11} and N_{01} can be established by the following argument. If we draw a line from each occupied site to its two neighboring sites, we will have drawn $2N$ lines. Also, in this process, we will have placed two lines between each nearest-neighbor 11 pair (one starting from each side) and one line between each 01 pair. Therefore,

$$2N = 2N_{11} + N_{01}. \tag{14–1}$$

A similar argument for empty sites gives

$$2(M - N) = 2N_{00} + N_{01}. \tag{14–2}$$

We see from Eqs. (14–1) and (14–2) that only one of N_{11}, N_{01}, and N_{00} is independent; we choose N_{01}. End effects are neglected in Eqs. (14–1) and (14–2) because $M \to \infty$.

A configuration or arrangement of N molecules on M sites with N_{01} pairs of type 01 will have an interaction potential energy

$$N_{11}w = \left(N - \frac{N_{01}}{2}\right)w,$$

according to Eq. (14–1). Suppose that there are altogether $g(N, M, N_{01})$ configurations with exactly N_{01} pairs of type 01. That is, suppose there are $g(N, M, N_{01})$ different ways in which N molecules can be distributed on M sites giving N_{01} pairs of type 01. The contribution of these configurations to Q is $g(N, M, N_{01})e^{-N_{11}w/kT}$, and the complete expression for Q is then

$$Q(N, M, T) = q^N \sum_{N_{01}} g(N, M, N_{01}) \exp\left[-\left(N - \frac{N_{01}}{2}\right) w/kT\right]$$

$$= (qe^{-w/kT})^N \sum_{N_{01}} g(N, M, N_{01})(e^{w/2kT})^{N_{01}}, \qquad (14\text{–}3)$$

where the sum is over all possible values of N_{01} for given N and M.

Having related Q formally to g, our next problem is to find an explicit expression for g. We might note at the outset that we must have for the *total* number of configurations with given N and M,

$$\sum_{N_{01}} g(N, M, N_{01}) = \frac{M!}{N!(M - N)!} \cdot \qquad (14\text{–}4)$$

In view of this relation, it is clear that Eq. (14–3) reduces, as it should, to Eq. (7–4) when $w = 0$.

Since we shall be using only the maximum term in the sum in Eq. (14–3), N, M, and N_{01} may all be regarded as very large numbers. For concreteness, suppose N_{01} is odd and that the site on the left of the linear array of sites is of type 1. For example:

1 1 1 ⋮ 0 0 ⋮ 1 ⋮ 0 ⋮ 1 ⋮ 0 0 ⋮ 1 1 ⋮ 0

$N = 7, \quad N_{01} = 7, \quad M = 13, \quad M - N = 6.$

Then there are $(N_{01} + 1)/2$ groups of 1's, $(N_{01} + 1)/2$ groups of 0's, and the left-hand group is a 1 group, while the right-hand group is a 0 group. These remarks follow from the fact that a 01 pair occurs at each boundary between a 1 group and a 0 group. Now consider the number of ways of arranging N 1's in $(N_{01} + 1)/2$ groups. Each 1 group must have at least one site of type 1 in it; thus the required number of arrangements is the number of ways of assigning the remaining $N - [(N_{01} + 1)/2]$ $(= X)$ 1's among the $(N_{01} + 1)/2$ $(= Y)$ groups, with no restriction on the number of these 1's per group. This number is

$$\frac{(X + Y - 1)!}{(Y - 1)!X!} = \frac{N!}{(N_{01}/2)![N - (N_{01}/2)]!},$$

where we have dropped unity compared with large numbers in the second expression. The corresponding number for the 0's is obtained by replacing N by $M - N$. Then g is twice the product of the above number for the 1's and the corresponding number for the 0's. The factor of two arises because the left group could as well be a 0 as a 1. But this factor is negligible, for we use only $\ln g$ and not g itself. Thus we have finally

$$g(N, M, N_{01}) = \frac{N!(M - N)!}{[N - (N_{01}/2)]![M - N - (N_{01}/2)]![(N_{01}/2)!]^2}. \tag{14–5}$$

This formula for g is now inserted in Eq. (14–3) for Q. But the sum is difficult, so we use the maximum-term method (Appendix II). Let $t(N_{01}, N, M, T) = g(e^{w/2kT})^{N_{01}}$ in Eq. (14–3). Then from the condition

$$\frac{\partial \ln t}{\partial N_{01}} = 0 = \frac{\partial \ln g}{\partial N_{01}} + \frac{w}{2kT} \tag{14–6}$$

and Eq. (14–5), we find

$$\frac{(\theta - \alpha)(1 - \theta - \alpha)}{\alpha^2} = e^{-w/kT}, \tag{14–7}$$

with

$$\theta = \frac{N}{M} \quad \text{and} \quad \alpha = \frac{N_{01}^*}{2M},$$

where N_{01}^* is the value of N_{01} giving the maximum term in the sum in Eq. (14–3). Equation (14–7) is a quadratic equation in α, and gives N_{01}^* as a function of N, M, and T:

$$\alpha = \frac{N_{01}^*}{2M} = \frac{2\theta(1 - \theta)}{\beta + 1}, \tag{14–8}$$

where

$$\beta = [1 - 4\theta(1 - \theta)(1 - e^{-w/kT})]^{1/2}.$$

The sign on the square root can be determined from the special case $w/kT = 0$. Before proceeding to find the thermodynamic functions of the system from Q, let us digress to note that Eq. (14–7) can be rewritten as

$$\frac{N_{11}^* N_{00}^*}{N_{01}^{*2}} = \frac{e^{-w/kT}}{4} \tag{14–9}$$

if we use Eqs. (14–1) and (14–2). This has the form of a chemical equilibrium quotient, as in Eq. (10–5), for the "reaction"

$$2(01) \rightleftarrows (11) + (00). \tag{14–10}$$

The "equilibrium constant" $e^{-w/kT}/4$ is consistent with the "partition functions" $q_{00} = q$, $q_{11} = qe^{-w/kT}$, and $q_{01} = 2q$ (configurational degeneracy). That is, $e^{-w/kT}/4 = q_{11}q_{00}/q_{01}^2$.

The chemical potential follows from

$$\ln Q = N \ln qe^{-w/kT} + \ln t(N_{01}^*, N, M, T) \tag{14–11}$$

and

$$-\frac{\mu}{kT} = \left(\frac{\partial \ln Q}{\partial N}\right)_{M,T} = \ln qe^{-w/kT} + \left(\frac{\partial \ln t}{\partial N}\right)_{N_{01}^*,M,T} + \left(\frac{\partial \ln t}{\partial N_{01}^*}\right)_{N,M,T} \left(\frac{\partial N_{01}^*}{\partial N}\right)_{M,T}.$$

The last term drops out because of Eq. (14–6). Then

$$-\frac{\mu}{kT} = \ln qe^{-w/kT} + \left(\frac{\partial \ln g}{\partial N}\right)_{N_{01}^*,M,T}.$$

From this and Eq. (14–5) we find

$$\lambda qe^{-w/kT} = \frac{1-\theta}{\theta}\left(\frac{\theta - \alpha}{1 - \theta - \alpha}\right), \tag{14–12}$$

where $\lambda = e^{\mu/kT}$ as usual. Finally, we eliminate α in Eq. (14–12) with the aid of Eq. (14–8), and obtain

$$y \equiv \lambda qe^{-w/kT} = \frac{\beta - 1 + 2\theta}{\beta + 1 - 2\theta}. \tag{14–13}$$

This equation gives λ as a function of θ and T. In the case of adsorption or binding, λ is (at least approximately) proportional to the equilibrium gas pressure or solute concentration [see, for example, Eq. (7–9)]. Hence this is the "adsorption isotherm." Equation (14–13) reduces to the Langmuir equation, (7–9), when $w/kT \to 0$. Figure 14–1 gives θ plotted against

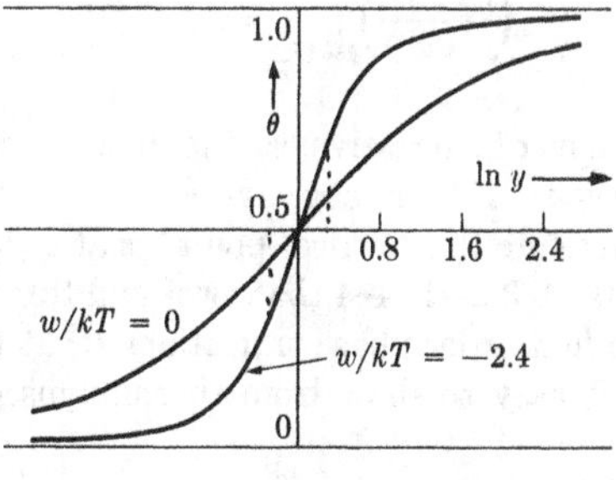

FIG. 14–1. Exact adsorption isotherm for one-dimensional lattice with nearest-neighbor interactions. The $w/kT = 0$ case is the Langmuir isotherm.

$\ln y$ for $w/kT = 0$ (Langmuir) and -2.4. The symmetry evident in Fig. 14–1 can be expressed analytically (Problem 14–1) by

$$y(\theta)y(1-\theta) = 1. \tag{14–14}$$

Next, we find the equation of state. We could use $\Phi = kT\,\partial \ln Q/\partial M$ (Eq. 7–7) but instead, with μ available, we employ

$$\frac{\Phi}{kT} = \frac{N\mu}{MkT} - \frac{A}{MkT} = \frac{N\mu}{MkT} + \frac{1}{M}\ln Q. \tag{14–15}$$

From Eqs. (14–8) and (14–11), we find

$$\frac{\Phi}{kT} = \theta \ln \frac{\beta - 1 + 2\theta}{\beta + 1 - 2\theta} + \frac{1}{M}\ln g(N, M, N_{01}^*) + \frac{2\theta(1-\theta)}{\beta + 1}\cdot\frac{w}{kT}\cdot$$

Finally, we use Eqs. (14–5) and (14–8) to eliminate N_{01}^* from $\ln g$. Lengthy algebra leads to

$$\frac{\Phi}{kT} = \ln \frac{\beta + 1}{\beta + 1 - 2\theta}\cdot \tag{14–16}$$

This reduces to Eq. (7–7) (Langmuir) when $w/kT \to 0$.

For a first-order phase transition to appear in this system, we must find the familiar van der Waals type of loop in the equation of state, (14–16), below a critical temperature. From the symmetry of Fig. 14–1, it is clear that if a critical point exists, the critical value of θ is 1/2. The usual equation-of-state p-v plot corresponds here to a Φ-$1/\theta$ plot (Problem 14–2). In a van der Waals loop, then, $\partial\Phi/\partial(1/\theta)$ will be positive at $\theta = 1/2$; or, $\partial\Phi/\partial\theta$ will be negative at $\theta = 1/2$. But we deduce from Eq. (14–16) that

$$\left(\frac{\partial \Phi/kT}{\partial \theta}\right)_{\theta=1/2} = 2e^{w/2kT}. \tag{14–17}$$

This quantity can never be negative, so a loop and first-order phase transition are not observed. If w is negative and $T \to 0$, the derivative approaches zero. Therefore the "critical point" is at $\theta_c = 1/2$, $T_c = 0$. We shall see in Sections 14–3 and 14–4 that two- and three-dimensional lattice systems do lead to first-order phase transitions (that is, with $T_c > 0$).

Incidentally, it is easy to show from thermodynamics (Problem 14–3) that

$$\left(\frac{\partial \mu/kT}{\partial \theta}\right)_T = \left(\frac{\partial \ln y}{\partial \theta}\right)_T = \frac{1}{\theta}\left(\frac{\partial \Phi/kT}{\partial \theta}\right)_T,$$

and therefore, in Fig. 14–1,

$$\left(\frac{\partial \theta}{\partial \ln y}\right)_{\theta=1/2} = \frac{e^{-w/2kT}}{4}.$$

This slope is steeper than the $w = 0$ slope if $w < 0$, but less steep if $w > 0$. If $w < 0$ and $T \to 0$, the slope approaches $+\infty$.

14–2 Elasticity of a linear polymer chain. In the first part of Section 7–4 we studied a linear chain of α and β units, with lengths of units l_α and l_β. We reconsider the same problem here, but now we allow interactions between nearest-neighbor units in the chain. The interaction energies (hydrogen bonds, for example) are, in obvious notation, $w_{\alpha\alpha}$, $w_{\alpha\beta}$, and $w_{\beta\beta}$. This model, and extensions of it, has been applied to the α-β transformation in fibrous proteins, and to the helix-random coil transformation in protein and nucleic acid molecules in solution (force on chain $= \tau = 0$). We shall, however, not go into the details of these applications.

Equation (7–57) is again the thermodynamic starting point. In place of the canonical ensemble partition function (7–58), we must use the equivalent of Eq. (14–3). We let $g(M_\alpha, M, N_{\alpha\beta})$ be the number of ways in which M_α α units can be distributed among M positions on the chain so that there are $N_{\alpha\beta}$ nearest-neighbor pairs of type $\alpha\beta$. This is obviously the same function as in Eq. (14–5), except for notation. In any one of these g configurations, the total interaction potential energy W is

$$\begin{aligned} W &= N_{\alpha\alpha}w_{\alpha\alpha} + N_{\alpha\beta}w_{\alpha\beta} + N_{\beta\beta}w_{\beta\beta} \\ &= -\frac{wN_{\alpha\beta}}{2} + M_\alpha w_{\alpha\alpha} + (M - M_\alpha)w_{\beta\beta}, \end{aligned} \tag{14–18}$$

where we have used Eqs. (14–1) and (14–2) and defined w by

$$w = w_{\alpha\alpha} + w_{\beta\beta} - 2w_{\alpha\beta}. \tag{14–19}$$

The energy w is the change in energy (or free energy if the w's are functions of T) in the process (14–10). This composite w plays the same role as w in the preceding section. Therefore the problem is not complicated in a fundamental way by the occurrence here of *three* different w's. Then, instead of Eq. (7–58), we have

$$\begin{aligned} Q(M_\alpha, M, T) &= j_\alpha^{M_\alpha} j_\beta^{M-M_\alpha} \sum_{N_{\alpha\beta}} g(M_\alpha, M, N_{\alpha\beta}) e^{-W/kT} \\ &= (j_\alpha e^{-w_{\alpha\alpha}/kT})^{M_\alpha} (j_\beta e^{-w_{\beta\beta}/kT})^{M-M_\alpha} \sum_{N_{\alpha\beta}} g(M_\alpha, M, N_{\alpha\beta}) (e^{w/2kT})^{N_{\alpha\beta}}. \end{aligned} \tag{14–20}$$

This is now in essentially the same form as Eq. (14–3). Equations (14–4) through (14–10) apply here without change (except notation).

For the force on the chain we find, just as in Eq. (14–13),

$$\frac{\tau(l_\beta - l_\alpha)}{kT} = \left(\frac{\partial \ln Q}{\partial M_\alpha}\right)_{M,T} = \ln\left[\left(\frac{\beta + 1 - 2\delta}{\beta - 1 + 2\delta}\right)\frac{j_\alpha e^{-w_{\alpha\alpha}/kT}}{j_\beta e^{-w_{\beta\beta}/kT}}\right], \quad (14\text{–}21)$$

where $\delta = M_\alpha/M$ (fraction of α units) and θ is replaced by δ in the function $\beta(\theta, T)$. Figure 14–1 is a plot of δ (ordinate) against

$$-\frac{\tau(l_\beta - l_\alpha)}{kT} + \ln\frac{j_\alpha e^{-w_{\alpha\alpha}/kT}}{j_\beta e^{-w_{\beta\beta}/kT}} \equiv \ln y. \quad (14\text{–}22)$$

A negative w means here that $\alpha\alpha$ and $\beta\beta$ pairs of units attract each other more than do $\alpha\beta$ pairs. This leads to a tendency for the system to split into two phases (predominantly α or predominantly β), a tendency which, however, cannot be realized in a one-dimensional system, as pointed out in the previous section. In Fig. 14–1, as $\tau \to \infty$, $\ln y \to -\infty$, and $\delta \to 0$ (all long units).

For free molecules (that is, $\tau = 0$), the equilibrium value of δ is determined by Eq. (14–21) with $\tau = 0$. This value of δ is a function of T, a dependence which has been investigated experimentally for the α helix-random coil transition in proteins by Doty and collaborators. The corresponding equation when the w's are all zero is (7–61).

14–3 Two-dimensional square lattice. Here we use the lattice gas language again, since the only practical application is to adsorption on a surface. The model is the same as in Section 14–1, except that the sites form a two-dimensional square lattice instead of a one-dimensional lattice. In particular, q is the partition function of an adsorbed molecule on a site, and w denotes the interaction energy between two molecules on nearest-neighbor sites. Some of the exact properties of this system have been found, especially by Onsager. These are confined mostly to the special condition $\theta = 1/2$, but by symmetry [just as in Eq. (14–14)], it can be deduced that $\theta_c = 1/2$ if a critical point exists, so this is the most interesting value of θ. Because of the advanced mathematical techniques needed for this problem, we shall confine ourselves to a brief summary, without proof, of some of the results obtained. References and more details will be found in S. M., Chapter 7. The same problem is treated by approximate methods in Sections 14–4 and 14–5.

The first result we mention is that the two-dimensional lattice gas can exist in two phases, a dilute phase ("gas") and a condensed phase ("liquid"), and below a critical temperature the two phases can be in equilibrium with each other. The transition from one phase to the other involves a latent

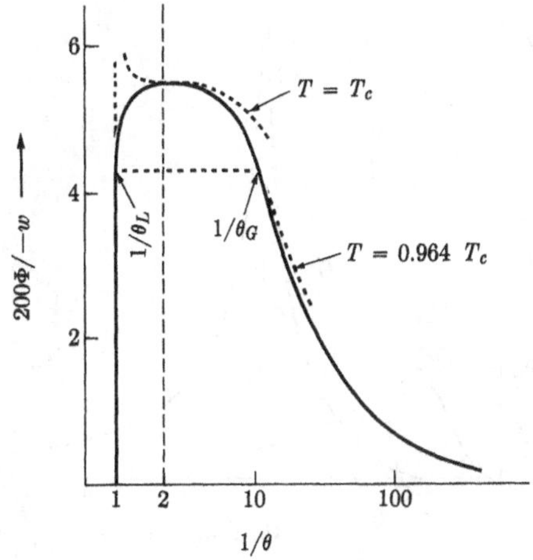

FIG. 14–2. Exact two-phase region for two-dimensional lattice gas with $w < 0$. Dotted curves are schematic isotherms.

heat. In other words, the model predicts a first-order phase transition under suitable conditions. The critical density is $\theta_c = 1/2$, and the critical temperature (first found by Kramers and Wannier) is given by $x_c = \sqrt{2} - 1 = 0.4142$, where $x = e^{w/2kT}$ and $x_c = e^{w/2kT_c}$. Thus w must be *negative* for a critical temperature to exist (Problem 14–4) and for a first-order phase transition to occur. A plot analogous to the usual p-v plot is given in Fig. 14–2 for $w < 0$. Inside the solid curve is the two-phase region. For a given temperature (i.e., value of x) below the critical temperature, the pressure at which the two phases are in equilibrium (i.e., the vapor pressure) is

$$\frac{\Phi}{kT} = \ln(1 + x^2) + \frac{1}{2\pi}\int_0^{\pi} \ln\{\tfrac{1}{2}[1 + (1 - k_1^2 \sin^2\varphi)^{1/2}]\}\, d\varphi, \qquad (14\text{–}23)$$

where

$$k_1 = \frac{4x(1 - x^2)}{(1 + x^2)^2}.$$

Actually, Eq. (14–23) is more general than just indicated: Φ is the pressure for *any* value of x (w positive or negative) at $\theta = 1/2$ (e.g., any point on the dashed line in Fig. 14–2). Thus, if $w = 0$ ($x = 1$), $\Phi/kT = \ln 2$, in agreement with Eq. (7–7).

Let θ_L be the value of θ, for given x, on the liquid side of the two-phase region (Fig. 14–2). Then for the same x, symmetry leads to $\theta_G = 1 - \theta_L$,

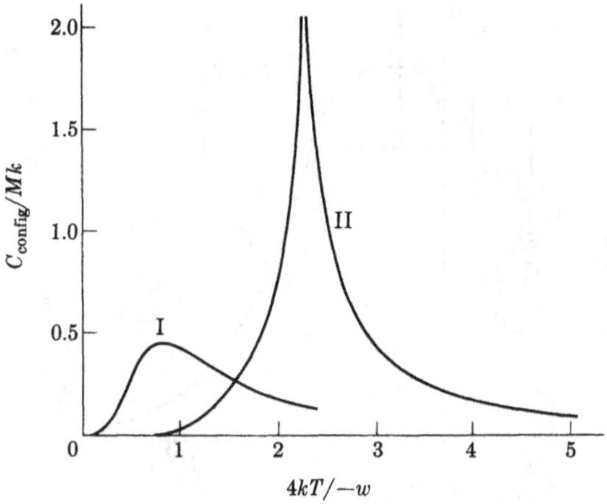

FIG. 14–3. Exact configurational heat capacities for one-dimensional (curve I) and two-dimensional (curve II) lattice gases.

where θ_G is the corresponding value of θ on the gas side. The dependence of θ_G on x is given by

$$\theta_G(x) = \frac{1}{2} - \frac{1}{2}\left[\frac{(1+x^2)(1-6x^2+x^4)^{1/2}}{(1-x^2)^2}\right]^{1/4}. \tag{14–24}$$

The contribution of nearest-neighbor interactions to the energy of the system is called the configurational energy, $E_{\text{config}} = N_{11}^* w$. We assume w is independent of T here. For $\theta = 1/2$, Onsager obtained explicit expressions, which we shall not write down, for E_{config} and the heat capacity $C_{\text{config}} = \partial E_{\text{config}}/\partial T$, involving elliptic integrals. Figure 14–3 shows C_{config} plotted against T. This is the configurational heat capacity along the dashed line in Fig. 14–2. As the temperature T_c is crossed, a singularity in C_{config} occurs. It is easy to show (Problem 14–5) from Section 14–1 that in one dimension ($\theta = 1/2$)

$$\frac{C_{\text{config}}}{Mk} = \left(\frac{w}{4kT}\operatorname{sech}\frac{w}{4kT}\right)^2. \tag{14–25}$$

This function is also plotted in Fig. 14–3; it does not have a singularity.

Many exact series expansions have been found, not only for the square lattice, but for other two- and three-dimensional lattices as well. As an example (see also Problem 14–17), we derive the second virial coefficient $B_2(T)$ in the expansion of Φ/kT in powers of θ for an arbitrary lattice with c nearest-neighbor sites to a given site (for example, $c = 4$ for a square lattice). The definition of the nth virial coefficient B_n is obvious from the expansion

$$\frac{\Phi}{kT} = \theta + B_2(T)\theta^2 + B_3(T)\theta^3 + \cdots. \tag{14–26}$$

We use the method suggested by Eq. (1–69), since it reduces the problem to a two-body problem. For an arbitrary lattice, Eqs. (14–1) through (14–3) become

$$cN = 2N_{11} + N_{01}, \tag{14–27}$$

$$c(M - N) = 2N_{00} + N_{01}, \tag{14–28}$$

$$Q(N, M, T) = (qe^{-cw/2kT})^N \sum_{N_{01}} g(N, M, N_{01})x^{N_{01}}, \tag{14–29}$$

where $x = e^{w/2kT}$ as before. The grand partition function is

$$\Xi = e^{\Phi M/kT} = 1 + Q(1, M, T)\lambda + Q(2, M, T)\lambda^2 + \cdots, \tag{14–30}$$

where $\lambda = e^{\mu/kT}$. When $N = 1$, the only possible value of N_{01} is c, and $g(1, M, c) = M$, since the one molecule can be on any of the M sites. Thus

$$Q(1, M, T) = qe^{-cw/2kT}Mx^c = Mq. \tag{14–31}$$

When $N = 2$, which is as far as we have to go for $B_2(T)$, N_{01} can have the values $2c$ or $2c - 2$. In the former case the two occupied sites are not nearest neighbors to each other, but in the latter case they are. Then

$$g(2, M, 2c) = \frac{M(M - c - 1)}{2}, \qquad g(2, M, 2c - 2) = \frac{Mc}{2}. \tag{14–32}$$

The first result, for example, follows from the fact that the first molecule can be placed at any one of M sites and then the second can be placed at any one of $M - (c + 1)$ sites to avoid "contact." Note that Eqs. (14–32) satisfy Eq. (14–4). Equations (14–29) and (14–32) give

$$Q(2, M, T) = q^2M\left(\frac{cx^{-2} + M - c - 1}{2}\right). \tag{14–33}$$

The grand partition function, (14–30), becomes then

$$\Xi = e^{\Phi M/kT} = 1 + Mz + M\,\frac{(cx^{-2} + M - c - 1)}{2}\,z^2 + \cdots, \tag{14–34}$$

where we define

$$z \equiv q\lambda. \tag{14–35}$$

If we take the logarithm of both sides of Eq. (14–34), we get

$$\frac{\Phi}{kT} = z + \frac{(cx^{-2} - c - 1)}{2} z^2 + \cdots. \tag{14–36}$$

Since Mz is of order M, the legitimacy of using the expansion of $\ln(1 + Mz + \cdots)$ here is questionable, to say the least. Actually, the result can be justified by another argument, which we give later in Section 15–1. Next, we use the thermodynamic equation $\overline{N}\, d\mu = M\, d\Phi$ (T constant), or

$$\theta = \frac{\overline{N}}{M} = \lambda \left(\frac{\partial \Phi/kT}{\partial \lambda}\right)_T = z \left(\frac{\partial \Phi/kT}{\partial z}\right)_T, \tag{14–37}$$

to obtain from Eq. (14–36),

$$\theta = z + (cx^{-2} - c - 1)z^2 + \cdots. \tag{14–38}$$

Since, by definition, z is proportional to λ, and $z \to \theta$ as $\theta \to 0$, it is appropriate to call z an "activity" in the chemical thermodynamic sense. The inverse of Eq. (14–38) is

$$z = \theta + (c + 1 - cx^{-2})\theta^2 + \cdots. \tag{14–39}$$

Finally, we use Eq. (14–39) to eliminate z from Eq. (14–36) and obtain Φ/kT in powers of θ, as in Eq. (14–26). We find

$$B_2(T) = \frac{c + 1 - cx^{-2}}{2} = \frac{c + 1 - ce^{-w/kT}}{2}. \tag{14–40}$$

This agrees with Eq. (7–7) when $w = 0$ ($x = 1$). The second virial coefficient can become negative if w/kT is sufficiently negative ($B_2 \to -cx^{-2}/2$). On the other hand, if w/kT is large and positive, $B_2 \to (c + 1)/2$ (two molecules cannot occupy nearest-neighbor sites).

The method we have used here can be extended to B_3, B_4, etc., and has very general applicability in statistical mechanics.* We shall encounter it again several times in this book (Chapters 15 and 19).

14–4 Bragg-Williams approximation. No two- or three-dimensional lattice statistics problem has as yet been given a treatment that is both complete and exact. Hence, approximate methods are useful. The present approximation is probably the simplest possible that retains the correct qualitative features. It is equivalent to the van der Waals approximation for imperfect gases and liquids (Chapter 16).

---

* T. L. Hill, *J. Chem. Phys.* **27**, 561 (1957).

In the Bragg-Williams approximation for a lattice gas with nearest-neighbor energy w, the configurational degeneracy and average nearest-neighbor interaction energy are both handled on the basis of a *random* distribution of molecules among sites [i.e., as if $w = 0$, as in Eq. (7–4)]. This is obviously an incorrect procedure, except in the limit as $T \to \infty$. Thus, instead of Eq. (7–4), we use

$$Q(N, M, T) = \frac{M! q^N e^{-\overline{N}_{11} w/kT}}{N!(M - N)!}, \tag{14–41}$$

where $\overline{N}_{11} w$ is the average interaction energy. We calculate $\overline{N}_{11}$ as follows: a molecule at a site has, on the average (random distribution), $c\theta = cN/M$ occupied nearest-neighbor sites next to it; therefore, $\overline{N}_{11} = (cN/M)(N/2) = cN^2/2M$, where the factor of two is inserted to avoid counting each 11 pair twice. The same partition function follows from Eq. (14–29) if we write

$$\sum_{N_{01}} g(N, M, N_{01}) x^{N_{01}} = x^{\overline{N}_{01}} \sum_{N_{01}} g = \frac{M! x^{\overline{N}_{01}}}{N!(M - N)!},$$

where $\overline{N}_{01} = cN(M - N)/M$ (random distribution).

The various thermodynamic properties may easily be derived (with w assumed constant) from Eq. (14–41). For example,

$$\ln Q = -\frac{A}{kT} = M \ln M - N \ln N - (M - N) \ln (M - N) + N \ln q - \frac{cN^2 w}{2MkT}; \tag{14–42}$$

$$\frac{S}{k} = \ln Q + T\left(\frac{\partial \ln Q}{\partial T}\right)_{N,M} = \ln \frac{M!}{N!(M - N)!} + N\left(\ln q + T\frac{d \ln q}{dT}\right); \tag{14–43}$$

$$\frac{\Phi}{kT} = \left(\frac{\partial \ln Q}{\partial M}\right)_{N,T} = \frac{cw\theta^2}{2kT} - \ln (1 - \theta) \tag{14–44}$$

$$= \theta + \left(\frac{1}{2} + \frac{cw}{2kT}\right)\theta^2 + \frac{1}{3}\theta^3 + \cdots; \tag{14–45}$$

$$\frac{\mu}{kT} = -\left(\frac{\partial \ln Q}{\partial N}\right)_{M,T} = \ln \frac{\theta e^{wc\theta/kT}}{(1 - \theta)q}; \tag{14–46}$$

$$y \equiv q\lambda e^{-cw/2kT} = \frac{\theta e^{cw(2\theta - 1)/2kT}}{1 - \theta}. \tag{14–47}$$

The entropy is seen to be the same as for the ideal lattice gas (Eq. 7–11). Equation (14–44) is the equation of state, and Eq. (14–46) or (14–47) is

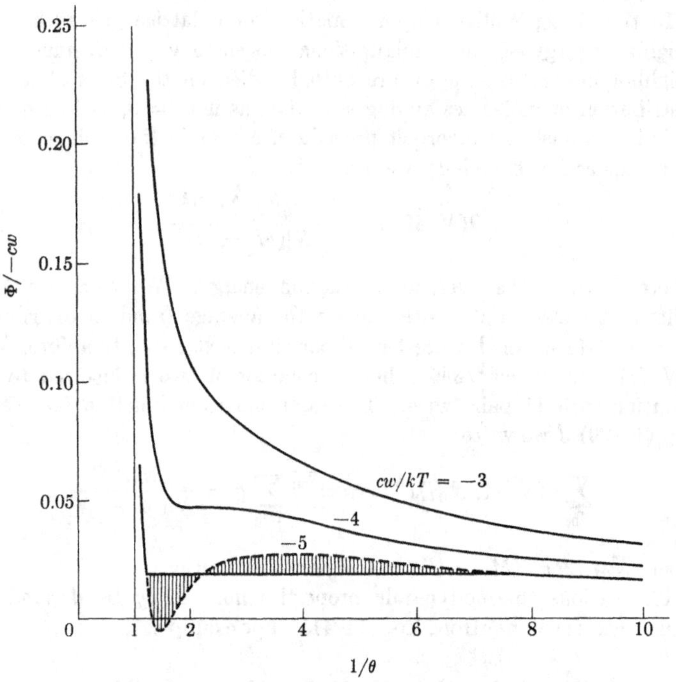

FIG. 14–4. Pressure-volume isotherms for a Bragg-Williams lattice gas. The curve labeled $cw/kT = -4$ is the critical curve.

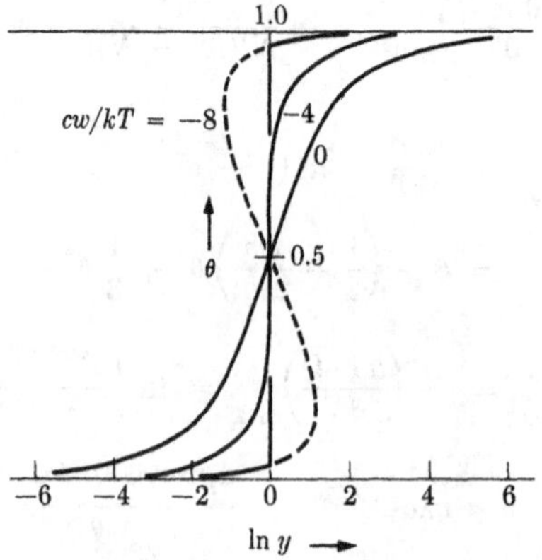

FIG. 14–5. Adsorption isotherm for Bragg-Williams lattice gas. The $cw/kT = 0$ case is the Langmuir isotherm.

the adsorption isotherm [see Eq. (7–9)]. The second virial coefficient in Eq. (14–45) agrees with the exact relation (14–40) in the limit as $w/kT \to 0$, as we should expect from the nature of the Bragg-Williams approximation. We have defined y above in such a way that the symmetry property (Eq. 14–14) $y(\theta)y(1-\theta) = 1$ holds (Problem 14–1). This definition of y is consistent with that in Eq. (14–13), where $c = 2$.

Both Eqs. (14–44) and (14–47) lead to loops (as in van der Waals equation) and to the prediction of a first-order phase transition (Figs. 14–4 and 14–5). To draw in the horizontal stable equilibrium path in Fig. 14–4, we may use either the thermodynamic equal-area theorem or the symmetry condition $\theta_L = 1 - \theta_G$. By symmetry, the critical point must be at $\theta_c = 1/2$. We can locate the critical temperature from the condition

$$\left(\frac{\partial \Phi/kT}{\partial \theta}\right)_{\theta=1/2} = 0. \tag{14–48}$$

We find $cw/kT_c = -4$. The exact value for a square lattice (Section 14–3) is $4w/kT_c = -7.05$. The Bragg-Williams theory incorrectly predicts a phase transition for a linear lattice ($c = 2$).

A number of phase transitions (vertical jumps) in adsorption isotherms, as in Fig. 14–5, have been observed experimentally. If we take (Problem 14–4), say, $w = -400$ cal·mole^{-1}, then for a two-dimensional hexagonal lattice we have* $T_c = 77$°K. At this temperature, however, the adsorbed phase is probably more nearly mobile than localized (Section 9–6) so a better model for the experimental systems is a two-dimensional gas-liquid transition (Sections 16–1 and 16–2). A system which is definitely localized, which shows a phase transition, and to which the Bragg-Williams theory has been applied successfully is the *absorption* of hydrogen gas (as atoms) by palladium metal.† The experimental critical temperature is $T_c =$ 568°K. We might then roughly estimate w, using $c = 0.6 \times 12 = 7.2$ ($\theta = 1$ corresponds to 0.6 H atom per Pd atom), from

$$w = -\frac{15.80kT_c}{6} \times \frac{6}{7.2} = -2500 \text{ cal·mole}^{-1}.$$

For consistency and convenience we are using throughout this chapter, except in Section 14–2, notation and language appropriate to a lattice gas. We could have used "magnetic language," instead, as in Section 12–4. To illustrate this, we turn next to the Bragg-Williams approximation for a ferromagnet or antiferromagnet.

---

* The exact value of $6w/kT_c$ for a hexagonal lattice is -15.80.

† See Fowler and Guggenheim, pp. 558–563, for details.

Section 12–4 should be reviewed for notation and thermodynamics. In a ferromagnet, the magnetic dipoles are associated with electron spin in atoms. Parallel spins or dipoles ($\uparrow\uparrow$ or $\downarrow\downarrow$) at nearest-neighbor positions in the lattice are assumed here to have an interaction energy $-J$, while antiparallel spins ($\uparrow\downarrow$) have an interaction energy $+J$. In Section 12–4, $J = 0$. These interaction energies are not simply dipole-dipole interactions, which are in fact negligibly weak by comparison, but are due instead to quantum-mechanical exchange forces, as in chemical bond theory. If J is positive, we have the ferromagnetic case, and if J is negative, we have the antiferromagnetic case.

The partition function $Q(M, T, H)$ of Eq. (12–40) has to be modified to take care of the nearest-neighbor interaction energies just referred to. For a given value of N (the number of $\downarrow$ dipoles) in the sum in Eq. (12–40), we insert a factor $e^{-W/kT}$, where W is the total interaction energy:

$$W = \overline{N}_{11}w_{11} + \overline{N}_{01}w_{01} + \overline{N}_{00}w_{00};$$

$$0 = \uparrow, \qquad 1 = \downarrow, \qquad w_{11} = w_{00} = -J, \qquad w_{01} = +J;$$

$$\overline{N}_{11} = \frac{cN^2}{2M}, \qquad \overline{N}_{01} = \frac{cN(M-N)}{M}, \qquad \overline{N}_{00} = \frac{c(M-N)^2}{2M}.$$

This gives

$$W = -\frac{cJ(2N-M)^2}{2M}. \tag{14–49}$$

After putting the factor $e^{-W/kT}$, with the above W, behind the summation sign in Eq. (12–40), we can rearrange as follows:

$$Q(M, T, H) = [qe^{(2mH+cJ)/2kT}]^M \sum_{N=0}^{M} \frac{M!e^{2cJN^2/MkT}[e^{-2(mH+cJ)/kT}]^N}{N!(M-N)!}. \tag{14–50}$$

Let us compare this with the grand partition function for the lattice gas, based on Eq. (14–41):

$$\Xi = \sum_N Q(N, M, T)\lambda^N = \sum_{N=0}^{M} \frac{M!e^{-cN^2w/2MkT}(q\lambda)^N}{N!(M-N)!}. \tag{14–51}$$

The summation in Eq. (14–50) clearly has the same form as Ξ in Eq. (14–51). Therefore, instead of rederiving results (Problem 14–6) for the magnetic problem, we can merely transcribe those already found for the lattice gas. For example, since

$$\overline{N} = \frac{\sum_N NQ(N, M, T)\lambda^N}{\sum_N Q(N, M, T)\lambda^N}, \tag{14–52}$$

the form of the sum in Eq. (14–51) suffices to determine $\overline{N}/M$ as a function of $-cw/2kT$ and $q\lambda$. This relationship has already been established in Eq. (14–47), using the canonical ensemble [which is the same as applying the maximum term method to Ξ in Eq. (14–51)]. Then, from the sum in Eq. (14–50) for the magnetic problem, $\overline{N}/M$ must be the *same function* as above, but with $2cJ/kT$ and $e^{-2(mH+cJ)/kT}$ as independent variables. Thus we have the correspondences

$$-\frac{cw}{2kT} \leftrightarrow \frac{2cJ}{kT} \qquad \text{or} \qquad w \leftrightarrow -4J, \tag{14–53}$$

$$y \equiv q\lambda e^{-cw/2kT} \leftrightarrow e^{-2(mH+cJ)/kT}e^{2cJ/kT} \qquad \text{or} \qquad y \leftrightarrow e^{-2mH/kT}. \tag{14–54}$$

The relation (14–53) also follows from Eq. (14–19). Therefore Fig. 14–5 is also a plot, for the magnetic problem, of $\overline{N}/M = (1/2)[1 - (I/mM)]$ (ordinate) against $-2mH/kT$ (abscissa).

The magnetic field strength is the analog of the chemical potential for a lattice gas, as pointed out at the end of Section 12–4. The related thermodynamic equations are

$$d(\Phi M) = d(kT \ln \Xi) = S\,dT + N\,d\mu + \Phi\,dM \qquad \text{(lattice gas)},$$

$$-dA = d(kT \ln Q) = S\,dT + I\,dH - \mu\,dM \qquad \text{(magnet)}.$$

When $J > 0$ (ferromagnetism), the critical (Curie) temperature is given by $cJ/kT_c = 1$. Below the critical temperature, "spontaneous magnetization" can exist. This is the magnetization I_s (Fig. 14–6) that a

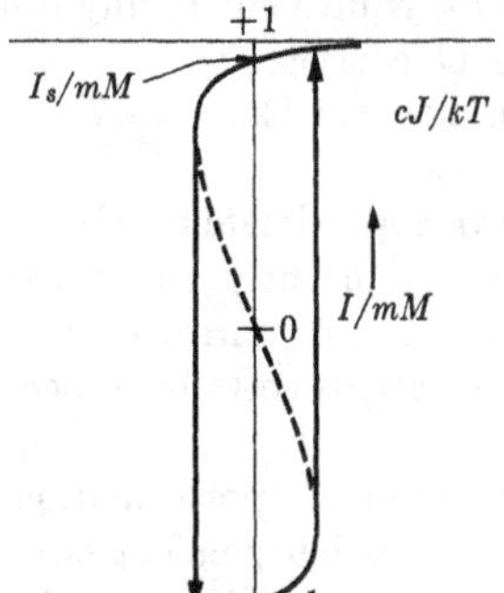

FIG. 14–6. Magnetization as a function of magnetic field for a Bragg-Williams ferromagnet. The metastable parts of the curve lead to the possibility of hysteresis. This is the same curve as $cw/kT = -8$ in Fig. 14–5.

ferromagnet ($T < T_c$) retains if a sample is placed in a magnetic field ($H > 0$) and then the field is removed ($H \to 0+$):

$$I_s = \lim_{H \to 0+} I(H). \tag{14-55}$$

In this state, even though $H = 0$, the majority of the spins are $\uparrow$, in the direction of the previously applied field. If $T \geq T_c$, $I_s = 0$.

In connection with any first-order phase transition, a hysteresis loop is possible, as illustrated in Fig. 14-6. Hysteresis is well known in magnetic and adsorption systems, for example.

As a final topic in this section, we show how the Bragg-Williams lattice gas equation (14-47) can be derived very simply from a one-site grand partition function ξ [compare Eq. (7-31)]. We select one particular site and regard it as being in the potential field of molecules occupying other sites in the lattice *at random*. Let θ be the fraction of other sites occupied. Then the average number of occupied sites nearest-neighbor to the particular site is $c\theta$. When the particular site is itself occupied, the potential energy of the occupying molecule in the field of its neighbors is $c\theta w$. Therefore, instead of $\xi = 1 + q\lambda$, as in Eq. (7-31), we have

$$\xi = 1 + qe^{-c\theta w/kT}\lambda. \tag{14-56}$$

The average occupation of the site is

$$\bar{s} = \lambda \left(\frac{\partial \ln \xi}{\partial \lambda} \right)_T = \frac{qe^{-c\theta w/kT}\lambda}{1 + qe^{-c\theta w/kT}\lambda}, \tag{14-57}$$

where θ is considered a fixed parameter in the differentiation. Since the particular site selected is equivalent to any other site, we have to impose the further condition of consistency, $\bar{s} = \theta$. With this substitution in Eq. (14-57), we again arrive at Eq. (14-47).

14-5 Quasi-chemical approximation. In this section we describe an approximation that is significantly better than the Bragg-Williams approximation but still not of unreasonable mathematical complexity. Further refinements of various sorts have been worked out, but we shall not pursue these here.

The essence of the present approximation is that pairs of nearest-neighbor sites are treated as independent of each other, though we know they are not, since they overlap. For example, in a square lattice:

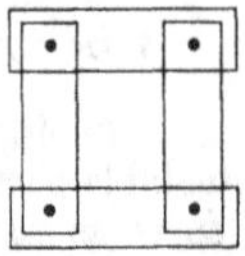

We start with Eq. (14–29) and use the assumed independence of pairs to approximate the function $g(N, M, N_{01})$. Each pair of sites can be occupied in four ways: 11, 01, 10, and 00. The total number of pairs of sites is $cM/2$. For given values of N, M, and N_{01}, the numbers of pairs of different types are [Eqs. (14–27) and (14–28)]:

$$\text{Number of 11 pairs} = N_{11} = \frac{cN}{2} - \frac{N_{01}}{2}.$$

$$\text{Number of 01 pairs} = \frac{N_{01}}{2}.$$

$$\text{Number of 10 pairs} = \frac{N_{01}}{2}.$$

$$\text{Number of 00 pairs} = N_{00} = \frac{c(M - N)}{2} - \frac{N_{01}}{2}.$$

If the pairs are independent of each other, each pair in the lattice can be assigned to one of the four categories above. The number of ways of doing this is

$$\omega(N, M, N_{01}) = \frac{(cM/2)!}{[(cN/2) - (N_{01}/2)]!\{[c(M - N)/2] - (N_{01}/2)\}![(N_{01}/2)!]^2}. \tag{14–58}$$

However, as it stands this cannot be set equal to $g(N, M, N_{01})$, for it will rather obviously not satisfy the condition (14–4). This is because many configurations counted in ω are actually impossible. For example, the two pairs of sites shown below cannot both be occupied in the manner indicated:

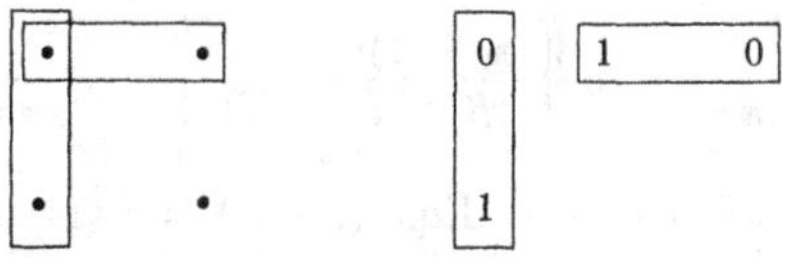

Thus ω overcounts the number of configurations. To take care of this, we must normalize ω:

$$g(N, M, N_{01}) = C(N, M)\omega(N, M, N_{01}), \tag{14–59}$$

$$\sum_{N_{01}} g = \frac{M!}{N!(M - N)!} = C(N, M) \sum_{N_{01}} \omega(N, M, N_{01}).$$

To find $C(N, M)$, we replace the ω sum by its maximum term. From

$\partial \ln \omega/\partial N_{01} = 0$, we obtain

$$\frac{N_{01}^*}{2} = \frac{cN(M-N)}{2M}, \qquad \omega(N, M, N_{01}^*) = \left[\frac{M!}{N!(M-N)!}\right]^c,$$

and therefore

$$C(N, M) = \left[\frac{M!}{N!(M-N)!}\right]^{1-c}. \tag{14-60}$$

Equations (14–58) through (14–60) provide our approximate expression for $g(N, M, N_{01})$. It should be noted that this argument accidentally gives the right answer for a one-dimensional lattice: put $c = 2$ in Eq. (14–59) and compare Eq. (14–5). Hence all the quasi-chemical relations below are exact in one dimension ($c = 2$).

From this point we proceed exactly as in Section 14–1, so we omit details (Problem 14–7). Equations (14–7) through (14–10) are again found, except that now $\alpha = N_{01}^*/cM$. From

$$\ln Q = N \ln qe^{-cw/2kT} + \ln t(N_{01}^*, N, M, T),$$

we deduce, as in Eqs. (14–11) through (14–13),

$$y \equiv \lambda q e^{-cw/2kT} = \left(\frac{1-\theta}{\theta}\right)^{c-1} \left(\frac{\theta - \alpha}{1 - \theta - \alpha}\right)^{c/2},$$

and

$$y = \left[\frac{(\beta - 1 + 2\theta)(1-\theta)}{(\beta + 1 - 2\theta)\theta}\right]^{c/2} \frac{\theta}{1-\theta}. \tag{14-61}$$

This is the adsorption isotherm. The symmetry condition $y(\theta)y(1-\theta) = 1$ is easy to verify. The equation of state is

$$\frac{\Phi}{kT} = \ln \left\{\left[\frac{(\beta+1)(1-\theta)}{(\beta+1-2\theta)}\right]^{c/2} \frac{1}{1-\theta}\right\}. \tag{14-62}$$

The qualitative behavior of Eqs. (14–61) and (14–62) is the same as in Figs. 14–5 and 14–4, respectively, for the Bragg-Williams theory. The critical point is at $\theta_c = 1/2$ and a temperature determined from Eq. (14–62):

$$\left(\frac{\partial \Phi/kT}{\partial \theta}\right)_{\theta=1/2} = 0 = c(x - 1) + 2,$$

or

$$x_c = e^{w/2kT_c} = \frac{c-2}{c}, \qquad \frac{cw}{kT_c} = 2c \ln \frac{c-2}{c}. \tag{14-63}$$

For $c = 4$ (square lattice), this gives $4w/kT_c = -5.54$, which is intermediate, as expected, between the Bragg-Williams value (−4) and the exact value (−7.05). As a check on Eq. (14–63), if we put $c = 2$, we find $T_c = 0$, as in Section 14–1. Incidentally, the Bragg-Williams result follows from Eq. (14–63) if we let $c \to \infty$ holding cw constant: $cw/kT_c \to -4$. Also, for example, the Bragg-Williams $y(\theta)$ (Eq. 14–47) can be derived from the quasi-chemical $y(\theta)$ (Eq. 14–61) in the same limit (Problem 14–8).

The quasi-chemical approximation was first used by Bethe in connection with the order-disorder transition in alloys (see Chapter 20). However, the mathematical identity of his quite different physical approach with the above equations was not obvious at first. Guggenheim soon introduced another method, equivalent to Bethe's and the combinatorial argument above, which was responsible for the choice of the term "quasi-chemical." We shall merely sketch Guggenheim's procedure.† The starting point is the intuitively reasonable quasi-chemical equilibrium relation, (14–9). [Incidentally, in the Bragg-Williams approximation, $w = 0$ in Eq. (14–9).] From this and the conservation equations (14–27) and (14–28), we derive Eq. (14–8) with $\alpha = N_{01}^*/cM$. The configurational energy is then

$$E_{\text{config}} = N_{11}^* w = \left(\frac{cN}{2} - \frac{N_{01}^*}{2}\right) w$$

$$= \frac{cM}{2}\left[\theta - \frac{2\theta(1-\theta)}{\beta + 1}\right] w. \qquad (14\text{–}64)$$

From this we can obtain A_{config} by integrating the thermodynamic equation

$$E_{\text{config}} = \left(\frac{\partial A_{\text{config}}/T}{\partial 1/T}\right)_{N,M}$$

between $1/T = 0$ (random distribution) and $1/T$. This integration is very tedious, incidentally. Then from A_{config} we get Φ and μ by differentiation with respect to M and N, respectively.

14–6 First-order phase transitions. In this section we make some comments of a general nature on first-order phase transitions (i.e., transitions with a latent heat). The exact treatment of Onsager and the Bragg-Williams and quasi-chemical approximations for a lattice gas provide us with examples. The familiar van der Waals equation of state for a fluid (Chapter 16) is another example. The Lennard-Jones and Devonshire (LJD) theory of liquids (Chapter 16) is still another.

The solid curve in Fig. 14–7(a) shows typical p-v experimental behavior below the critical temperature. The dashed portions represent meta-

† See Fowler and Guggenheim, pp. 437–438, 441–443.

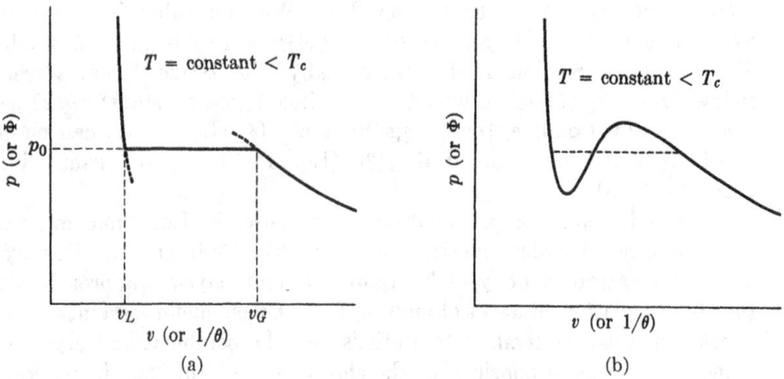

FIG. 14–7. (a) Typical experimental p-v isotherm behavior below the critical temperature. (b) Corresponding curve from approximate theory with restraint of homogeneity.

stable states sometimes observed. If we start with dilute gas (p small) and then increase the pressure, when $p = p_0$ (and $v = v_G$) is reached, another phase—liquid with $v = v_L$—suddenly appears in the system. The sharpness of the appearance of a new phase is the feature of particular theoretical interest. In this respect, a phase transition is quite different from a chemical transition (i.e., chemical reaction), for in the chemical case the equilibrium composition varies in a smooth and continuous way with change in a variable (e.g., total pressure or temperature). The essential difference is that there is a *cooperative* aspect to a phase transition which is missing in a chemical reaction. Thus in a dilute gas reaction $A \rightleftarrows B$, the probability that any given molecule is in state A (or state B) is independent of the state of the other molecules in the gas. But, in a phase transition, if A = molecule in dilute phase and B = molecule in dense phase, there is a tendency for a large number of molecules to switch as a group from state A to state B because the molecules in state B can *stabilize each other* (hence the term "cooperative") through intermolecular attractions. Unlike the chemical reaction example, the tendency for the conversion of a particular A into a B is *not* independent of the state of other molecules; rather, the conversion is aided by the presence of other B molecules. Thus a phase transition resembles a landslide or autocatalytic process.

An example of the distinction between a smooth (chemical) transition and a sudden (phase) transition can be seen in the magnetic system discussed in Sections 12–4 and 14–4. In Section 12–4, there are no interactions between magnetic spins, i.e., the spins behave independently. We therefore have a "chemical" type of equilibrium between states $\uparrow$ and $\downarrow$. If

θ is the fraction of dipoles or spins in state $\downarrow$ and $1 - \theta$ is the fraction in state $\uparrow$, then Eq. (12–46) can be written (in the notation of Chapter 10)

$$\frac{\theta}{1-\theta} = K = e^{-\Delta F^\circ/kT} = e^{-2mH/kT}.$$

That is, K is the equilibrium constant and $\Delta F^\circ = 2mH$ is the energy change in the process: state $\uparrow \rightarrow$ state $\downarrow$. If we vary the magnetic field or temperature, K will change smoothly and the transition from predominantly one state (at equilibrium) to predominantly the other will take place smoothly. On the other hand, if we introduce a negative interaction energy between nearest-neighbor spins in the same state [$J > 0$ in Eqs. (14–49) through (14–55)], the spins tend to switch states *cooperatively* instead of independently and a sudden instead of smooth transition will take place at $H = 0$ if H is varied holding T constant at $T < T_c$ (Fig. 14–6).

The occurrence of a sharp phase-transition point can also be understood from Fig. 14–8. Curve G shows the typical dependence of the chemical potential of a dilute gas on pressure at constant temperature (Eq. 4–25). Curve L represents the chemical potential of the liquid. This curve is almost flat because of the relatively small volume per molecule in the liquid state; i.e., $(\partial\mu/\partial p)_T = v$. The stable phase at any p is the one with lower μ. Because of the quite different slopes, the curves intersect at a sharply defined pressure p_0 (the vapor pressure at the temperature under consideration). Thus, if we start with $p < p_0$, only the gas phase is present, for its chemical potential is lower. On increasing p, the liquid phase appears very suddenly at $p = p_0$, the crossing point of the μ-p curves. Both phases are present at $p = p_0$, and only liquid is present when $p > p_0$. The dashed curves again correspond to metastable states.

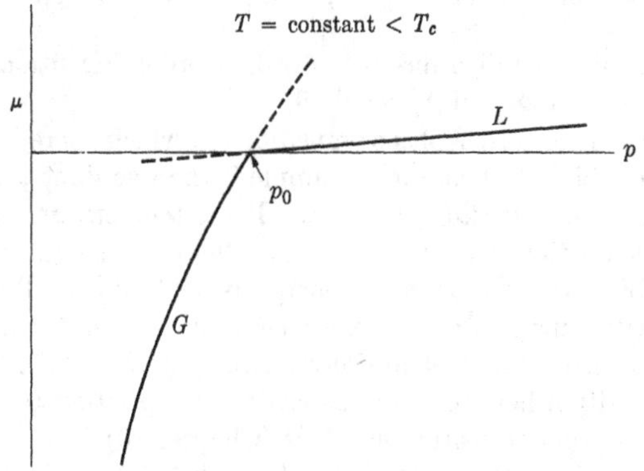

FIG. 14–8. Chemical potential as a function of pressure (T constant), showing liquid-gas transition.

Just as for a chemical equilibrium, we can interpret in a general way the relative stability of the gas (dilute) or liquid (dense) phase by observing that the gas phase is favored by a relatively high entropy and that the liquid is favored by a relatively low (intermolecular potential) energy. The dense phase is always more stable at low temperature or high pressure, and the dilute phase is more stable at high temperature or low pressure. At the transition point ($p = p_0, \mu_G = \mu_L$), the energy and entropy effects just balance each other.

From the Bragg-Williams, quasi-chemical, van der Waals, or LJD (Chapter 16) theories, we obtain a continuous loop, as in Fig. 14–7(b), instead of a three-branched p-v curve such as in Fig. 14–7(a). Why this discrepancy? A loop is always encountered in a theory which (a) uses the partition function Q and (b) introduces implicitly or explicitly the restraint of uniform density through the extent of the system. It is not possible under this restraint for two phases of different density to exist together in the system, as required at $p = p_0$ in Fig. 14–7(a). The loop in Fig. 14–7(b) is the p-v curve of an artificial system *forced* to be homogeneous (one density) under all conditions. A three-branched curve as in Fig. 14–7(a) will result if Q is evaluated exactly (all configurations being represented, including those in which the density is different in different regions) or even if Q is evaluated approximately but all possible densities are allowed in any small region of the system. It should be added that mathematically sharp corners occur in the curve only in the limit of an infinitely large system. If the system is large but finite, the corners will be rounded off somewhat.

It is easy to see that the partition functions Ξ and Δ can never lead to a loop, regardless of the nature of the approximation used in Q. This follows from the fluctuation equations (2–10) and (2–14): $(\partial p/\partial v)_T$ cannot be positive.

A much more detailed discussion of first-order transitions will be found in S. M., Section 28 and Appendix 9.

So-called higher-order phase transitions in which there is no latent heat are also possible. A theoretical example is the two-dimensional lattice gas of Section 14–3, *restricted to* $\theta = 1/2$. If the temperature of this system is varied, a transition occurs at $T = T_c$: there is a singularity in the heat capacity (Fig. 14–3) though the energy is continuous. The same kind of heat-capacity curve is found experimentally and/or theoretically in, for example, a ferromagnet or antiferromagnet at $H = 0$, a binary solution with composition held fixed at the critical composition (Chapter 20), and the order-disorder transition in alloys (Chapter 20).

For a review of the whole subject of phase transitions, the reader is referred to Temperley (Supplementary Reading list).

Problems

14–1. Prove the symmetry property $y(\theta)y(1-\theta) = 1$ for Eqs. (14–13) and (14–47). (Pages 240 and 249.)

14–2. Calculate a few points and draw approximate curves for Φ/kT against θ and $1/\theta$ for an ideal lattice gas, Eq. (7–7). (Page 240.)

14–3. Derive the thermodynamic equation

$$\left(\frac{\partial\mu/kT}{\partial\theta}\right)_T = \frac{1}{\theta}\left(\frac{\partial\Phi/kT}{\partial\theta}\right)_T.$$

Note the connection with Problem 7–13. (Page 240.)

14–4. Calculate T_c for a two-dimensional (square) lattice gas, using $x_c = 0.4142 = e^{w/2kT_c}$ and $w = -400$ cal·mole^{-1}. (Pages 243 and 249.)

14–5. Derive Eq. (14–25) for the heat capacity of a one-dimensional lattice gas with $\theta = 1/2$. (Page 244.)

14–6. Use Eqs. (12–38) and (14–50) to derive the relation between I and H for a Bragg-Williams ferromagnet. (Page 250.)

14–7. Fill in the details in the derivation of the quasi-chemical equations, (14–60) through (14–63). (Page 254.)

14–8. Derive the Bragg-Williams $y(\theta)$ (Eq. 14–47) from the quasi-chemical $y(\theta)$ (Eq. 14–61), using the limit $c \to \infty$ (cw constant). (Page 255.)

14–9. Apply the (quasi-chemical) method of Section 14–5 to a one-dimensional lattice gas with a first-neighbor interaction energy w_1 and a second-neighbor interaction energy w_2. Instead of assuming independent pairs of sites, assume independent triplets.

14–10. Introduce second-neighbor, and higher, interactions into the Bragg-Williams lattice gas treatment.

14–11. Derive equations for S and E for a Bragg-Williams lattice gas, considering w a function of temperature, $w(T)$.

14–12. When H_2 gas at pressure p is absorbed by Pd metal, the absorbed hydrogen is in the form of atoms. Use the equilibrium condition μ_{H_2} (gas) $= 2\mu_H$ (absorbed) to deduce the absorption isotherm $p(\theta)$ using the Bragg-Williams approximation.

14–13. Expand the one-dimensional Φ/kT in powers of θ in Eq. (14–16) and verify Eq. (14–40) for B_2 when $c = 2$.

14–14. Define $P(N) = Q(N, M, T)\lambda^N$ for a Bragg-Williams or quasi-chemical lattice gas. $P(N)$ is proportional to the probability that an open system (λ, M, T given) will contain N molecules. Discuss the form of the function $P(N)$ when $w < 0$, $T < T_c$ and: (a) $\lambda = q^{-1}e^{cw/2kT}$ (i.e., $\ln y = 0$ in Fig. 14–5); (b) λ different from the value in (a) but still within the extent of the loop (Fig. 14–5); and (c) λ outside of the extent of the loop.

14–15. Apply the argument of Eqs. (14–56) and (14–57) to a pair of nearest-neighbor sites instead of to a single site. That is, treat the pair exactly [see Eq. (7–32)], and assume all other sites in the lattice are occupied at random. Show that, in this approximation,

$$y(\theta) = \frac{2\theta x^{(c-1)(2\theta-1)}}{(x^2+1-\beta^2)^{1/2}+(1-2\theta)x}.$$

Is this result exact in one dimension? Does it satisfy $y(\theta)y(1-\theta) = 1$? Does it reduce to the Bragg-Williams approximation when $w/kT \rightarrow 0$?

14–16. Consider the partition function for a lattice gas,

$$\Delta(N, \Phi, T) = \sum_{M=N}^{\infty} Q(N, M, T)\eta^M$$
$$= Q(N, N, T)\eta^N\left[1 + \frac{Q(N, N+1, T)}{Q(N, N, T)}\eta + \cdots\right],$$

where $\eta = e^{-\Phi/kT}$. This expansion is valid at high pressures where there are very few empty sites ($\theta \rightarrow 1$). Use the same general procedure as in Eqs. (14–30) through (14–38) to obtain an exact expansion of $(1-\theta)/\theta$ in powers of η through the η^2 term.

14–17. Obtain the Bragg-Williams $\ln Q$, Eq. (14–42), by starting with the exact Eq. (14–29) and making use of the high-temperature expansion $x = 1 + (w/2kT) + \cdots$. [This method, due to Kirkwood, can be extended to higher terms; see Eqs. (20–30) through (20–32).]

Supplementary Reading

FOWLER and GUGGENHEIM, Chapter 10.
NEWELL, G. F., and MONTROLL, E. W., *Revs. Mod. Phys.* **25**, 353 (1953).
S. M., Section 28, Chapter 7, and Appendix 9.
TEMPERLEY, H. N. V., *Changes of State.* London: Cleaver-Hume, 1956.
TER HAAR, Chapter 12.

CHAPTER 15

IMPERFECT GASES

This chapter is concerned with gases which are dilute, but not so dilute that we can ignore intermolecular forces altogether. The first two sections, in which our main object is to relate thermodynamic virial coefficients to intermolecular forces, are the basic ones. The rest of the chapter is devoted to a few special topics. A much more complete treatment of this subject will be found in S. M., Chapter 5.

The recurring theme throughout this chapter is that the application of the grand partition function to a dilute system such as an imperfect gas makes possible the reduction of a many-body problem in statistical mechanics to one-body, two-body, etc., problems.

15–1 Virial expansion for a one-component gas. The equation of state of a sufficiently dilute gas is $p = \rho kT$. This is a universal law, the same for all gases. At higher concentrations the equation of state (experimental or theoretical) can be put in the form

$$\frac{p}{kT} = \rho + B_2(T)\rho^2 + B_3(T)\rho^3 + \cdots, \tag{15–1}$$

known as a virial expansion. The $B_n(T)$ are called virial coefficients. These coefficients are in general different for different gases and depend in particular on intermolecular forces. Roughly speaking, when the gas is dense enough so that pairs of molecules spend appreciable amounts of time near each other, the term in B_2 must be introduced. Interactions between three molecules involve B_3, etc. The series (15–1) converges, when $T < T_c$, for values of ρ up to $\rho = 1/v_G$ (Fig. 14–7), at which point liquid begins to appear in the system and there is obviously a singularity in the function $p(\rho)$. When $T > T_c$, the series converges for all ρ.

From the Taylor expansion

$$\frac{p}{\rho kT} = 1 + \left(\frac{\partial p/\rho kT}{\partial \rho}\right)_{T,\rho=0}\rho + \frac{1}{2!}\left(\frac{\partial^2 p/\rho kT}{\partial \rho^2}\right)_{T,\rho=0}\rho^2 + \cdots, \tag{15–2}$$

we have the thermodynamic relations

$$B_n(T) = \frac{1}{(n-1)!}\left(\frac{\partial^{n-1} p/\rho kT}{\partial \rho^{n-1}}\right)_{T,\rho=0} \qquad (n = 2, 3, \ldots). \tag{15–3}$$

This emphasizes the fact that the virial coefficients are properties of the gas in the limit of zero density, $\rho = 0$. This feature of the B_n will be confirmed by our statistical-mechanical expressions below.

First we give a quite general statistical-mechanical argument which is valid for any one-component gas that possesses a virial expansion [not all gases do, incidentally; e.g., a plasma of ionized hydrogen atoms (see Chapter 18)]. The results will be applicable, for example, to polyatomic gases, degenerate (quantum) gases (Chapter 22), etc. After deriving general relations, we shall turn specifically and in more detail to a classical monatomic gas (Section 15–2).

The grand partition function is most convenient here, for reasons already explained [see Eqs. (1–69) and (14–26)]. We have

$$\Xi(\lambda, V, T) = e^{pV/kT} = \sum_{N\geq 0} Q_N(V, T)\lambda^N = 1 + \sum_{N\geq 1} Q_N\lambda^N, \tag{15–4}$$

where

$$Q_N(V, T) \equiv Q(N, V, T), \qquad \lambda = e^{\mu/kT}.$$

In Eq. (15–4), we have put $Q_0 = 1$ since, when $N = 0$, the system has only one state, and this with energy $E = 0$ [see Eq. (1–29)]. Equation (15–4) is a power series in λ, the absolute activity. It proves very convenient later if we define a new activity, z, proportional to λ and having the property that $z \to \rho$ as $\rho \to 0$ [as in Eq. (14–35)]. To find the desired connection between z and λ, we use the limit $\lambda \to 0$ in Eq. (15–4):

$$\ln \Xi = \frac{pV}{kT} = Q_1\lambda + \cdots,$$

$$\overline{N} = \lambda\left(\frac{\partial \ln \Xi}{\partial \lambda}\right)_{V,T} = Q_1\lambda = \frac{pV}{kT}\cdot$$

Then clearly $z = Q_1\lambda/V$, since $\rho = \overline{N}/V$. The partition function Q_1 (one molecule in V) is the same quantity as q in Eq. (3–10). If we put zV/Q_1 in place of λ in Eq. (15–4), the result is

$$\Xi = 1 + \sum_{N\geq 1}\left(\frac{Q_N V^N}{Q_1^N}\right) z^N.$$

To simplify notation, we define $Z_N(V, T)$ by the relation

$$\frac{Z_N(V, T)}{N!} = \frac{Q_N(V, T)\,V^N}{Q_1(V, T)^N}\cdot \tag{15–5}$$

In particular, $Z_1 = V$. In classical statistical mechanics, Z_N turns out, with this definition, to be the configuration integral [see, for example, Eq. (6–22)]. We shall verify this statement in the next section. Then, finally,

$$\Xi(\lambda, V, T) = e^{pV/kT} = 1 + \sum_{N \geq 1} \frac{Z_N(V, T)}{N!} z^N. \tag{15–6}$$

Now we take the logarithm of both sides of Eq. (15–6), expand the logarithm on the right, divide by V, and obtain an expansion for p in powers of z:

$$\frac{p}{kT} = \sum_{j \geq 1} b_j(T) z^j, \tag{15–7}$$

where

$$\begin{aligned} 1!Vb_1 &= Z_1 = V, \qquad b_1 = 1, \\ 2!Vb_2 &= Z_2 - Z_1^2, \\ 3!Vb_3 &= Z_3 - 3Z_1Z_2 + 2Z_1^3, \end{aligned} \tag{15–8}$$

etc. A general relation is available* relating the b_j to the Z_N. To convert the z expansion, (15–7), into a ρ expansion, as in Eq. (15–1), we use the equation

$$\overline{N} = z\left(\frac{\partial \ln \Xi}{\partial z}\right)_{V,T} \quad \text{or} \quad \rho = z\left(\frac{\partial p/kT}{\partial z}\right)_T. \tag{15–9}$$

This gives

$$\rho = \sum_{j \geq 1} j b_j(T) z^j. \tag{15–10}$$

If we invert this series to get z as a power series in ρ, we can substitute $z(\rho)$ into Eq. (15–7) and obtain the required virial expansion. To invert the series, we use for simplicity a straightforward algebraic method, though more elegant procedures are available for this purpose. We substitute

$$z = \rho + a_2\rho^2 + a_3\rho^3 + \cdots \tag{15–11}$$

in Eq. (15–10) to obtain an identity in ρ. Equating coefficients of like powers of ρ on the two sides of the equation, we find

$$\begin{aligned} a_2 &= -2b_2, \\ a_3 &= -3b_3 - 4a_2b_2 = -3b_3 + 8b_2^2, \end{aligned}$$

* See, for example, S. M., Eq. (23.44).

etc. As a final step, we then put Eq. (15–11) (with the a's just deduced) in Eq. (15–7) and get

$$\frac{p}{kT} = \rho + B_2(T)\rho^2 + B_3(T)\rho^3 + \cdots, \tag{15–12}$$

where

$$B_2 = -b_2, \qquad B_3 = 4b_2^2 - 2b_3, \qquad \text{etc.} \tag{15–13}$$

In retrospect, we see, above, that we have found the virial coefficients B_n in terms of the b_j, the b_j in terms of the Z_N, and the Z_N in terms of the Q_N. We note further that to calculate B_2, we need only Q_1 and Q_2; to calculate B_3, we need Q_1, Q_2, and Q_3; etc. Now $Q_1(V, T)$ is the partition function for a single molecule in a box of volume V; $Q_2(V, T)$ is the partition function for two molecules in V; etc. Thus even though the actual gas has, say, 10^{20} molecules in it, we have reduced the calculation of B_2 to one- and two-molecule problems in quantum mechanics (i.e., we need the energy levels for one and two molecules in V to compute Q_1 and Q_2), etc. In view of the comments at the beginning of this section, this result is intuitively reasonable. Thus we expect that binary interactions become significant when the B_2 term is needed in the equation of state, (15–12), and this is confirmed by our finding that B_2 depends on the properties of at most *two* molecules in the volume V. That no additional molecules are in V in our computation of Q_1 and Q_2 (for B_2) corresponds to the fact, mentioned in connection with Eq. (15–3), that the virial coefficients are properties of the gas at *zero* macroscopic density. That is, B_2 depends on binary interactions in a vacuum, B_3 on ternary interactions in a vacuum, etc.

At this point we digress to examine one step in the above argument which happens to lead to correct results but which needs justification. This is the step (15–6) → (15–7). [We employed the same procedure in Eq. (14–36) and in finding the connection between z and λ following Eq. (15–4).] The difficulty is that the expansion $\ln(1 + x) = x - (1/2)x^2 + \cdots$, which we have made use of, is valid only if $x^2 < 1$; but in $1 + Vz + \cdots$ in Eq. (15–6), Vz is of order $\overline{N}$. The following argument avoids this complication. We start with the desired form of expansion, (15–7), and work backward to find the connection between the b_j and the Z_N. That is, at the outset, the b_j in (15–7) are undetermined coefficients. From the expansion (15–7), we form the function

$$e^{pV/kT} = \exp\left(V \sum_{j\geq 1} b_j z^j\right) = \prod_{j\geq 1} e^{Vb_j z^j}$$

$$= \prod_{j\geq 1}\left[\sum_{m_j\geq 0} \frac{1}{m_j!}\,(Vb_j)^{m_j} z^{jm_j}\right]. \tag{15–14}$$

This is a power series in z. The coefficient of z^N in the series is

$$\sum_{\mathbf{m}} \left[\prod_{j=1}^{N} \frac{(Vb_j)^{m_j}}{m_j!} \right], \tag{15-15}$$

where the sum is over all sets $\mathbf{m} = m_1, \ldots, m_N$ satisfying the condition

$$\sum_{j=1}^{N} jm_j = N.$$

On equating coefficients of z^N in Eqs. (15-6) and (15-14), we have, then, that $Z_N/N!$ is equal to the expression (15-15). The first few relations are

$$\begin{aligned} Z_1 &= Vb_1 = V, \\ \tfrac{1}{2}Z_2 &= Vb_2 + \tfrac{1}{2}(Vb_1)^2, \\ \tfrac{1}{6}Z_3 &= Vb_3 + (Vb_1)(Vb_2) + \tfrac{1}{6}(Vb_1)^3. \end{aligned} \tag{15-16}$$

These are easily seen to be equivalent to Eqs. (15-8), which give the b_j explicitly in terms of the Z_N.

We return now to (15-12) and related equations and derive a few more general expressions. The fugacity f of an imperfect gas is, like z, proportional to λ, but with a proportionality constant such that $f \to p$ as $\rho \to 0$. That is, $f \to \rho kT$ as $\rho \to 0$; hence the connection between f and z is $f = zkT$. Thus Eq. (15-11) gives f/kT as a power series in ρ. The inverse of Eq. (15-7) would give f/kT as a power series in p/kT (Problem 15-1). Also, if we define an activity coefficient γ by the relation $z = \gamma\rho$, then, from Eq. (15-11),

$$\gamma = \frac{z}{\rho} = 1 + a_2\rho + a_3\rho^2 + \cdots. \tag{15-17}$$

An alternative and more elegant expression for γ can be found as follows. From $z = Q_1\lambda/V$ and $z = \gamma\rho$, we have

$$\frac{\mu(\rho, T)}{kT} = \ln\left(\frac{V}{Q_1}\right) + \ln \rho + \ln \gamma(\rho, T), \tag{15-18}$$

where V/Q_1 is a function of T only, since Q_1 is proportional to V (see Problem 3-1, where q has the same meaning as Q_1 here). If we now integrate the thermodynamic equation

$$d\left(\frac{\mu}{kT}\right) = v\, d\left(\frac{p}{kT}\right) = \frac{1}{\rho}\left(\frac{\partial p/kT}{\partial \rho}\right)_T d\rho \qquad (T \text{ constant}),$$

using the virial expansion (15–12) for p/kT, we find

$$\frac{\mu}{kT} = \text{constant} + \ln \rho + \sum_{k \geq 1} \left(\frac{k+1}{k}\right) B_{k+1}\rho^k. \tag{15–19}$$

From the definition of z and the limit $\rho \to 0$, we see that the integration constant is $\ln (V/Q_1)$. Therefore, from Eqs. (15–18) and (15–19),

$$\ln \gamma = -\sum_{k \geq 1} \beta_k(T)\rho^k, \tag{15–20}$$

where we have defined β_k by

$$\beta_k = -\left(\frac{k+1}{k}\right) B_{k+1}. \tag{15–21}$$

Incidentally, it should be noted that Eqs. (15–7), (15–10), (15–12), and (15–20) are all thermodynamic expansions that can be interrelated by purely thermodynamic operations. Hence the connections obtained above between the b_j, B_n, and β_k are essentially thermodynamic in origin. Molecular theory enters when we relate any of these coefficients to the Z_N or Q_N, as in Eqs. (15–8).

A comment on the orders of magnitude in Eqs. (15–8) may be helpful. From Eq. (15–7), we have that

$$z = O\left(\frac{\overline{N}}{V}\right), \qquad \frac{p}{kT} = O\left(\frac{\overline{N}}{V}\right), \qquad b_j = O\left(\frac{1}{\rho^{j-1}}\right) = O\left(\frac{V^{j-1}}{\overline{N}^{j-1}}\right).$$

Now let us write the equation for b_j in (15–8) in the form

$$\frac{j!\,Vb_j}{V^j} = \frac{j!\,b_j}{V^{j-1}} = \frac{Z_j - \cdots}{V^j}.$$

From the definition (15–5), we see that $Z_j = O(V^j)$. Therefore each separate term (only one is shown) on the right of the above equation is of order unity, but the complete right side is of order $1/\overline{N}^{j-1}$, since this is the order of the left side. Thus the b_j in Eq. (15–8) represent small terms left over after cancellation of the major contributions to the right-hand side.

15–2 One-component classical monatomic gas. The special case we discuss in this section is a one-component classical monatomic gas with an intermolecular potential energy assumed pairwise additive. The reader should review Eqs. (6–15) through (6–22) and Appendix IV, where basic equations and criteria for the use of classical statistics are developed.

We observe first that for this system $Q_1 = q = V/\Lambda^3$, according to Eq. (4–10). Then, in Eq. (15–5), $Q_N V^N/Q_1{}^N = Q_N \Lambda^{3N}$. Thus, on comparing Eqs. (15–5) and (6–21), we see that the two Z_N's have the same relation to Q_N. Hence, in this special case, Z_N of Section 15–1 is just the configuration integral defined by Eq. (6–22). The first three configuration integrals are

$$Z_1 = \int_V d\mathbf{r}_1 = V,$$

$$Z_2 = \iint_V e^{-u(r_{12})/kT}\, d\mathbf{r}_1\, d\mathbf{r}_2, \tag{15–22}$$

$$Z_3 = \iiint_V e^{-[u(r_{12})+u(r_{13})+u(r_{23})]/kT}\, d\mathbf{r}_1\, d\mathbf{r}_2\, d\mathbf{r}_3,$$

where $d\mathbf{r} = dx\, dy\, dz$, $r_{12} = |\mathbf{r}_2 - \mathbf{r}_1|$, and $u(r)$ is the intermolecular pair potential (Appendix IV) which, in principle at least, can be calculated from quantum mechanics. Beginning with Z_3, the assumption of pairwise additivity in the potential energy $U(\mathbf{r}_1, \mathbf{r}_2, \mathbf{r}_3)$ appears. That is,

$$U = u(r_{12}) + u(r_{13}) + u(r_{23}).$$

From Eqs. (15–8) and (15–13), the second virial coefficient of a classical monatomic gas is

$$B_2(T) = -b_2 = -\frac{1}{2V}(Z_2 - Z_1^2)$$

$$= -\frac{1}{2V}\iint_V [e^{-u(r_{12})/kT} - 1]\, d\mathbf{r}_1\, d\mathbf{r}_2. \tag{15–23}$$

Since $u(r_{12})$ goes rapidly to zero for intermolecular distances r_{12} greater than, say, 15 or 20 A, the integrand in Eq. (15–23) is nonzero only when the elements of volume $d\mathbf{r}_1$ and $d\mathbf{r}_2$ are close to each other. For this reason we change variables from $\mathbf{r}_1$ and $\mathbf{r}_2$ to $\mathbf{r}_1$ and $\mathbf{r}_{12} = \mathbf{r}_2 - \mathbf{r}_1$ (position of molecule 2 relative to position of 1 as origin). Integration over $\mathbf{r}_{12}$ leads to a result that is independent of the location of $d\mathbf{r}_1$, except when $d\mathbf{r}_1$ is in a region of negligible extent (V is macroscopic) within a distance of order 20 A from the walls. Then

$$B_2(T) = -\frac{1}{2V}\int_V d\mathbf{r}_1 \int_V [e^{-u(r_{12})/kT} - 1]\, d\mathbf{r}_{12}$$

$$= -\frac{1}{2}\int_0^\infty [e^{-u(r)/kT} - 1] 4\pi r^2\, dr, \tag{15–24}$$

where we have put $d\mathbf{r}_{12} = 4\pi r^2\, dr$, and the upper limit $r = \infty$ can be used, since the only contributions to the integral come in the first 20 A or so.

For a (hypothetical) gas of hard spheres,

$$\begin{aligned} u(r) &= +\infty \qquad r < a \\ &= 0 \qquad r \geq a, \end{aligned} \tag{15-25}$$

and

$$B_2 = \frac{2\pi a^3}{3}, \tag{15-26}$$

where a is the distance of closest approach of the centers of two spheres, or the diameter of one sphere. Thus B_2 is four times the volume of a sphere.

If we use the much more realistic Lennard-Jones potential, (IV-1), for $u(r)$, the second virial coefficient becomes

$$B_2(T) = -2\pi r^{*3} \int_0^\infty \left[\exp\left(\frac{2\epsilon}{kT} y^{-6} - \frac{\epsilon}{kT} y^{-12} \right) - 1 \right] y^2 \, dy, \tag{15-27}$$

where $y = r/r^*$. This result predicts that for all molecules with an intermolecular potential of the Lennard-Jones form, B_2/r^{*3} is a universal function of kT/ϵ. This is another example of a law of corresponding states. Thus, if experimental values of B_2 as a function of T for a given gas are plotted by adjusting both horizontal and vertical scales (thereby determining values of ϵ and r^* as given in Table IV-1), then the experimental curve can be made to coincide with the theoretical curve of B_2/r^{*3} against kT/ϵ calculated from Eq. (15-27). This is illustrated in Fig. 15-1 for A, Ne, N_2, and CH_4. (The two latter gases behave effectively as monatomic gases as far as the equation of state is concerned.)

The Boyle temperature T_B is the temperature at which $B_2 = 0$. The theoretical value, from Eq. (15-27), proves to be $kT_B/\epsilon = 3.42$. At temperatures below T_B, B_2 is negative owing to the predominant effect of the potential well in $u(r)$ on the integral for B_2. At sufficiently high temperatures, on the other hand, the potential well gets "washed out" ($\epsilon/kT \to 0$) and the short-range repulsion dominates. Hence, B_2 is positive, as for hard spheres. At very high temperatures, the effective "hard-sphere diameter" decreases with increasing T, because the molecules can approach each other more closely when they are more energetic. Therefore $B_2(T)$ passes through a maximum.

Turning now to B_3, Eq. (15-13) gives B_3 in terms of b_2^2 and b_3. Define $x_{ij} = e^{-u(r_{ij})/kT}$, and recall that

$$b_2 = \frac{1}{2V} \iint_V (x_{12} - 1) \, d\mathbf{r}_1 \, d\mathbf{r}_2 = \frac{1}{2} \int_V (x_{12} - 1) \, d\mathbf{r}_{12}.$$

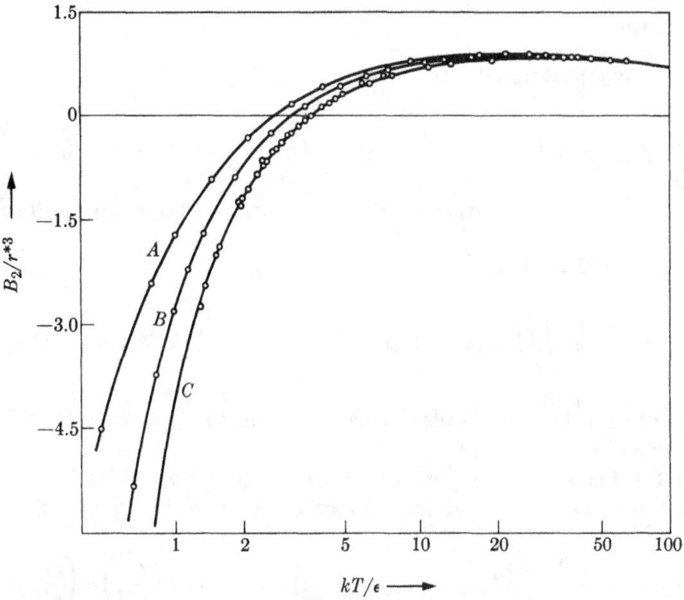

FIG. 15–1. Reduced second virial coefficient (B_2/r^{*3}) as a function of reduced temperature (kT/ϵ). Curve C is the classical curve calculated from Eq. (15–27). The experimental points on curve C are a mixture of points for A, Ne, N_2, and CH_4. Curves A and B are the calculated quantum curves (Section 22–5) for He and H_2, respectively, with corresponding experimental points also shown.

Now we notice that

$$\begin{aligned}\iiint_V (x_{12} - 1)(x_{13} - 1)\, d\mathbf{r}_1\, d\mathbf{r}_2\, d\mathbf{r}_3 &= V\left[\int_V (x_{12} - 1)\, d\mathbf{r}_{12}\right]\left[\int_V (x_{13} - 1)\, d\mathbf{r}_{13}\right] \\ &= 4b_2^2 V. \end{aligned} \tag{15–28}$$

This gives us an expression for b_2^2. For b_3, we have from Eq. (15–8),

$$6Vb_3 = \iiint_V (x_{12}x_{13}x_{23} - x_{12} - x_{13} - x_{23} + 2)\, d\mathbf{r}_1\, d\mathbf{r}_2\, d\mathbf{r}_3. \tag{15–29}$$

For symmetry, we combine Eqs. (15–28) (using the first line) and (15–29) in the form

$$3B_3V = 12b_2^2V - 6b_3V$$

$$= \iiint_V [(x_{12} - 1)(x_{13} - 1) + (x_{12} - 1)(x_{23} - 1) + (x_{13} - 1)(x_{23} - 1) - (x_{12}x_{13}x_{23} - x_{12} - x_{13} - x_{23} + 2)]\, d\mathbf{r}_1\, d\mathbf{r}_2\, d\mathbf{r}_3.$$

After cancellation, this gives

$$B_3(T) = -\frac{1}{3V}\iiint_V (x_{12} - 1)(x_{13} - 1)(x_{23} - 1)\, d\mathbf{r}_1\, d\mathbf{r}_2\, d\mathbf{r}_3. \quad (15\text{–}30)$$

A fairly long calculation leads, for hard spheres, to $B_3 = (5/8)B_2^2$, where B_2 is given by Eq. (15–26).

For the Lennard-Jones potential, we first change variables to $\mathbf{r}_1$, $\mathbf{r}_{12}$, $\mathbf{r}_{13}$ and integrate over $\mathbf{r}_1$, giving a factor V. Then, as in Eq. (15–27),

$$B_3 = -\frac{r^{*6}}{3}\iiint_V (x_{12} - 1)(x_{13} - 1)(x_{23} - 1)\, d\left(\frac{\mathbf{r}_{12}}{r^{*3}}\right) d\left(\frac{\mathbf{r}_{13}}{r^{*3}}\right). \quad (15\text{–}31)$$

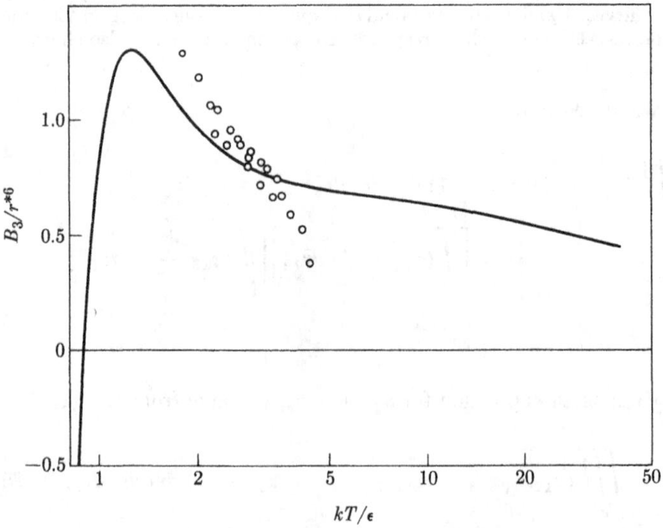

FIG. 15–2. B_3/r^{*6} as a function of kT/ϵ, calculated from the classical equation (15–31). The experimental points are a mixture of points for A, N_2, and CH_4.

This gives B_3/r^{*6} as a universal function of kT/ϵ. Figure 15-2 shows this function, calculated numerically from Eq. (15-31) (after rearrangement for computational convenience). The experimental values of B_3 in the figure have been "reduced," using value of ϵ and r^* obtained from the second virial coefficient (i.e., by fitting the theoretical curve in Fig. 15-1). The agreement with experiment is only fair. Possible contributions to the discrepancy are: (1) error in the experimental values of B_3; (2) lack of effective spherical symmetry in N_2; (3) approximate nature of the Lennard-Jones potential; and (4) nonadditivity of the intermolecular potential of three molecules.

The dependence of B_2/r^{*3} and B_3/r^{*6} on kT/ϵ only, for the Lennard-Jones potential, is in fact a general result for all virial coefficients [see Eq. (15-38) below] if pairwise additivity of the potential is assumed for B_3, B_4, ... In this case the complete equation of state (15-12) can be written, using $v = 1/\rho$,

$$\frac{pv}{kT} = 1 + \left(\frac{r^{*3}}{v}\right)\varphi_2\left(\frac{kT}{\epsilon}\right) + \left(\frac{r^{*3}}{v}\right)^2\varphi_3\left(\frac{kT}{\epsilon}\right) + \cdots$$

$$= \varphi\left(\frac{v}{r^{*3}}, \frac{kT}{\epsilon}\right). \tag{15-32}$$

It is clear that not only the Lennard-Jones potential but any two-parameter potential which is of the form

$$u(r) = (\text{energy parameter}) \times h\left(\frac{r}{\text{distance parameter}}\right) \tag{15-33}$$

and which is assumed pairwise additive, will lead to the law of corresponding states for imperfect gases, (15-32).

Equations (15-32) and (15-33) imply that (subscript c refers to critical point)

$$\frac{v_c}{r^{*3}} = c_1, \qquad \frac{kT_c}{\epsilon} = c_2, \qquad \frac{p_c r^{*3}}{\epsilon} = c_3,$$
$$\frac{p_c v_c}{kT_c} = c_4, \qquad \frac{T_B}{T_c} = c_5, \tag{15-34}$$

where $c_1, \ldots, c_5$ are dimensionless constants, the same for all gases obeying Eq. (15-33) with the same function h. Table 15-1 furnishes a test of all but the last of these relations for simple molecules, using experimental critical constants reduced by values of ϵ and r^* from the experimental second virial coefficient. The constancy of the "constants" is very good but not excellent. We can conclude that, to a rather good approximation, the intermolecular potential for effectively spherical molecules satisfies (15-33) and is pairwise additive.

TABLE 15–1

REDUCED CRITICAL CONSTANTS

	$\frac{v_c}{r^{*3}}$	$\frac{kT_c}{\epsilon}$	$\frac{p_c r^{*3}}{\epsilon}$	$\frac{p_c v_c}{kT_c}$
Ne	2.35	1.25	0.157	0.295
A	2.23	1.26	0.164	0.290
Xe	2.05	1.31	0.187	0.293
N_2	2.09	1.33	0.185	0.291
O_2	1.90	1.31	0.201	0.292
CH_4	2.09	1.29	0.178	0.288
Average	2.12	1.29	0.179	0.292

Equations (15–34) and Table 15–1 provide a method* for the estimation, from critical constants, of r^* and ϵ for other molecules that are of the same type but which are not in the table. That is, we can use the relations

$$\epsilon = \frac{kT_c}{1.29},$$
$$r^{*3} = \frac{v_c}{2.12}, \qquad \text{or} \qquad r^{*3} = \frac{0.179\epsilon}{p_c} = \frac{0.139kT_c}{p_c}. \tag{15–35}$$

There is a similarity in the form of Eqs. (15–23) and (15–30) for B_2 and B_3 which can be shown to be general in the case of pairwise additivity. We give the result but not the proof. But first we digress to introduce "cluster diagrams." Define $f_{ij} = x_{ij} - 1$. The function $f_{ij} = e^{-u(r_{ij})/kT} - 1$ is nonzero only when r_{ij} is small. The molecules 1, 2 are said to form a "cluster" when f_{12} occurs as the integrand in an integral over $d\mathbf{r}_1$ and $d\mathbf{r}_2$. The term is appropriate because the integrand is nonzero only when the two molecules are near each other. Equation (15–23) provides an example. Three molecules 1, 2, 3 form a cluster when any of the following integrands occur (nonzero only when all three molecules are near each other): $f_{12}f_{13}f_{23}$, $f_{12}f_{23}$, $f_{12}f_{13}$, $f_{13}f_{23}$. Equation (15–30) is an example. The integrands in Eqs. (15–23) and (15–30) can be represented schematically by "cluster diagrams":

(15–36)

* T. L. HILL, *J. Chem. Phys.* **16**, 399 (1948).

A line between two circles (molecules) i and j means that a factor f_{ij} is present in the integrand. The other cluster diagrams for three molecules are:

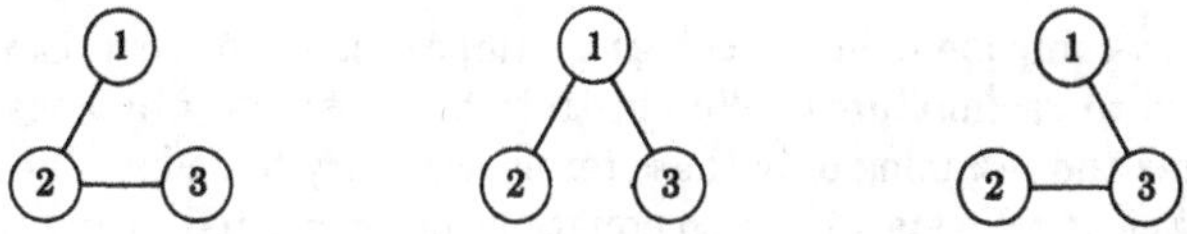

(15-37)

Now Eq. (15-21) relates B_{k+1} to β_k. The general result* referred to at the beginning of this paragraph is that

$$\beta_k = \frac{1}{k!V} \int \underset{V}{\cdots} \int S'_{1,2,\ldots,k+1} \, d\mathbf{r}_1 \, d\mathbf{r}_2 \ldots d\mathbf{r}_{k+1}, \tag{15-38}$$

where $S'_{1,2,\ldots,k+1}$ is the sum of all different products of f's that connect molecules 1, 2, . . . , $k + 1$ in a "doubly connected" cluster diagram. In a doubly connected diagram, between each pair of molecules there are at least two entirely independent paths which do not cross at any circle. In addition, a cluster of two molecules is considered doubly connected, for classification purposes, and is sometimes written . Thus the diagrams (15-36) are doubly connected, but those in (15-37) are only singly connected. Hence $S'_{1,2} = f_{12}$ and $S'_{1,2,3} = f_{12}f_{13}f_{23}$. This is consistent with Eqs. (15-23) and (15-30). However, for β_3 and B_4, there are ten terms in $S'_{1,2,3,4}$ corresponding to the following ten doubly connected diagrams:

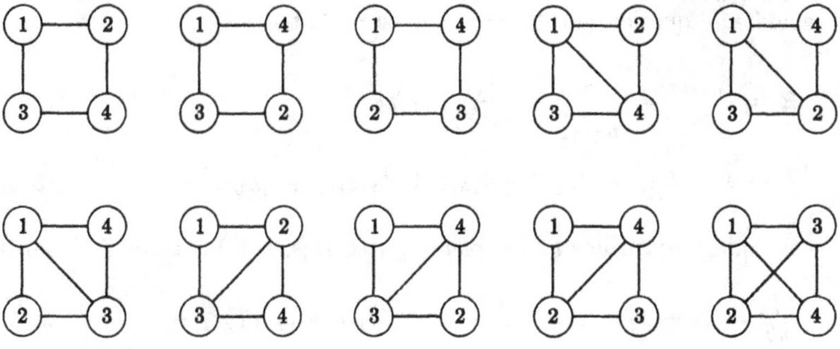

(15-39)

---

* See S. M., Chapter 5, for proof.

Usually β_k is referred to as an "irreducible cluster integral," and b_j is called a "cluster integral." Cluster diagrams were first introduced by Mayer, starting from the canonical ensemble.*

15–3 Two-component imperfect gas. Imperfect gas theory can easily be extended to gas mixtures. We illustrate this possibility by considering a binary gas and retaining only those terms necessary to deduce the second virial coefficient. We start off with equations not restricted to any particular kind of gas (classical, quantum, polyatomic, etc.). The grand partition function is

$$\Xi(\lambda_1, \lambda_2, V, T) = e^{pV/kT} = \sum_{N_1, N_2 \geq 0} Q_{N_1N_2}(V, T)\lambda_1^{N_1}\lambda_2^{N_2}. \tag{15–40}$$

We define z_1, z_2, and $Z_{N_1N_2}$ by the equations

$$z_1 = \frac{Q_{10}\lambda_1}{V}, \qquad z_2 = \frac{Q_{01}\lambda_2}{V}, \tag{15–41}$$

$$\frac{Z_{N_1N_2}}{N_1!N_2!} = \frac{Q_{N_1N_2}V^{N_1+N_2}}{Q_{10}^{N_1}Q_{01}^{N_2}}. \tag{15–42}$$

The coefficients Q_{10}/V and Q_{01}/V in Eqs. (15–41) are functions of T only. The definitions (15–41) and (15–42) will prove below to have the desired properties that (a) $z_1 \to \rho_1$ and $z_2 \to \rho_2$ when ρ_1 and $\rho_2 \to 0$, and (b) $Z_{N_1N_2}$ becomes the configuration integral in classical statistical mechanics. Equation (15–40) can now be rewritten as

$$\begin{aligned}\Xi = e^{pV/kT} &= \sum_{N_1, N_2 \geq 0} \frac{Z_{N_1N_2}}{N_1!N_2!} z_1^{N_1} z_2^{N_2} \\ &= 1 + Vz_1 + Vz_2 + \tfrac{1}{2}Z_{20}z_1^2 + Z_{11}z_1z_2 + \tfrac{1}{2}Z_{02}z_2^2 + \cdots.\end{aligned} \tag{15–43}$$

On taking logarithms and expanding [see Eqs. (15–14) through (15–16)],

$$\frac{p}{kT} = z_1 + z_2 + b_{20}(T)z_1^2 + b_{11}(T)z_1z_2 + b_{02}(T)z_2^2 + \cdots, \tag{15–44}$$

where

$$\begin{aligned} b_{10} = b_{01} = 1, &\qquad 2!0!Vb_{20} = Z_{20} - V^2, \\ 1!1!Vb_{11} = Z_{11} - V^2, &\qquad 0!2!Vb_{02} = Z_{02} - V^2. \end{aligned} \tag{15–45}$$

* See Mayer and Mayer, Chapter 13.

From

$$\rho_i = z_i \left(\frac{\partial p/kT}{\partial z_i} \right)_{T, z_j}, \qquad i = 1, 2,$$

we have

$$\begin{aligned} \rho_1 &= z_1 + 2b_{20}z_1^2 + b_{11}z_1z_2 + \cdots, \\ \rho_2 &= z_2 + 2b_{02}z_2^2 + b_{11}z_1z_2 + \cdots, \end{aligned} \tag{15–46}$$

and the inverses (Problem 15–2)

$$\begin{aligned} z_1 &= \rho_1 - 2b_{20}\rho_1^2 - b_{11}\rho_1\rho_2 + \cdots, \\ z_2 &= \rho_2 - 2b_{02}\rho_2^2 - b_{11}\rho_1\rho_2 + \cdots. \end{aligned} \tag{15–47}$$

These last equations also provide series in the activity coefficients $\gamma_1 = z_1/\rho_1$ and $\gamma_2 = z_2/\rho_2$.

Substitution of Eqs. (15–47) in Eq. (15–44) gives the desired virial expansion

$$\frac{p}{kT} = \rho_1 + \rho_2 + B_{20}(T)\rho_1^2 + B_{11}(T)\rho_1\rho_2 + B_{02}(T)\rho_2^2 + \cdots, \tag{15–48}$$

where

$$B_{20} = -b_{20}, \qquad B_{11} = -b_{11}, \qquad B_{02} = -b_{02}. \tag{15–49}$$

The coefficients B_{20} and B_{02} are just the second virial coefficients of the two *pure* gases (Section 15–1); B_{11} is new and depends on the properties of two molecules in the volume V, one of each species.

For a classical monatomic gas mixture, we have from Eq. (6–25) that

$$Q_{N_1N_2} = \frac{Z_{N_1N_2}}{N_1!N_2!\Lambda_1^{3N_1}\Lambda_2^{3N_2}}, \tag{15–50}$$

where

$$Z_{N_1N_2} = \int_V e^{-U_{N_1N_2}/kT}\, d\{N_1\}\, d\{N_2\}, \tag{15–51}$$

and $d\{N_1\}$ means $d\mathbf{r}_1 \ldots d\mathbf{r}_{N_1}$ for species 1, etc. Comparison with Eq. (15–42) shows that, in this special case, $Z_{N_1N_2}$ defined in Eq. (15–42) is the classical configuration integral, (15–51). Then, in Eq. (15–48), B_{20} and B_{02} are given by Eq. (15–24) for a pure gas [using the appropriate $u(r)$ for each gas] and

$$\begin{aligned} B_{11} = -b_{11} &= -\frac{1}{V}(Z_{11} - V^2) \\ &= -\int_0^\infty [e^{-u_{11}(r)/kT} - 1]4\pi r^2\, dr. \end{aligned} \tag{15–52}$$

In this equation, u_{11} is the intermolecular potential between one molecule of each type. Information about u_{11} can be deduced from very accurate measurements on the two pure gases and on the mixture, since only B_{11} in Eq. (15–48) is not a property of a pure gas. If a Lennard-Jones potential is used for each of the three pair interactions, a good approximation for the parameters of u_{11}, in the absence of other information, is

$$\epsilon_{11} = (\epsilon_1\epsilon_2)^{1/2}, \qquad r_{11}^* = \tfrac{1}{2}(r_1^* + r_2^*), \tag{15–53}$$

where ϵ_1, ϵ_2, r_1^*, and r_2^* refer to the pure gases 1 and 2. The first relation follows from the simple theory of dispersion forces, and the second would be correct for hard spheres.

The above equations provide a summary of the statistical-mechanical relations that determine the thermodynamic properties of a slightly imperfect binary gas mixture. The (quantum-mechanical) interactions between the three kinds of pairs of molecules are the basic quantities appearing in these equations.

Next we investigate briefly an example of a chemical equilibrium occurring in an imperfect gas mixture. It will be recalled that Chapter 10 was restricted to chemical reactions occurring in ideal gases. Suppose that in the binary mixture 1, 2 the dissociation (or association) equilibrium,

$$\textcircled{2} \rightleftarrows 2\textcircled{1},$$

as in Section 10–2, takes place. The equilibrium condition is $\mu_2 = 2\mu_1$, or $\lambda_2 = \lambda_1^2$. Then from the definitions of z_1 and z_2 in Eq. (15–41),

$$\frac{z_2 V}{Q_{01}} = \frac{z_1^2 V^2}{Q_{10}^2},$$

or

$$\frac{z_1^2}{z_2} = \frac{(Q_{10}/V)^2}{(Q_{01}/V)} = K(T). \tag{15–54}$$

Here $K(T)$ is the same kind of equilibrium constant as in Eq. (10–6), and the Q's have the same meaning as the q's in (10–6). However, the equilibrium constant is equal to a quotient of activities rather than to a quotient of concentrations. The fact that z_1^2/z_2 is a function of T only, $K(T)$, is of course a purely thermodynamic result. But the relation of $K(T)$ to the Q's is statistical-mechanical.

We can also write

$$K(T) = \frac{z_1^2}{z_2} = \frac{\rho_1^2\gamma_1^2}{\rho_2\gamma_2}, \tag{15–55}$$

where γ_1 and γ_2 are activity coefficients, determined by Eqs. (15–47). Thus,

$$K(T) = \frac{\rho_1^2}{\rho_2} [1 + (b_{11} - 4b_{20})\rho_1 + \cdots]. \tag{15–56}$$

The expansion in brackets here gives the first-order correction term, associated with gas imperfection, to the ratio ρ_1^2/ρ_2 [which is exactly equal to $K(T)$ only in the limit as $\rho_1, \rho_2 \to 0$]. The linear term in ρ_2 is omitted because it is of order ρ_1^2. This expansion is thermodynamic in origin if we regard the b_{ij} as being determined from the experimental equation of state, (15–48), and the relations (15–49). But Eqs. (15–45) for the b_{ij} are of a molecular, not thermodynamic, nature.

A special case of the equilibrium ② ⇄ 2①, above, is the formation of dimers, or clusters of two molecules, in a slightly imperfect one-component gas. Of course, from a thermodynamic point of view, any dimer formation, however defined, is automatically taken care of by the second virial coefficient B_2 and need not be discussed explicitly. But for some purposes it is profitable to adopt the alternative but necessarily equivalent point of view that dimers exist and are in equilibrium with monomers. We shall say something about the definition of a dimer below. Higher clusters (trimers, etc.) can be included in the treatment, but for simplicity we confine ourselves to monomers and dimers (i.e., the gas is assumed very dilute). Incidentally, we are referring here to "real" or "physical" clusters, not the "mathematical clusters" of Section 15–2. A much more detailed treatment of clusters of both kinds will be found in S. M., Chapter 5.

The gas is considered an equilibrium mixture of two species: monomers, ①, and dimers, ②. The partition function of one monomer in V is Q_{10}; of two monomers in V is Q_{20} (the two monomers can interact with each other); and of one dimer in V is Q_{01}. The concentration of molecules in the gas, from the one-component point of view, is

$$\rho = \rho_1 + 2\rho_2, \tag{15–57}$$

where

$$\frac{\rho_1^2}{\rho_2} = K(T) = \frac{Q_{10}^2}{VQ_{01}}. \tag{15–58}$$

The extra term given in Eq. (15–56) is not involved in B_2, so we drop it here. From the above two equations we find

$$\rho = \rho_1 + \frac{2\rho_1^2}{K},$$

or, for the concentrations of monomers and dimers,

$$\rho_1 = \rho - \frac{2}{K}\rho^2, \tag{15-59}$$

$$\rho_2 = \frac{\rho^2}{K}, \tag{15-60}$$

to terms in ρ^2. If we substitute Eqs. (15–59) and (15–60) in Eq. (15–48) for a binary gas mixture, we get for the equation of state

$$\frac{p}{kT} = \rho + \left(-\frac{1}{K} - b_{20}\right)\rho^2 + \cdots, \tag{15-61}$$

where

$$-\frac{1}{K} - b_{20} = -\frac{1}{2V}\left[\frac{2V^2(Q_{20} + Q_{01})}{Q_{10}^2} - V^2\right]. \tag{15-62}$$

This should be the second virial coefficient B_2 of the one-component gas. From Section 15–1,

$$B_2 = -\frac{1}{2V}\left(\frac{2V^2Q_2}{Q_1^2} - V^2\right). \tag{15-63}$$

Since Q_1 and Q_{10} have the same meaning, we conclude from (15–62) and (15–63) that, for self-consistency, we must have $Q_2 = Q_{20} + Q_{01}$. But this is just what we should expect since, from the monomer-dimer point of view, two molecules in V can exist in two sets of states: (a) two monomers; and (b) one dimer. As far as thermodynamics is concerned, the splitting of Q_2 into two parts, Q_{20} and Q_{01}, is quite arbitrary; it depends on the definition of "dimer" (but the same B_2 is found in any case, provided $Q_{20} + Q_{01} = Q_2$). Hence the numbers of monomers and dimers calculated by Eqs. (15–59) and (15–60) are also arbitrary. The most obvious classification for the states of Q_2 is to include the "bound" states (in quantum or classical mechanics) in Q_{01} and all others in Q_{20}. The reader interested in pursuing this subject further should consult S. M., Section 27.

15–4 Imperfect gas near a surface. In this section and the next we consider a one-component imperfect gas in an external field. Here the external field is a short-range one provided by a solid surface, as, for example, in Eq. (7–2). The present treatment provides a relatively exact (compared to Chapter 7) approach to the problem of physical adsorption of gases on solids. We shall merely sketch the basic equations here. Further details are available elsewhere.*

---

* S. M., Appendix 10, where additional references are given. Also T. L. Hill, *J. Phys. Chem.* **63**, 456 (1959).

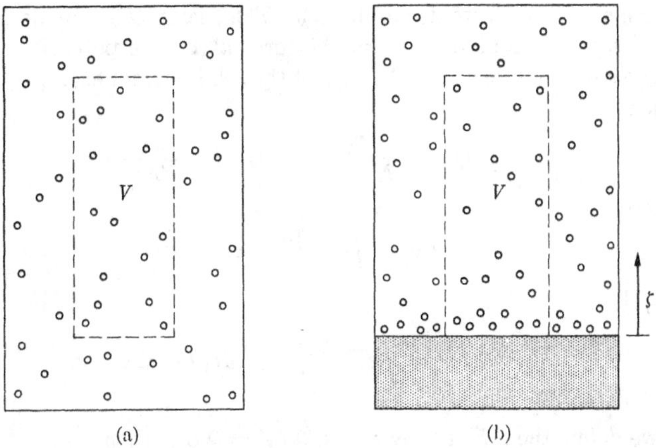

FIG. 15–3. Gas in volume V (a) in the absence and (b) in the presence of an adsorbing surface.

The system of interest is a one-component gas in a volume V in the absence (Fig. 15–3a) and in the presence (Fig. 15–3b) of an adsorbing surface of area $\mathcal{A}$. In both cases the gas has a chemical potential μ (or activity z) and temperature T. If $\overline{N}$ is the average number of molecules in the system with solid present and if $\overline{N}^0$ is the average number with solid absent, then the number of adsorbed molecules, by Gibbs' surface excess definition, is $\overline{N} - \overline{N}^0$. The basic thermodynamic equation for the system with the solid present is [compare Eq. (7–6)]

$$dE = T\,dS - \varphi\,d\mathcal{A} - p\,dV + \mu\,dN, \tag{15–64}$$

where p is the gas pressure far from the surface. Then it follows that

$$\Xi(\mu, V, \mathcal{A}, T) = e^{(pV+\varphi\mathcal{A})/kT} = \sum_N Q_N(V, \mathcal{A}, T)\lambda^N. \tag{15–65}$$

In the absence of the solid, we use the notation

$$\Xi^0(\mu, V, T) = e^{pV/kT} = \sum_N Q_N^0(V, T)\lambda^N, \tag{15–66}$$

where Ξ^0 and Q_N^0 have the same meanings as Ξ and Q_N in Section 15–1.

Although we could pursue this problem using the general formulation of Section 15–1, let us investigate instead the specific case of a classical monatomic gas. The potential energy of the gas (when the solid is present) is the sum of the usual (Section 15–2) intermolecular potential energy U_N^0 (solid absent) and the potential energy of the gas molecules in the field of the solid, which we denote by U_N^*. The energy U_N^* is made up of a sum of N separate contributions of the form $u^*(\mathbf{r})$, the potential energy of a single gas molecule at $\mathbf{r}$ in the field of the solid [see, for example, Eq. (7–2)]. Then

$$Q_N^0 = \frac{Z_N^0}{N!\Lambda^{3N}}, \qquad Q_N = \frac{Z_N}{N!\Lambda^{3N}}, \tag{15–67}$$

where

$$Z_N^0 = \int_V e^{-U_N^0/kT}\, d\mathbf{r}_1 \cdots d\mathbf{r}_N \tag{15–68}$$

and

$$Z_N = \int_V e^{-(U_N^0+U_N^*)/kT}\, d\mathbf{r}_1 \cdots d\mathbf{r}_N. \tag{15–69}$$

If we define the activity by $z = Q_1^0\lambda/V = \lambda/\Lambda_3$, then

$$\Xi^0 = \sum_N \frac{Z_N^0}{N!} z^N, \qquad \Xi = \sum_N \frac{Z_N}{N!} z^N. \tag{15–70}$$

We now follow the same argument as in Section 15–1, and find

$$\frac{pV}{kT} = V \sum_{j\geq 1} b_j^0 z^j, \qquad \frac{pV + \varphi\mathfrak{a}}{kT} = V \sum_{j\geq 1} b_j z^j \tag{15–71}$$

$$\overline{N}^0 = V \sum_{j\geq 1} j b_j^0 z^j, \qquad \overline{N} = V \sum_{j\geq 1} j b_j z^j$$

$$\overline{N} - \overline{N}^0 = V \sum_{j\geq 1} j(b_j - b_j^0) z^j, \tag{15–72}$$

where the b's are related to the corresponding Z's by Eq. (15–8) (but $Z_1 \neq V$, $b_1 \neq 1$), and $p(z, T)$ is the pressure either in the absence of or far from the surface. Equation (15–72) is the "adsorption isotherm" giving the amount adsorbed as a function of the activity z (or fugacity, $f = zkT$). At very low gas pressures (the Henry's law region), $p/kT = z$ and the number of molecules adsorbed is

$$\overline{N} - \overline{N}^0 = V(b_1 - b_1^0)z = V(b_1 - 1)\frac{p}{kT},$$

where

$$V(b_1 - 1) = Z_1 - V = \int_V [e^{-u^*(\mathbf{r})/kT} - 1]\, d\mathbf{r}.$$

If u^* is a function of the distance ζ from the surface only [as in Eq. (7–2) and Problem 7-1], we have

$$\frac{\overline{N} - \overline{N}^0}{\mathcal{A}} = \frac{p}{kT} \int_0^\infty [e^{-u^*(\zeta)/kT} - 1]\, d\zeta. \tag{15-73}$$

The integral determines the relative extent of binding of different molecules on a surface, and this in turn depends on $u^*(\zeta)$. As is typical of a Henry's law constant (Chapter 19), the constant is determined by the interaction of just one "solute" (gas) molecule with the "pure solvent" (solid surface). Equation (15-73) has been applied successfully to experimental data by Freeman and Halsey.†

It is left to the reader to obtain explicit expressions (in terms of u^* and u) for the quadratic terms in the expansion of $\overline{N} - \overline{N}^0$ in powers of z or p/kT.

From Eqs. (15-71) and (15-72), the surface pressure in the limit as $p \to 0$ is given by

$$\frac{\varphi \mathcal{A}}{kT} = V(b_1 - b_1^0)z = \overline{N} - \overline{N}^0, \tag{15-74}$$

which has the form of the equation of state of a two-dimensional ideal gas. Note from Eq. (15-73) that it is possible for $\overline{N} - \overline{N}^0$ and φ to be negative.

Incidentally, if we subtract

$$dE^0 = T\, dS^0 - p\, dV + \mu\, dN^0$$

from Eq. (15-57), we get the basic thermodynamic equation, in terms of surface excesses,

$$dE_s = T\, dS_s - \varphi\, d\mathcal{A} + \mu\, dN_s, \tag{15-75}$$

where

$$E_s = E - E^0, \qquad S_s = S - S^0, \qquad N_s = N - N^0.$$

This is the analog of Eq. (7-6).

15-5 Imperfect gas in an electric field. The discussion here is a continuation of that in Sections 12-1 and 12-2, which should be reviewed.

Section 12-3 was concerned with a dilute gas of independent molecules in an electric field. In an imperfect gas or liquid, the molecules are not independent and the problem becomes very involved. Each molecule is not only in the field (D) of the external charges, but is also in the field of the other molecules. These intermolecular interactions are more complicated than usual because (a) dipole-dipole forces have a relatively long

† M. P. Freeman and G. D. Halsey, Jr., *J. Phys. Chem.* **59**, 181 (1955).

range, and (b) both the external field and the field due to other molecules induce an additional dipole in a given molecule, which in turn contributes to the interaction of the molecule with other molecules. Superimposed on these electrostatic interactions are the usual van der Waals attractions and repulsions (themselves ultimately, of course, of electrostatic origin).

Although the properties of a dielectric liquid in an electric field are extremely difficult to analyze without approximation, use of the grand ensemble and low-density expansions make it possible to treat an imperfect gas in a straightforward way. We confine ourselves here to slightly imperfect gases (up to the second virial coefficient), and even in this case we develop general equations only. Detailed treatment of realistic special cases or models involves rather complicated calculations.*

The grand partition function, (12–15), can be written

$$\Xi = 1 + Q_1(V, T, D)\lambda + Q_2(V, T, D)\lambda^2 + \cdots. \qquad (15\text{–}76)$$

Then

$$\frac{pV}{kT} = \ln \Xi = Q_1\lambda + (Q_2 - \tfrac{1}{2}Q_1^2)\lambda^2 + \cdots \qquad (15\text{–}77)$$

and, from Eq. (12–17) (note the definition of M_N),

$$\overline{M} = M_1Q_1\lambda + (Q_2M_2 - Q_1^2M_1)\lambda^2 + \cdots. \qquad (15\text{–}78)$$

Also, from Eq. (12–9),

$$\overline{N} = \lambda\left(\frac{\partial \ln \Xi}{\partial \lambda}\right)_{V,T,D} = Q_1\lambda + (2Q_2 - Q_1^2)\lambda^2 + \cdots. \qquad (15\text{–}79)$$

For a very dilute gas (p, ρ, or $\lambda \to 0$), only the leading terms in Eqs. (15–77) through (15–79) need be retained. For a slightly imperfect gas, for which a second virial coefficient is required in the equation of state, Eqs. (15–77) through (15–79) indicate that we have to consider the properties of only *two molecules* in the volume V and external field D. Of course, in this case, interactions (including dipole-dipole) between the two molecules, and the influence of D on the interactions, have to be taken into account.

From Eqs. (15–77) and (15–79), we can obviously derive, just as in Section 15–1, the virial expansion

$$\frac{p}{kT} = \rho + \sum_{n\geq 2} B_n(T, D)\rho^n, \qquad (15\text{–}80)$$

* See A. D. Buckingham, *J. Chem. Phys.* **23**, 2370 (1955); A. D. Buckingham and J. A. Pople, *Trans. Faraday Soc.* **51**, 1029, 1179 (1955).

where the B_n and Q_N are related by the same formal expressions as in Section 15–1. Also, we can show* (Problem 15–3) that the virial expansion for the polarization is

$$\overline{P} = \frac{\overline{M}}{V} = M_1 \rho - kT \sum_{n \geq 2} \left(\frac{1}{n-1} \right) \left(\frac{\partial B_n}{\partial D} \right)_T \rho^n. \qquad (15\text{–}81)$$

Problems

15–1. Obtain an expansion for f/kT (f = fugacity) in powers of p/kT up to the cubic term. (Page 265.)

15–2. Deduce the expansions (15–47) for a binary mixture from (15–46). (Page 275.)

15–3. Use Eqs. (15–77) through (15–79) to verify the coefficient of ρ^2 in the polarization expansion, (15–81). (Page 283.)

15–4. Estimate values of ϵ and r^* for ethane from $T_c = 305.2°\text{K}$ and $p_c = 48.8$ atm.

15–5. The intermolecular potential for monatomic molecules,

$$u(r) = -\frac{\alpha}{r^6} + \beta e^{-r/\gamma},$$

where α, β, and γ are positive constants, is often used. (a) If the system (gas) obeys the law of corresponding states, what conclusion can be drawn about the independence or dependence of α, β, and γ on each other? (b) If this potential is made to agree with the Lennard-Jones potential for large r and in the location of the bottom of the well ($u = -\epsilon$ = minimum at $r = r^*$), what are the connections between α, β, and γ and r^* and ϵ? Calculate α, β, and γ for argon.

15–6. Sketch a plot of the function $e^{-u(r)/kT} - 1$ for hard spheres and for the Lennard-Jones potential (at several different temperatures).

15–7. Find B_2 at high temperatures for the potential

$$\begin{aligned} u(r) &= +\infty & r &< r^* \\ &= -\epsilon \left(\frac{r^*}{r} \right)^6 & r &\geq r^*. \end{aligned}$$

That is, for $r \geq r^*$, replace $e^{-u/kT}$ by $1 - (u/kT)$.

15–8. Show, using an integration by parts, that

$$B_2 = -\frac{1}{6kT} \int_0^\infty r \frac{du(r)}{dr} e^{-u(r)/kT} 4\pi r^2 \, dr \qquad (15\text{–}82)$$

is equivalent to Eq. (15–24).

---

* T. L. Hill, *J. Chem. Phys.* **28**, 61 (1958). A number of other details are also given in this paper.

15–9. Extend the series of Section 15–1 one more term.

15–10. Show that the expansion of E/NkT in powers of ρ for a classical monatomic gas starts off as

$$\frac{E}{NkT} = \frac{3}{2} + \frac{\rho}{2kT}\int_0^\infty u(r)e^{-u(r)/kT}4\pi r^2\,dr + \cdots, \tag{15–83}$$

and that a more general relation is

$$\frac{E}{NkT} = \frac{3}{2} + \frac{T}{\rho}\sum_{j\geq 2}\frac{db_j}{dT}z^j. \tag{15–84}$$

15–11. For a classical monatomic gas, put S/Nk at given ρ and T in the form

$$\frac{S}{Nk} = \frac{S}{Nk}\text{ (ideal gas, same }\rho\text{ and }T) + C(T)\rho + \cdots.$$

Find $C(T)$ in terms of $u(r)$. Show that for a finite temperature, C is always negative (as might be expected intuitively). Find the asymptotic form of C at high temperatures. Use the $u(r)$ in Problem 15–7 to calculate C at high temperatures.

15–12. Use Problem 15–4, critical constants for methane, and Eqs. (15–53) to estimate ϵ_{11} and r_{11}^* for the ethane-methane interaction.

15–13. Extend Eqs. (15–43) through (15–49) to include the third virial coefficient.

15–14. If Eq. (15–48) for a binary mixture is put in the form

$$\frac{p}{kT} = \rho + B_2\rho^2 + \cdots, \qquad \rho = \rho_1 + \rho_2,$$

show that

$$B_2(x_1, T) = B_{20}(T)x_1^2 + B_{11}(T)x_1(1 - x_1) + B_{02}(T)(1 - x_1)^2,$$

where x_1 is the mole fraction of component 1.

15–15. Consider the isomeric equilibrium ② ⇄ ① in a binary gas mixture. Find the analog of Eq. (15–56). If the equation of state is written as in Problem 15–14, show that

$$B_2(T) = \frac{B_{20}K^2 + B_{11}K + B_{02}}{(1 + K)^2}.$$

15–16. Define the surface concentration $\Gamma = (\overline{N} - \overline{N}^0)/\mathcal{A}$. Extend Eq. (15–74) to read

$$\frac{\varphi}{kT} = \Gamma + B_2(T)\Gamma^2 + \cdots,$$

and find an expression for B_2 in terms of u^* and u.

15–17. Relate the thermodynamic functions E_s, S_s, φ, and N_s in Eq. (15–75) to partition functions.

Supplementary Reading

Fowler and Guggenheim, Chapter 7.
Hirschfelder, Curtiss, and Bird, Chapter 3.
Mayer and Mayer, Chapter 13.
S. M., Chapter 5.

CHAPTER 16

APPROXIMATE CELL AND HOLE THEORIES OF THE LIQUID STATE

The general series expansion approach of the previous chapter is exact but in practice can be applied only to dilute gases. The treatment of dense gases requires higher virial coefficients, and it will be recalled that an investigation of the virial coefficient B_n involves an n-body problem. Although the series method runs into computational difficulties when applied to dense gases, it breaks down altogether when liquid is present in the system; that is, there is a singularity in the function $p(\rho)$ at the condensation point, as already explained in connection with Eq. (15–1).

Therefore we have to turn to new techniques in studying the liquid state. Any exact approach to this problem involves, in one form or another, the treatment of a system containing of the order of, say, 10^{20} molecules. For example, in the canonical ensemble, the integral (6–22) with $N = O(10^{20})$ must be evaluated. Because of this possibly insuperable difficulty,* *approximate* theories of the liquid state are of great interest. We discuss approximate cell and hole theories in this chapter, and the distribution function method in the next chapter. We restrict ourselves to a classical monatomic system with pairwise additive potential energy. Section 6–2 gives the criteria for the use of classical statistics. As usual, our object is to provide an introduction to the subject and not to review the latest refinements.

We begin by discussing the van der Waals equation of state. This theory is of interest because of its simplicity, its prediction of a phase transition, and its resemblance to the Bragg-Williams lattice gas approximation (Section 14–4).

16–1 The van der Waals equation of state. If no forces act between the molecules of the system,

$$Q = \frac{q^N}{N!}, \qquad q = \frac{V}{\Lambda^3}, \tag{16–1}$$

as we found in Eq. (4–12). Here the partition function q pertains to a

---

* High-speed computing machines seem to be the only hope in this connection. Model systems with of the order of several hundred particles have been examined so far. See, for example, B. J. Alder and T. E. Wainwright, *J. Chem. Phys.* **31,** 459 (1959), where other references are given.

single particle moving in a volume V which is potential-free. In order to derive the van der Waals equation, we assume that in a fluid of interacting molecules, each molecule moves, independently, in a uniform potential field provided by the other molecules, these being distributed in a random manner (hence the analogy with the Bragg-Williams theory). The potential energy between a pair of molecules is taken to be (see Problem 15–7)

$$\begin{aligned} u(r) &= +\infty \qquad r < r^* \\ &= -\epsilon\left(\frac{r^*}{r}\right)^6 \qquad r \geq r^*. \end{aligned} \tag{16–2}$$

This interaction potential has a "hard-sphere core" and the usual r^{-6} attractive term for large r. The minimum occurs at $u = -\epsilon$, $r = r^*$.

Because of the assumed independence of the molecules, we still have $Q = q^N/N!$ but we make two modifications in q, both arising from intermolecular forces. First, because of the hard-sphere core in each intermolecular interaction, not all of the volume V is available for the motion of a given molecule. We therefore replace V by a "free volume" V_f. Second, we insert a Boltzmann factor $e^{-\varphi/2kT}$ to take care of the intermolecular potential field in which the given molecule is moving. The energy φ, a function of N/V, is the potential energy of interaction between any one molecule and all others in the system. The factor of two is inserted in the exponent because each pair interaction has to be shared between two molecules in counting up the total potential energy. Thus we write

$$Q = \frac{q^N}{N!}, \qquad q = \frac{V_f e^{-\varphi/2kT}}{\Lambda^3}. \tag{16–3}$$

Now we turn to the explicit forms to be used for φ and V_f. In the neighborhood of a particular molecule, the density of other molecules will be zero between $r = 0$ (location of the specified molecule) and $r = r^*$, because of the hard core, and will be constant at the value N/V from $r = r^*$ to $r = \infty$, because of the assumed random distribution of molecules. (Actually, the distribution is not random and the density not constant—see Chapter 17.) Between r and $r + dr$, for $r \geq r^*$, there are $(N/V)\cdot 4\pi r^2\,dr$ other molecules. The potential energy of interaction between each of these and the central molecule at $r = 0$ is given by Eq. (16–2). Therefore

$$\begin{aligned} \varphi &= -\int_{r^*}^{\infty} \epsilon\left(\frac{r^*}{r}\right)^6 \frac{N}{V}\, 4\pi r^2\, dr \\ &= -\frac{2a_v N}{V}, \qquad a_v \equiv \frac{2\pi\epsilon r^{*3}}{3}. \end{aligned} \tag{16–4}$$

As the particular molecule of interest wanders through the volume V, a volume $(4\pi/3)r^{*3}$ is excluded to it by each other molecule in the system. However, we have to divide this quantity by two for the same reason as above: the excluded volume arises from an intermolecular pair interaction, (16–2), and only half of the effect can be assigned to a given molecule. This argument will be confirmed by a direct calculation of B_2 below. Therefore we write

$$V_f = V - Nb, \qquad b = \frac{2\pi r^{*3}}{3}. \tag{16–5}$$

Use of the volume Nb in Eq. (16–5) implies nonoverlapping, i.e., additivity, of excluded volumes. This can only be correct in the limit of low densities, but we use this expression for V_f at all densities, as an approximation.

Equations (16–3) through (16–5) now furnish us with a complete canonical ensemble partition function from which the various thermodynamic properties can be deduced. For example (Problem 16–1),

$$-\frac{A}{kT} = \ln Q = N \ln\left(\frac{V - Nb}{\Lambda^3}\right) + \frac{N^2 a_v}{VkT} - N \ln N + N, \tag{16–6}$$

and

$$\left(p + \frac{N^2 a_v}{V^2}\right)(V - Nb) = NkT. \tag{16–7}$$

Equation (16–7) is the van der Waals equation of state. Thus the constants a_v and b, defined in Eqs. (16–4) and (16–5), are the usual van der Waals constants, but here they are given expression in terms of the parameters ϵ and r^* of the intermolecular potential function. The subscript is included in a_v to avoid confusion with a in Eq. (16–24).

If we expand Eq. (16–7) in powers of $1/v = \rho = N/V$, we obtain the virial expansion

$$\frac{p}{kT} = \rho + \left(b - \frac{a_v}{kT}\right)\rho^2 + b^2\rho^3 + b^3\rho^4 + \cdots. \tag{16–8}$$

The second virial coefficient is $B_2 = b - (a_v/kT)$. If we multiply Eq. (16–8) by b and let $\theta = \rho b$ (the maximum density corresponds to $\theta = 1$),

$$\frac{pb}{kT} = \theta + \left(1 - \frac{\epsilon}{kT}\right)\theta^2 + \theta^3 + \theta^4 + \cdots. \tag{16–9}$$

This is very similar to Eq. (14–45) in the Bragg-Williams theory.

The second virial coefficient, calculated from Eq. (15–24) and the potential (16–2), agrees in the limit of high temperatures (see Problem 15–7) with B_2 in Eq. (16–8). This is to be expected, since the random molecular distribution assumed in calculating φ is approached at high temperatures, and the excluded volume correction in V_f is correct if the density is low enough so that only pair interactions need be considered.

The fact that the van der Waals equation exhibits critical and phase-transition behavior as in Figs. 14-4 and 14-7(b) is well known. This is, of course, the reason for our interest in the equation: it is mathematically very simple, but still predicts a first-order gas-liquid transition. The critical constants, from Eq. (16-7), are

$$v_c = 3b, \qquad T_c = \frac{8a_v}{27bk}, \qquad p_c = \frac{a_v}{27b^2}, \tag{16-10}$$

or, in terms of the parameters ϵ and r^*,

$$\frac{v_c}{r^{*3}} = 2\pi = 6.28, \qquad \frac{kT_c}{\epsilon} = \frac{8}{27} = 0.296,$$
$$\frac{p_c r^{*3}}{\epsilon} = \frac{1}{18\pi} = 0.0177, \qquad \frac{p_c v_c}{kT_c} = \frac{3}{8} = 0.375. \tag{16-11}$$

The individual values of the critical constants in (16-11) are in very poor agreement with experiment (Table 15-1). Hence we cannot take the van der Waals model seriously in a quantitative sense, though it is useful for qualitative purposes.

Because of the form of the potential (16-2) [see Eq. (15-33)], we expect the van der Waals equation to obey the law of corresponding states. This is easily verified if we rewrite Eq. (16-7) as

$$\frac{pv}{kT} = \frac{(v/b)}{(v/b) - 1} - \frac{a_v}{(v/b)bkT}. \tag{16-12}$$

Now $b = 2\pi r^{*3}/3$ and $a_v/bkT = \epsilon/kT$. Therefore, pv/kT is a universal function of v/r^{*3} and kT/ϵ, as in Eq. (15-32).

The two-dimensional van der Waals equation is of considerable interest as an approximate equation of state for an adsorbed monolayer. By the same reasoning as above (Problem 16-2), we find the equation of state

$$\left(\varphi + \frac{N^2 a'}{\mathcal{A}^2}\right)(\mathcal{A} - Nb') = NkT, \tag{16-13}$$

where

$$a' = \frac{\pi \epsilon r^{*2}}{4}, \qquad b' = \frac{\pi r^{*2}}{2}. \tag{16-14}$$

If the critical properties of the same gas [with $u(r)$ given by Eq. (16-2)] are compared in three dimensions and in two dimensions, we find $T_c(3 \text{ dim})/T_c(2 \text{ dim}) = 2$ (Problem 16-3). This result has been confirmed approximately in several adsorption systems.

16–2 Cell theories of liquids. In a cell theory of liquids, we imagine the volume V divided up into a lattice of N cells, with one molecule in each cell. The motion of each molecule, within its cell and in the potential field of its neighbors, is assumed independent of the motion of the other molecules. This resembles the Einstein model for a crystal (the reader should review the beginning of Section 5–1), except that we shall be using classical mechanics here and the potential is not necessarily parabolic. The conversion of a cell model for a crystal into a cell model for a liquid is accomplished rather artificially by the introduction of the so-called "communal entropy"—absent in a crystal, present in a gas, and assumed also present in a liquid. Let us digress at this point to explain what is meant by the term "communal entropy."

In a crystal each molecule is confined in a "cell" or "cage" formed by its nearest neighbors, and is only very rarely involved in excursions outside the cell. At the other extreme, in a dilute gas, each molecule is quite free to wander over the *entire* volume V of the container. Because of this additional freedom, gas molecules are said to have "communal entropy" not possessed by molecules in a crystal. The liquid state is intermediate in nature, and it is not at all obvious to what extent the liquid state possesses communal entropy. Originally, Hirschfelder, Stevenson, and Eyring* assumed that the liquid state had essentially the complete communal entropy and that the communal entropy therefore appeared on melting as a large part of the entropy of fusion. This view was later criticized by O. K. Rice.† It can safely be said that the situation with regard to the communal entropy in the liquid state remains obscure even at present, although Kirkwood has given the concept rigorous formal definition.

We now consider the simplest possible illustration of communal entropy. Suppose that we have a system of N monatomic molecules, without intermolecular forces, in a volume V. Then, as in Eq. (16–1),

$$A = -kT \ln Q = -kT \ln \frac{V^N}{N!\Lambda^{3N}} = -NkT \ln \frac{ve}{\Lambda^3}, \tag{16–15}$$

and

$$S = Nk \ln \frac{ve^{3/2}}{\Lambda^3} + Nk. \tag{16–16}$$

On the other hand, suppose, by the use of hypothetical partitions, that the volume V is divided up into N cells, each of volume $v = V/N$, and that each cell is occupied by a molecule which is restricted to move inside the

* J. Hirschfelder, D. Stevenson, and H. Eyring, *J. Chem. Phys.* **5**, 896 (1937).

† O. K. Rice, *J. Chem. Phys.* **6**, 476 (1938).

cell. Again we assume that there are no intermolecular forces. From the communal-entropy point of view this situation resembles that in a crystal. The configuration integral is now v^N instead of V^N; also, the factor $1/N!$ in Eq. (16–15) is omitted, since the molecules are now distinguishable (the cells can be labeled). Hence

$$A = -kT \ln Q = -kT \ln \frac{v^N}{\Lambda^{3N}} = -NkT \ln \frac{v}{\Lambda^3}, \qquad (16\text{–}17)$$

and

$$S = Nk \ln \frac{ve^{3/2}}{\Lambda^3}. \qquad (16\text{–}18)$$

Equations (16–15) and (16–16) may be compared with Eqs. (16–17) and (16–18). It is clear that if we start with the system divided up into cells and then remove the partitions, the system acquires in this process the additional "communal" entropy, $\Delta S = Nk$.

Now suppose that we take into account, very roughly, the intermolecular forces in the cell model described above (for a crystal) by assuming that each molecule moves in a field of constant potential φ within its cell, where φ arises from the interaction of a given molecule with all the other molecules of the system. The only effect this will have on thermodynamic properties is to raise the energy of the system by $N\varphi/2$. Equation (16–17) becomes

$$A = -NkT \ln \frac{v}{\Lambda^3} + \frac{N\varphi}{2}. \qquad (16\text{–}19)$$

As a next step, suppose the interaction potential between a given molecule and all other molecules is not assumed constant, but rather is a function of position in the cell, $\varphi(\mathbf{r})$, where the origin $\mathbf{r} = 0$ is located at the minimum in φ. For example, owing to intermolecular repulsions, we might expect φ to become very large in a condensed phase when the confined molecule approaches its neighbors at the edge of its cell. The probability of observing the central molecule in a given element of volume is no longer uniform throughout the cell but must be obtained from a Boltzmann factor. As a result of this, the "effective" or "free" volume through which the central molecule can move is reduced to a value less than v. In fact, the cell configuration integral leading to Eq. (16–17),

$$v = \int_\Delta d\mathbf{r},$$

where Δ represents the cell, must now be replaced by

$$v_f = \int_\Delta e^{-[\varphi(\mathbf{r})-\varphi(0)]/kT}\, d\mathbf{r}, \qquad (16\text{–}20)$$

where v_f is the "free" or "effective" volume. Equation (16–19) is then modified to read

$$A = -NkT \ln \frac{v_f}{\Lambda^3} + \frac{N\varphi(0)}{2}. \qquad (16\text{–}21)$$

In this picture we would expect $\varphi(0)$ to be a function of ρ or v, and v_f to be a function of v and T. This equation is very closely related to Eq. (5–3) for an Einstein crystal: v_f/Λ^3 corresponds to q^3 in Eq. (5–3). The difference is that (a) v_f/Λ^3 is a *classical* partition function, and (b) the potential well inside the cell is *not* assumed to be parabolic in Eq. (16–20) (i.e., we are not limited to small vibrations), as it is in the Einstein model.

Finally, to make Eq. (16–21) applicable to a liquid, we replace v_f by $v_f e$, just as v in Eq. (16–17) is replaced by ve in Eq. (16–15). That is, we assume the liquid has communal entropy since, in a liquid, a given molecule is not confined to a particular cell but can wander over the entire volume V. As already mentioned above, this is a strictly intuitive step which is found on more careful analysis to be inaccurate. Thus, for a liquid,

$$A = -NkT \ln \frac{v_f(v, T)e}{\Lambda^3} + \frac{N\varphi(0; v)}{2}. \qquad (16\text{–}22)$$

Before going on to a particular application of Eq. (16–22), we note that the van der Waals free energy, Eqs. (16–3) and (16–6), although not based on a cell model, can be put in the form of Eq. (16–22) with

$$v_f(v) = \frac{V_f}{N} = v - b, \qquad \varphi(0; v) = -\frac{2a_v}{v}. \qquad (16\text{–}23)$$

The van der Waals v_f is a function of v but not T.

We now turn to the cell theory of Lennard-Jones and Devonshire* (LJD), based on Eq. (16–22) as a starting point. Many refinements of this theory have been worked out, but they have not changed the position of the LJD theory as the basic prototype for cell theories of liquids.

In the LJD model, each molecule moves within its cell in the potential field of its nearest neighbors assumed fixed at the centers of their cells. To simplify the problem, the c nearest neighbors are treated as uniformly "smeared" over a spherical surface of radius a, where a is the distance between the centers of nearest-neighbor cells. Then

$$a^3 = \gamma v, \qquad (16\text{–}24)$$

where γ is a numerical constant (not an activity coefficient here) depend-

---

* J. E. LENNARD-JONES and A. F. DEVONSHIRE, *Proc. Roy. Soc.* (*London*), **A163**, 53 (1937); **A165**, 1 (1938).

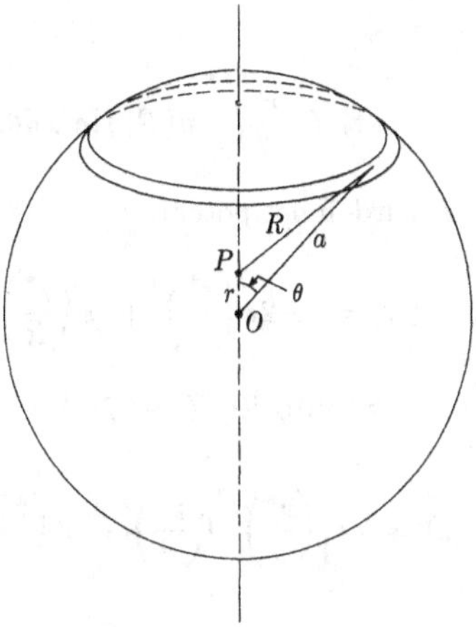

FIG. 16-1. Cell geometry for the Lennard-Jones and Devonshire theory.

ing on the geometry of the lattice. For a face-centered cubic lattice, $\gamma = \sqrt{2}$ and $c = 12$. Our task is to calculate (Eq. 16-20)

$$v_f = \int_\Delta e^{-\psi(\mathbf{r})/kT} 4\pi r^2 \, dr, \tag{16-25}$$

where

$$\psi(\mathbf{r}) = \varphi(\mathbf{r}) - \varphi(0). \tag{16-26}$$

In Fig. 16-1, the central molecule is at P, a distance r from the center of the cell. The area of the ring shown on the surface of the sphere is

$$2\pi a^2 \sin\theta \, d\theta.$$

The number of "smeared" nearest neighbors in this area is

$$c \cdot \frac{2\pi a^2 \sin\theta \, d\theta}{4\pi a^2} = \frac{c}{2} \sin\theta \, d\theta,$$

and the potential energy of interaction between the molecule at P and the neighbors in the ring is

$$u(R) \cdot \frac{c}{2} \sin\theta \, d\theta,$$

where

$$R^2 = r^2 + a^2 - 2ar\cos\theta.$$

Hence the total energy of interaction between the molecule at P and all of its c neighbors is

$$\varphi(r) = \frac{c}{2}\int_0^\pi u(R)\sin\theta\,d\theta, \tag{16–27}$$

where we use the Lennard-Jones potential

$$u(R) = -2\epsilon\left(\frac{r^*}{R}\right)^6 + \epsilon\left(\frac{r^*}{R}\right)^{12}. \tag{16–28}$$

We substitute Eq. (16–28) into Eq. (16–27), and obtain, on carrying out the integration,

$$\psi(r) = \varphi(r) - \varphi(0) = c\epsilon\left[\left(\frac{r^*}{a}\right)^{12} l\left(\frac{r^2}{a^2}\right) - 2\left(\frac{r^*}{a}\right)^6 m\left(\frac{r^2}{a^2}\right)\right], \tag{16–29}$$

where

$$\varphi(0) = c\epsilon\left[-2\left(\frac{r^*}{a}\right)^6 + \left(\frac{r^*}{a}\right)^{12}\right], \tag{16–30}$$

$$l(x) = (1 + 12x + 25.2x^2 + 12x^3 + x^4)(1 - x)^{-10} - 1, \tag{16–31}$$

$$m(x) = (1 + x)(1 - x)^{-4} - 1. \tag{16–32}$$

If we define

$$v^* = \frac{v}{a^3}\,r^{*3} = \frac{r^{*3}}{\gamma}, \tag{16–33}$$

Eqs. (16–29) and (16–30) become

$$\psi(r) = \varphi(r) - \varphi(0) = c\epsilon\left[\left(\frac{v^*}{v}\right)^4 l\left(\frac{r^2}{a^2}\right) - 2\left(\frac{v^*}{v}\right)^2 m\left(\frac{r^2}{a^2}\right)\right], \tag{16–34}$$

$$\varphi(0) = c\epsilon\left[-2\left(\frac{v^*}{v}\right)^2 + \left(\frac{v^*}{v}\right)^4\right]. \tag{16–35}$$

When v/v^* is small ($< \sim 1.6$), $\psi(r)$ has a minimum at $r = 0$ and rises rapidly as r increases. For large v/v^*, as should be expected from the physical model, $\psi(r)$ has a low maximum at $r = 0$, a minimum near $r = a - r^*$, and rises rapidly when $r > a - r^*$.

The potential $\psi(r)$ in Eq. (16–34) is substituted in Eq. (16–25), and we find (putting $y = r^2/a^2$)

$$v_f = 2\pi a^3 g, \tag{16–36}$$

$$g = \int_0^s \exp\left\{-\frac{c\epsilon}{kT}\left[\left(\frac{v^*}{v}\right)^4 l(y) - 2\left(\frac{v^*}{v}\right)^2 m(y)\right]\right\} y^{1/2}\,dy.$$

The upper limit s in the integral is determined by the "boundary condition" that, when $c\epsilon/kT = 0$ (ideal gas),

$$v_f = v = 2\pi a^3 \int_0^s y^{1/2}\,dy,$$

or

$$s = \left(\frac{3}{4\pi\gamma}\right)^{2/3}.$$

Actually, the results for the critical and liquid regions are quite insensitive to the choice of s.

Equations (16–22), (16–35), and (16–36) now determine all the thermodynamic properties of the system. For example, for the equation of state, we find (Problem 16–4)

$$\frac{pv}{kT} = 1 - \frac{2c\epsilon}{kT}\left[\left(\frac{v^*}{v}\right)^2 - \left(\frac{v^*}{v}\right)^4\right] + \frac{4c\epsilon}{kT}\left[\left(\frac{v^*}{v}\right)^4 \frac{g_l}{g} - \left(\frac{v^*}{v}\right)^2 \frac{g_m}{g}\right], \tag{16–37}$$

where

$$g_l = \int_0^s \exp\left\{-\frac{c\epsilon}{kT}\left[\left(\frac{v^*}{v}\right)^4 l(y) - 2\left(\frac{v^*}{v}\right)^2 m(y)\right]\right\} y^{1/2} l(y)\,dy \tag{16–38}$$

and

$$g_m = \int_0^s \exp\left\{-\frac{c\epsilon}{kT}\left[\left(\frac{v^*}{v}\right)^4 l(y) - 2\left(\frac{v^*}{v}\right)^2 m(y)\right]\right\} y^{1/2} m(y)\,dy. \tag{16–39}$$

The quantities g, g_l, and g_m have to be calculated by numerical integration for each pair of values of v^*/v and $c\epsilon/kT$. Extensive tables of g, g_l, g_m, and several thermodynamic properties have been published by Hirschfelder, Curtiss, and Bird (Supplementary Reading list). The tables given actually include the contribution of second- and third-neighbor shells, but this refinement turns out to have little effect on thermodynamic properties. We can summarize the results as follows:

(1) Since pv/kT above is a universal function of v/v^* and kT/ϵ, the law of corresponding states is obeyed.

(2) A plot of pv^*/kT against v/v^*, for different fixed values of kT/ϵ, shows typical critical behavior and loops of the van der Waals type. The critical constants (for $\gamma = \sqrt{2}$ and $c = 12$) are

$$\frac{v_c}{r^{*3}} = 1.25, \qquad \frac{kT_c}{\epsilon} = 1.30,$$
$$\frac{p_c r^{*3}}{\epsilon} = 0.614, \qquad \frac{p_c v_c}{kT_c} = 0.590. \tag{16–40}$$

Comparison with the experimental values in Table 15–1 shows that the LJD theory predicts the critical temperature very well, but is not so successful on the other critical constants [though it is certainly an improvement over the van der Waals results, (16–11)].

(3) The expansion of pv/kT in powers of v^*/v has no linear term, so that the second virial coefficient is zero. This is a consequence of the fact that, in this model, each cell has exactly one molecule in it at all densities; at low densities, with this restriction, binary molecular interactions (responsible for B_2) become of negligible importance.

We can conclude that the LJD cell theory is completely unsatisfactory for a dilute gas and only fairly good in the critical region. But from the nature of the model, we would expect it to be quite successful, and it is, at very high pressures where the assumption of exactly one molecule per cell is reasonable.

Incidentally, the two-dimensional theory is included in the LJD references on p. 292. We shall quote just the one result that T_c (3 dim)/ T_c(2 dim) $= 1.89$, compared with the van der Waals value 2.00.

16–3 Hole theories of liquids. A rather obvious refinement of cell theories of the LJD type is to relax the restriction of exactly one molecule per cell. The most general approach would be to allow 0, 1, 2, . . . molecules in a cell. (If the intermolecular potential $u(r)$ is assumed to have a hard-sphere core for sufficiently small r, there would be a *maximum* possible number of molecules in a cell of fixed volume.) Much work along these lines has been done, but we shall not review this area here. Unfortunately, these refinements do not improve the LJD theory a great deal.

A special case of generalized cell theories arises when the cells are chosen small enough so that the number of molecules per cell can be only zero or one. In this case, it is customary to speak of a "hole theory" of liquids—an empty cell being a "hole." Strictly speaking, this situation is possible only with a hard-sphere core in $u(r)$; but in practice, since the Lennard-Jones potential rises so fast as $r \to 0$, this point can be ignored (except at very high temperatures). We consider here an approximate hole theory of liquids, due to Cernuschi and Eyring, which is in fact identical with the quasi-chemical approximation for a three-dimensional lattice gas (Section 14–5). In the lattice gas, a "site" is a cell and an "empty site" is a hole.

We make the following specific assignments of parameters in Section 14–5. The lattice is assumed close-packed, so $c = 12$. The Lennard-Jones potential $u(r)$ (with parameters ϵ and r^*) is used, and we choose the lattice spacing (nearest-neighbor distance between sites) as r^*. This is perhaps the most natural choice, but is somewhat arbitrary. (A refinement would be to select the lattice spacing so as to minimize A for given

N, V, T.) Then, in Section 14–5, $w = -\epsilon$. The partition function q refers to the motion of a single molecule in its cell, and $q = v_f/\Lambda^3$ in the notation of Section 16–2. Clearly v_f should be different for each different number and arrangement of nearest-neighbor molecules, but as an approximation we assume v_f is evaluated when all ($c = 12$) neighbors are present. Since the nearest-neighbor distance r^* is fixed in this model, q is a function of T only. Because we shall consider just the equation of state, which does not depend on $q(T)$, we need not pursue this question further [but an obvious evaluation of $q(T)$, if needed, would be that offered by Eq. (16–36) with $a = r^*$].

The equation of state is (14–62), if we put $c = 12$ and

$$\Phi = \frac{pV}{M} = \frac{pr^{*3}}{\sqrt{2}}.$$

Then, from Eq. (14–63), the critical constants are (Problem 16–5)

$$\frac{v_c}{r^{*3}} = \sqrt{2} = 1.414, \qquad \frac{kT_c}{\epsilon} = 2.74, \qquad \frac{p_c r^{*3}}{\epsilon} = 0.663,$$
$$\frac{p_c v_c}{kT_c} = 0.342. \tag{16–41}$$

Table 16–1 brings together the critical constants of Table 15–1 and Eqs. (16–11), (16–40), and (16–41). The hole theory of Cernuschi and Eyring is, on the whole, not as good as the LJD theory in predicting critical constants.

Equation (14–62) can be rewritten as $pv/kT = (1/\theta) \ln \{\ \}$. Since $\theta = r^{*3}/\sqrt{2}\,v$ and $w = -\epsilon$, pv/kT is a universal function of v/r^{*3} and kT/ϵ. Hence this model also obeys the law of corresponding states.

TABLE 16–1

REDUCED CRITICAL CONSTANTS

	$\frac{v_c}{r^{*3}}$	$\frac{kT_c}{\epsilon}$	$\frac{p_c r^{*3}}{\epsilon}$	$\frac{p_c v_c}{kT_c}$
Experimental	2.12	1.29	0.179	0.292
van der Waals	6.28	0.296	0.0177	0.375
LJD	1.25	1.30	0.614	0.590
Cernuschi-Eyring	1.41	2.74	0.663	0.342

16–4 Law of corresponding states. We have by this time encountered several special cases of the law of corresponding states in gases and liquids. We give here a brief argument showing how this law can be deduced in a general way.*

Consider the canonical ensemble partition function for a classical monatomic gas, Eq. (6–21). Actually, the argument below applies also, approximately, to diatomic and polyatomic gases if the angular dependence of the intermolecular force is relatively unimportant so that the translational partition function is separable from rotation, etc., or approximately so (in other words, if these molecules are effectively spherically symmetrical). This rules out polar molecules, hydrogen bonding molecules, etc. From thermodynamics, we know that $-A/NkT$, being an intensive quantity, must be a function of v and T only. Then $N^{-1}\ln Q_N$ is a function of v and T only. Therefore, from Eq. (6–21), $N^{-1}\ln(Z_N/N!)$ is a function of v and T only, call it $\psi_1(v, T)$. That is,

$$\ln\left(\frac{Z_N}{N!}\right) = N\psi_1(v, T),$$

or

$$\frac{Z_N}{N!} = e^{\psi_1 N} = \psi_2(v, T)^N, \tag{16–42}$$

where $\psi_2 = e^{\psi_1}$. Now we can rewrite Eq. (6–22) as

$$Z_N = r^{*3N}\int_V e^{-U/kT}\, d\left(\frac{\mathbf{r}_1}{r^{*3}}\right)\cdots d\left(\frac{\mathbf{r}_N}{r^{*3}}\right), \tag{16–43}$$

where

$$\frac{U}{kT} = \sum_{1\leq i<j\leq N}\frac{u(r_{ij})}{kT} = \sum_{ij}\frac{\epsilon}{kT}h\left(\frac{r_{ij}}{r^*}\right),$$

if we assume a pairwise additive, two-parameter, intermolecular potential of the form (15–33), in which ϵ and r^* are energy and distance parameters (not necessarily identical with those in the Lennard-Jones potential) characteristic of each species of molecule. The function h is assumed the same for all gases under consideration. Then Z_N, from Eq. (16–43), can be written

$$Z_N = r^{*3N}\psi_3\left(\frac{\epsilon}{kT}, \frac{V}{r^{*3}}, N\right), \tag{16–44}$$

* The original argument of this type is due to K. S. Pitzer, *J. Chem. Phys.* **7**, 583 (1939).

where ψ_3 is the same function for all molecules. Next, on comparing Eqs. (16–42) and (16–44), we see that ψ_3 must in fact have the form

$$\psi_3 = N!\Psi\left(\frac{\epsilon}{kT}, \frac{v}{r^{*3}}\right)^N,$$

where Ψ is also the same function for all molecules. This result gives us

$$Z_N = r^{*3N}N!\Psi\left(\frac{\epsilon}{kT}, \frac{v}{r^{*3}}\right)^N = Q_N N!\Lambda^{3N},$$

or

$$-\frac{A}{NkT} = \frac{1}{N}\ln Q_N = \ln\left[\frac{r^{*3}\Psi(\epsilon/kT, v/r^{*3})}{\Lambda^3}\right]. \qquad (16\text{–}45)$$

For the equation of state, we have then

$$\frac{pv}{kT} = -\frac{v}{r^{*3}}\left(\frac{\partial A/NkT}{\partial v/r^{*3}}\right)_T = \frac{v}{r^{*3}}\left(\frac{\partial \ln \Psi}{\partial v/r^{*3}}\right)_{\epsilon/kT}. \qquad (16\text{–}46)$$

Therefore pv/kT is a universal function of v/r^{*3} and kT/ϵ, as was to be proved.

In summary, then, we expect the law of corresponding states to be obeyed by those monatomic (or effectively monatomic, as far as translational motion goes) classical gases whose intermolecular potentials are pairwise additive and of the form (15–33) (all with the same function h).

In Figs. 15–1 and 15–2 and in Table 15–1 we have seen experimental evidence for the existence of this law of corresponding states. A much more extensive summary and analysis of experimental data is given by Guggenheim.*

Quantum fluids (e.g., H_2 and He) require an extension of the above law of corresponding states. This will be considered in Chapter 22.

---

* E. A. Guggenheim, *Thermodynamics*. 3rd ed. Amsterdam: North-Holland, 1957. See pp. 165–172.

PROBLEMS

16–1. Derive equations for μ, E/NkT, and S/Nk as functions of v and T for a van der Waals fluid. Compare these with ideal gas equations. (Page 288.)

16–2. Derive Eqs. (16–13) and (16–14) for a two-dimensional van der Waals fluid. (Page 289.)

16–3. Show that $T_c\,(3\text{ dim})/T_c\,(2\text{ dim}) = 2$ for a van der Waals fluid. (Page 289.)

16–4. Deduce the LJD equation of state from Eqs. (16–22), (16–35), and (16–36). (Page 295.)

16–5. Verify the numerical values in (16–41) for the reduced critical constants of the approximate hole theory of liquids considered in Section 16–3. (Page 297.)

16–6. Find T_B and T_B/T_c (T_B = Boyle temperature) for a van der Waals fluid.

16–7. Find the second virial coefficient for the hole theory of Section 16–3.

16–8. Show that if pv/kT is a universal function of v/r^{*3} and kT/ϵ, then pv/kT is also a universal function of v/v_c and T/T_c. The law of corresponding states is often expressed in the latter way.

16–9. An ideal gas is adsorbed on a surface as a two-dimensional van der Waals fluid. Derive the adsorption isotherm.

16–10. Two simple equations of state often used are:

$$\text{Berthelot:}\quad \left(p + \frac{a_v}{Tv^2}\right)(v - b) = kT$$

and

$$\text{Dieterici:}\quad p(v - b) = kTe^{-a_v/kTv}.$$

Find the second and third virial coefficients in each case. If each of these equations is considered to have been derived from Eqs. (16–3), find φ.

16–11. If the curvature at the bottom of the LJD potential well is used to determine the frequency ν of an Einstein model, find ν as a function of c, ϵ, a, r^*, and m (mass). Also, find the condition under which the LJD potential well has a maximum at $r = 0$ instead of a minimum.

SUPPLEMENTARY READING

FOWLER and GUGGENHEIM, Chapter 8.
HIRSCHFELDER, CURTISS, and BIRD, Chapter 4.
PRIGOGINE, Chapter 7.
SLATER, Chapter 12.
S. M., Chapter 8.

CHAPTER 17

DISTRIBUTION FUNCTIONS IN CLASSICAL MONATOMIC FLUIDS

An alternative approach to the theory of liquids that has received much attention is the distribution-function method. In this chapter we give an introduction to the subject, using arguments that are more intuitive and transparent than would be the case in a formal and completely rigorous treatment. This type of discussion seems particularly appropriate here in view of the fact that a detailed survey of the formal type is already available in S. M., Chapter 6.

For simplicity we restrict the discussion to a classical monatomic fluid with a pairwise additive intermolecular potential energy, though the applicability of the distribution-function method is by no means limited to this special case. In fact, as will be intuitively obvious, in systems of effectively spherically symmetrical polyatomic molecules, the radial distribution function plays the same role as in monatomic systems. The corresponding function in systems of polyatomic molecules that are not spherically symmetrical is obtained by integrating a more general, angular dependent, distribution function over all rotational orientations.

17–1 Radial distribution function. Consider a one-component fluid with number density ρ and temperature T. Imagine that an observer is stationed on one particular molecule as it moves through the fluid, and that he makes observations on the number of molecules found at different distances from the ("central") molecule on which he is sitting. Since the system is a fluid, the distribution of other molecules around the central molecule will be found, on the average, to be spherically symmetrical. The mean number of molecules observed in an infinitesimal element of volume $d\tau$ at a distance r from the central molecule will not be, in general, just $\rho\, d\tau$, because the presence of the central molecule itself perturbs its immediate environment. For example, if the molecules of the fluid are hard spheres of diameter a, then *no* other molecule would be observed at distances $r < a$. Or, if the intermolecular potential is of the Lennard-Jones type, there would be a tendency for other molecules to accumulate at a distance of about $r = r^*$. In any case, at large distances, where the influence of the central molecule has died out, the mean number of molecules $\rho\, d\tau$ in $d\tau$ would be approached.

In general, we specify the mean number of molecules observed in $d\tau$ by $\rho g(r)\, d\tau$, where $g(r)$, called the radial distribution function, is the factor by which the mean "local density" $\rho g(r)$ at r deviates from the bulk density ρ. Because all real molecules become effectively "hard" for r sufficiently small, $g \to 0$ as $r \to 0$. Also, as indicated above, $g \to 1$ as $r \to \infty$.

The radial distribution function $g(r)$ depends on density and temperature. Therefore we shall sometimes write it as $g(r, \rho, T)$.

The function $g(r)$ cannot be obtained by direct thermodynamic measurement, but fortunately it can be deduced by a nonthermodynamic experimental method: x-ray diffraction. The observed diffraction pattern is a sum of interference effects from pairs of molecules. The radial distribution function determines the relative weight [proportional to $4\pi r^2 g(r)$] in the sum given to different intermolecular distances. From the observed pattern, one can calculate backward to deduce $g(r)$. Figure 17–1 shows typical experimental curves in gas and liquid states. The peaks in the liquid

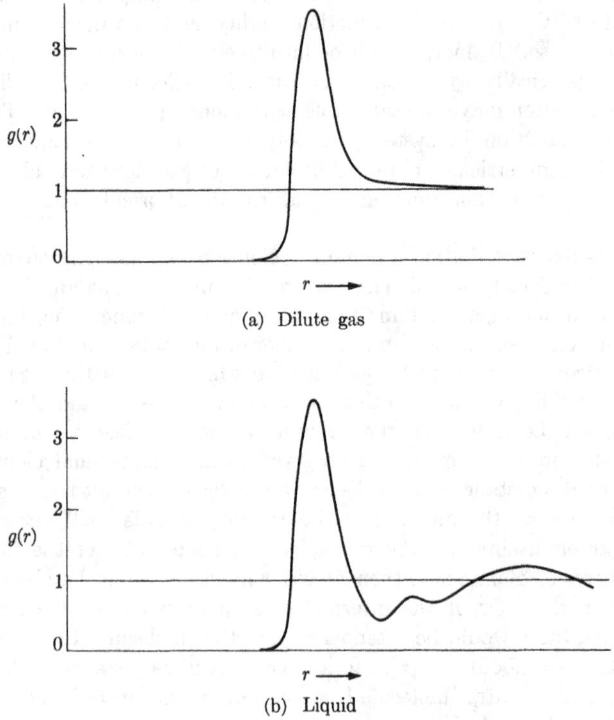

FIG. 17–1. Typical experimental radial distribution functions for (a) dilute gas and (b) liquid. In both cases the high peak occurs near $r = r^*$ (Lennard-Jones 6–12 potential).

curve are smeared-out remnants of relatively sharp peaks in $g(r)$ in the solid state (obtained by averaging over different directions outward from the central molecule, since spherical symmetry is lacking). These peaks in $g(r)$ for the solid correspond, as r increases, to first neighbors, second neighbors, etc. Thus the liquid has a certain amount of short-range order, about each molecule, which is a residue of the long-range order in the corresponding crystal.

The best single experimental paper available on this subject is an extensive one on argon.*

The radial distribution function is of interest in statistical mechanics primarily because the thermodynamic functions of the fluid can be expressed in terms of it, as we shall see in Section 17–2. Because of this, it becomes part of the task of statistical mechanics to provide the necessary framework from which $g(r)$ can be calculated theoretically, at least in principle. This topic is the subject of Section 17–3. Comparison of theoretical $g(r)$'s with experiment can be made either directly with experimental $g(r)$'s (from x-ray work) or indirectly with experimental thermodynamic properties of the fluid.

17–2 Relation of thermodynamic functions to $g(r)$. The simplest connection to establish concerns the internal energy E. It follows immediately from Eqs. (1–35) and (6–21) that

$$E = \tfrac{3}{2}NkT + \overline{U}. \tag{17–1}$$

The first term is the mean kinetic energy, and the second term is the mean potential energy. It is very easy to express $\overline{U}$ in terms of $g(r)$. Consider any molecule as the "central" molecule. The total intermolecular potential energy between the central molecule and other molecules in the fluid at distances between r and $r + dr$ is

$$u(r) \cdot \rho g(r) \cdot 4\pi r^2\, dr.$$

Then $\overline{U}$ is obtained by integrating over all values of r and multiplying by $N/2$, since any of the N molecules might be "central." The factor of two is inserted so that each pair interaction is counted only once. Thus

$$\frac{E}{NkT} = \frac{3}{2} + \frac{\rho}{2kT}\int_0^\infty u(r) g(r, \rho, T) 4\pi r^2\, dr. \tag{17–2}$$

If, for a gas, we suppose $g(r, \rho, T)$ expanded in powers of ρ as

$$g(r, \rho, T) = g_0(r, T) + \rho g_1(r, T) + \rho^2 g_2(r, T) + \cdots, \tag{17–3}$$

* A. Eisenstein and N. S. Gingrich, *Phys. Rev.* **62**, 261 (1942).

we see immediately from Eq. (15–83) that

$$g_0(r, T) = \lim_{\rho \to 0} g(r, \rho, T) = e^{-u(r)/kT}. \tag{17–4}$$

This is just what we should expect, for in the limit as $\rho \to 0$, only one molecule at a time need be considered in the neighborhood of the central molecule, and its spatial probability distribution will be determined by the Boltzmann factor (17–4) (see Fig. 17–1a). This, incidentally, is equivalent to saying that the second virial coefficient involves pair interactions only. At higher densities, many molecules are in the vicinity of the central molecule, and the interactions between them influence the distribution about the central molecule.

The expansion (17–3) is, of course, not valid for a liquid. Equation (17–2) must be used as it stands.

Next, we consider the pressure in a fluid. Although there are several ways to do this, we calculate the pressure as the force per unit area which the molecules on one side of a mathematical surface S in the fluid (Fig. 17–2) exert on the molecules on the other side. There are two contributions to this force: the first, p_K, is associated with momentum transport and the second, p_U, with intermolecular forces. Since the momentum distribution is independent of the existence of intermolecular forces (see Section 6–4), the contribution of momentum transport to the pressure is just the same as in an ideal gas at the same density and temperature, namely, $p_K = \rho kT$. So we have to consider only the second contribution, p_U.

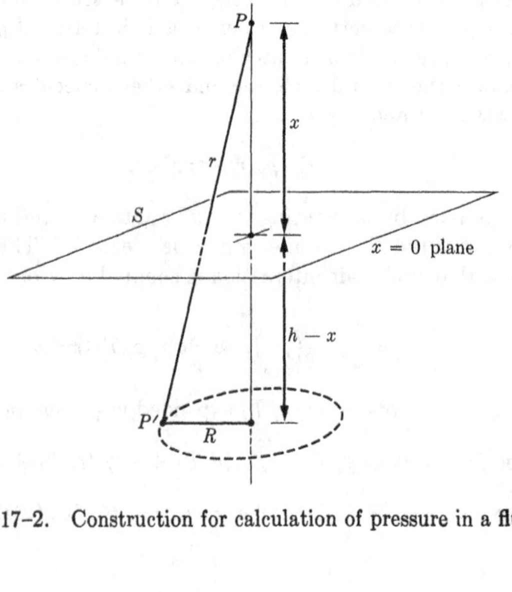

FIG. 17–2. Construction for calculation of pressure in a fluid.

In Fig. 17-2, the surface S is the plane $x = 0$. The force which a molecule at P a distance x from S exerts on a molecule at P' on the other side of S is $-u'(r) = -du(r)/dr$. The force normal to S is $-u'(r)h/r$. The mean number of molecules in the ring (radius R) with values of R between R and $R + dR$ and values of h between h and $h + dh$ is $\rho g(r) 2\pi R\, dR\, dh$. The normal force exerted by the one molecule at P on the molecules in the ring is then

$$-u'(r)\,\frac{h}{r}\cdot \rho g(r) 2\pi R\, dR\, dh. \tag{17-5}$$

Now we change independent variable from R to r. Since $r^2 = h^2 + R^2$, $r\,dr = R\,dR$. Hence (17-5) becomes

$$-2\pi\rho u'(r) g(r)\, dr\, h\, dh.$$

Then the normal force exerted by the molecule at P on *all* molecules on the opposite side of S ($x \leq 0$) is $-2\pi\rho I_1(x)$, where

$$I_1(x) = \int_x^\infty h\, dh \int_h^\infty u'(r) g(r)\, dr. \tag{17-6}$$

Finally, we will get p_U if we add up the contributions $-2\pi\rho I_1(x)$ of all molecules in a cylinder of unit cross-sectional area and with axis perpendicular to S, extending from $x = 0$ to $x = \infty$. The number of molecules in this cylinder between x and $x + dx$ is $\rho\, dx$. Therefore

$$p_U = -2\pi\rho^2 \int_0^\infty I_1(x)\, dx. \tag{17-7}$$

Equation (17-7) can be simplified. An integration by parts (Problem 17-1) reduces $I_1(x)$ in (17-6) to

$$I_1(x) = -\frac{x^2}{2}\int_x^\infty u'(r) g(r)\, dr + \frac{1}{2}\int_x^\infty r^2 u'(r) g(r)\, dr. \tag{17-8}$$

This result is substituted in Eq. (17-7), and two further integrations by parts are carried out. We find (Problem 17-2)

$$p_U = -2\pi\rho^2(-\tfrac{1}{6} + \tfrac{1}{2}) \int_0^\infty r^3 u'(r) g(r)\, dr, \tag{17-9}$$

where the two fractions result from the respective terms in (17-8). Finally, $p = p_K + p_U$, and hence

$$\frac{p}{kT} = \rho - \frac{\rho^2}{6kT}\int_0^\infty r\,\frac{du(r)}{dr}\, g(r, \rho, T) 4\pi r^2\, dr. \tag{17-10}$$

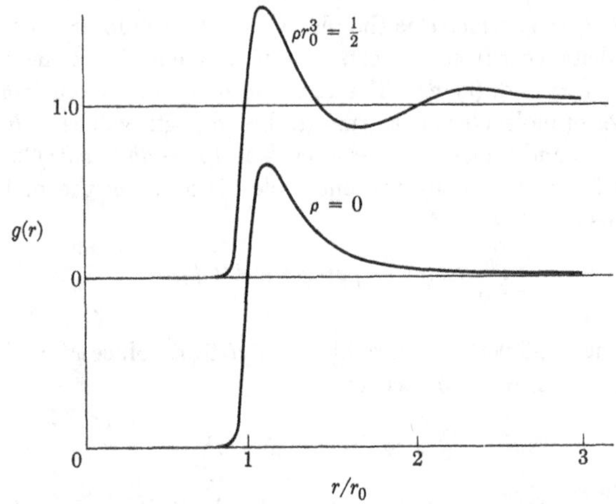

FIG. 17–3. Radial distribution functions for a gas calculated from the Lennard-Jones 6–12 potential (r_0 is defined in Appendix IV).

Equation (17–10) is valid for a liquid or gas. If, for a gas, we introduce the expansion (17–3) and then compare with Eq. (15–82), we confirm that $g_0 = e^{-u(r)/kT}$. Comparison of Eq. (17–10) with Eq. (15–31) for B_3 gives an expression for g_1, but a simpler procedure is to use Eq. (17–20) (Problem 17–3). Although g_0 has only one peak (at $r = r^*$), as seen in Fig. 17–1(a), with the inclusion of just one more term (that is, using $g = g_0 + \rho g_1$) a second peak appears in $g(r)$ (Fig. 17–3).

The last thermodynamic function we consider is the chemical potential. This, together with E and p for given N, V and T, determines completely all the thermodynamic properties of the system (for example, A and S). Of course, if $g(r, \rho, T)$ is known over the entire temperature range from T to $T = \infty$, then A (and hence S and μ) can be found by integrating

$$\left(\frac{\partial A/T}{\partial 1/T}\right)_{N,V} = E$$

from T to $T = \infty$, using Eq. (17–2) for E and the limit $g \to 1$ when $T \to \infty$ (random distribution). But this is not a very practical suggestion since ordinarily $g(r)$ is not available as a function of T.

Instead, we introduce the idea, due to Onsager and Kirkwood, of a coupling parameter ξ which can vary from $\xi = 0$ to $\xi = 1$. This is called a "charging parameter" in electrolyte theory (Chapter 18). We imagine that the strength of the interaction between a particular central molecule and all other molecules in the fluid is reduced from its normal value by a factor ξ. [That is, $u(r)$ for these interactions is replaced by

$\xi u(r)$ or, with the Lennard-Jones potential, ϵ is replaced by $\xi\epsilon$.] This is often expressed by the statement that the central molecule is "coupled" to the remaining molecules to the degree ξ. All other molecules interact normally with each other. Then the radial distribution function about the central molecule will be different than usual, and is denoted by $g(r, \rho, T; \xi)$. The ordinary $g(r)$ is $g(r; 1)$. Of course, there is no way to measure $g(r; \xi)$ directly, but theoretical equations determining $g(r; \xi)$ can be derived (Section 17–3).

We write the Helmholtz free energy of the fluid as $A = A' + A''$, where A' is the hypothetical Helmholtz free energy the system would have (same N, V, T) if there were no intermolecular forces, and A'' is the contribution of the intermolecular forces to A. Then

$$\mu = \left(\frac{\partial A}{\partial N}\right)_{V,T} = \mu' + \mu'', \tag{17–11}$$

where

$$\mu' = \left(\frac{\partial A'}{\partial N}\right)_{V,T} = kT \ln \Lambda^3 + kT \ln \rho, \tag{17–12}$$

$$\mu'' = \left(\frac{\partial A''}{\partial N}\right)_{V,T} = kT \ln \gamma. \tag{17–13}$$

Equation (17–12) follows from Eq. (4–23) or (15–18). The activity coefficient γ is defined, as in Chapter 15, by $z = \lambda/\Lambda^3 = \gamma\rho$. Since N is a very large number, we can write

$$\mu'' = \left(\frac{\partial A''}{\partial N}\right)_{V,T} = A''(N, V, T) - A''(N - 1, V, T). \tag{17–14}$$

Because of the relation between the Helmholtz free energy and work in thermodynamics, we may conclude from Eq. (17–14) that μ'' is the isothermal, reversible work that has to be done on the system *against intermolecular forces* in order to add one more molecule to the system. More exactly, we mean by the above language that μ'' is the work done on the system (V and T constant) in passing from an initial state containing $N - 1$ molecules fully "coupled" with each other and 1 molecule not coupled with any of the others (that is, $\xi = 0$) to a final state with all molecules fully coupled ($\xi = 1$). The complete chemical potential (μ) is then this work (μ'') plus a contribution (μ') which depends only on number density and temperature and not on the presence or absence of intermolecular forces.

We choose the one molecule that is orginally uncoupled ($\xi = 0$) as the central molecule. For an arbitrary intermediate value of ξ, the radial distribution function about the central molecule is $g(r, \rho, T; \xi)$. The potential energy of interaction of the central molecule with another molecule at r is $\xi u(r)$. Then $u(r)\, d\xi$ is the work that must be done on the system, because of this one interaction, if ξ is increased by $d\xi$. Thus the work done on the system when ξ is increased by $d\xi$, because of interactions between the central molecule and all molecules between r and $r + dr$, is

$$u(r)\, d\xi \cdot \rho g(r;\, \xi) \cdot 4\pi r^2\, dr.$$

The total work μ'' is just the integral of this expression over r and over ξ from $\xi = 0$ to $\xi = 1$. Therefore

$$\mu = kT \ln \Lambda^3 + kT \ln \rho + kT \ln \gamma(\rho, T), \qquad (17\text{–}15)$$

$$kT \ln \gamma = \rho \int_0^1 \int_0^\infty u(r) g(r, \rho, T;\, \xi) 4\pi r^2\, dr\, d\xi. \qquad (17\text{–}16)$$

This is our final expression for the chemical potential.

From the discussion following Eq. (17–4), we expect that

$$\lim_{\rho \to 0} g(r, \rho, T;\, \xi) = e^{-\xi u(r)/kT}. \qquad (17\text{–}17)$$

If we substitute (17–17) for g in Eq. (17–16), and integrate over ξ, we obtain a result that agrees with Eqs. (15–20) and (15–24).

17–3 Integral equation for $g(r;\, \xi)$. In the preceding section we have derived equations relating thermodynamic functions of a fluid to the radial distribution function. In this section we obtain an integral equation that determines $g(r;\, \xi)$ theoretically. The argument we use is intuitive in nature, but correct.* A rigorous argument requires a detailed discussion beyond the scope of this book (see S. M., Chapter 6).

Consider a central molecule whose position, without loss of generality, can be regarded as fixed at $\mathbf{r}_2$ (Fig. 17–4). We define the local chemical

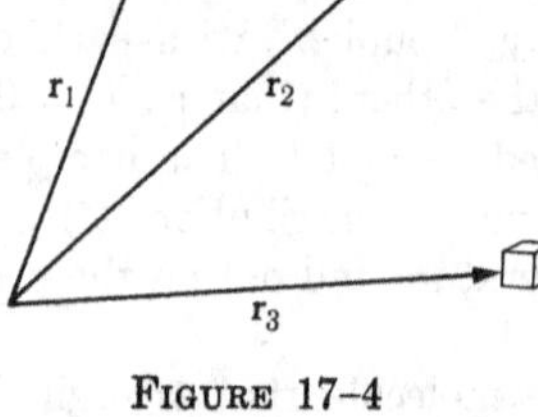

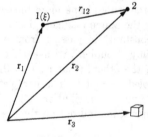

FIGURE 17–4

* T. L. HILL, *J. Chem. Phys.* **30**, 1521 (1959); *J. Phys. Chem.* **61**, 548 (1957).

potential $\mu(r_{12})$ at $\mathbf{r}_1$, a distance r_{12} from $\mathbf{r}_2$, as the work against intermolecular forces necessary to "charge up" or "couple" a new molecule at $\mathbf{r}_1$ plus $kT \ln \Lambda^3$ plus kT times the logarithm of the local number density at $\mathbf{r}_1$. This is by analogy with Eqs. (17–12) and (17–13). Clearly, when $r_{12} \to \infty$, the molecule at $\mathbf{r}_2$ has no influence on the coupling process at $\mathbf{r}_1$, and $\mu(\infty) = \mu$. For convenience we rewrite Eq. (17–15) as

$$\mu = \mu(\infty) = kT \ln \Lambda^3 + kT \ln \rho + \rho \int_0^1 \int_V u(r_{13}) g(r_{13};\,\xi)\, d\mathbf{r}_3\, d\xi. \tag{17–18}$$

That is, we use $d\mathbf{r}_3$ to represent an element of volume in the neighborhood of $\mathbf{r}_1$.

Because the system is at equilibrium, we assume that the local chemical potential $\mu(r_{12})$, defined above, has a constant value everywhere. This is an extension to the molecular level of the thermodynamic principle of constancy of the chemical potential in phase equilibria. That is, $\mu(r_{12}) = \mu(\infty) = \mu$. This is the intuitive step in the present argument, which, however, as indicated above, can be justified.

Next, we write down an explicit expression for $\mu(r_{12})$. First we note that the mean number density at $\mathbf{r}_1$ is $\rho g(r_{12})$, which we use in place of ρ in the concentration term, (17–12). To calculate the work against intermolecular forces in coupling a new molecule at $\mathbf{r}_1$, we have to introduce a new, higher-order, distribution function $g^{[3]}(\mathbf{r}_1, \mathbf{r}_2, \mathbf{r}_3;\,\xi)$. This is defined by stating that the mean number density at $\mathbf{r}_3$ when a molecule is fixed at $\mathbf{r}_2$ and when a partially coupled (ξ) molecule is fixed at $\mathbf{r}_1$ is $\rho g^{[3]}$. (The significance of the superscript is that three molecules are involved; the radial distribution function is often written with a superscript two.) The potential energy of interaction of the molecule at $\mathbf{r}_1$ with those in $d\mathbf{r}_3$ is, then, $\xi u(r_{13}) \cdot \rho g^{[3]}\, d\mathbf{r}_3$. The corresponding work required to increase ξ by $d\xi$ is $u(r_{13})\, \rho g^{[3]}\, d\mathbf{r}_3\, d\xi$. Also, the work involved in the interaction with the molecule at $\mathbf{r}_2$ is $u(r_{12})\, d\xi$. On integrating over $\mathbf{r}_3$ and ξ, we obtain for $\mu(r_{12})$,

$$\mu(r_{12}) = kT \ln \Lambda^3 + kT \ln \rho g(r_{12}) + u(r_{12}) + \rho \int_0^1 \int_V u(r_{13}) g^{[3]} d\mathbf{r}_3\, d\xi. \tag{17–19}$$

We now equate $\mu(\infty)$ in Eq. (17–18) with $\mu(r_{12})$, and deduce

$$-kT \ln g(r_{12}) = u(r_{12}) + \rho \int_0^1 \int_V u(r_{13}) [g^{[3]}(\mathbf{r}_1, \mathbf{r}_2, \mathbf{r}_3;\,\xi) - g(r_{13};\,\xi)]\, d\mathbf{r}_3\, d\xi. \tag{17–20}$$

This result is not quite general enough, for it pertains to $g(r_{12}; 1)$ but not to $g(r_{12}; \xi)$. To remedy this shortcoming, we use the same type of argument as above, but add a molecule at $\mathbf{r}_1$ finally coupled only to the extent ξ instead of $\xi = 1$. This resembles adding to a fluid of one species ($\xi = 1$) a molecule of a second species (ξ). Therefore we digress to briefly consider a binary solution.

Let the species be α and β, and let the number densities be ρ_α and ρ_β. Consider two infinitesimal elements of volume, $d\mathbf{r}_\alpha$ and $d\mathbf{r}_\beta$. The probability that an α molecule is in $d\mathbf{r}_\alpha$ is $\rho_\alpha\, d\mathbf{r}_\alpha$ (since $d\mathbf{r}_\alpha \to 0$, this is also the mean number of α molecules in $d\mathbf{r}_\alpha$). If an α molecule is in $d\mathbf{r}_\alpha$, let the probability that a β molecule is in $d\mathbf{r}_\beta$ be $\rho_\beta g_{\alpha\beta}(r_{12}, \rho_\alpha, \rho_\beta, T)\, d\mathbf{r}_\beta$. Then the probability that an α molecule is in $d\mathbf{r}_\alpha$ and a β molecule is in $d\mathbf{r}_\beta$ is the product of the two probabilities, $\rho_\alpha \rho_\beta g_{\alpha\beta}\, d\mathbf{r}_\alpha\, d\mathbf{r}_\beta$. This result has to be symmetrical with respect to α and β; so $g_{\alpha\beta}$ is not only, as above, the radial distribution function for β molecules about a central α molecule, but is also the radial distribution function for α molecules about a central β molecule.

Suppose we have a solution in which ρ_β is very small, $\rho_\beta \to 0$, so that in effect each β has an environment made up entirely of α molecules. Now if we add one more β molecule to the solution, by the argument leading to Eq. (17–15) [see also Eqs. (18–17) through (18–19)],

$$\mu_\beta = kT \ln \Lambda_\beta^3 + kT \ln \rho_\beta + \rho_\alpha \int_0^1 \int_0^\infty u_{\alpha\beta}(r) g_{\alpha\beta}(r, \rho_\alpha, T; \xi) 4\pi r^2\, dr\, d\xi, \qquad (17\text{–}21)$$

where ξ refers to the β molecule. The last term is very closely related and leads immediately to an expression for the Henry's law constant of the solute β in the solvent α (Problem 19–14).

We return to our generalization of Eq. (17–20). Consider a hypothetical "binary solution" containing a fluid of number density ρ ("solvent") and a very few ("solute") molecules of the same type but with coupling parameter ξ. Let the "solute" number density be ρ_ξ, where $\rho_\xi \to 0$. A "solvent" molecule is fixed at $\mathbf{r}_2$, and we add a new "solute" molecule to the system at $\mathbf{r}_1$. If $\mathbf{r}_1$ is far from $\mathbf{r}_2$, $r_{12} \to \infty$, then, just as in Eq. (17–21),

$$\mu_\xi(\infty) = kT \ln \Lambda^3 + kT \ln \rho_\xi + \rho \int_0^\xi \int_V u(r_{13}) g(r_{13}; \xi')\, d\mathbf{r}_3\, d\xi'. \qquad (17\text{–}22)$$

But if r_{12} is finite,

$$\mu_\xi(r_{12}) = kT \ln \Lambda^3 + kT \ln \rho_\xi g(r_{12}; \xi) + \xi u(r_{12}) + \rho \int_0^\xi \int_V u(r_{13}) g^{[3]}(\xi')\, d\mathbf{r}_3\, d\xi'. \qquad (17\text{–}23)$$

Then from $\mu_\xi(\infty) = \mu_\xi(r_{12})$,

$$-kT \ln g(r_{12}; \xi) = \xi u(r_{12}) + \rho \int_0^\xi \int_V u(r_{13})[g^{[3]}(\mathbf{r}_1, \mathbf{r}_2, \mathbf{r}_3; \xi') - g(r_{13}; \xi')] \, d\mathbf{r}_3 \, d\xi'. \quad (17\text{–}24)$$

This is our final result, a generalization of Eq. (17–20), first derived by Kirkwood (but not by this kind of argument).

Equation (17–24) relates a two-particle ($n = 2$) distribution function, g, to a three-particle ($n = 3$) function, $g^{[3]}$. If we choose two fixed molecules instead of one (as above), an equation analogous to (17–24) can be derived relating an $n = 3$ function to an $n = 4$ function (Problem 17–4). This procedure can be carried on indefinitely, giving us a hopelessly complex set of N interlocking integral equations. We should, in fact, expect a very involved result of this kind, because the original configuration integral over $3N$ coordinates (Eq. 6–22) seems impossible to handle analytically and exactly, and there is no reason to believe that any alternative approach (e.g., the distribution-function method) will provide a solution which is fundamentally any simpler.

In view of this situation, Kirkwood suggested an intuitively attractive approximation, called the "superposition approximation," which serves to "close" the set of integral equations. The approximation is to assume that the influences of the fixed molecules at $\mathbf{r}_1$ (with ξ) and $\mathbf{r}_2$ on the mean density at $\mathbf{r}_3$ are independent of each other. Analytically, the approximation is

$$g^{[3]}(\mathbf{r}_1, \mathbf{r}_2, \mathbf{r}_3; \xi) = g(r_{13}; \xi) g(r_{23}). \quad (17\text{–}25)$$

If we put this in Eq. (17–24), we have

$$-kT \ln g(r_{12}; \xi) = \xi u(r_{12}) + \rho \int_0^\xi \int_V u(r_{13}) g(r_{13}; \xi')[g(r_{23}) - 1] \, d\mathbf{r}_3 \, d\xi'. \quad (17\text{–}26)$$

This is an integral equation in $g(r; \xi)$, called the Kirkwood equation. It has been solved numerically for a fluid of hard spheres and for the Lennard-Jones potential (see S. M., Chapter 6), and thermodynamic functions have been calculated from the solutions. An alternative but essentially equivalent integral equation, due to Born, Green and Yvon, has also been solved numerically. The results in both cases are qualitatively completely satisfactory, but only moderately accurate quantitatively, as should be expected since the superposition approximation is used.

When applied to an imperfect gas, the superposition approximation can be shown (Problem 17–5) to be exact for B_3 but not B_4.

17–4 Formal definition of distribution functions. In this section we show how distribution functions are related to the configuration integral. In a more exhaustive discussion (e.g., in S. M., Chapter 6), we would use these relations as a starting point.

Let us refer back to the paragraph preceding Eq. (17–21), and apply the argument to a one-component fluid. Then we have that

$$\rho^2 g^{(2)}(\mathbf{r}_1, \mathbf{r}_2)\, d\mathbf{r}_1\, d\mathbf{r}_2$$

is the probability that any one molecule of the fluid will be found in $d\mathbf{r}_1$ and any second molecule in $d\mathbf{r}_2$, where $g^{(2)}$ is the radial distribution function (the same as g above). To be more general, let us define $g^{(n)}(\mathbf{r}_1, \ldots, \mathbf{r}_n)$ by the statement that $\rho^n g^{(n)}\, d\mathbf{r}_1 \ldots d\mathbf{r}_n$ is the probability that any one molecule will be in $d\mathbf{r}_1$, another in $d\mathbf{r}_2$, etc. We now relate this probability to the configuration integral.

As in Eq. (6–28),

$$\frac{e^{-H/kT}\, dq_x\, dp_x}{\int e^{-H/kT}\, dq_x\, dp_x}$$

is the probability that the system will be found in the classical state $dq_x dp_x$. We integrate over the momenta and find that

$$\frac{e^{-U/kT}\, d\mathbf{r}_1 \ldots d\mathbf{r}_N}{Z}$$

is the probability that molecule 1 is in $d\mathbf{r}_1, \ldots,$ and that molecule N is in $d\mathbf{r}_N$. The probability that molecule 1 is in $d\mathbf{r}_1$, molecule 2 in $d\mathbf{r}_2, \ldots,$ and molecule n in $d\mathbf{r}_n$, irrespective of the configuration of the remaining $N - n$ molecules, is then

$$\frac{d\mathbf{r}_1 \ldots d\mathbf{r}_n \int \underset{V}{\cdots} \int e^{-U/kT}\, d\mathbf{r}_{n+1} \ldots d\mathbf{r}_N}{Z}. \tag{17–27}$$

Now, the probability that *any* molecule is in $d\mathbf{r}_1$, any second molecule in $d\mathbf{r}_2, \ldots,$ and any nth molecule in $d\mathbf{r}_n$, irrespective of the positions of the other molecules, is just (17–27) multiplied by $N!/(N-n)!$, since any one of N molecules can be in $d\mathbf{r}_1$, any one of $N - 1$ in $d\mathbf{r}_2, \ldots,$ and any one of $N - n + 1$ in $d\mathbf{r}_n$. This probability is just what we called $\rho^n g^{(n)}\, d\mathbf{r}_1 \ldots d\mathbf{r}_n$, above. So we have

$$\rho^n g^{(n)}(\mathbf{r}_1, \ldots, \mathbf{r}_n) = \frac{N!}{(N-n)!} \frac{\int \underset{V}{\cdots} \int e^{-U/kT}\, d\mathbf{r}_{n+1} \ldots d\mathbf{r}_N}{Z}. \tag{17–28}$$

For many purposes, when n is a small number, we can use

$$\frac{N!}{\rho^n(N-n)!} = V^n\left[1 + O\left(\frac{1}{N}\right)\right]$$

and

$$g^{(n)}(\mathbf{r}_1, \ldots, \mathbf{r}_n) = \frac{V^n \int \underset{V}{\cdots} \int e^{-U/kT}\, d\mathbf{r}_{n+1} \ldots d\mathbf{r}_N}{Z}. \tag{17–29}$$

Note that $g^{(n)}$ treats all the molecules in the set of n molecules equivalently. On the other hand, $g^{[3]}$ and $g^{[4]}$, introduced elsewhere in this chapter [see Eq. (17–19) and Problems 17–4 and 17–12], single out one molecule for special consideration.

Let us define a quantity $w^{(n)}$ by the equation

$$g^{(n)} = e^{-w^{(n)}/kT}. \tag{17–30}$$

The physical significance of $w^{(n)}$ can be seen as follows. We substitute (17–30) for $g^{(n)}$ in Eq. (17–28), take the logarithm of both sides of the resulting equation, and then differentiate with respect to the position of one of the n molecules 1, . . . , n, say molecule i. This gives

$$-\nabla_i w^{(n)} = \frac{\int \underset{V}{\cdots} \int e^{-U/kT}(-\nabla_i U)\, d\mathbf{r}_{n+1} \ldots d\mathbf{r}_N}{\int \underset{V}{\cdots} \int e^{-U/kT}\, d\mathbf{r}_{n+1} \ldots d\mathbf{r}_N}. \tag{17–31}$$

Now $-\nabla_i U$ is the force, f_i, acting on molecule i (for any given configuration $\mathbf{r}_1, \ldots, \mathbf{r}_N$). Therefore the right-hand side of Eq. (17–31) is the *mean* force $\bar{f}_i^{(n)}$, acting on i, averaged over all configurations of molecules $n+1, \ldots, N$ (the set of molecules $1, 2, \ldots, n$ being in a fixed configuration $\mathbf{r}_1, \ldots, \mathbf{r}_n$). Thus,

$$\bar{f}_i^{(n)} = -\nabla_i w^{(n)}. \tag{17–32}$$

This tells us that $w^{(n)}$ is the potential whose negative gradient gives the mean force acting on any one of the molecules 1, . . . , n. Therefore $w^{(n)}$ is called the potential of mean force.

As a specific example, suppose we start with two molecules far apart in a fluid with given N, V, T, so that $g^{(2)} = 1$ and $w^{(2)} = 0$. We then move the two molecules together reversibly in the thermodynamic sense (i.e., extremely slowly) to within a distance r. The force against which work has to be done in this process is $\bar{f}^{(2)}$, and the work done on the system is $w^{(2)}(r)$. In the limit as $\rho \to 0$, the two molecules are in effect in a vacuum, and $w^{(2)}(r)$ is just $u(r)$. But at a finite density, the work $w^{(2)}(r)$

is influenced by the presence of other molecules via statistical averaging (Eq. 17–31). In fact, Eq. (17–20) gives a general expression for $w^{(2)}(r_{12})$, since $-kT \ln g = w^{(2)}$.

To extend the idea of distribution functions to open systems, we start with the probability

$$\frac{N!}{(N-n)!} \cdot \frac{d\mathbf{r}_1 \ldots d\mathbf{r}_n \int \cdots_V \int e^{-U_N/kT} \, d\mathbf{r}_{n+1} \ldots d\mathbf{r}_N}{Z_N} \tag{17–33}$$

for a closed system with N molecules that any molecule is in $d\mathbf{r}_1, \ldots,$ and any other one in $d\mathbf{r}_n$. (The subscript N is inserted here and below to indicate a closed system with N molecules; we have omitted the subscript N until now in this chapter to simplify the notation.) This is the probability referred to preceding Eq. (17–28). In the notation of Chapter 15, the probability that an open system contains N molecules is

$$P_N = \frac{\lambda^N Q_N}{\Xi} = \frac{z^N Z_N}{N!\Xi}. \tag{17–34}$$

We therefore multiply (17–33) by (17–34) and sum over N to get the probability that molecules are in $d\mathbf{r}_1, \ldots, d\mathbf{r}_n$, irrespective of the locations of other molecules, in an open system (i.e., irrespective of N). We define $g^{(n)}$ in an open system by equating this probability to $\rho^n g^{(n)} \, d\mathbf{r}_1 \ldots d\mathbf{r}_n$, where $\rho = \overline{N}/V$. Thus we have that

$$\rho^n g^{(n)}(\mathbf{r}_1, \ldots, \mathbf{r}_n) = \frac{1}{\Xi} \sum_{N \geq n} \frac{z^N}{(N-n)!} \int \cdots_V \int e^{-U_N/kT} \, d\mathbf{r}_{n+1} \ldots d\mathbf{r}_N. \tag{17–35}$$

Here the physical significance of $g^{(n)}$ is the same as $g_N^{(n)}$ for a closed system with $N = \overline{N}$. The two $g^{(n)}$'s differ only by a term of order $1/\overline{N}$. Although the open system $g^{(n)}$ appears to be more complicated, it has a number of advantages in theoretical work.

We can define a $w^{(n)}$ for an open system by the relation (17–30) and again show (Problem 17–6) that $w^{(n)}$ is a potential of mean force. (This time the force is averaged not only over all configurations of molecules $n + 1, \ldots, N$ but also over the number of molecules N.) The physical significance of $w^{(n)}$ is the same as $w_{\overline{N}}^{(n)}$ for a closed system. In the example above [preceding Eq. (17–33)], $w^{(2)}(r)$ is the reversible work done on an *open* system (μ, V, T fixed) in bringing together two molecules from $r = \infty$ to r.

17–5 Surface tension. In this section we illustrate the use of distribution functions in an inhomogeneous region. The system we choose is a plane interface between a one-component gas and liquid. Our object is

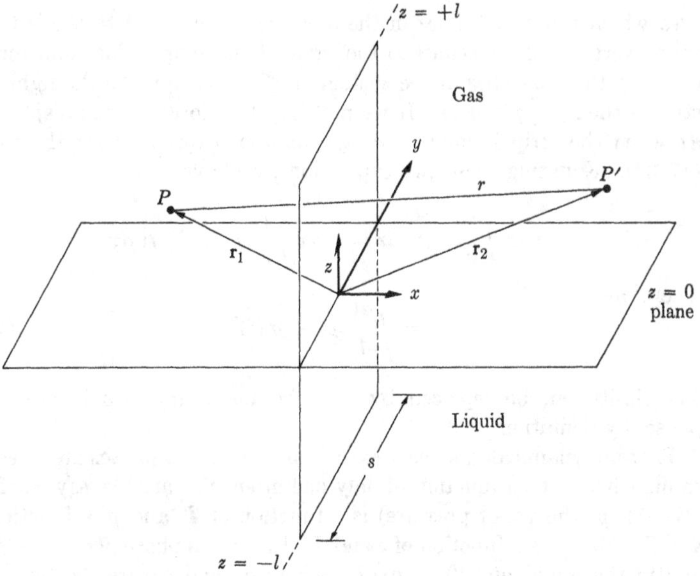

FIG. 17-5. Construction for calculation of surface tension.

to relate the surface tension to $g^{(2)}$. We use an argument, due to Kirkwood and Buff,* which is a generalization of that used for the pressure in Eq. (17-10).

Consider a hypothetical macroscopic strip of zero thickness, width s, and height $2l$, immersed in the fluid and extending from bulk liquid to bulk gas through and perpendicular to the interface (Fig. 17-5). The xy-plane is parallel to the interface and located in the transition region (the exact location is immaterial). We include in the thermodynamic system here only that part of the fluid in the area $s \times 2l$ and to the left of the strip ($x \leq 0$). Now imagine the strip moved parallel to the surface through an infinitesimal distance dx. Both the volume and interfacial area of the system change, so the work done by the system is

$$p\,dV - \gamma\,d\mathcal{A} = p \cdot 2ls\,dx - \gamma \cdot s\,dx, \tag{17-36}$$

where γ is the surface tension (*not* an activity coefficient) and $\mathcal{A}$ the surface area.

The mean number density in this system is a function of z, $\rho(z)$ (z is not to be confused with the activity). We define $p'(z)$ as the force per unit

* J. G. KIRKWOOD and F. P. BUFF, *J. Chem. Phys.* **17**, 338 (1949).

area which those molecules in the area $s \times dz$ at z and to the left of the strip exert on all molecules to the right of the strip. The total force exerted by the molecules in the system ($x \leq 0$) on those to the right of the strip is then $\int_{-l}^{+l} p'(z) s\, dz$. If we multiply this force by the displacement dx when the strip is moved, we get another expression for the work in (17–36). Equating these two expressions, we have

$$s\, dx \int_{-l}^{+l} p'(z)\, dz = s\, dx \int_{-l}^{+l} p\, dz - \gamma s\, dx.$$

Therefore,

$$\gamma = \int_{-l}^{+l} [p - p'(z)]\, dz. \tag{17–37}$$

The limits can be replaced by $\pm\infty$ because $p'(z) \to p$ in either bulk phase, by definition.

Thermodynamically, since γ exists only when two phases are in equilibrium, γ has to be a function of only one intensive variable, say T. In Eq. (17–37), p (the vapor pressure) is a function of T, and p' a function of z and T. Also, ρ is a function of z and T. In the gas phase $\rho(z, T) \to \rho_G(T)$, and in the liquid $\rho(z, T) \to \rho_L(T)$, where ρ_G and ρ_L are the bulk phase values.

The remaining part of the problem is to express $p'(z)$ in terms of $g^{(2)}$ (which we denote by g hereafter). Consider a molecule at the point P ($\mathbf{r}_1$), $x_1, y_1 = 0, z_1$, and another at P' ($\mathbf{r}_2$), x_2, y_2, z_2, as in Fig. 17–5, where $x_1 \leq 0$ and $x_2 \geq 0$. The normal (i.e., x-component of the) force exerted by the molecule at P on the molecule at P' is $-(x_2 - x_1)u'(r)/r$, where $r = |\mathbf{r}_2 - \mathbf{r}_1|$. The mean number density at $\mathbf{r}_2$ when a molecule is fixed at $\mathbf{r}_1$ is $\rho(z_2)g(z_1, z_2, r)$, which defines the distribution function g. Since g depends on z_1 and z_2 as well as on r, this pair distribution function is not a "radial" distribution function. At large separations r, $g \to 1$; that is, $g(z_1, z_2, \infty) = 1$. The mean number of molecules in $d\mathbf{r}_2$, when a molecule is fixed at $\mathbf{r}_1$, is $\rho(z_2)g(z_1, z_2, r)\, d\mathbf{r}_2$; and the normal force exerted on these molecules by the molecule at $\mathbf{r}_1$ is

$$-\frac{(x_2 - x_1)u'(r)}{r} \cdot \rho(z_2)g(z_1, z_2, r)\, d\mathbf{r}_2.$$

The normal force exerted on *all* molecules with $x_2 \geq 0$ is then

$$-\int_{-\infty}^{+\infty}\int_{-\infty}^{+\infty}\int_{0}^{\infty} \left(\frac{x_2 - x_1}{r}\right) u'(r)\rho(z_2)g(z_1, z_2, r)\, dx_2\, dy_2\, dz_2, \tag{17–38}$$

which we denote by $-G(x_1, z_1)$. Finally, the normal force exerted on all molecules with $x_2 \geq 0$ by all molecules in the area $s\, dz_1$ at z_1 and to the left of the strip is

$$-s\,dz_1 \int_{-\infty}^{0} G(x_1, z_1)\rho(z_1)\,dx_1.$$

To obtain the contribution of intermolecular forces to $p'(z_1)$ (a force per unit area), we divide this last result by the area $s\,dz_1$. In addition, just as in Eq. (17–10), there is a contribution $\rho(z_1)kT$ owing to momentum transport. Therefore

$$p'(z_1) = \rho(z_1)kT - \rho(z_1) \int_{-\infty}^{0} G(x_1, z_1)\,dx_1, \tag{17–39}$$

where $-G$ is given by (17–38).

We change variables from $\mathbf{r}_2$ to $\mathbf{r}_{12}$, that is, from x_2, y_2, z_2 to $x_{12} = x_2 - x_1$, $y_{12} = y_2$, $z_{12} = z_2 - z_1$. Of course, $r^2 = x_{12}^2 + y_{12}^2 + z_{12}^2$. Then

$$G(x_1, z_1) = \int_{-\infty}^{+\infty}\int_{-\infty}^{+\infty}\int_{-x_1}^{+\infty} \left(\frac{x_{12}}{r}\right) u'(r)\rho(z_{12} + z_1)g(z_1, z_{12}, r)\,dx_{12}\,dy_{12}\,dz_{12}.$$

Next, we perform an integration by parts on the integral in Eq. (17–39). The integrand in the resulting integral over x_1 is an even function of x_1. For symmetry, then, we extend the integration from $-\infty$ to $+\infty$, multiply by 1/2, and also replace x_1 by x_{12} as the dummy variable of integration. This gives the final result

$$p'(z_1) = \rho(z_1)kT - \tfrac{1}{2}\rho(z_1) \iiint_{-\infty}^{+\infty} \left(\frac{x_{12}^2}{r}\right) \frac{du(r)}{dr}\,\rho(z_{12} + z_1)g(z_1, z_{12}, r)\,dx_{12}\,dy_{12}\,dz_{12}, \tag{17–40}$$

which can be put into Eq. (17–37) for γ.

It is not difficult to see (Problem 17–7) that Eq. (17–40) reduces to Eq. (17–10), as it should, in either bulk phase.

In applying Eq. (17–40), we encounter essentially the same difficulty as in the theory of homogeneous fluids (Section 17–3). That is, the "singlet" distribution function $\rho(z)$ can be related to the pair function $g^{(2)}$ (or g) by an integral equation (Problem 17–8), the pair function to a triplet function by another integral equation, etc. Exact solution of this set of equations to obtain $\rho(z_1)$ and $g(z_1, z_{12}, r)$ for use, for example, in Eq. (17–40) is impossible.

Numerical calculations have been made, however, on two approximations: (1) Fowler's approximation,* in which an obviously unrealistic sharp discontinuity is assumed in the density at the interface (Problem

* Kirkwood and Buff, *loc. cit.*

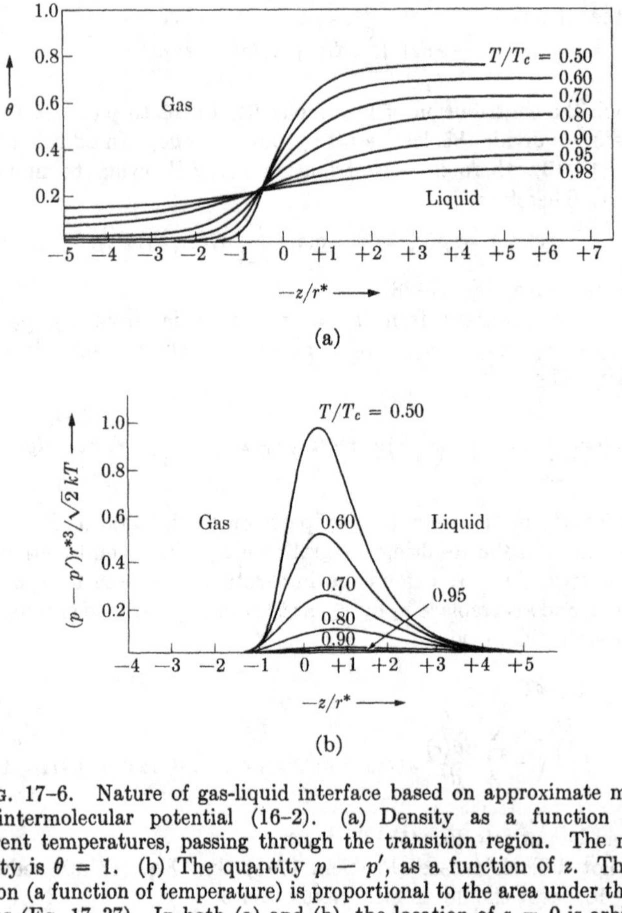

FIG. 17–6. Nature of gas-liquid interface based on approximate model and the intermolecular potential (16–2). (a) Density as a function of z, at different temperatures, passing through the transition region. The maximum density is $\theta = 1$. (b) The quantity $p - p'$, as a function of z. The surface tension (a function of temperature) is proportional to the area under the $p - p'$ curves (Eq. 17–37). In both (a) and (b), the location of $z = 0$ is arbitrary for each curve.

17–9); and (2) Hill's approximation,* which presumably gives a correct qualitative picture of the transition region (Fig. 17–6) but which is based, nevertheless, on a very crude model. Approximation (1), together with an *experimental* radial distribution function for bulk liquid argon, leads to a value of γ at 90°K for argon that differs from the experimental value (11.9 ergs·cm^{-2}) by 25%; while approximation (2) gives a value of γ that is off by a factor of two.

---

* T. L. HILL, *J. Chem. Phys.* **20**, 141 (1952).

Problems

17–1. Choose $u = \int_h^\infty$ and $dv = h\,dh$, integrate (17–6) by parts, and verify (17–8). (Page 305.)

17–2. Deduce p_U in Eq. (17–9) from Eqs. (17–7) and (17–8), using two integrations by parts: (a) choose $dv = x^2\,dx$ with the first integral in (17–8), and (b) choose $dv = dx$ with the second integral in (17–8). (Page 305.)

17–3. At low densities, $\rho \to 0$, we can put, in Eq. (17–20),

$$g^{[3]}(\xi) \to e^{-u(r_{23})/kT} e^{-\xi u(r_{13})/kT},$$

$$g(\xi) \to e^{-\xi u(r_{13})/kT}.$$

Show, then, that g_1 in the expansion (17–3) is given by

$$g_1(r_{12}, T) = e^{-u(r_{12})/kT} \int [e^{-u(r_{13})/kT} - 1][e^{-u(r_{23})/kT} - 1]\,d\mathbf{r}_3. \quad (17\text{–}41)$$

(Page 306.)

17–4. Extend the argument leading to Eq. (17–24) to derive an integral equation for $g^{[3]}$ in terms of $g^{[4]}$, where $\rho g^{[4]}(\mathbf{r}_1, \ldots, \mathbf{r}_4; \xi)$ is the mean number density at $\mathbf{r}_4$ when molecules are fixed at $\mathbf{r}_2$ and $\mathbf{r}_3$ and a partially coupled (ξ) molecule is fixed at $\mathbf{r}_1$. (Page 311.)

17–5. Prove that the superposition approximation gives the correct third virial coefficient. (Page 311.)

17–6. Show that $w^{(n)}$ is a potential of mean force in an open system. (Page 314.)

17–7. Show that $p'(z_1)$ in Eq. (17–40) reduces to p in Eq. (17–10) in a bulk phase. (Page 317.)

17–8. Use the method of Eq. (17–15) to derive the following integral equation for the surface-tension problem (assume the local chemical potential has the same value at every point):

$$kT \ln \rho(z_1) = kT \ln \rho_\alpha + \int_0^1 \int_V u(r)[\rho_\alpha g_\alpha(r; \xi) - \rho(z_{12} + z_1) g(z_1, z_{12}, r; \xi)]\,d\mathbf{r}_{12}\,d\xi, \quad (17\text{–}42)$$

where α refers to either bulk phase. (Page 317.)

17–9. Imagine that a sample of bulk liquid is divided into two parts by a plane of area $\mathcal{A}$, and that the two parts are separated reversibly by gradually

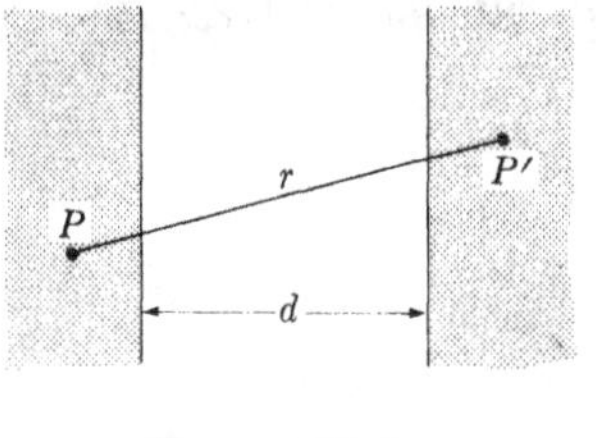

Figure 17–7

increasing d (Fig. 17–7) from $d = 0$ to $d = \infty$. Assume that the (vapor) space between the two liquid parts has essentially zero density, and that each liquid part remains homogeneous right up to the vapor region, where there is a sharp discontinuity in density (from the liquid density ρ to $\rho = 0$). If a molecule is fixed at P (Fig. 17–7), the number density at P' is assumed to be $\rho g(r)$, where $g(r)$ is the bulk liquid radial distribution function. We can now calculate the normal force between the molecule at P and the other half of the liquid [compare Eq. (17–7)], and then we can calculate the total work necessary to pull the two liquid parts away from each other. This work is set equal to $2\gamma\mathcal{A}$, since the amount of new surface area formed is $2\mathcal{A}$. Derive, in this way, the equation

$$\gamma = \frac{\pi\rho^2}{8}\int_0^\infty r^4 \frac{du(r)}{dr}\, g(r, \rho, T)\, dr, \tag{17–43}$$

originally due to Fowler. This can also be shown to be a special case of Eqs. (17–37) and (17–40). (Page 318.)

17–10. Show schematically what you would expect the functions $4\pi r^2 g(r)$ and $g(r)$ to look like for a monatomic solid with cubic close packing [$g(r)$ here is averaged over all directions outward from the central atom]. The successive numbers of neighbors are 12, 6, 24, ... at distances a, $\sqrt{2}a$, $\sqrt{3}a$, ... From these curves guess the qualitative form of $g(r)$ for the liquid state of the same substance.

17–11. Apply Eq. (17–21) to the case where the "solvent" is a dilute gas α. Show that this equation is consistent with Eq. (15–47),

$$z_\beta/\rho_\beta = 1 - b_{11}\rho_\alpha + \cdots.$$

17–12. Devise a formal definition of $g^{[3]}$ analogous to Eq. (17–28) and find the relation between $g^{[3]}$ and $g^{(3)}$.

17–13. Investigate the possible application of the law of corresponding states to surface tension.

Supplementary Reading

De Boer, J., *Repts. Progr. Phys.* **12,** 305 (1949).

Green, H. S., *Molecular Theory of Fluids.* Amsterdam: North-Holland, 1952.

Hirschfelder, Curtiss, and Bird, Chapters 4 and 5.

Ono, S., and Kondo, S., *Handbuch der Physik,* Vol. 10. Berlin: Springer, 1960.

S. M., Chapter 6.

CHAPTER 18

DILUTE ELECTROLYTE SOLUTIONS AND PLASMAS

Since a great many accounts of the Debye-Hückel theory of electrolytes (based on the linearized Poisson-Boltzmann equation) are available, in Section 18–1 we shall give only a rather short discussion of this topic. The reader who desires a more detailed treatment should consult Fowler and Guggenheim, Chapter 9. The rest of the chapter is devoted to a less conventional approach to the problem, based on Kirkwood's theory of solutions.* The distribution-function methods of Section 17–3 are extended in Section 18–2 to solutions in general and, in Section 18–3, to electrolyte solutions in particular. It is shown in Section 18–3 that the Debye-Hückel limiting law can be deduced from the Kirkwood solution theory as well as by the usual (Debye-Hückel) method of Section 18–1.

18–1 Debye-Hückel theory. The model we consider here and in Section 18–3 is the following. The volume V is *completely* filled with a dielectric *continuum* of dielectric constant ϵ which, for simplicity, is *not* considered a molecular species of the system. Immersed in the continuum (see Fig. 18–1) are "hard" monatomic ions of diameter a, N_i of species i. Thus the system is in effect a "gas" mixture of ions in a continuous dielectric medium. If the continuum is a vacuum, $\epsilon = 1$ and the system is a real gas mixture (i.e., a plasma). The charge on an ion of species i is $z_i\mathbf{e}$, where $\mathbf{e}$ is the charge on a proton. In view of our continuum assumption (see Fig. 18–1), the interionic potential energy for an ij pair is

$$\begin{aligned} u_{ij}(r) &= +\infty \qquad r < a \\ &= \frac{z_i z_j \mathbf{e}^2}{\epsilon r} \qquad r \geq a. \end{aligned} \tag{18–1}$$

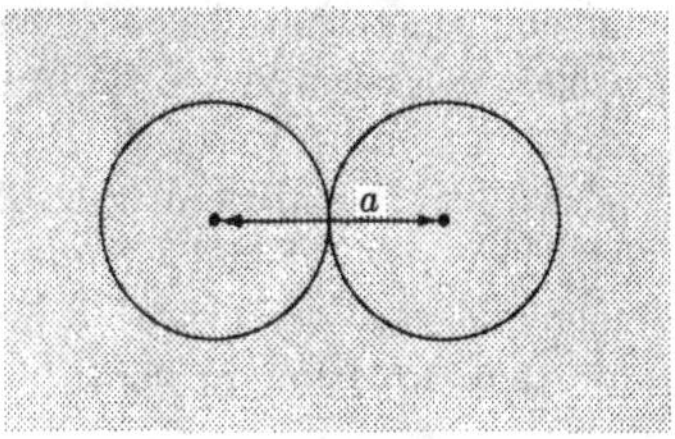

FIG. 18–1. Model for "hard" ions immersed in a dielectric continuum.

Of course, real ions are not hard spheres, and a real solvent does not behave like a continuum when r is small (say, $r < 5a$). Therefore Eq. (18–1) is only an approximation. However, this equa-

* J. G. KIRKWOOD, *J. Chem. Phys.* **3**, 300 (1935).

tion becomes exact, in effect, in the limit of very dilute solutions, for then only large values of r are significant.

We denote the concentration (number density) of species i by $\rho_i = N_i/V$. In order for the solution to be electrically neutral, we must have $\sum_i \rho_i z_i = 0$ (which we call the neutrality condition). If the solution is not neutral, the excess charge will collect on the surface, and the properties of the system will depend on the shape of container.

If one is interested in temperature and pressure effects on an electrolyte solution, the solvent must be treated explicitly as a molecular species (as in Fowler and Guggenheim, Chapter 9). In this case, however, not much real progress can be made without giving up Eq. (18–1) for small values of r (say $r < 5a$) in favor of a detailed study of the behavior of solvent molecules around a pair of ions. We avoid all such complications in our brief treatment by adopting the continuum model already described.

The basic thermodynamic equation for our system is

$$dE = T\,dS - p\,dV + \sum_i \mu_i\,dN_i. \tag{18–2}$$

The pressure p is the pressure of the "gas" mixture of ions. In reality, of course, if the continuum is a solvent and not a vacuum, p is the *osmotic* pressure of the electrolyte (this will be confirmed in Sections 19–1 and 19–2). In a plasma ($\epsilon = 1$), p is the total pressure.

Our object in this section is to give the Debye-Hückel derivation of the ionic activity coefficient γ_i for an arbitrary species i in a very dilute solution. In the limit as $\rho_j \to 0$ (all j), we would have for the chemical potential (Eq. 4–23) of species i,

$$\mu_i = kT \ln \Lambda_i^3 + kT \ln \rho_i \qquad (\text{all } \rho_j \to 0). \tag{18–3}$$

Of course, if the molecular nature of the solvent were being taken into account, Λ_i^3 would be multiplied by a quantity (a function of pressure and temperature) closely related to the Henry's law constant and determined by the nature of the interaction between one ion and pure solvent [see Eq. (17–21) and Chapter 19]. We are interested in the first-order correction to Eq. (18–3), when the solution is not quite dilute enough for that equation to hold:

$$\mu_i = kT \ln \Lambda_i^3 + kT \ln \rho_i + kT \ln \gamma_i. \tag{18–4}$$

Perhaps the first thought to come to mind is that we have here, in effect, a dilute gaseous mixture of ions and hence that the methods of Chapter 15 on imperfect gases are applicable. Thus, all we would need for the first-order correction mentioned above would be to calculate second virial coefficients as in Section 15–3. However, if we substitute the coulombic potential energy, (18–1), in Eq. (15–52), we see [on writing $e^{-u/kT} = 1 - (u/kT)$] that the integral diverges because of the upper limit. In fact, the integral will diverge not only for the coulombic potential r^{-1} but also for r^{-2} and r^{-3}. Convergence begins with r^{-4} (in the van der Waals interaction, we have r^{-6}). The same is true for the higher virial coefficients. The physical reason for the divergence is that with forces of sufficiently long range, it is not possible to treat a dilute system in terms of, successively, binary interactions, ternary interactions, etc., as the concentration increases. Instead, many-body or "collective" interactions appear even at the first departure from the dilute solution behavior of Eq. (18–3).

Thus, a virial expansion does not exist for systems made up of particles interacting by long-range forces (pair potential energy $\propto r^{-n}$, $n \leq 3$); other methods must be used.

We now turn to the Debye-Hückel argument, which leads to exact *limiting* expressions (at very low concentrations) but which is otherwise approximate. We consider the spherically symmetrical (on the average) neighborhood around one particular ("central") ion of type j (as it moves through the system; or we can regard it as fixed). Let the mean electrostatic potential at a distance r from the j ion be $\psi_j(r)$. The zero for $\psi_j(r)$ is chosen at $r = \infty$. Also, let the mean charge density at r be $n_j(r)$. Then Poisson's equation in electrostatics states that

$$\nabla^2\psi_j(r) = -\frac{4\pi n_j(r)}{\epsilon} \qquad (r \geq a). \tag{18–5}$$

Because of the presence of the central j ion, $\psi_j(r)$ will be nonzero for finite r [$\psi_j(r)$ will have the same sign as z_j], and also $n_j(r)$ will be nonzero for finite r [$n_j(r)$ will have a sign opposite to that of z_j because of a net accumulation of ions of opposite charge]. The region over which $\psi_j(r)$ and $n_j(r)$ differ significantly from zero is called the ion atmosphere of the central j ion. As $r \to \infty$, $n_j(r) \to 0$ because of the neutrality condition.

The general plan is to convert Eq. (18–5) into a tractable differential equation in $\psi_j(r)$, to solve the differential equation, and finally to use $\psi_j(r)$ thus found to calculate the activity coefficient γ_j by a charging process (as in Chapter 17).

From Section 17–3, we have that the mean concentration of ions of type i a distance r from a central j ion is

$$\rho_i^j(r) = \rho_i g_{ij}(r) = \rho_i e^{-w_{ij}(r)/kT}, \tag{18–6}$$

where $w_{ij}(r)$ is the potential of mean force. Since, with our model, no ion can be closer to another than $r = a$, $\rho_i^j(r) = 0$ and $w_{ij}(r) = +\infty$ for $r < a$. At $r = \infty$, $w_{ij} = 0$, $g_{ij} = 1$, and the concentration of i ions is the bulk concentration ρ_i. The first approximation we make in the Debye-Hückel theory is to assume that $\rho_i^j(r)$ is determined by a simple Boltzmann factor,

$$\rho_i^j(r) = \rho_i e^{-z_i \mathbf{e}\psi_j(r)/kT} \qquad (r \geq a). \tag{18–7}$$

This is analogous to assuming that in an electrochemical equilibrium between two bulk phases α and β, with a potential difference $\psi = \psi^\alpha - \psi^\beta$,

$$\rho_i^\alpha = \rho_i^\beta e^{-z_i \mathbf{e}\psi/kT},$$

with *neglect of activity coefficients*. This analogy will be pursued more explicitly in Section 18–3. Comparison with Eq. (18–6) shows that the assumption (18–7) also amounts to setting $w_{ij}(r) = z_i \mathbf{e}\psi_j(r)$ for $r \geq a$. Exact expressions for $w_{ij}(r)$ and $z_i\mathbf{e}\psi_j(r)$ will be compared in Section 18–3. Although Eq. (18–7) is in general not exact, it proves to be correct in the limit of very dilute solutions.

Equation (18–7) makes it possible to write Eq. (18–5) as a differential equation in $\psi_j(r)$ only. The charge density at r is

$$n_j(r) = \mathbf{e}\sum_i z_i \rho_i^j(r) \tag{18–8}$$

$$= \mathbf{e}\sum_i z_i \rho_i e^{-z_i \mathbf{e}\psi_j(r)/kT} \qquad (r \geq a). \tag{18–9}$$

Then substitution of Eq. (18–9) for $n_j(r)$ in Eq. (18–5) leads to the so-called Poisson-Boltzmann differential equation in $\psi_j(r)$. This equation is nonlinear and very difficult to solve in general (except by numerical or series methods). But much effort to solve the equation is probably not justified in any case, since Eq. (18–7) and therefore Eq. (18–9) are approximations in the first place.

To make further progress, we limit the discussion to cases in which $z_i \mathbf{e}\psi_j(r)/kT \ll 1$ for the important values of r. This is the second Debye-Hückel approximation; it allows us to linearize the Poisson-Boltzmann equation. After solving the linearized equation, we can come back to see what is meant by "important values of r" and for what electrolyte concentrations linearization is a good approximation. It will turn out that this approximation, like the first one, is justified if the solution is sufficiently dilute.

Therefore we expand the exponential in Eq. (18–9) up to the linear term. The Poisson-Boltzmann equation becomes

$$\nabla^2\psi_j = -\frac{4\pi\mathbf{e}}{\epsilon}\sum_i z_i\rho_i\left(1 - \frac{z_i\mathbf{e}\psi_j}{kT}\right),$$

or

$$\frac{1}{r}\frac{d^2(r\psi_j)}{dr^2} = \kappa^2\psi_j \qquad (r \geq a), \tag{18–10}$$

where

$$\kappa^2 = \frac{4\pi e^2}{\epsilon kT}\sum_i \rho_i z_i^2. \tag{18–11}$$

The leading term in the expansion does not contribute to Eq. (18–10) because of the neutrality condition. The summation in Eq. (18–11) is twice the "ionic strength," often used in thermodynamics (and expressed in moles·liter^{-1}). Thus $\kappa \propto$ (ionic strength)$^{1/2}$ and $\kappa \to 0$ as the solution approaches infinite dilution. When we rewrite Eq. (18–10) in the familiar form

$$\frac{d^2(r\psi_j)}{dr^2} = \kappa^2(r\psi_j),$$

the solution is obviously

$$r\psi_j = C_1 e^{-\kappa r} + C_2 e^{+\kappa r}.$$

Since we know that $\psi_j \to 0$ as $r \to \infty$, we have to choose $C_2 = 0$. Therefore

$$\psi_j(r) = \frac{C_1 e^{-\kappa r}}{r} \qquad (r \geq a). \tag{18–12}$$

To evaluate C_1 we must consider the situation inside the sphere $r = a$ (Fig. 18–1), where no charges can be present except the charge $z_j e$ at $r = 0$. Laplace's equation is therefore satisfied inside the sphere. Let φ_j be the inside electrostatic potential. Then

$$\nabla^2\varphi_j = \frac{1}{r}\frac{d^2(r\varphi_j)}{dr^2} = 0 \qquad (r < a).$$

Successive integrations give

$$\frac{d(r\varphi_j)}{dr} = C_3$$

and

$$\varphi_j(r) = C_3 + \frac{C_4}{r} \qquad (r < a). \tag{18–13}$$

When $r \to 0$, φ_j must approach the coulombic potential $z_j e/\epsilon r$. Therefore $C_4 = z_j e/\epsilon$. To evaluate the remaining two constants, C_1 and C_3, we use the two boundary conditions at $r = a$:

$$\begin{aligned} \psi_j(a) &= \varphi_j(a), \\ \psi_j'(a) &= \varphi_j'(a). \end{aligned} \tag{18–14}$$

Equations (18–12) through (18–14) give

$$C_1 = \frac{z_j e e^{\kappa a}}{\epsilon(1 + \kappa a)}, \qquad C_3 = -\frac{z_j e \kappa}{\epsilon(1 + \kappa a)}.$$

Therefore

$$\psi_j(r) = \frac{z_j e e^{-\kappa(r-a)}}{\epsilon r(1 + \kappa a)} \qquad (r \geq a). \tag{18–15}$$

We also note that, in Eq. (18–13), C_4/r is the contribution to the potential inside the sphere owing to the central j ion itself, and hence C_3 must be the contribution of all the ions outside the sphere (i.e., the contribution of the ion atmosphere). We denote the latter potential by $\varphi_j^{\text{atm}}(r)$. This quantity is a constant. Hence, at $r = 0$, where the central charge $z_j e$ is located, we have

$$\varphi_j^{\text{atm}}(0) = C_3 = -\frac{z_j e \kappa}{\epsilon(1 + \kappa a)}. \tag{18–16}$$

We can now use Eq. (18–16) to calculate the electrostatic contribution to γ_j. The connection is established by extending the argument of Eqs. (17–11) through (17–14) to solutions [in fact, we have already applied it to a binary solution in Eq. (17–21)]. We have

$$\mu_j = \mu_j' + \mu_j'', \tag{18–17}$$

$$\mu_j' = \left(\frac{\partial A'}{\partial N_j}\right)_{N_{\alpha \neq j}, V, T} = kT \ln \Lambda_j^3 + kT \ln \rho_j, \tag{18–18}$$

$$\mu_j'' = \left(\frac{\partial A''}{\partial N_j}\right)_{N_{\alpha \neq j}, V, T} = kT \ln \gamma_j, \tag{18–19}$$

where $kT \ln \gamma_j$ is the isothermal reversible work, against intermolecular forces, required to add one more molecule of species j to the system. For simplicity, we calculate the electrostatic contribution to $kT \ln \gamma_j$ using the work of a charging process, and then add on separately the hard-sphere interaction contribution, since the latter is already available from Chapter 15. To compute the electrostatic work of adding one more j ion, we introduce the charging (coupling) parameter ξ_j on the j ion. For an arbitrary value of ξ_j, the j ion has a charge $\xi_j z_j e$, and hence

$$\varphi_j^{\text{atm}}(0; \xi_j) = -\frac{\xi_j z_j e \kappa}{\epsilon(1 + \kappa a)}.$$

If we bring in an increment of charge $z_j e\, d\xi_j$ to the j ion, the work done against the ion atmosphere (i.e., against the electrostatic contribution to the intermolecular forces) is $\varphi_j^{\text{atm}}(0;\, \xi_j) z_j e\, d\xi_j$. Therefore,

$$\int_0^1 \varphi_j^{\text{atm}}(0;\, \xi_j) z_j e \; d\xi_j = -\frac{z_j^2 e^2 \kappa}{2\epsilon(1 + \kappa a)} \tag{18–20}$$

is the electrostatic contribution to $kT \ln \gamma_j$.

For the hard-sphere contribution to $kT \ln \gamma_j$ (which would remain even if we put each $z_j = 0$) in a dilute solution, we merely have to extend Eq. (15–47) for a binary gas mixture to an arbitrary number of components. In the binary case [and in the notation of Eq. (15–47)] we have from Eqs. (15–24), (15–25), and (15–52),

$$\begin{aligned} \ln \gamma_2 &= B_{11}\rho_1 + 2B_{02}\rho_2 + \cdots \\ &= \frac{4\pi a^3}{3}(\rho_1 + \rho_2) + \cdots. \end{aligned}$$

For any number of components, then, we obviously have

$$\ln \gamma_j = \frac{4\pi a^3}{3} \sum_i \rho_i \qquad (\text{all } j). \tag{18–21}$$

Finally, then, for the electrolyte solution,

$$\ln \gamma_j = -\frac{z_j^2 e^2 \kappa}{2\epsilon kT(1 + \kappa a)} + \frac{4\pi a^3}{3} \sum_i \rho_i. \tag{18–22}$$

This is a statistical-mechanical equation for the activity coefficient of a single ionic species. Of course, in thermodynamics we can measure activity coefficients only for neutral salts, so only "neutral" combinations of γ_j's from Eq. (18–22) (e.g., $\gamma_+\gamma_-$, $\gamma_{++}\gamma_-^2$, etc.) can be checked experimentally. It is well known that excellent agreement is achieved between experimental activity coefficients and the limiting form of Eq. (18–22) *at high dilutions*, but not always with Eq. (18–22) itself. At high dilutions, we have the so-called Debye-Hückel limiting law,

$$\ln \gamma_j \to -\frac{z_j^2 e^2 \kappa}{2\epsilon kT} \qquad (\kappa \to 0). \tag{18–23}$$

We note that $\ln \gamma_j$ is proportional to the square of the charge on a j ion and to the square root of ionic strength.

Equation (18–22) has the disadvantage of reflecting the approximate nature of our choice of $u_{ij}(r)$ in (18–1). It is therefore not surprising that Eq. (18–22) is not always in satisfactory agreement with experiment. On the other hand, Eq. (18–23) does *not* involve a. In fact, it is rather obvious on physical grounds that Eq. (18–23) must depend on the validity of Coulomb's law, $u_{ij}(r) = z_i z_j e^2/\epsilon r$, only for $r \gg a$, because in a very dilute solution the ions are in general far apart. But we may have confidence that Coulomb's law (with ϵ) will become exact for large r; therefore we expect Eq. (18–23) to be an exact limiting law provided that our approximations, (a) $w_{ij}(r) \rightarrow z_i e\psi_j(r)$, and (b) $z_i e\psi_j(r)/kT \ll 1$, become valid in a very dilute solution. The fact that the Debye-Hückel limiting law is verified experimentally indicates that this is indeed the case. On the theoretical side, we shall confirm (b) below and (a) in Section 18–3.

We can examine more carefully the point above about the importance of $r \gg a$ for dilute solutions by investigating the (net) charge density distribution (the "ion atmosphere") about the central j ion. From Eqs. (18–10) and (18–15),

$$n_j(r) = -\frac{\epsilon\kappa^2\psi_j}{4\pi} = -\frac{z_j e\kappa^2 e^{-\kappa(r-a)}}{4\pi r(1 + \kappa a)}. \tag{18–24}$$

That is, $n_j(r) \propto e^{-\kappa r}/r$. Note that $n_j(r)$ has a sign opposite to z_j, as expected. The net amount of charge between r and $r + dr$ is proportional to $r^2 n_j(r)$, or $re^{-\kappa r} \equiv f(r)$. The function $\kappa f(r)$ is plotted in Fig. 18–2. This figure shows the form and extent of the ion atmosphere. In par-

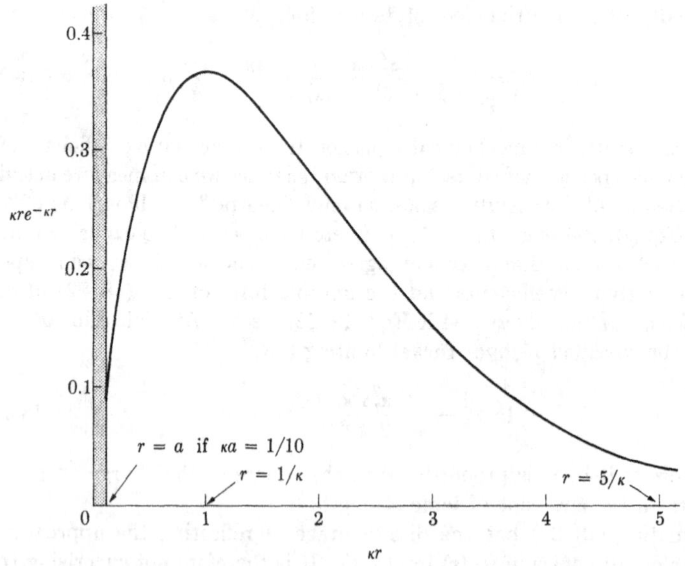

FIG. 18–2. Relative net amount of charge between r and $r + dr$ in ion atmosphere around a particular ion in a dilute electrolyte solution.

ticular, the maximum in $f(r)$ occurs at $r = 1/\kappa$, as is easy to verify (Problem 18-1). Thus the most important region of the ion atmosphere is in the neighborhood of $r = 1/\kappa$. This region moves to larger values of r as the solution becomes more dilute ($\kappa \to 0$). This confirms the fact that the behavior of u_{ij} near $r = a$ is unimportant for the limiting law.

To obtain some feeling for the orders of magnitude involved, we can calculate (Problem 18-2) that in an aqueous 1-1 electrolyte solution at 25°C ($\epsilon = 78.5$),

$$\frac{1}{\kappa} = \frac{3.04}{c^{1/2}} \quad \text{(in A)}, \tag{18-25}$$

where c is the concentration of salt in moles·liter^{-1}. Thus, if $c = 0.01$ moles·l^{-1}, $1/\kappa = 30.4$ A. By comparison, a is of order 3 or 4 A, so that $1/\kappa = O(10a)$ or $\kappa a = O(10^{-1})$ at $c = 0.01$ moles·l^{-1}. At about this dilution, then, we should expect the limiting law, (18-23), to begin to be rather accurate [again assuming approximations (a) and (b) above to be valid], and this is found to be the case. If $c = 10^{-4}$ moles·l^{-1}, $1/\kappa = 304$ A, etc.

At this point, we are in a position to go back to the question of the conditions under which the linear approximation, (b), made in writing Eq. (18-10), is legitimate. We require that

$$\frac{z_i \mathrm{e} \psi_j(r)}{kT} \ll 1.$$

As we have seen, the important values of r are those in the neighborhood of $r = 1/\kappa$. Therefore we put $r = 1/\kappa$ in Eq. (18-15), and obtain (dropping factors of order unity)

$$\frac{\mathrm{e}^2 \kappa}{\epsilon kT} \ll 1$$

as the necessary condition for the validity of the approximation. This is the same as

$$\frac{1}{\kappa} \gg \frac{\mathrm{e}^2}{\epsilon kT}. \tag{18-26}$$

For water at 25°C, this gives the condition $1/\kappa \gg 7.1$ A (Problem 18-3) (that is, $1/\kappa \gg 2a$). Hence we come to the conclusion that for $c < 0.01$ moles·l^{-1} (1-1 electrolyte), where $1/\kappa > 30$ A, the linear approximation should begin to be a good approximation. For $c < 10^{-3}$ or 10^{-4} moles·l^{-1}, it should be an excellent approximation.

The Debye-Hückel argument itself cannot provide any theoretical justification for approximation (a) in a dilute solution [$w_{ij}(r) \to z_i \mathrm{e} \psi_j(r)$

as $\kappa \to 0$], as it does for approximation (b) [see (18–26)]. However, as already mentioned, there is indirect experimental evidence supporting both (a) and (b). We shall return to (a) in Section 18–3. But let us note here what the Debye-Hückel theory has to say about $w_{ij}(r)$. From Eq. (18–15) we have

$$\begin{aligned} w_{ij}(r) &= +\infty \qquad (r < a) \\ &= z_i e \psi_j(r) = \frac{z_i z_j e^2 e^{-\kappa(r-a)}}{\epsilon r(1 + \kappa a)} \qquad (r \geq a). \end{aligned} \tag{18–27}$$

In very dilute solutions, $\kappa a \to 0$ and

$$w_{ij}(r) \to \frac{z_i z_j e^2 e^{-\kappa r}}{\epsilon r}. \tag{18–28}$$

As implied by earlier remarks, Eq. (18–28) proves to be exact, as a limiting law, but Eq. (18–27) is an approximation.

Incidentally, since by definition (see $g_{\alpha\beta}$ in Section 17–3) $w_{ij}(r) = w_{ji}(r)$, we can deduce the symmetry condition $z_i\psi_j(r) = z_j\psi_i(r)$ from approximation (a), $w_{ij} = z_i e\psi_j$, which was used in arriving at the Poisson-Boltzmann equation, (18–5) and (18–9). This symmetry condition is seen to be satisfied by Eq. (18–15), which is a solution of the *linearized* Poisson-Boltzmann equation. But when the complete (nonlinear) Poisson-Boltzmann equation is used, it turns out that the solution of this equation does *not* satisfy the symmetry condition in general. Thus we can be sure that approximation (a) is incorrect except when the electrolyte solution is dilute enough (Eq. 18–26) to justify linearizing the exponential in Eq. (18–7).

Finally, we turn to a consideration of thermodynamic functions (Problem 18–4) derivable from the Debye-Hückel limiting law, (18–23). From Eqs. (18–2), (18–4), and (18–23), we have

$$\frac{F}{kT} = \frac{1}{kT}\sum_i N_i\mu_i = \sum_i N_i \ln \Lambda_i^3 \rho_i + \frac{F_{\text{el}}}{kT}, \tag{18–29}$$

where

$$\frac{F_{\text{el}}}{kT} = -\frac{e^2\kappa}{2\epsilon kT}\sum_i N_i z_i^2 = -\frac{\kappa^3 V}{8\pi}. \tag{18–30}$$

We use F_{el} to represent the electrostatic contribution to F. To find the (osmotic) pressure, we employ Eq. (18–29) and the thermodynamic relationship,

$$V\left(\frac{\partial p}{\partial V}\right)_{\mathbf{N},T} = \left(\frac{\partial F}{\partial V}\right)_{\mathbf{N},T} = -\frac{kT}{V}\sum_i N_i + \frac{e^2\kappa}{4\epsilon V}\sum_i N_i z_i^2.$$

We multiply this equation by dV/V, integrate from V to ∞, and obtain

$$\frac{p}{kT} = \sum_i \rho_i - \frac{\kappa^3}{24\pi}. \tag{18–31}$$

The Helmholtz free energy is then

$$\frac{A}{kT} = \frac{F}{kT} - \frac{pV}{kT} = \sum_i N_i \ln\left(\frac{\Lambda_i^3 \rho_i}{e}\right) + \frac{A_{\mathrm{el}}}{kT}, \tag{18–32}$$

where

$$\frac{A_{\mathrm{el}}}{kT} = -\frac{\mathbf{e}^2\kappa}{3\epsilon kT}\sum_i N_i \mathbf{z}_i^2 = -\frac{\kappa^3 V}{12\pi}. \tag{18–33}$$

As a check on Eq. (18–33), we can make use of a different charging process, one in which we start with all ions discharged and then charge all ions at the same rate up to the final charge. Thus there is a single charging parameter ξ. In this process, if carried out reversibly with T, V, and the N_i held constant,

$$\Delta A = A_{\mathrm{el}}(\xi = 1) - A_{\mathrm{el}}(\xi = 0) = A_{\mathrm{el}}(\xi = 1) = W_{\mathrm{el}},$$

where W_{el} is the work done on the system in the charging process. When all ions are charged to the degree ξ, we have from Eq. (18–16) that

$$\begin{aligned}\varphi_j^{\mathrm{atm}}(0;\,\xi) &= -\frac{\xi \mathbf{z}_j \mathbf{e}\kappa(\xi)}{\epsilon} \\ &= -\frac{\xi^2 \mathbf{z}_j \mathbf{e}\kappa}{\epsilon},\end{aligned}$$

since $\kappa(\xi) = \xi\kappa$ according to Eq. (18–11). Then the work necessary to bring up an additional charge $\mathbf{z}_j\mathbf{e}\,d\xi$ to one j ion is $\varphi_j^{\mathrm{atm}}(0;\,\xi)\mathbf{z}_j\mathbf{e}\,d\xi$. Hence the total work necessary to charge up all the ions is

$$\begin{aligned}W_{\mathrm{el}} = A_{\mathrm{el}}(\xi = 1) &= \int_0^1 \sum_j N_j \varphi_j^{\mathrm{atm}}(0;\,\xi)\mathbf{z}_j\mathbf{e}\,d\xi \\ &= -\frac{\mathbf{e}^2\kappa}{3\epsilon}\sum_j N_j \mathbf{z}_j^2,\end{aligned}$$

in agreement with Eq. (18–33).

18–2 Kirkwood theory of solutions. In this section we extend the methods of Section 17–3 to solutions of monatomic molecules. As in that section, we use [in Eq. (18–37)] an intuitive argument (not employed by

Kirkwood*) to avoid involvement in the detailed formal properties of distribution functions. The results we find here will then be applied in the next section to electrolyte solutions in particular.

We begin with Eqs. (18–17) through (18–19), which apply to any solution of monatomic molecules:

$$\mu_i = kT \ln \Lambda_i^3 + kT \ln \rho_i + kT \ln \gamma_i, \qquad (18\text{–}34)$$

where $kT \ln \gamma_i$ is the work W_i, against intermolecular forces, required to add one i molecule to the system. We now derive an equation for W_i. The average number of molecules of species s in the element of volume $d\mathbf{r}_s$ in the neighborhood of a partially coupled molecule of species i is

$$\rho_s e^{-w_{is}(r_{is},\xi_i)/kT} \, d\mathbf{r}_s, \qquad (18\text{–}35)$$

where w_{is} is the potential of mean force. The contribution to W_i, arising from interactions with these molecules, when ξ_i is increased by $d\xi_i$, is (18–35) multiplied by $u_{is}(r_{is})\, d\xi_i$. Therefore, if we integrate over $\mathbf{r}_s$ and ξ_i, and sum over s, we have

$$W_i = kT \ln \gamma_i = \sum_s \rho_s \int_0^1 \int_V u_{is} e^{-w_{is}(\xi_i)/kT} \, d\mathbf{r}_s \, d\xi_i. \qquad (18\text{–}36)$$

This is the generalization of Eq. (17–16) for solutions.

Consider an element of volume $d\mathbf{r}_i$ near a fixed molecule of species j and the equilibrium between i molecules in $d\mathbf{r}_i$ and i molecules at a very large distance from the fixed j molecule. We can write

$$\begin{aligned} \mu_i(\infty) &= kT \ln \Lambda_i^3 + kT \ln \rho_i + kT \ln \gamma_i \\ &= \mu_i(r_{ij}) = kT \ln \Lambda_i^3 + kT \ln \rho_i^j(r_{ij}) + W_i^j(r_{ij}), \end{aligned} \qquad (18\text{–}37)$$

where ρ_i^j is the mean concentration of i molecules in $d\mathbf{r}_i$, and W_i^j represents the work that has to be done against intermolecular forces in order to add an i molecule to the system at $\mathbf{r}_i$.

The average number of s molecules in $d\mathbf{r}_s$, when a j molecule is fixed at $\mathbf{r}_j$ and a partially coupled i molecule is at $\mathbf{r}_i$, is

$$\rho_s e^{-w_{ij}^s(\mathbf{r}_{ij},\mathbf{r}_{sj},\xi_i)/kT} \, d\mathbf{r}_s,$$

where w_{ij}^s is a potential of mean force defined by

$$g^{[3]}(\mathbf{r}_i, \mathbf{r}_j, \mathbf{r}_s; \xi_i) = e^{-w_{ij}^s/kT}, \qquad (18\text{–}38)$$

---

* J. G. Kirkwood, *J. Chem. Phys.* **3**, 300 (1935).

in the notation of Section 17-3. The physical significance of w_{ij}^s is the following: it is the reversible work which must be done to move an s molecule from infinity through the solution up to the position $\mathbf{r}_s$, when a partially coupled (ξ_i) i molecule is fixed at $\mathbf{r}_i$ and a j molecule is fixed at $\mathbf{r}_j$ during the process. In the superposition approximation, (17-25), we have

$$w_{ij}^s(\xi_i) = w_{is}(\xi_i) + w_{js}, \tag{18-39}$$

but this is not true in general, because the effects of the molecular environments of the i and j molecules on the s molecule are not simply additive but perturb each other. In the limit as $\rho_i \to 0$ (all i), that is, a very dilute gas,

$$w_{ij}^s(\xi_i) \to \xi_i u_{is} + u_{js},$$

and (18-39) is correct (if the intermolecular potential energy is itself pairwise additive).

The total work of coupling i at $\mathbf{r}_i$ is then

$$W_i^j(r_{ij}) = u_{ij}(r_{ij}) + \sum_s \rho_s \int_0^1 \int_V u_{is} e^{-w_{ij}^s(\xi_i)/kT} \, d\mathbf{r}_s \, d\xi_i. \tag{18-40}$$

We also have the relation

$$\rho_i^j(r_{ij}) = \rho_i e^{-w_{ij}(r_{ij})/kT}. \tag{18-41}$$

On combining Eqs. (18-36), (18-37), (18-40), and (18-41), we obtain

$$w_{ij}(r_{ij}) = u_{ij}(r_{ij}) + \sum_s \rho_s \int_0^1 \int_V u_{is} [e^{-w_{ij}^s(\xi_i)/kT} - e^{-w_{is}(\xi_i)/kT}] \, d\mathbf{r}_s \, d\xi_i. \tag{18-42}$$

This is the generalization of Eq. (17-20) to solutions.

We can deduce a more general equation than (18-42), for $w_{ij}(\xi_i)$, in exactly the same way as we obtained Eq. (17-24) from Eq. (17-20), but we omit this derivation. The final result can easily be written down on inspection of Eqs. (17-20), (17-24), and (18-42).

Also, we can extend the above argument without difficulty to successively larger groups of fixed molecules. For example, if we couple a new molecule of species i near fixed j and k molecules, we find (Problem 18-5)

$$w_{jk}^i = w_{ij} + u_{ik} + \sum_s \rho_s \int_0^1 \int_V u_{is} [e^{-w_{ijk}^s(\xi_i)/kT} - e^{-w_{ij}^s(\xi_i)/kT}] \, d\mathbf{r}_s \, d\xi_i. \tag{18-43}$$

Finally, we note that Eq. (17-10) for the pressure of a monatomic fluid can easily be extended to fluid mixtures (Problem 18-6):

$$\frac{p}{kT} = \sum_i \rho_i - \frac{1}{6kT} \sum_{i,j} \rho_i \rho_j \int_0^\infty r_{ij} u'_{ij}(r_{ij}) g_{ij}(r_{ij}) 4\pi r_{ij}^2 \, dr_{ij}, \qquad (18\text{–}44)$$

where $g_{ij} = e^{-w_{ij}/kT}$ is a function of all the ρ's and T.

18–3 Electrolyte solutions. In this section we apply the equations of Section 18–2 to the idealized electrolyte model described in the first paragraph of Section 18–1. Here again the solvent, if any, is considered a nonmolecular continuum. We first write down general equations and then observe that the Debye-Hückel limiting relations furnish a solution of these general equations in the high dilution limit. This provides a more firm statistical-mechanical foundation for the Debye-Hückel limiting law than the Debye-Hückel argument (Section 18–1) itself.

We begin with Eqs. (18–37). It is now convenient for us to write W_i^j in the form

$$W_i^j(r_{ij}) = kT \ln \gamma_i^j(r_{ij}) + z_i \mathrm{e} \psi_j(r_{ij}) \qquad (r_{ij} > a), \qquad (18\text{–}45)$$

where $\psi_j(r_{ij})$ is the mean electrostatic potential a distance r_{ij} from a fixed j ion [with $\psi_j(\infty) = 0$], and where γ_i^j is a "local activity coefficient," defined by Eq. (18–45) in terms of W_i^j and ψ_j. The point of separating W_i^j in this way is that now Eqs. (18–37) give

$$\rho_i^j(r_{ij}) = \rho_i e^{-z_i \mathrm{e} \psi_j(r_{ij})} \left[\frac{\gamma_i}{\gamma_i^j(r_{ij})}\right] \qquad (r_{ij} > a), \qquad (18\text{–}46)$$

which (a) is analogous to an electrochemical phase equilibrium *with activity coefficients included* [see the discussion following Eq. (18–7)], and (b) furnishes the explicit correction factor by which the approximation (18–7) has to be multiplied to make it exact. Of course, the introduction of γ_i^j does not provide anything new and can be avoided, if desired, in view of Eq. (18–41):

$$e^{-w_{ij}(r_{ij})/kT} = \frac{e^{-z_i \mathrm{e} \psi_j(r_{ij})/kT} \gamma_i}{\gamma_i^j(r_{ij})} \qquad (r_{ij} > a). \qquad (18\text{–}47)$$

Qualitatively, we would expect $\gamma_i^j(r_{ij}) \neq \gamma_i$ for finite r_{ij} since (a) the mean local ionic strength at r_{ij} is different than in the bulk solution ($r_{ij} = \infty$), and (b) the atmosphere around an i ion at r_{ij} is not spherically symmetrical as it is at $r_{ij} = \infty$. The exact Poisson-Boltzmann equation is

$$\nabla^2 \psi_j(r) = -\frac{4\pi \mathrm{e}}{\epsilon} \sum_i z_i \rho_i e^{-w_{ij}(r)/kT} \qquad (r > a) \qquad (18\text{–}48)$$

$$= -\frac{4\pi \mathrm{e}}{\epsilon} \sum_i z_i \rho_i e^{-z_i \mathrm{e} \psi_j(r)/kT} \left[\frac{\gamma_i}{\gamma_i^j(r)}\right] \qquad (r > a). \qquad (18\text{–}49)$$

An explicit expression for $\psi_j(r_{ij})$, the mean electrostatic potential at $\mathbf{r}_i$ a distance r_{ij} from a fixed j ion, may be written as follows. The contribution to ψ_j owing to an s ion at $\mathbf{r}_s$ is $z_s\mathrm{e}/\epsilon r_{is}$. The contribution of all s ions in $d\mathbf{r}_s$ is then

$$\frac{z_s\mathrm{e}}{\epsilon r_{is}}\,\rho_s e^{-w_{sj}(r_{sj})/kT}.$$

Hence the total potential at $\mathbf{r}_i$, arising from the fixed j ion and its ion atmosphere, is

$$\psi_j(r_{ij}) = \frac{z_j\mathrm{e}}{\epsilon r_{ij}} + \sum_s \rho_s \int_V \frac{z_s\mathrm{e}}{\epsilon r_{is}}\, e^{-w_{sj}/kT}\, d\mathbf{r}_s. \tag{18-50}$$

This has to be the solution of the exact Poisson-Boltzmann equation, (18–48), but it is expressed in terms of the (in general) unknown functions w_{sj}.

The electrostatic potential at $r_{ij} = 0$, when the central j ion is charged to the degree ξ_j and owing to all ions other than j itself, is, from Eq. (18–50),

$$\psi_j^{\text{atm}}(0; \xi_j) = \sum_s \rho_s \int_{V'} \frac{z_s\mathrm{e}}{\epsilon r_{sj}}\, e^{-w_{sj}(\xi_j)/kT}\, d\mathbf{r}_s, \tag{18-51}$$

where V' means that the region $r_{sj} < a$ is excluded from the integration since no ions (other than j) are present there ($w_{sj} = +\infty$ for $r_{sj} < a$). If, in Eq. (18–36), we introduce Eq. (18–51) and use (18–1) for u_{is}, we have

$$kT \ln \gamma_i = \lambda_i + \int_0^1 z_i\mathrm{e}\psi_i^{\text{atm}}(0; \xi_i)\, d\xi_i, \tag{18-52}$$

where λ_i is a small term arising from the region $r_{is} < a$ [see, for example, Eq. (18–22)]. This is a generalized version of Eq. (18–20), used in the Debye-Hückel theory.

Equation (18–42) for w_{ij} is converted into an equation applicable to electrolytes by substitution of Eq. (18–1) for u_{ij} and u_{is}. An explicit equation for $\ln \gamma_i^j/\gamma_i$ in Eq. (18–47) can then be written down, using Eq. (18–42) for w_{ij} and Eq. (18–50) for ψ_j (Problem 18–7).

We now verify that the Debye-Hückel limiting expressions for a very dilute solution are consequences of the equations of the present section and Section 18–2. We seek a (limiting) solution of the series of equations of which Eqs. (18–42) and (18–43) are the first two members, using Eq. (18–1) for u_{ij}. We try a solution of the form

$$w_{ij}^s(\xi_i) = w_{is}(\xi_i) + w_{sj}, \tag{18-53}$$

$$w_{ijk}^s(\xi_i) = w_{is}(\xi_i) + w_{sj} + w_{sk}, \tag{18-54}$$

etc. (i.e., superposition), where

$$w_{ij}(r_{ij}) = \frac{\mathbf{z}_i\mathbf{z}_j\mathbf{e}^2 e^{-\alpha r_{ij}}}{\epsilon r_{ij}}. \tag{18–55}$$

Here α is a parameter whose value is to be determined. For arbitrary ξ_i, we replace $\mathbf{z}_i$ by $\mathbf{z}_i\xi_i$. Further, we expand all exponentials, $e^{-w_{ij}/kT}$, and retain only linear terms, pending establishment of the conditions that render this step legitimate (i.e., $w_{ij}/kT \ll 1$). We ignore small terms associated with the region $r < a$, because these prove to be negligible (as in the Debye-Hückel limiting law) under conditions such that the solution we find is valid. Substitution of Eqs. (18–53), (18–54), and $e^{-w_{ij}/kT} = 1 - (w_{ij}/kT)$ in Eq. (18–43) and in all higher members of the series, reduces these equations to Eq. (18–42), which, therefore, is the only one we need consider.

Equation (18–42) reads (after integrating over ξ_i)

$$w_{ij}(r_{ij}) = \frac{\mathbf{z}_i\mathbf{z}_j\mathbf{e}^2}{\epsilon r_{ij}} + \sum_s \rho_s \int_V \frac{\mathbf{z}_i\mathbf{z}_s\mathbf{e}^2}{\epsilon r_{is}}\left(-\frac{w_{sj}}{kT}\right) d\mathbf{r}_s. \tag{18–56}$$

Next, we substitute Eq. (18–55) in Eq. (18–56), factor out $\mathbf{z}_i\mathbf{z}_j\mathbf{e}^2/\epsilon$, and write $r_{ij} = r$, $r_{is} = u$ and $r_{js} = v$:

$$\frac{e^{-\alpha r}}{r} = \frac{1}{r} - \frac{\mathbf{e}^2}{\epsilon kT}\sum_s \rho_s\mathbf{z}_s^2 \int_V \frac{e^{-\alpha v}}{uv}\, d\mathbf{r}_s. \tag{18–57}$$

In Fig. 18–3, rotation of the element of area $dx\,dy$ about the x-axis sweeps out $d\mathbf{r}_s$, so that $d\mathbf{r}_s = 2\pi y\,dx\,dy$. We transform coordinates x and y to u and v, where

$$u^2 = x^2 + y^2, \qquad v^2 = y^2 + (r - x)^2.$$

The Jacobian of this transformation gives

$$dx\,dy = \frac{uv}{yr}\,du\,dv,$$

or

$$d\mathbf{r}_s = \frac{2\pi uv}{r}\,du\,dv.$$

The limits of integration we use are

$$0 \le v \le \infty,$$
$$|r - v| \le u \le r + v.$$

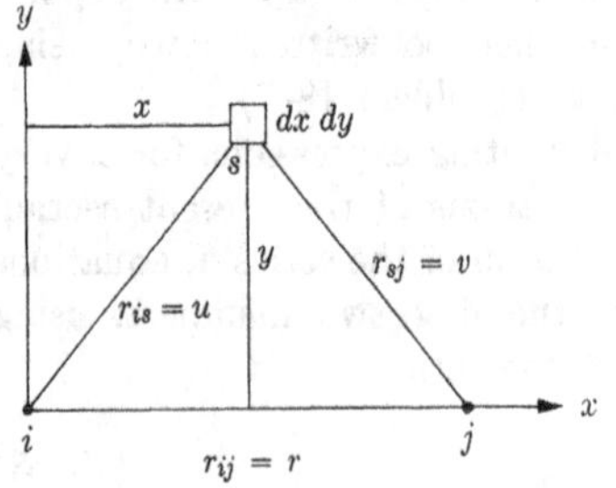

FIGURE 18–3

Equation (18-57) becomes, then,

$$\frac{e^{-\alpha r}}{r} = \frac{1}{r} - \frac{2\pi e^2}{\epsilon r kT} \sum_s \rho_s z_s^2 \iint, \tag{18-58}$$

where

$$\iint = \int_0^\infty e^{-\alpha v}\, dv \int_{|r-v|}^{r+v} du = \frac{2}{\alpha^2}(1 - e^{-\alpha r}).$$

Thus Eq. (18-58) is an identity [that is, Eqs. (18-53) through (18-55) are a solution of Eqs. (18-42), (18-43), etc.] *if*

$$\alpha^2 = \frac{4\pi e^2}{\epsilon kT} \sum_s \rho_s z_s^2,$$

that is, if $\alpha = \kappa$ (Eq. 18-11). Equation (18-55) turns out to be, then, just the Debye-Hückel limiting potential of mean force, (18-28).

From Eq. (18-55) with $\alpha = \kappa$, we find, as in the Debye-Hückel theory, that the ion atmosphere has an extent of order $r = 1/\kappa$ and hence that the linearization of $e^{-w_{ij}/kT}$ is legitimate if the solution is dilute enough to satisfy (18-26).

Let us confirm that the remaining Debye-Hückel results are also obtained. (a) When we put Eq. (18-55) (with $\alpha = \kappa$) in Eq. (18-50) for ψ_j, we find, using Eq. (18-58), that

$$\psi_j(r) = \frac{z_j e}{\epsilon r} e^{-\kappa r}.$$

Therefore, $w_{ij} = z_i e \psi_j$. However, on comparing Eqs. (18-42) and (18-50), we see that this is a limiting or asymptotic relationship. In general, $w_{ij} \neq z_i e \psi_j$ (see also Problem 18-7). (b) Equation (18-51) gives

$$\psi_j^{\text{atm}}(0; \xi_j) = -\frac{\xi_j z_j e \kappa}{\epsilon},$$

and hence, from Eq. (18-52),

$$kT \ln \gamma_i = -\frac{z_i^2 e^2 \kappa}{2\epsilon}.$$

(c) Finally, if we put Eqs. (18-1) and (18-55) (with $\alpha = \kappa$) in Eq. (18-44) for p/kT, we obtain (Problem 18-8) Eq. (18-31) again. Note that the integral in Eq. (18-44) converges because the potential of mean force, (18-55), is a short-range potential if $\kappa > 0$ (the ions i and j are both screened by ion atmospheres). But if we try to calculate second virial coefficients from Eq. (18-44) by taking the limit $\kappa \to 0$ in g_{ij} [see the paragraph following Eq. (17–10)], w_{ij} in (18–55) approaches u_{ij} and the integral diverges (the ions i and j are no longer screened by ion atmospheres). In summary: Eq. (18–44) can be used in electrolyte theory, but we cannot expand g_{ij} in this equation as a power series in the concentrations for this would lead to a virial expansion that diverges.

Important, but advanced, recent papers on electrolyte theory have been published by Mayer,* Kirkwood and Poirier,† and Meeron.‡

Problems

18–1. Show that the most probable value of r in the Debye-Hückel ion atmosphere, (18–24), is $1/\kappa$. Find also the mean value of r, $\bar{r}$, in terms of κ. (Page 329.)

18–2. Verify Eq. (18–25) for $1/\kappa$ in water at 25°C. (Page 329.)

18–3. Show that the condition (18–26) amounts to $1/\kappa \gg 7.1$ A for water at 25°C. (Page 329.)

18–4. Verify the details in the derivation of the thermodynamic functions (18–29) through (18–33). (Page 330.)

18–5. Derive Eq. (18–43) for w^i_{jk}. (Page 333.)

18–6. Extend the derivation of Eq. (17–10) to the pressure of a monatomic fluid mixture. (Page 333.)

18–7. Show that

$$kT \ln\left[\frac{\gamma^j_i(r_{ij})}{\gamma_i}\right] = \text{small terms} + \sum_s \rho_s \int_0^1 \int_{V'} \frac{z_i z_s e^2}{\epsilon r_{is}}$$

$$\times \left[e^{-w^s_{ij}(\xi_i)/kT} - e^{-w_{sj}/kT} - e^{-w_{is}(\xi_i)/kT}\right] dr_s\, d\xi_i. \qquad (18\text{–}59)$$

Verify that $\gamma^j_i/\gamma_i = 1$ in the Debye-Hückel limiting law concentration region. (Page 335.)

18–8. Deduce the Debye-Hückel limiting expression for p/kT, (18–31), from Eqs. (18–44) and (18–55). (Page 337.)

18–9. Consider a charge $z_j e$ at the center of a sphere of radius a and with dielectric constant ϵ_{in}. Outside the sphere is a dilute electrolyte of point ions in a medium of dielectric constant ϵ. Use the linear Poisson-Boltzmann equation outside the sphere. Show that the work done in charging up the sphere (keeping the electrolyte fully charged) against the ion atmosphere and dielectrics, is

$$\frac{1}{2}\frac{z_j^2 e^2}{\epsilon a}\left(\frac{1}{1 + \kappa a} - \frac{\epsilon}{\epsilon_{\text{in}}}\right).$$

---

* J. E. Mayer, *J. Chem. Phys.* **18**, 1426 (1950).

† J. G. Kirkwood and J. C. Poirier, *J. Phys. Chem.* **58**, 591 (1954).

‡ E. Meeron, *J. Chem. Phys.* **28**, 630 (1958).

18–10. This is the same as Problem 18–9 except that the charge $z_j e$ is smeared uniformly over the *surface* of the sphere. Show that the work here is

$$\frac{1}{2}\,\frac{z_j^2 e^2}{\epsilon a(1+\kappa a)}.$$

18–11. This is the same as Problem 18–9 except that the charge $z_j e$ is smeared uniformly throughout the *volume* of the sphere. Show that the work here is

$$\frac{1}{2}\,\frac{z_j^2 e^2}{\epsilon a}\left(\frac{1}{1+\kappa a}+\frac{\epsilon}{5\epsilon_{\text{in}}}\right).$$

18–12. Show from Eq. (18–24) that the integral of the charge density $n_j(r)$ in the Debye-Hückel theory, over all space, gives $-z_j e$, as expected.

18–13. Show that in the Debye-Hückel ion atmosphere, (18–24), the fraction of net charge between a and $2a$, in the limit as $\kappa a \to 0$, is $3(\kappa a)^2/2$.

18–14. Use Eq. (18–23) to calculate γ_{++} in a 2–2 aqueous electrolyte at 25°C and at a concentration 10^{-4} moles·l^{-1}.

18–15. Find the expression for $1/\kappa$, equivalent to (18–25), which is applicable for (a) a 2–2 electrolyte, (b) a 3–3 electrolyte, and (c) a 1–1 electrolyte in solvents with (*i*)$\epsilon = 10$ and (*ii*)$\epsilon = 1$ (solvent = vacuum; i.e., a plasma). In each case find the value of c giving $1/\kappa = 30$ A.

18–16. Calculate and compare the magnitude of terms in Eqs. (18–22) and (18–31) for a 1–1 aqueous electrolyte at 25°C and a concentration 10^{-3} moles·l^{-1}. Take $a = 3$ A.

18–17. Verify the evaluation of the integral in Eq. (18–58).

18–18. Show that Eq. (18–36) leads to

$$\ln \gamma_2 = B_{11}\rho_1 + 2B_{02}\rho_2 + \cdots$$

for a dilute binary gas mixture, a result used in Eq. (18–21) (obtained from Chapter 15).

Supplementary Reading

FOWLER and GUGGENHEIM, Chapter 9.

HARNED, H. S., and OWEN, B. B., *Physical Chemistry of Electrolytic Solutions*, 3rd ed., New York: Reinhold, 1958.

KIRKWOOD, J. G., *J. Chem. Phys.* **3**, 300 (1935); *Chem. Revs.* **19**, 275 (1936).

LANDAU and LIFSHITZ, Sections 74 and 91.

CHAPTER 19

DILUTE LIQUID SOLUTIONS

The main thesis of the present chapter is that the expansion methods of imperfect gas theory (Chapter 15) can be applied to a liquid solution which is dilute with respect to at least one ("solute") component, provided that the interaction between solute molecules is a short-range one (energy $\propto r^{-n}$, $n > 3$, for large r). The technique is to reduce the many-body problem to, successively, (a) the interaction of one solute molecule with pure solvent (this determines the Henry's law constant, for example), (b) the interaction between two solute molecules immersed in pure solvent, etc. This procedure is *not* possible, as we saw in Chapter 18, with the long-range forces in an electrolyte solution.

The same expansion methods can be applied to gaseous and solid solutions, but these cases are less important than liquid solutions.

The expansions encountered in a particular solution theory depend on the choice of independent variables and ensemble. The expansions of the different theories may be interconverted by suitable thermodynamic manipulations, so any formally exact theory is in principle equivalent to any other one. But, generally speaking, if one is interested in the expansion of a given dependent variable as a function of a certain set of independent variables, the simplest result will be obtained if the ensemble is suitably chosen to yield the expansion directly—instead of indirectly by thermodynamic operations on series obtained from another ensemble. Hence there is some point in discussing more than one dilute solution theory.

We shall emphasize in this chapter, rather arbitrarily, the McMillan-Mayer theory, which is particularly suitable for an investigation of osmotic systems and the osmotic pressure. Section 19–1 contains a general discussion of this theory, while Section 19–2 is devoted to particular applications. For contrast, we consider in Section 19–3 a solution theory which is especially appropriate for solutions at fixed pressure and temperature with molality as composition variable. This is a very common choice of independent variables in practical solution thermodynamics.

An important alternative approach to solution theory, which we shall not include, is that due to Kirkwood and Buff.*

---

* J. G. Kirkwood and F. P. Buff, *J. Chem. Phys.* **19**, 774 (1951).

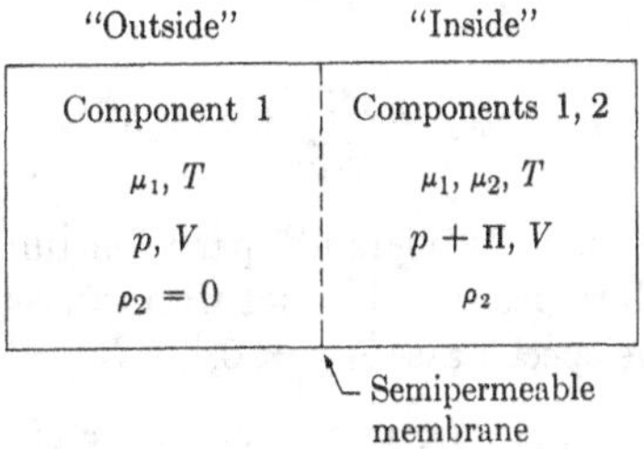

FIG. 19–1. Osmotic equilibrium system. The membrane passes component 1 (solvent) but not component 2 (solute).

19–1 McMillan-Mayer solution theory. We begin by considering a two-component solution which may contain monatomic or polyatomic molecules, and which may obey classical or quantum statistics. Component 1 is the solvent and component 2 the solute, the dilute component. We assume short-range forces between solute molecules, as already explained above. Our object is to choose partition functions in such a way that we can develop a treatment here for "solute in solvent" which exactly parallels the treatment of "gas in vacuum" in Section 15–1. For this reason, the reader should review Section 15–1. The derivation we give* is chosen for its relative simplicity, rather than for its generality.†

The two-component solution referred to above (the "inside" solution of Fig. 19–1) is in osmotic equilibrium with pure solvent ("outside" in Fig. 19–1). That is, a semipermeable membrane, permeable to solvent only, separates solvent from solution. The outside system has specified values of the chemical potential μ_1 and temperature T. The values of μ_1 and T fix the pressure p. The "inside" solution is also at temperature T and contains solvent at μ_1 and solute at μ_2. The values of T, μ_1, and μ_2 then determine, for example, the inside pressure $p + \Pi$ and solute number density ρ_2. Π is the osmotic pressure—the extra pressure needed on the inside (in the presence of solute which tends to lower μ_1) to give μ_1 the same value inside as outside.

The grand partition function for the "inside" solution is

$$e^{(p+\Pi)V/kT} = \sum_{N_1,N_2\geq 0} Q_{N_1,N_2}(V, T)\lambda_1^{N_1}\lambda_2^{N_2}$$

$$= \sum_{N_2\geq 0} \Psi_{N_2}(\mu_1, V, T)\lambda_2^{N_2}, \qquad (19\text{–}1)$$

* T. L. HILL, *J. Chem. Phys.* **30**, 93 (1959).

† More general and complicated derivations will be found in S.M., Chapter 6, and in McMillan and Mayer (Supplementary Reading list).

where

$$\Psi_{N_2} = \sum_{N_1 \geq 0} Q_{N_1,N_2}\lambda_1^{N_1}. \tag{19-2}$$

Here $\Psi_{N_2}(\mu_1, V, T)$ is a "semigrand" partition function for a system at V and T, open with respect to 1 but not with respect to 2 (Problem 19–1). We note that in the special case $N_2 = 0$,

$$\Psi_0 = \sum_{N_1 \geq 0} Q_{N_1}\lambda_1^{N_1} = e^{pV/kT}, \tag{19-3}$$

which is the grand partition function for the outside system. We shall be interested in small values of N_2, that is, Ψ_1, Ψ_2, Ψ_3, etc., for solvent systems (μ_1, V, T) with only one, two, three, etc., solute molecules immersed in the solvent. We now have to digress to introduce activities, the Henry's law constant, etc.

If, as in Section 15–3, we define activities z_1 and z_2 by Eqs. (15–41), then $z_1 \rightarrow \rho_1$ and $z_2 \rightarrow \rho_2$ in the limit $\rho_1, \rho_2 \rightarrow 0$ (ideal gas mixture). In this limit

$$\mu_1 = \mu_1^0(T) + kT \ln p_1,$$

$$\mu_2 = \mu_2^0(T) + kT \ln p_2,$$

where

$$\mu_1^0(T) = kT \ln \left(\frac{V}{kTQ_{10}}\right), \qquad \mu_2^0(T) = kT \ln \left(\frac{V}{kTQ_{01}}\right). \tag{19-4}$$

Examples will be found in Eqs. (4–26), (8–40), and Problem 9–3.

It is more convenient for the present problem, however, to define a solute activity $\mathfrak{z}_2$, proportional to z_2, but with the property that $\mathfrak{z}_2 \rightarrow \rho_2$ when $\rho_2 \rightarrow 0$ *in the inside solution* (i.e., with μ_1 and T fixed). In this limit we have pure solvent at μ_1 and T, and not an ideal gas mixture, so z_2 does not approach ρ_2 but approaches instead, say, $\gamma_2^0\rho_2$, which defines $\gamma_2^0(\mu_1, T)$. This is a limiting value of the activity coefficient γ_2 defined by $z_2 = \gamma_2\rho_2$ (Eq. 15–47). Hence the connection between $\mathfrak{z}_2$ and z_2 is $z_2 = \gamma_2^0(\mu_1, T)\mathfrak{z}_2$. We shall encounter explicit expressions for γ_2^0 below.

The inside solution in the limit $\rho_2 \rightarrow 0$ is often referred to as the "infinitely dilute solution." When $\rho_2 = 0$, the inside solution is the same as the solvent (outside).

The fugacity of the solute in the solution (at any composition) is defined as usual by $f_2 = z_2kT$. (That is, if the solute here is in equilibrium with an ideal gas phase in which the partial pressure of component 2 is p_2, then f_2 in solution is equal to p_2 in the ideal gas, since z_2 is the same in the two phases at equilibrium.) Now we define the Henry's law constant $k_2(\mu_1, T)$ for the solute in the solvent at μ_1, T by the statement that $f_2 \rightarrow k_2x_2$ as $x_2 \rightarrow 0$, where x_2 is the mole fraction of the solute. In this limit,

$x_2 \to \rho_2 v_1(\mu_1, T)$, where v_1 is the volume per molecule of pure solvent at μ_1, T. Therefore, in the limit as $\rho_2 \to 0$, we have

$$z_2 = \gamma_2^0 \rho_2 = \frac{k_2 \rho_2 v_1}{kT},$$

or

$$k_2(\mu_1, T) = \frac{\gamma_2^0(\mu_1, T)kT}{v_1(\mu_1, T)}, \tag{19–5}$$

which gives k_2 in terms of γ_2^0.

Returning to the activity $\mathfrak{z}_2$, we shall verify below that the definition

$$\mathfrak{z}_2 \equiv \frac{\lambda_2 \Psi_1}{\Psi_0 V} \tag{19–6}$$

gives, as required, $\mathfrak{z}_2 \to \rho_2$ as $\rho_2 \to 0$ in the solution. Accepting this definition tentatively, we find that

$$z_2 = \frac{Q_{01}\lambda_2}{V} = \gamma_2^0 \mathfrak{z}_2 = \frac{\gamma_2^0 \lambda_2 \Psi_1}{\Psi_0 V},$$

or

$$\gamma_2^0 = \frac{Q_{01}\Psi_0}{\Psi_1} = \frac{Ve^{-\mu_2^0/kT}\Psi_0}{kT\Psi_1}. \tag{19–7}$$

Since γ_2^0 is a function of μ_1 and T only and Q_{01} is proportional to V, Ψ_1/Ψ_0 must also be proportional to V. We see that γ_2^0 depends on the properties of one solute molecule in a vacuum (i.e., on Q_{01}) and on the nature of the interaction between one solute molecule and pure solvent (i.e., on Ψ_1/Ψ_0). We shall examine a special case of Eq. (19–7) in Problem 19–2. Equation (19–7) gives for k_2,

$$k_2 = \frac{Q_{01}\Psi_0 kT}{\Psi_1 v_1}. \tag{19–8}$$

From the relations

$$\frac{Q_{01}\lambda_2}{V} = z_2 \to \rho_2 \qquad \text{(component 2 in vacuum)},$$

$$\frac{Q_{01}\lambda_2}{\gamma_2^0 V} = \mathfrak{z}_2 \to \rho_2 \qquad \text{(component 2 in solvent)},$$

we see that Q_{01}/γ_2^0 plays the role of an "effective" partition function for one solute molecule in a volume V filled by solvent at μ_1 and T (analogous to Q_{01} for one molecule of component 2 in a volume V otherwise unoccupied). The factor $1/\gamma_2^0$ thus takes care of the "contact" of the solute molecule with solvent. This gives some indication of the physical significance of γ_2^0 (see also Problem 19–2).

Another interpretation of γ_2^0 is the following:

$$\frac{\mathfrak{z}_2}{z_2} = \frac{\rho_2 \text{ (solute in infinitely dilute solution)}}{\rho_2 \text{ (solute in infinitely dilute gas)}} = \frac{Q_{01}/\gamma_2^0 V}{Q_{01}/V}$$

$$= \frac{1}{\gamma_2^0(\mu_1, T)} = K(\mu_1, T).$$

Comparison with Eq. (10–6) shows that $1/\gamma_2^0 = K(\mu_1, T)$ is the "equilibrium constant" for the process

$$\text{solute in gas} \rightleftarrows \text{solute in solution.}$$

The numerical value of γ_2^0 depends on how a single solute molecule "likes" (including both energy and entropy effects) being in a vacuum relative to pure solvent.

We now return to the expansion (19–1). To simplify the notation we use N for N_2 (number of solute molecules), replace λ_2 by $\mathfrak{z}_2$ (Eq. 19–6) and Ψ_N by Z_N^*, where $Z_N^*(\mu_1, V, T)$ is defined by

$$Z_N^* = \frac{N!\Psi_N\Psi_0^{N-1}V^N}{\Psi_1^N}. \tag{19–9}$$

Note that $Z_1^* = V$. The definition (19–9) is constructed so that Z_N^* plays the same formal role here for N solute molecules immersed in the solvent as Z_N does in Section 15–1 for N molecules in a vacuum (one-component gas). This will become more apparent as we proceed. Then Eq. (19–1) becomes

$$e^{(p+\Pi)V/kT} = \Psi_0 + \Psi_0 \sum_{N\geq 1} \frac{Z_N^*}{N!} \mathfrak{z}_2^N.$$

Using Eq. (19–3), we have

$$e^{\Pi V/kT} = 1 + \sum_{N\geq 1} \frac{Z_N^*(\mu_1, V, T)}{N!} \mathfrak{z}_2^N. \tag{19–10}$$

This equation is seen to be formally identical with Eq. (15–6) for a one-component gas: Π replaces p, Z_N^* replaces Z_N, and $\mathfrak{z}_2$ replaces z. (We still have to check later that $\mathfrak{z}_2 \to \rho_2$ as $\rho_2 \to 0$ in the solution, just as $z \to \rho$ as $\rho \to 0$ in the gas.) The solute is thus treated as a quasi-one-component system, with the solvent playing an implicit background role through its influence on Z_N^*. We shall be able to put this in more physical terms below for a special case. Incidentally, this point of view is obviously the same as that adopted for the solvent, without formal justification, in Chapter 18 (on electrolyte theory).

Equation (15–6) is a special case of Eq. (19–10): if the "solvent" is a vacuum, then the outside pressure p in Fig. 19–1 is zero and Π is the total pressure of the gas (solute) on the inside.

We can now manipulate Eq. (19–10) just as we did Eq. (15–6). On taking the logarithm of Eq. (19–10) and expanding [see also the discussion of Eqs. (15–14) through (15–16)], we find

$$\frac{\Pi}{kT} = \sum_{j\geq 1} b_j^*(\mu_1, T)\mathfrak{z}_2^j, \tag{19–11}$$

where the b_j^* are related to the Z_N^* by Eqs. (15–8). From the thermodynamic equation [see (1–65)]

$$\overline{N}_2 = \left[\frac{\partial(p + \Pi)V}{\partial\mu_2}\right]_{\mu_1,T,V},$$

we have (p is determined by μ_1 and T only and hence is constant)

$$\rho_2 = \frac{\overline{N}_2}{V} = \mathfrak{z}_2\left(\frac{\partial\Pi/kT}{\partial\mathfrak{z}_2}\right)_{\mu_1,T} = \sum_{j\geq 1} jb_j^*(\mu_1, T)\mathfrak{z}_2^j, \tag{19–12}$$

just as in Eq. (15–10). We finally confirm here the property already used above that $\mathfrak{z}_2 \to \rho_2$ as $\rho_2 \to 0$, since $b_1^* = 1$. Then, as in Eq. (15–12), we have the osmotic pressure virial expansion

$$\frac{\Pi}{kT} = \rho_2 + \sum_{n\geq 2} B_n^*(\mu_1, T)\rho_2^n, \tag{19–13}$$

where the B_n^* are related to the b_j^* by Eqs. (15–13). In the limit $\rho_2 \to 0$, $\Pi/kT \to \rho_2$.

From the general thermodynamic equation, for a binary solution,

$$N_1\,d\mu_1 + N_2\,d\mu_2 = -S\,dT + V\,dp,$$

we have in our special case and notation,

$$\overline{N}_2\,d\mu_2 = V\,d(p + \Pi) = V\,d\Pi \qquad (\mu_1,\ T \text{ constant}).$$

Therefore

$$\left(\frac{\partial\mu_2/kT}{\partial\rho_2}\right)_{\mu_1,T} = \frac{1}{\rho_2}\left(\frac{\partial\Pi/kT}{\partial\rho_2}\right)_{\mu_1,T}. \tag{19–14}$$

We substitute Eq. (19–13) here for Π/kT, carry out the differentiation, multiply by $d\rho_2$, and finally integrate. The result is

$$\frac{\mu_2}{kT} = \text{constant} + \ln\rho_2 + \sum_{k\geq 1}\left(\frac{k+1}{k}\right)B_{k+1}^*\rho_2^k, \tag{19–15}$$

which is the analog of Eq. (15–19). To evaluate the integration constant, we recall that in the limit $\rho_2 \to 0$,

$$z_2 = \frac{Q_{01}e^{\mu_2/kT}}{V} = \gamma_2^0 \mathfrak{z}_2 \to \gamma_2^0 \rho_2.$$

From this relation we find that

$$\text{constant} = \ln \frac{V\gamma_2^0}{Q_{01}} = \ln \frac{V\Psi_0}{\Psi_1} = \frac{\mu_2^0}{kT} + \ln k_2 v_1.$$

Therefore Eq. (19–15) can be written

$$\frac{\mu_2}{kT} = \ln \frac{V\Psi_0}{\Psi_1} + \ln \rho_2 + \ln \bar{\gamma}_2, \tag{19–16}$$

where

$$\ln \bar{\gamma}_2 = -\sum_{k\geq 1} \beta_k^*(\mu_1, T)\rho_2^k, \tag{19–17}$$

$$\beta_k^* = -\left(\frac{k+1}{k}\right) B_{k+1}^*. \tag{19–18}$$

The activity coefficient $\bar{\gamma}_2$ (the bar does not mean an average value) is obviously defined by the equation $\mathfrak{z}_2 = \rho_2\bar{\gamma}_2$ and has the property, desirable for a practical activity coefficient, that $\bar{\gamma}_2 \to 1$ as $\rho_2 \to 0$ (μ_1 and T held constant). The relation between γ_2 and $\bar{\gamma}_2$ is

$$z_2 = \gamma_2\rho_2 = \gamma_2^0\mathfrak{z}_2 = \gamma_2^0\bar{\gamma}_2\rho_2,$$

or

$$\bar{\gamma}_2 = \frac{\gamma_2}{\gamma_2^0}. \tag{19–19}$$

The treatment above parallels completely that in Section 15–1 for a one-component imperfect gas. Here, when $\rho_2 \to 0$, interaction between one molecule and pure solvent is responsible for the term $\ln (V\Psi_0/\Psi_1)$ in Eq. (19–16) for μ_2 but does not influence the "equation of state," $\Pi = \rho_2 kT$; at higher concentrations, B_2^*, which depends on the properties of a pair of solute molecules in pure solvent, contributes to μ_2 and Π/kT; etc. The quantities Z_N^*, b_j^*, B_n^*, β_k^* have a significance for solute molecules in solvent which is identical with that of Z_N, b_j, B_n, and β_k for gas molecules in vacuum.

The above equations are necessarily quite formal because of their generality. To give a better physical understanding of their significance, we now turn to the simplest special case, a classical binary solution of monatomic molecules, and introduce the potential of mean force.

Classical binary solution of monatomic molecules. We first have to extend Eq. (17-35) for $g^{(n)} = e^{-w^{(n)}/kT}$ to a binary system. We let $\{n_1\}$ represent the spatial positions $\mathbf{r}_1, \ldots, \mathbf{r}_{n_1}$ of a set of n_1 molecules of component 1, and let $d\{n_1\}$ represent $d\mathbf{r}_1 \ldots d\mathbf{r}_{n_1}$. Then, as in Eq. (17-33), the probability that, in a *closed* binary system with numbers of molecules N_1 and N_2, any n_1 molecules of component 1 will be found in $d\{n_1\}$ and any n_2 of component 2 in $d\{n_2\}$ is

$$\frac{N_1!}{(N_1 - n_1)!} \frac{N_2!}{(N_2 - n_2)!} \times \frac{d\{n_1\}\, d\{n_2\} \int_V e^{-U_{N_1N_2}/kT}\, d\{N_1 - n_1\}\, d\{N_2 - n_2\}}{Z_{N_1N_2}}. \tag{19-20}$$

The probability that an *open* system characterized by z_1, z_2, V, T contains the numbers of molecules N_1, N_2 is

$$\frac{z_1^{N_1} z_2^{N_2} Z_{N_1N_2}}{N_1! N_2! \Xi}. \tag{19-21}$$

If we take the product of (19-20) and (19-21) and sum over N_1 and N_2, we have the probability that any n_1 molecules of 1 are in $d\{n_1\}$ and any n_2 of 2 are in $d\{n_2\}$, in an open system. We denote this probability by $\rho_1^{n_1}\rho_2^{n_2} g^{(\mathbf{n})}\, d\{n_1\}\, d\{n_2\}$, which defines the distribution function $g^{(\mathbf{n})}$ ($\mathbf{n}$ refers to the set of numbers n_1, n_2—a general notational system we have been using throughout most of this book), a function of z_1, z_2, and T as well as of $\{n_1\}$ and $\{n_2\}$. Thus we have

$$\rho_1^{n_1}\rho_2^{n_2} g^{(\mathbf{n})} = \frac{1}{\Xi} \sum_{\substack{N_1 \geq n_1 \\ N_2 \geq n_2}} \frac{z_1^{N_1} z_2^{N_2}}{(N_1 - n_1)!(N_2 - n_2)!} \times \int_V e^{-U_{N_1N_2}/kT}\, d\{N_1 - n_1\}\, d\{N_2 - n_2\}, \tag{19-22}$$

which is the generalization of Eq. (17-35). Since N_1 and N_2 are merely dummy summation indices, we can change notation by putting N_1 for $N_1 - n_1$, $N_1 + n_1$ for N_1, etc. This gives

$$g^{(\mathbf{n})}(\mathbf{z}, \{\mathbf{n}\}, T) = \frac{\gamma_1^{n_1}\gamma_2^{n_2}}{\Xi} \sum_{\mathbf{N} \geq 0} \frac{z_1^{N_1} z_2^{N_2}}{N_1! N_2!} \int_V \exp\left(-U_{\mathbf{N}+\mathbf{n}}/kT\right) d\{\mathbf{N}\}. \tag{19-23}$$

where $\mathbf{z}$ refers to the activity set z_1, z_2.

Now we define $w^{(\mathbf{n})}$ by

$$g^{(\mathbf{n})} = \exp\left[-w^{(\mathbf{n})}/kT\right], \tag{19-24}$$

substitute this expression in Eq. (19-23), and integrate over $\{\mathbf{n}\}$. The result is

$$\int_V \exp[-w^{(\mathbf{n})}/kT]\,d\{\mathbf{n}\} = \frac{\gamma_1^{n_1}\gamma_2^{n_2}\sum_{\mathbf{N}\geq 0}(z_1^{N_1}z_2^{N_2}/N_1!N_2!)Z_{\mathbf{N}+\mathbf{n}}}{\sum_{\mathbf{N}\geq 0}(z_1^{N_1}z_2^{N_2}/N_1!N_2!)Z_{\mathbf{N}}}. \qquad (19\text{–}25)$$

As in Section 17–4, $w^{(\mathbf{n})}$ is the potential of the mean force acting between the set of molecules $\mathbf{n}$ located at $\{\mathbf{n}\}$ in the solution characterized by $\mathbf{z}$, V, T. We are particularly interested in the special case of a set of N solute molecules only, $\mathbf{n} = 0, N$, in the limit as $\mathbf{z} \rightarrow z_1, 0$ (i.e., N solute molecules immersed in pure solvent at z_1, T—the outside system in Fig. 19–1). In this limit, only terms with $N_2 = 0$ contribute in the sums of Eq. (19–25), so we find

$$\int_V e^{-w^{(N)}/kT}\,d\{N\} = \frac{\gamma_2^{0^N}\sum_{N_1\geq 0}(z_1^{N_1}/N_1!)Z_{N_1N}}{\sum_{N_1\geq 0}(z_1^{N_1}/N_1!)Z_{N_1}}. \qquad (19\text{–}26)$$

We shall make use of this result presently.

If we substitute Eq. (19–7) in Eq. (19–9), we obtain

$$Z_N^* = \frac{N!\Psi_N V^N}{\Psi_0}\left(\frac{\Psi_0}{\Psi_1}\right)^N = \frac{N!\Psi_N \gamma_2^{0^N}\Lambda_2^{3N}}{\Psi_0}. \qquad (19\text{–}27)$$

Also, from Eq. (19–2), we have

$$\Psi_N = \frac{1}{\Lambda_2^{3N}N!}\sum_{N_1\geq 0}\frac{z_1^{N_1}Z_{N_1N}}{N_1!}. \qquad (19\text{–}28)$$

Using Eq. (19–28), we eliminate Ψ_N and Ψ_0 from Eq. (19–27) and observe that Z_N^* is the same as the right side of Eq. (19–26). Therefore,

$$Z_N^* = \int_V e^{-w^{(N)}/kT}\,d\{N\}. \qquad (19\text{–}29)$$

Equation (19–29) should be compared with Z_N for a classical one-component monatomic gas (Section 15–2):

$$Z_N = \int_V e^{-U_N/kT}\,d\{N\}. \qquad (19\text{–}30)$$

We see that, for classical monatomic systems† at least, the analogy between the formal significance of Z_N^*, b_j^*, etc., for a dilute solution and Z_N, b_j, etc., for a dilute gas, already emphasized above, extends even further: U_N in Eq. (19–30) is the potential of the force between N gas

† Actually the analogy between $w^{(N)}$ and U_N is much more general. It extends to systems of polyatomic molecules of any complexity in which vibration is separable and in which rotation (internal and external) and translation are classical. See S. M., Section 40.

molecules in a vacuum, while $w^{(N)}$ in Eq. (19–29) is the potential of the (mean) force between N solute molecules in the pure solvent. Thus the *physical* significance, in terms of force or work, of $w^{(N)}$ and U_N in the respective cases is identical.

This analogy makes it possible to write down immediately, by inspection, a number of equations for dilute solutions. For example, from Eq. (15–24) for a one-component gas, we have

$$B_2^*(z_1, T) = -\tfrac{1}{2}\int_0^\infty [e^{-w^{(2)}(r, z_1, T)/kT} - 1]4\pi r^2\, dr. \qquad (19\text{–}31)$$

Similarly, from Eq. (15–30), we have an equation for $B_3^*(z_1, T)$. It should be emphasized here, though, that the assumption

$$U_3 = u_{12} + u_{13} + u_{23}, \qquad (19\text{–}32)$$

used in Eq. (15–30), is in a different class than

$$w^{(3)} = w_{12} + w_{13} + w_{23}, \qquad (19\text{–}33)$$

needed for B_3^*. Equation (19–32) is a quantum-mechanical approximation, and a good one in general, but Eq. (19–33) is statistical mechanical (the w's are functions of z_1 and T) in origin and not necessarily a very good approximation. Equation (19–33) is in fact the superposition approximation of Chapters 17 and 18. Another example is Eq. (17–10):

$$\frac{\Pi}{kT} = \rho_2 - \frac{\rho_2^2}{6kT}\int_0^\infty rw'(r, T, z_1, 0)e^{-w(r,T,z_1,z_2)/kT}4\pi r^2\, dr, \qquad (19\text{–}34)$$

where $w = w^{(2)}$ and here the solution is not necessarily dilute. The superposition approximation (pairwise additivity) has been made use of in $w(z_1, 0)$ but not in $w(z_1, z_2)$. The extension of Eq. (19–34) to a multicomponent solute is obvious from Eq. (18–44). It is this extension which is in reality involved in Problem 18–8 for the Debye-Hückel p (actually, Π), if the dielectric is not a vacuum.

In view of the fact that the solvent does not play an explicit role in Eq. (19–29) and in view of the interpretation of $w^{(N)}$ as the potential of mean force between solute molecules immersed in the outside solution (solvent) of Fig. 19–1, it is physically obvious (and can easily be confirmed†) that whether the solvent is made up of one component or many components, formal relations such as Eq. (19–29) and those between Z_N^*, b_j^*, B_n^*, etc., are unchanged. A multicomponent solvent influences the system indirectly or implicitly through its influence on $w^{(N)}$, which is now a function of T and one z for each solvent species (we denote this

† See S. M., Section 40.

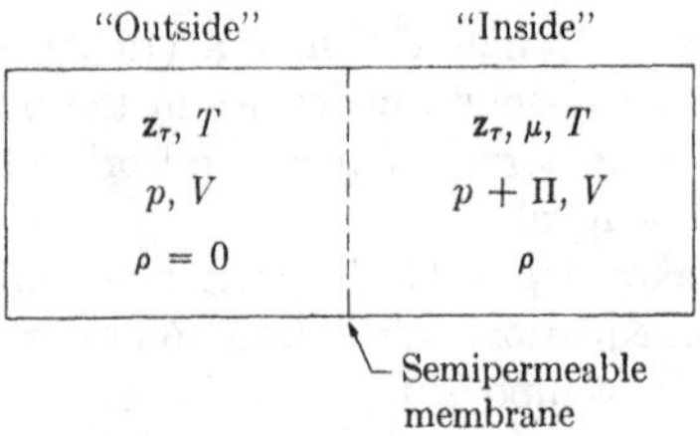

FIG. 19–2. Osmotic equilibrium system. The membrane passes all solvent species (subscript τ) but not the solute (no subscript).

set of solvent z's by $\mathbf{z}_\tau$). For example, Eqs. (19–13) and (19–17) become

$$\frac{\Pi}{kT} = \rho + \sum_{n \geq 2} B_n^*(\mathbf{z}_\tau, T)\rho^n, \tag{19–35}$$

$$\ln \bar{\gamma} = \sum_{k \geq 1} \left(\frac{k+1}{k}\right) B_{k+1}^*(\mathbf{z}_\tau, T)\rho^k, \tag{19–36}$$

where ρ is the number density of the single solute ("nondiffusible") species on the "inside" in Fig. 19–2 and $\bar{\gamma}$ is its activity coefficient. All of the solvent species can pass through the semipermeable membrane in Fig. 19–2. The second virial coefficient is

$$B_2^*(\mathbf{z}_\tau, T) = -\tfrac{1}{2}\int_0^\infty \{\exp[-w(r, \mathbf{z}_\tau, T)/kT] - 1\} 4\pi r^2 \, dr, \tag{19–37}$$

where $w(r)$ is the reversible work necessary to bring two solute molecules together from $r = \infty$ to r in the solvent at $\mathbf{z}_\tau$, T (i.e., in the outside solution).

19–2 Applications of the McMillan-Mayer theory. We consider here a few examples of the calculation of B_2^* for a solute in a solvent (which may be multicomponent). B_2^* then determines the first departures from limiting behavior in Π/kT and $\ln \bar{\gamma}$ [Eqs. (19–35) and (19–36)], for example.

Hard spheres. Suppose the nondiffusible species (i.e., the solute) in an osmotic equilibrium can be represented approximately by hard spheres of diameter a. Spherical protein molecules, under some conditions, might be an example.† That is,

$$\begin{aligned} w(r) &= +\infty \qquad r < a \\ &= 0 \qquad\;\; r \geq a, \end{aligned} \tag{19–38}$$

† The many "internal" degrees of freedom—vibration and rotation—of such a molecule can be ignored in calculating virial coefficients.

as in Eq. (15–25). Then we find, as in Eqs. (15–26) and (15–30), that

$$B_2^* = \frac{2\pi a^3}{3}, \qquad B_3^* = \tfrac{5}{8}B_2^{*2}. \tag{19–39}$$

If we let v_m be the volume occupied by the molecule (sphere), $B_2^* = 4v_m$.

Considerable work has been done† on the calculation of B_2^* for "hard" molecules which are not spheres. Of course, in this case a generalization of Eq. (19–37) has to be used (Problem 19–3) which involves integration over rotational orientations as well as over r. It is found that B_2^*/v_m is always greater than 4 for nonspherical particles and that, for highly anisometric particles (e.g., "needles" or "pancakes"), B_2^*/v_m is approximately equal to the ratio of the long to the short dimension. Solutions of such particles will therefore be "nonideal" at much lower molar concentrations than spheres of the same volume.

The calculation of B_2^* for polymer and polyelectrolyte molecules will be deferred to Chapter 21.

Donnan membrane equilibrium. A Donnan equilibrium is an osmotic equilibrium in which the nondiffusible species (solute) is charged and all other species, including at least two which are charged and of opposite sign, can pass through the membrane.

Necessarily, if there is a charged nondiffusible species, then for electrical neutrality at least one other charged species must be present on the "inside." If there is only one other charged species (of sign opposite to the solute) on the inside, it too will be confined to the inside, even though it could under other conditions pass through the membrane, for otherwise the outside solution would not be electrically neutral. (Even extremely small departures from neutrality would be resisted by the creation of very large electrostatic potentials.) The uncharged solvent can, of course, be on both sides. In this case, then, we have an electrolyte solution on the inside and solvent (usually water) on the outside. This is not a Donnan equilibrium. The osmotic pressure Π here is just that (called p) of Chapter 18. We cannot use a virial expansion of Π because $w(r)$ $(=w^{(2)})$ for a pair of ions immersed in the outside system (solvent) has the coulombic form, $1/r$ (Eq. 18–1), and hence the virial coefficients diverge. Incidentally, the reason why the coulombic potential energy (18–1) is denoted by $u(r)$ in Chapter 18 and by $w(r)$ here is that we are now treating the solvent as a molecular component of the system: Eq. (18–1) is an approximation for the work necessary to bring two ions together, calculated from a *mean* force, that is, a force averaged over all configurations of the solvent molecules.

---

† See B. H. Zimm, *J. Chem. Phys.* **14,** 164 (1946); L. Onsager, *Ann. N. Y. Acad. Sci.* **51,** 627 (1949); Hirschfelder, Curtiss, and Bird, pp. 183–187.

To achieve a Donnan equilibrium, we must have, besides the charged solute on the inside, at least two other charged (opposite to each other) species on the inside. Then both of these latter species can also be on the outside, with electrical neutrality achieved on both sides. An example would be P^+ (protein, say), Na^+, Cl^-, and H_2O on the inside, and Na^+, Cl^-, and H_2O on the outside. The "solvent" here (i.e., the outside system) *is itself an electrolyte solution.* Equations (19–35) and (19–36) are applicable. We can use a virial expansion approach because $w^{(N)}(\mathbf{z}_T, T)$ for a set of N charged solute molecules refers to the set immersed in the outside (electrolyte) solution. That is: the solute molecules in the solvent are screened by ion atmospheres; $w(r)$ has the form $e^{-\kappa r}/r$ (κ refers to the outside solution), at least approximately; hence, the virial coefficients do not diverge.

The traditional approach to this problem is the method of Donnan in which a membrane potential is introduced. One of the more satisfying aspects of the present statistical-mechanical method is that this operationally questionable concept† does not enter the discussion.

As a first example of a Donnan equilibrium, let all ions (including the solute) be treated as point charges ($a = 0$), let the charge on a solute ion be‡ $\mathsf{z}\mathbf{e}$, and let us use the Debye-Hückel limiting $w(r)$ (Eq. 18–28) for a pair of solute ions immersed in the outside solution:

$$w(r) = \frac{\mathsf{z}^2\mathbf{e}^2 e^{-\kappa r}}{\epsilon r}, \tag{19–40}$$

where κ depends on the *outside* ionic strength. We note that intensive properties of the system are completely determined by the concentration ρ of solute on the inside, by T, and by the dielectric constant and ionic strength on the outside. Examples of such intensive properties are Π, $\ln \bar{\gamma}$, and ρ_k^i/ρ_k^0 for a diffusible ion of type k (Problem 19–4), where ρ_k^i is the inside concentration and ρ_k^0 is the outside concentration.

To calculate B_2^*, we put Eq. (19–40) for $w(r)$ in Eq. (19–37), after expanding the exponential in the latter equation up to the term in $(w/kT)^2$. This suffices to give us B_2^* up to the linear term in $\alpha \equiv \mathbf{e}^2\kappa/\epsilon kT$, which is as far as we are justified in going [see Eq. (18–26)]. Elementary integrations then give

$$B_2^* = \frac{\mathsf{z}^2}{2\sum} - \frac{\mathsf{z}^4\alpha}{8\sum}, \tag{19–41}$$

---

† See E. A. GUGGENHEIM, *Thermodynamics.* Amsterdam: North-Holland, 1957; pp. 374–381.

‡ Note the difference between the symbols z (charge number) and $\mathbf{z}$ (activity set). The distinction will also be clear from the context.

where

$$\sum = \sum_k \rho_k^0 z_k^2.$$

The first term on the right can be shown to be a consequence only of the fact that the Debye-Hückel $w(r)$ is consistent with electrical neutrality in the outside solution (Problem 19–5). The second term is probably† not the exact linear term in α, because higher terms in $w(r)$ [i.e., Eq. (19–40) gives the leading term in an expansion] may also contribute linear terms in α to B_2^*.

As a second example of a Donnan equilibrium, let us calculate B_2^* for the potential of mean force between solute molecules (in the solvent),

$$\begin{aligned} w(r) &= +\infty & r &< a \\ &= \frac{\mathsf{z}^2\mathsf{e}^2 e^{-\kappa(r-a)}}{\epsilon r(1+\kappa a)} & r &\geq a, \end{aligned} \tag{19–42}$$

where κ again refers to the outside solution. This $w(r)$ would be appropriate as an approximate potential in two cases. (1) All ions, including solute ions of charge ze, have a diameter a (Eq. 18–27). (2) The solute ions are relatively large spheres of diameter a, each with total charge ze smeared uniformly over the surface; the relatively small diffusible ions are treated as point charges (Problem 18–10). One finds in either case,

$$B_2^* = \frac{2\pi a^3}{3} + \frac{\mathsf{z}^2}{2\sum} - \frac{\mathsf{z}^4 \alpha}{8(1+\kappa a)^2 \sum}. \tag{19–43}$$

The first term on the right is the hard sphere term (Eq. 19–39), the second is the "neutrality" term, and the third is definitely not the exact linear term in α because Eq. (19–42) is not even an exact leading term for $w(r)$. However, Eq. (19–43) should be a useful approximation for α not too large.

Solute with binding equilibrium. A relatively complicated but instructive and important (especially in physical biochemistry) application of the McMillan-Mayer theory is to an osmotic system in which the solute molecules can bind (adsorb) one of the diffusible (solvent) species. The most important example is the binding of hydrogen ions by solutes such as proteins, nucleic acids, polyelectrolytes, etc. But in other systems, other ions or molecules, in the solvent, may also be bound to solute molecules. Some of the questions that arise for this kind of system are the following: How is the binding equilibrium (e.g., the titration curve, if H^+ is being bound) influenced by solute concentration? How does the osmotic

---

† See T. L. Hill, *Faraday Soc. Disc.* **21**, 31 (1956); *J. Phys. Chem.* **61**, 548 (1957).

pressure second virial coefficient depend on the concentration of the species being bound? How does the potential of mean force between a pair of solute molecules (at infinite dilution, $\rho \to 0$) depend on the concentration of the species being bound?

The present problem is a generalization of Section 7–3, which should be reviewed. In Section 7–3 we were concerned with binding of one species on another in an ideal gas mixture. Here we have intersolute interactions and the presence of solvent to complicate matters.

To keep the notation from getting too complex, we consider the following special case: (1) there is only one solute species; (2) only molecules of one solvent species can be bound on solute molecules; and (3) the potential of average force $w^{(\mathbf{N})}$ between a set of solute molecules depends on the number of bound molecules on each of the solute molecules in the set and on the location of the center of mass of each of the solute molecules, but *not* on the manner of distribution of bound molecules among the binding sites of each solute molecule nor on the rotational orientation of the solute molecules. All of these restrictions can, however, be removed.*

For convenience we refer to the solvent species that can be bound as A (adsorbate). Each solute molecule can bind up to m A molecules, $s = 0, 1, \ldots, m$. There are thus a total of $m + 1$ solute "subspecies," depending on the value of s. The partition function of a single solute molecule with s A molecules bound to it (an s-solute), in a box of volume V at T, is denoted by $q(s)$. We let λ_s be the absolute activity of s-solute molecules and λ be the absolute activity of A molecules. The activity z_s for an s-solute is defined as usual (Eq. 15–41) by

$$z_s \equiv \frac{q(s)\lambda_s}{V}. \tag{19–44}$$

Then $z_s \to \rho_s$ if *all* species become infinitely dilute (ideal gas mixture). The activities of all solvent species other than A are represented by the set $\mathbf{z}_T$. The outside solution (solvent) in the osmotic equilibrium is therefore characterized by $\mathbf{z}_T$, λ, and T; and the inside solution is characterized by $\mathbf{z}_T$, λ, T, and the total (all subspecies) solute concentration $\rho(=\sum_s \rho_s)$.

The general method we use is: (a) write an expression for $\Pi V/kT$ that is a generalization of Eq. (19–10), regarding the solute as multicomponent ($m + 1$ "different" solutes); (b) introduce in this expression the fact that actually the solute subspecies are not independent of each other but are interrelated by binding equilibria; and finally (c) develop series expansions, etc., treating the solute as a single (composite) component.

---

* J. G. Kirkwood and J. B. Shumaker, *Proc. Nat. Acad. Sci.* **38,** 855, 863 (1952); T. L. Hill, *J. Chem. Phys.* **23,** 623, 2270 (1955).

The rather obvious generalization, which we shall not prove,* of Eqs. (19–10) and (19–29) to a multicomponent solute in an osmotic equilibrium, is

$$e^{\Pi V/kT} = \sum_{\mathbf{N}\geq 0} \left[\prod_s \frac{(z_s/\gamma_s^0)^{N_s}}{N_s!} \right] \int_V \exp\left[-w^{(\mathbf{N})}/kT\right] d\{\mathbf{N}\}, \qquad (19\text{–}45)$$

where the sum is over all solute sets $\mathbf{N} = N_0, N_1, \ldots, N_m$, and where γ_s^0 and $w^{(\mathbf{N})}$ are functions of $\mathbf{z}_\tau$, λ, and T. As explained in Section 19–1, γ_s^0 depends on the interaction of one s-solute molecule with solvent (outside solution), and $w^{(\mathbf{N})}$ depends on the interaction between a set of $\mathbf{N}$ solute molecules immersed in the solvent (outside solution).

Let $\overline{N}_i$ be the mean number of i-solute molecules in the inside solution. Then, as in the equation preceding (19–12),

$$\overline{N}_i = z_i \left(\frac{\partial (p + \Pi) V/kT}{\partial z_i} \right)_{z_{s\neq i}, \mathbf{z}_\tau, \lambda, T, V} = z_i \left(\frac{\partial \Pi V/kT}{\partial z_i} \right)_{z_{s\neq i}, \mathbf{z}_\tau, \lambda, T, V}$$

$$= \frac{1}{e^{\Pi V/kT}} \sum_{\mathbf{N}\geq 0} N_i \left[\prod_s \frac{(z_s/\gamma_s^0)^{N_s}}{N_s!} \right] \int_V \exp\left[-w^{(\mathbf{N})}/kT\right] d\{\mathbf{N}\}. \qquad (19\text{–}46)$$

To conform with the notation in Section 7–3, we let $\overline{M}$ be the total number of solute molecules of all subspecies and let $\overline{N}$ be the total number of A molecules bound to these solute molecules. Then

$$\overline{M} = \sum_{i=0}^{m} \overline{N}_i, \qquad \overline{N} = \sum_{i=1}^{m} i\overline{N}_i. \qquad (19\text{–}47)$$

Hence we get expressions for $\overline{M}$ and $\overline{N}$ from Eq. (19–46) by replacing N_i, behind the summation sign, by $\sum_i N_i$ and $\sum_i iN_i$, respectively.

At this point we recognize explicitly the fact that the solute subspecies are interconnected by binding equilibria:

$$s\text{-solute} \rightleftarrows 0\text{-solute} + sA,$$

$$\mu_s = \mu_0 + s\mu,$$

$$\lambda_s = \lambda_0 \lambda^s, \qquad z_s = \frac{q(s)\lambda_0\lambda^s}{V}. \qquad (19\text{–}48)$$

This means that only one of $\lambda_0, \lambda_1, \ldots, \lambda_m$ is independent, and we choose λ_0 (0-solute; solute with no A bound). We substitute Eq. (19–48) for z_s

* See: Eqs. (15–43) and (15–51); S. M., Section 40; and Problem 19–15.

in Eq. (19–45) and obtain

$$e^{\Pi V/kT} = \sum_{\mathbf{N}\geq 0}\left[\prod_s \frac{(H_s\lambda_0\lambda^s)^{N_s}}{N_s!}\right]\int_V \exp\left[-w^{(\mathbf{N})}/kT\right] d\{\mathbf{N}\}, \quad (19\text{–}49)$$

where

$$H_s = \frac{q(s)}{V\gamma_s^0}. \quad (19\text{–}50)$$

The physical significance of H_s (a function of T, λ, $\mathbf{z}_\tau$) for an s-solute molecule is evident from the paragraph following Eq. (19–8): $q(s)/\gamma_s^0$ is the effective partition function of an s-solute molecule immersed in the outside solution; $q(s)$ itself is discussed following Eq. (7–46).

Note that if we perform the operation $\lambda_0\partial(\Pi V/kT)/\partial\lambda_0$ on Eq. (19–49), we get just $\sum_s \overline{N}_s$ [see Eqs. (19–46) and (19–47)]. Therefore

$$\overline{M} = \lambda_0\left(\frac{\partial \Pi V/kT}{\partial\lambda_0}\right)_{\mathbf{z}_\tau,\lambda,T,V}. \quad (19\text{–}51)$$

This shows that if we regard all solute molecules (of whatever subspecies) as a single component, then μ_0 should be considered the chemical potential of the solute (thus μ_0 has the same significance as μ' in Section 7–3). Next, we observe that the purely formal operation $\lambda\partial(\Pi V/kT)/\partial\lambda$ on Eq. (19–49), ignoring the actual dependence of the γ_s^0 and $w^{(\mathbf{N})}$ on λ, gives $\sum_s s\overline{N}_s$. Hence,

$$\overline{N} = \lambda\left(\frac{\partial \Pi V/kT}{\partial\lambda}\right)_{\lambda_0,T,V,\gamma^0,w}. \quad (19\text{–}52)$$

If we regard Eq. (19–49) as a power series in λ_0, the solute absolute activity, we will have equations analogous to those in Section 19–1 for an osmotic equilibrium with a single solute. Here, however, the various coefficients will be more complicated because they involve averaging over the different solute subspecies. For simplicity, let us go only as far as the second virial coefficient.* If we consider only those sets $\mathbf{N}$ containing a total of 0, 1, or 2 solute molecules, Eq. (19–49) becomes

$$e^{\Pi V/kT} = 1 + \lambda_0 V\sum_s H_s\lambda^s + \frac{\lambda_0^2}{2!}\sum_{s,s'}(H_s\lambda^s)(H_{s'}\lambda^{s'}) \times \int_V e^{-w_{ss'}/kT}\, d\mathbf{r}_s\, d\mathbf{r}_{s'} + \cdots, \quad (19\text{–}53)$$

* For higher terms, see T. L. HILL, *J. Chem. Phys.* **23**, 623, 2270 (1955).

where the double sum includes terms with $s = s'$, and $w_{ss'}$ is the potential of mean force between one s-solute molecule and one s'-solute molecule, immersed in the outside solution. We define a solute activity $\mathfrak{z}$ by

$$\mathfrak{z} = \lambda_0 \sum_s H_s \lambda^s \tag{19-54}$$

so that $\mathfrak{z} \to \rho(=\overline{M}/V)$ as $\rho \to 0$. Then, just as in Eq. (19-10),

$$e^{\Pi V/kT} = 1 + \frac{Z_1'}{1!}\mathfrak{z} + \frac{Z_2'}{2!}\mathfrak{z}^2 + \cdots, \tag{19-55}$$

where

$$Z_1' = V,$$

$$Z_2' = \frac{\sum_{s,s'} (H_s\lambda^s)(H_{s'}\lambda^{s'}) \int_V e^{-w_{ss'}/kT}\, d\mathbf{r}_s\, d\mathbf{r}_{s'}}{(\sum_s H_s\lambda^s)^2}. \tag{19-56}$$

It is necessary to digress briefly to introduce the radial distribution function $g(r)$ for solute molecules at infinite dilution, $\rho \to 0$ (i.e., in the outside solution). As in the paragraph preceding Eq. (17-21),

$$\rho_s \rho_{s'}\, e^{-w_{ss'}/kT}\, d\mathbf{r}_s\, d\mathbf{r}_{s'}$$

is the probability that an s-solute molecule is in $d\mathbf{r}_s$ and an s'-solute molecule is in $d\mathbf{r}_{s'}$, in the limit $\rho \to 0$. In this limit, Eq. (19-46) gives

$$\overline{N}_i \to \frac{z_i V}{\gamma_i^0} = \frac{q(i)\lambda_0\lambda^i}{\gamma_i^0} = V\lambda_0 H_i \lambda^i,$$

or

$$\rho_i = \lambda_0 H_i \lambda^i,$$
$$\rho = \sum_i \rho_i = \lambda_0 \sum_i H_i \lambda^i. \tag{19-57}$$

Therefore the probability above becomes

$$\frac{\rho^2 (H_s\lambda^s)(H_{s'}\lambda^{s'}) e^{-w_{ss'}/kT}\, d\mathbf{r}_s\, d\mathbf{r}_{s'}}{(\sum_s H_s\lambda^s)^2}.$$

If we sum this expression over s and s', we have the total probability that any solute molecule (any subspecies) is in $d\mathbf{r}_s$ and any other is in $d\mathbf{r}_{s'}$. If we regard the solute as a single component, this same probability is $\rho^2 g(r)\, d\mathbf{r}_s\, d\mathbf{r}_{s'}$, where $g(r)$ is the solute radial distribution function. On comparing the two expressions, we have

$$g(r) = e^{-w(r)/kT} = \frac{\sum_{s,s'} (H_s\lambda^s)(H_{s'}\lambda^{s'})e^{-w_{ss'}(r)/kT}}{(\sum_s H_s\lambda^s)^2}, \tag{19-58}$$

where $w(r)$ is the potential of mean force for a pair of solute molecules in the outside solution, properly averaged over the different solute subspecies (Problem 19–6). Since λ is proportional to the activity and approximately proportional to the concentration of the A molecules, Eq. (19–58) contains the dependence of $w(r)$ or $g(r)$ on this activity or concentration. Incidentally, in Eq. (19–58), $w_{ss'}$ is also a function of λ since A is a component of the solvent ($w_{ss'}$ depends on r, $\mathbf{z}_r$, λ, T), but this will generally be a relatively unimportant effect.

We note that $e^{-w(r)/kT}$ in Eq. (19–58) has the formal appearance of a one-component (A) grand partition function (Eq. 7–27) for A molecules bound on a pair of solute molecules a distance r apart divided by the grand partition function with $r = \infty$. This result might have been anticipated by analogy with the following thermodynamic relations for such a system:

$$dE = T\,dS + DW_{\text{on}} + \mu\,dN,$$

$$d(A - \mu N) = DW_{\text{on}} \qquad (T, \mu \text{ constant}),$$

$$-\Delta(\mu N - A) = -kT \ln \frac{\Xi(r)}{\Xi(\infty)} = w(r) \qquad (T, \mu \text{ constant}),$$

where DW_{on} and $w(r)$ are reversible work done on the system, the latter being the work (potential of mean force) necessary to bring the two solute molecules together from $r = \infty$.

We return now to Eq. (19–56), which, with the aid of Eq. (19–58), becomes

$$Z'_2 = \int_V e^{-w/kT}\, d\{2\}. \tag{19-59}$$

This has the same form as Eq. (19–29); furthermore, w has the same physical significance (potential of mean force) in the two equations, despite the complication here of subspecies averaging. This analogy proves in fact to be general; that is, for Z'_N.

From Eqs. (19–10) and (19–55), it is clear then that we can define a group of quantities Z'_N, b'_j, β'_k, B'_n (all of which depend on the properties of small groups of solute molecules in the outside solution) in complete analogy with Z^*_N, b^*_j, etc. For example, Eqs. (19–35) through (19–37) hold here for Π/kT and $\ln \bar{\gamma}$ provided we replace B^*_n by $B'_n(\mathbf{z}_r, \lambda, T)$ and w by $w(r, \mathbf{z}_r, \lambda, T)$ in these equations.

Next, let us derive an equation for $\overline{N}/\overline{M}$, the mean number of A molecules bound per solute molecule. Specifically, we want to express $\overline{N}/\overline{M}$ as a power series in ρ. From Eq. (19–52),

$$\frac{\overline{N}}{V} = \left(\frac{\partial \Pi/kT}{\partial \ln \lambda}\right)_{\lambda_0,T,\gamma^0,w} = \frac{\partial}{\partial \ln \lambda} \sum_{j\geq 1} b'_j \mathfrak{z}^j$$

$$= \left(\sum_{j\geq 1} j b'_j \mathfrak{z}^{j-1}\right) \lambda_0 \sum_s sH_s\lambda^s + \sum_{j\geq 2} \left(\frac{\partial b'_j}{\partial \ln \lambda}\right)_{T,\gamma^0,w} \mathfrak{z}^j.$$

Then

$$\frac{\overline{N}/V}{\rho} = \frac{\overline{N}}{\overline{M}} = \frac{\sum_s sH_s\lambda^s}{\sum_s H_s\lambda^s} + \frac{1}{\rho}\sum_{j\geq 2} \left(\frac{\partial b'_j}{\partial \ln \lambda}\right)_{T,\gamma^0,w} \mathfrak{z}^j.$$

Using the series $\mathfrak{z}(\rho)$ (Eq. 15–11) and Eqs. (15–13) relating the B'_n and b'_j, we find

$$\frac{\overline{N}}{\overline{M}} = \frac{\sum_s sH_s\lambda^s}{\sum_s H_s\lambda^s} - \rho \left(\frac{\partial B'_2}{\partial \ln \lambda}\right)_{T,\gamma^0,w} - \tfrac{1}{2}\rho^2 \left(\frac{\partial B'_3}{\partial \ln \lambda}\right)_{T,\gamma^0,w} - \cdots. \tag{19–60}$$

In the limit $\rho \to 0$, we get, as expected, the same result as Eqs. (7–29) and (7–51) [except that, because of the presence of the solvent, $q(s)/\gamma_s^0$ replaces $q(s)$]. The terms in ρ take care of the effect of solute concentration on the amount of binding. We can understand the linear term in ρ, qualitatively, as follows. If, for example, adsorbing A molecules on a pair of solute molecules increases the repulsion between the solute molecules (i.e., $\partial B'_2/\partial \ln \lambda > 0$), then, when two solute molecules are brought together from $r = \infty$, they will desorb some A molecules. But this is essentially what happens when the concentration of solute is increased: pairs of solute molecules spend more time near each other, so $\overline{N}/\overline{M}$ decreases.

An explicit expression for $\partial B'_2/\partial \ln \lambda$ in Eq. (19–60) is easily shown to be (Problem 19–7)

$$\frac{\sum_s sH_s\lambda^s}{V(\sum_s H_s\lambda^s)^2} \times \left[\frac{\sum_{s,s'} (H_s\lambda^s)(H_{s'}\lambda^{s'})\int_{ss'}}{\sum_s H_s\lambda^s} - \frac{\sum_{s,s'} (s+s')(H_s\lambda^s)(H_{s'}\lambda^{s'})\int_{ss'}}{2\sum_s sH_s\lambda^s}\right], \tag{19–61}$$

where $\int_{ss'}$ is the integral in Eq. (19–56).

As a final topic, we consider an important special case. Suppose that the solute and A molecules are charged and that the potential of mean force between a pair of solute subspecies a distance r apart is proportional to the product of the charges on the two molecules:

$$\frac{w_{ss'}(r)}{kT} = \mathbf{z}_s\mathbf{z}_{s'}f(r). \tag{19–62}$$

For example, we might have (Eq. 19–42)

$$f(r) = \frac{\mathbf{e}^2 e^{-\kappa(r-a)}}{\epsilon r(1+\kappa a)kT}\cdot \tag{19–63}$$

Let the charge number of a 0-solute be $\mathbf{z}_0$ and that of an A molecule be $\mathbf{z}$. Then $\mathbf{z}_s = \mathbf{z}_0 + s\mathbf{z}$ for an s-solute. A typical case would be a protein molecule with $\mathbf{z} = +1$ ($A = H^+$), $\mathbf{z}_0$ negative, and $\mathbf{z}_m$ positive. We expand $\exp(-w_{ss'}/kT)$ in Eq. (19–58) up to the quadratic term:

$$e^{-w/kT} = 1 - \left\langle\frac{w_{ss'}}{kT}\right\rangle + \frac{1}{2}\left\langle\left(\frac{w_{ss'}}{kT}\right)^2\right\rangle - \cdots, \tag{19–64}$$

where

$$\left\langle\left(\frac{w_{ss'}}{kT}\right)^n\right\rangle = \frac{\sum_{s,s'}(H_s\lambda^s)(H_{s'}\lambda^{s'})(w_{ss'}/kT)^n}{(\sum_s H_s\lambda^s)^2}\cdot \tag{19–65}$$

These are averages, it will be noted, in which the probabilities of s and s' are "unperturbed" or independent of each other [compare Eq. (13–19) and Problem 14–17]. We put Eq. (19–62) into Eq. (19–64) and obtain

$$e^{-w/kT} = 1 - (\bar{\mathbf{z}})^2 f(r) + \tfrac{1}{2}(\overline{\mathbf{z}^2})^2 f(r)^2 - \cdots, \tag{19–66}$$

where

$$\overline{\mathbf{z}^n} = \frac{\sum_s H_s\lambda^s \mathbf{z}_s^n}{\sum_s H_s\lambda^s}\cdot \tag{19–67}$$

Then from Eq. (19–66),

$$\frac{w(r)}{kT} = (\bar{\mathbf{z}})^2 f(r) - \tfrac{1}{2}[\overline{\mathbf{z}^2} - (\bar{\mathbf{z}})^2][\overline{\mathbf{z}^2} + (\bar{\mathbf{z}})^2]f(r)^2 + \cdots. \tag{19–68}$$

The leading term in Eq. (19–68) gives the potential of mean force between a pair of solute molecules a distance r apart which would obtain if the binding of A molecules on one solute molecule were uninfluenced by the binding on the other (i.e., $\bar{\mathbf{z}}$ is the average charge number of an *isolated* solute molecule in the solvent). Actually, the binding on the two solute molecules is not independent (Problem 19–8), and we expect that the perturbation of one by the other will always be such as to lower the potential of mean force. This is confirmed by the second term in Eq. (19–68), which is always negative.

The coefficients in Eq. (19–68) can also be expressed in terms of $\bar{s}$ and $\overline{s^2}$, where the averaging here is the same as in Eq. (19–67) [thus $\bar{s}$

is equal to $\overline{N}/\overline{M}$ in Eq. (19–60) in the limit as $\rho \to 0$—that is, isolated solute molecules]:

$$\bar{z} = z_0 + \bar{s}z, \qquad \overline{z^2} - \bar{z}^2 = z^2[\overline{s^2} - (\bar{s})^2]. \tag{19–69}$$

Also, it is easy to show that (Problem 19–9)

$$\overline{s^2} - (\bar{s})^2 = \lambda \left(\frac{\partial \bar{s}}{\partial \lambda}\right)_{T,\gamma^0}. \tag{19–70}$$

Consider the special case in which λ (or the concentration of A) is chosen so that $\bar{z} = 0$; that is, two solute molecules far from each other ($r = \infty$) in the outside solution would have zero charge on the average. An equivalent statement is that λ is chosen so that the average charge on a solute molecule is zero ("isoionic point") in the limit as $\rho \to 0$. Then, from Eq. (19–68),

$$\frac{w(r)}{kT} = -\frac{(\overline{z^2})^2 f(r)^2}{2}. \tag{19–71}$$

This equation tells us that if the two solute molecules are brought together to finite r, then w will be negative. The reason for this is the following: when the (fluctuating) charge number z_s on one of the two solute molecules happens to be, say, negative, this will increase the probability of the other charge number $z_{s'}$ being positive [$w_{ss'}$ will be negative and hence this ss' combination gets extra weight in Eq. (19–58)]. Thus there is a correlation between the fluctuating charges on the two solute molecules tending to favor charges on them of opposite sign. This is closely analogous to (a) the net attraction between rotating dipolar molecules in a gas which arises from a correlation between molecular orientations favoring those mutual arrangements with negative potential energy, and to (b) the origin of London dispersion forces (Appendix IV).

Incidentally, for simplicity we have assumed from the outset that $w_{ss'}$ depends only on the total numbers of A molecules bound, s and s', and not on how these molecules are distributed among solute sites. It is clear, however, that if the A molecules are charged, the interaction $w_{ss'}$ will involve not only the total charges we have been concerned with but also electric moments depending on particular A distributions. In this more general situation, then, there will be a contribution to a negative $w(r)$ in Eq. (19–71) not only from total charge correlation but also from a correlation between electric moments associated with the distribution of A molecules among available sites on the two solute molecules.

The simplest model to which we can apply Eq. (19–71) is the following: each solute molecule has m independent and equivalent sites for binding A molecules; q is the partition function for one A molecule bound on one site; and γ^0 is the same for all s (we neglect the dependence of γ^0

on λ). Then (see Problem 7-9)

$$H_s = \frac{1}{V\gamma^0} \cdot \frac{m!q^s}{s!(m-s)!}, \qquad \bar{s} = \frac{mq\lambda}{1+q\lambda},$$

$$\overline{z^2} = z^2[\overline{s^2} - (\bar{s})^2] = z^2\lambda \frac{\partial \bar{s}}{\partial \lambda} = \frac{z^2 mq\lambda}{(1+q\lambda)^2} \tag{19-72}$$

$$= \frac{z^2 m}{2 + q\lambda + (q\lambda)^{-1}}. \tag{19-73}$$

The form (19–73) is essentially the same as Eq. (11) of Kirkwood and Shumaker.

Finally, we calculate B_2', for use in Π/kT or $\ln \bar{\gamma}$, from the $w_{ss'}$ of Eqs. (19–62) and (19–63). Comparison of Eq. (19–66) with the expansion of $e^{-w/kT}$ from Eq. (19–42) shows that B_2' is given here by Eq. (19–43) but with z^2 replaced by $(\bar{z})^2$ and z^4 replaced by $(\overline{z^2})^2$. In the special case $\bar{z} = 0$ ("isoionic" at $\rho = 0$), B_2' can be written

$$B_2' = \frac{2\pi a^3}{3} - \frac{\pi(\overline{z^2})^2 e^4}{2(\epsilon kT)^2(1+\kappa a)^2\kappa}, \tag{19-74}$$

where κ refers to the outside solution. The first term is positive and the second is negative. At sufficiently low ionic strengths, the second term predominates (Problem 19–10). Equation (19–74) is the same as Eq. (17) of Kirkwood and Shumaker.*

19–3 Constant pressure solution theory.† The solution theories of McMillan and Mayer and Kirkwood and Buff are based on the grand canonical ensemble and open system distribution functions. The natural composition variable is the concentration, since the system is at constant volume. These theories are formally exact and necessarily equivalent through suitable thermodynamic manipulations. In the present section we discuss an alternative, rigorous solution theory designed to yield directly thermodynamic functions expressed in a particularly practical form. For example, for a binary solution, the chemical potentials and partial molal volumes, entropies, and heat contents can be developed as power series in the molality of the solute, with coefficients which depend on properties of the solvent (and small sets of solute molecules) at the

* See S. N. Timasheff, *et al.*, *J. Am. Chem. Soc.* **79**, 782 (1957) for an experimental confirmation.

† See T. L. Hill, *J. Am. Chem. Soc.* **79**, 4885 (1957); *J. Chem. Phys.* **30**, 93 (1959).

same pressure and temperature as the solution. The pressure (instead of the volume) is held fixed at the outset, and hence molality is a natural composition variable. One can substitute mole fraction for molality as composition variable, but we shall not discuss this possibility here.

We restrict the treatment to binary solutions (1 = solvent, 2 = solute) and use a method which is closely analogous to that in the first part of Section 19–1. In Section 19–1, to prepare the solution we start with pure solvent at μ_1 and T and add solute, holding μ_1 and T constant (this is accomplished by use of a semipermeable membrane). The pressure changes as we add solute (the increase in pressure is Π, the osmotic pressure). Here, on the other hand, we start with the same pure solvent but choose p and T as independent variables (instead of μ_1 and T). We then add solute holding p and T constant. In this latter case, the solvent chemical potential changes. Hence the change in solvent chemical potential plays a role analogous to that of Π in Section 19–1.

To obtain the desired independent variables, we use an ensemble apparently first introduced by Stockmayer (in a study of the relation between light scattering and composition fluctuations):

$$\Gamma(N_1, p, T, \mu_2) = e^{-N_1\mu_1/kT} = \sum_{N_2 \geq 0} \Delta_{N_2}(N_1, p, T)e^{N_2\mu_2/kT}, \quad (19\text{–}75)$$

where

$$\Delta_{N_2} = \sum_V Q(N_1, N_2, V, T)e^{-pV/kT}. \quad (19\text{–}76)$$

The partition function Δ_{N_2} is the p, T, $\mathbf{N}$ partition function already encountered several times [e.g., Eqs. (1–87) and (1–91)]. The partition function Γ is a "semigrand" partition function for a system, at p and T, open with respect to 2 but not with respect to 1. The right side of Eq. (19–75) is seen to be a power series in the absolute activity of the solute, $\lambda_2 = e^{\mu_2/kT}$, with coefficients which depend on properties of the solvent (N_1, p, T) containing small numbers (N_2) of solute molecules.

For convenience, we replace the absolute activity λ_2 by a more practical activity a_2, proportional to λ_2, but defined in such a way that (as will be seen below) $a_2 \to m_2$ as $m_2 \to 0$, where $m_2 = \overline{N}_2/N_1$. We shall refer to m_2 as the "molality" of the solute, though this differs from the conventional molality, $1000\, m_2/W_1$, by a constant (W_1 is the molecular weight of the solvent).

The substitution of a_2 for λ_2 in Eq. (19–75) gives, after dividing by the leading term, Δ_0,

$$\frac{\Gamma}{\Delta_0} = 1 + \sum_{N \geq 1} \frac{X_N}{N!} a_2^N, \quad (19\text{–}77)$$

where

$$X_N = \frac{N!\,\Delta_N\,\Delta_0^{N-1}N_1^N}{\Delta_1^N}, \tag{19–78}$$

$$a_2 = \frac{\lambda_2\,\Delta_1}{\Delta_0 N_1}. \tag{19–79}$$

We note that $X_1 = N_1$. The logarithm of the quotient

$$\frac{\Delta_N\,\Delta_0^{N-1}}{\Delta_1^N} = e^{-\Delta F_N/kT} \tag{19–80}$$

in Eq. (19–78) has the physical significance of a Gibbs free energy change, as indicated, since $F = -kT \ln \Delta$ in general. In Eq. (19–80), ΔF_N is the free energy change (non-pV work done on the system by the surroundings) for the process

$$N \text{ systems with } N_1,\, N_2 = 1,\, p,\, T \rightarrow$$

$$\begin{cases} 1 \text{ system with } N_1,\, N_2 = N,\, p,\, T \\ \qquad\qquad + \\ N - 1 \text{ (solvent) systems with } N_1,\, N_2 = 0,\, p,\, T. \end{cases}$$

We can manipulate Eq. (19–77) just as we did Eqs. (15–6) (imperfect gas) and (19–10) (osmotic system). We first note that

$$N_1\mu_1(p, T, 0) = -kT \ln \Delta_0,$$

where $\mu_1(p, T, 0)$ is the chemical potential of the pure solvent. Then if we define

$$\mu_1'(p, T, m_2) = \mu_1(p, T, m_2) - \mu_1(p, T, 0),$$

we have (Eq. 15–7)

$$-\frac{\mu_1'(p, T, a_2)}{kT} = \frac{1}{N_1}\ln\frac{\Gamma}{\Delta_0} = \sum_{j\geq 1} \theta_j(p, T)a_2^j, \tag{19–81}$$

where

$$\begin{aligned} 1!N_1\theta_1 &= X_1 = N_1, \qquad \theta_1 = 1, \\ 2!N_1\theta_2 &= X_2 - X_1^2, \\ 3!N_1\theta_3 &= X_3 - 3X_1X_2 + 2X_1^3, \qquad \text{etc.} \end{aligned} \tag{19–82}$$

From the Gibbs-Duhem equation,

$$a_2\left[\frac{\partial(-\mu_1'/kT)}{\partial a_2}\right]_{p,T} = m_2, \tag{19–83}$$

we have

$$m_2(p, T, a_2) = \sum_{j\geq 1} j\theta_j(p, T)a_2^j. \tag{19–84}$$

The inverse of Eq. (19–84), in logarithmic form (Eq. 15–20), is

$$\ln \gamma_2'(p, T, m_2) = -\sum_{k\geq 1} \delta_k(p, T)m_2^k, \tag{19–85}$$

where γ_2' (the solute activity coefficient) $= a_2/m_2$, and where

$$\begin{aligned} \delta_1 &= 2\theta_2, \\ \delta_2 &= 3\theta_3 - 6\theta_2^2, \end{aligned} \tag{19–86}$$

etc. Finally, if we replace a_2 by m_2 as independent variable in Eq. (19–81), by use of Eq. (19–85) we find

$$-\frac{\mu_1'(p, T, m_2)}{kT} = m_2 + \sum_{n\geq 2} C_n(p, T)m_2^n, \tag{19–87}$$

where

$$C_n = -\frac{n-1}{n}\delta_{n-1}. \tag{19–88}$$

Equation (19–87) is the formal equivalent of the virial expansion of an imperfect gas. Equations (19–85) and (19–87) give essentially the expansions of the two chemical potentials in powers of the molality.

Dilute solution. In a dilute solution $\gamma_2' \to 1$ and $a_2 \to m_2$, according to Eq. (19–85). Hence, from Eq. (19–79),

$$\mu_2 = kT \ln\left(\frac{N_1 \Delta_0}{\Delta_1}\right) + kT \ln m_2 \qquad (m_2 \to 0). \tag{19–89}$$

In the notation of Section 19–1, we also have

$$\begin{aligned} \mu_2 &= \mu_2^0(T) + kT \ln f_2 & \text{(19–90)} \\ &= \mu_2^0(T) + kT \ln k_2(p, T)x_2 \qquad (x_2 \to 0), & \text{(19–91)} \end{aligned}$$

where f_2 = fugacity of solute, x_2 = mole fraction of solute, k_2 = Henry's law constant, and μ_2^0 = chemical potential of solute gas at unit fugacity. Since $x_2 \to m_2$ as $m_2 \to 0$, comparison of Eqs. (19–89) and (19–91) yields, for the Henry's law constant,

$$k_2 = \frac{N_1 \Delta_0 e^{-\mu_2^0/kT}}{\Delta_1}. \tag{19–92}$$

If we write

$$\frac{k_2}{N_1} = \frac{\Delta_0 e^{-\mu_2^0/kT}}{\Delta_1} = e^{-\Delta F/kT}, \tag{19–93}$$

then ΔF is the Gibbs free energy change for the process

$$\text{System with } N_1,\ N_2 = 1,\ p,\ T \rightarrow \begin{cases} \text{system (solvent) with } N_1,\ N_2 = 0,\ p,\ T \\ \qquad + \\ \text{1 molecule of solute in gas at } f_2 = 1. \end{cases}$$

Of course, when the solution is not dilute, $f_2 = k_2 m_2 \gamma_2'$ in Eq. (19–90), and m_2 is replaced by $m_2 \gamma_2'$ in Eq. (19–89).

For the solvent, we have in general

$$-\frac{\mu_1'}{kT} = -\ln \frac{f_1}{f_1^0}, \tag{19–94}$$

where f_1^0 is the fugacity of the pure solvent. For a dilute solution (Raoult's law)

$$-\frac{\mu_1'}{kT} = -\ln(1 - x_2) \rightarrow x_2 \rightarrow m_2, \tag{19–95}$$

in agreement with Eq. (19–87).

Osmotic pressure. We omit series expansions (Problem 19–11) for the partial molal volumes, heat contents, and entropies,† and the corresponding extensive properties, but we consider the osmotic pressure briefly. Suppose we have osmotic equilibrium between the solution at p, T, m_2 and the pure solvent at $p - \Pi$, T:

$$\mu_1(p, T, m_2) = \mu_1(p - \Pi, T, 0). \tag{19–96}$$

But

$$\mu_1(p, T, 0) - \mu_1(p - \Pi, T, 0) = \int_{p-\Pi}^{p} v_1(p', T)\, dp'. \tag{19–97}$$

Therefore, from Eqs. (19–87), (19–96), and (19–97),

$$\frac{1}{kT} \int_{p-\Pi}^{p} v_1(p', T)\, dp' = m_2 + \sum_{n \geq 2} C_n(p, T) m_2^n. \tag{19–98}$$

This equation determines Π as a function of p, T, and m_2. If the pure solvent is incompressible, the left side becomes $\Pi v_1/kT$.

Example. Consider the following simple model. The solvent is an *inert* incompressible fluid of volume $V_0 = N_1 v_1$, whose only role is to

† See T. L. Hill, *J. Am. Chem. Soc.* **79,** 4885 (1957).

provide a suspension medium for solute molecules. The solute molecules are monatomic and interact with each other. These interactions are characterized by imperfect-gas type virial coefficients $B_2^*(T)$, $B_3^*(T)$, etc. The configuration integral for solute molecules, Z_N^*, can be written in terms of the virial coefficients as follows [Eqs. (15–13) and (15–16)]:

$$\begin{aligned} Z_1^* &= V, \\ Z_2^* &= -2VB_2^* + V^2, \\ Z_3^* &= -3VB_3^* + 12VB_2^{*2} - 6V^2B_2^* + V^3. \end{aligned} \tag{19–99}$$

The solution is also assumed incompressible, with volume

$$V = V_0 + N_2 v_2.$$

For the pure solvent we write

$$Q(N_1, 0, V) = Q_0\delta(V - V_0),$$

where $\delta(V - V_0)$ is the Dirac δ-function (introduced because of incompressibility). In general

$$Q(N_1, N, V) = \frac{Q_0\delta[V - (V_0 + Nv_2)]Z_N^*(V)}{N!\Lambda_2^{3N}}, \tag{19–100}$$

where $Z_N^*/N!\Lambda_2{}^{3N}$ is the canonical ensemble partition function of the solute molecules, with the Z_N^* given by Eqs. (19–99). Then, from Eq. (19–76),

$$\Delta_N = \frac{e^{-p(V_0+Nv_2)/kT}Q_0Z_N^*(V_0 + Nv_2)}{N!\Lambda_2^{3N}}$$

and

$$e^{-\Delta F_N/kT} = \frac{Z_N^*(V_0 + Nv_2)}{N!(V_0 + v_2)^N}.$$

For example, to terms in N_1^{-2},

$$e^{-\Delta F_2/kT} = \frac{1}{2}\left[1 + 2\frac{v_2}{V_0} - 2\frac{B_2^*}{V_0} - \left(\frac{v_2}{V_0}\right)^2 + \cdots\right],$$

$$\begin{aligned} e^{-\Delta F_3/kT} = \frac{1}{6}\Bigg[1 + 6\frac{v_2}{V_0} - 6\frac{B_2^*}{V_0} + 6\left(\frac{v_2}{V_0}\right)^2 \\ + 12\left(\frac{B_2^*}{V_0}\right)^2 - 18\frac{v_2B_2^*}{V_0^2} - 3\frac{B_3^*}{V_0^2} + \cdots\Bigg], \end{aligned}$$

and

$$\theta_2 = \frac{v_2}{v_1} - \frac{B_2^*}{v_1} = -C_2,$$

$$\theta_3 = \frac{3}{2}\left(\frac{v_2}{v_1}\right)^2 + 2\left(\frac{B_2^*}{v_1}\right)^2 - 3\,\frac{v_2B_2^*}{v_1^2} - \frac{1}{2}\,\frac{B_3^*}{v_1^2}, \tag{19–101}$$

$$C_3 = \left(\frac{v_2}{v_1}\right)^2 - 2\,\frac{v_2B_2^*}{v_1^2} + \frac{B_3^*}{v_1^2}.$$

The osmotic pressure is given by

$$\frac{\Pi v_1}{kT} = -\frac{\mu_1'}{kT},$$

which can be shown without difficulty (Problem 19–12) to be equivalent to

$$\frac{\Pi}{kT} = \rho_2 + B_2^*\rho_2^2 + B_3^*\rho_2^3 + \cdots, \tag{19–102}$$

as expected [see Eq. (19–13)].

Relation to McMillan-Mayer theory. The coefficients B_n^* of Section 19–1 and the C_n in this section are properties of the pure solvent under the same conditions: μ_1, T, and $p(\mu_1, T)$ or p, T, and $\mu_1(p, T)$. We expect therefore that general relations exist between the B_n^* and the C_n [Eqs. (19–101) refer to a special case only]. From thermodynamics we find†

$$2C_2v_1 = 2B_2^* - \bar{v}_2^0 + \alpha_1v_1 \tag{19–103}$$

and a similar but more complicated relation between C_3 and B_3^*, where $\bar{v}_2^0$ is the value of $\bar{v}_2$ (partial molal volume) in the limit as $m_2 \to 0$; α_1 (Problem 19–4) is the value of $(\partial\rho_1/\partial\rho_2)_{T,\mu_1}$ for the inside solution in the limit as $\rho_2 \to 0$; and v_1 is the volume per molecule in the pure solvent. In the example immediately above, $\alpha_1 = -v_2/v_1$, $\bar{v}_2^0 = v_2$, and therefore

$$C_2v_1 = B_2^* - v_2,$$

in agreement with Eq. (19–101).

---

† T. L. Hill, *J. Chem. Phys.* **30**, 93 (1959).

PROBLEMS

19–1. Find the characteristic thermodynamic function associated with the partition function Ψ_{N_2} in Eq. (19–2) and its relation to other thermodynamic functions [as in Eq. (1–73), for example]. (Page 342.)

19–2. Replace the sum in Ψ_0 and Ψ_1 by a single term with $N_1 = \overline{N}_1$ and show that, for monatomic molecules,

$$\frac{1}{\gamma_2^0} = \frac{\int_V e^{-U_{\overline{N}_1,1}/kT}\, d\{\overline{N}_1\}}{\int_V e^{-U_{\overline{N}_1}/kT}\, d\{\overline{N}_1\}} \tag{19–104}$$

[see Eq. (19–20) for notation]. Give a physical interpretation of this equation. (Page 343.)

19–3. Use Eqs. (8–28), (15–5), and (15–8) to show that for, say, a pair of "hard" right circular cylinders,

$$-2VB_2^* = \frac{1}{(4\pi)^2}\int (e^{-w/kT} - 1) \sin\theta_1\, d\theta_1\, d\varphi_1 \sin\theta_2\, d\theta_2\, d\varphi_2\, d\mathbf{r}_1\, d\mathbf{r}_2,$$

where w is a function of r_{12}, θ_1, θ_2, φ_1, and φ_2 ($w = +\infty$ for cylinders overlapping and $w = 0$ otherwise). (Page 351.)

19–4. It can be shown† that the inside/outside concentration ratio of a diffusible species k in an osmotic equilibrium is given by

$$\frac{\rho_k^i}{\rho_k^0} = 1 + b_{11}(k)\rho + \cdots, \tag{19–105}$$

where $b_{11}(k)$ is a cluster integral of the form (15–52) involving the potential of mean force $w_{11}(k)$, in the outside solution, between one solute molecule and one molecule of species k. (a) If the solute ions are spheres of diameter a and total surface charge $\mathbf{z}\mathbf{e}$, if all other ions are treated as point charges (see Problem 18–10), and if an ion of species k has a charge $\mathbf{z}_k\mathbf{e}$, then $w_{11}(k)$ is given by Eq. (19–42) with a replaced by $a/2$ and $\mathbf{z}^2$ by $\mathbf{z}\mathbf{z}_k$. Show that

$$b_{11}(k) = -\frac{\pi a^3}{6} - \frac{\mathbf{z}\mathbf{z}_k}{\sum} + \frac{\mathbf{z}^2\mathbf{z}_k^2\alpha}{4[1 + (\kappa a/2)]^2 \sum}.$$

(b) If the solute binds A ions, show that

$$b_{11}(k) = -\frac{\pi a^3}{6} - \frac{\bar{\mathbf{z}}\mathbf{z}_k}{\sum} + \frac{\overline{\mathbf{z}^2}\mathbf{z}_k^2\alpha}{4[1 + (\kappa a/2)]^2 \sum},$$

where the averages are defined by Eq. (19–67). (c) Show that in a two-component osmotic system, Eq. (19–105) leads to $\alpha_1 = b_{11}/v_1$, where α_1 is the quantity appearing in Eq. (19–103). (Pages 352 and 368.)

---

† T. L. HILL, *J. Am. Chem. Soc.* **80**, 2923 (1958).

19–5. Show that the term in z^2 in Eq. (19–41) is a consequence of (a) the condition of electrical neutrality in the outside solution, and (b) $w_{ij} = z_i z_j f(r)$ for an ij pair of ions in this solution, where $f(r)$ is arbitrary. (Page 353.)

19–6. Show that the $w(r)$ defined by Eq. (19–58) is the potential of the mean force between two solute molecules if $w_{ss'}(r)$ is the potential of mean force between an s-solute and an s'-solute. (Page 358.)

19–7. Show that (19–61) is equal to $\partial B_2'/\partial \ln \lambda$. (Page 359.)

19–8. In the special case (19–62) through (19–70), $\bar{s}$ is the mean value of s for a pair of solute molecules separated by $r = \infty$ in the outside solution. (a) Let $\bar{\mathfrak{s}}$ be the mean value of s for a finite separation r. Show that

$$\bar{\mathfrak{s}} = \bar{s} - z\bar{z}[\overline{s^2} - (\bar{s})^2]f(r) + \cdots$$

and give a physical interpretation of the result. (b) Let $\hat{s}$ be the mean value of s on one solute molecule of a pair if the other solute molecule is at r and has a fixed charge number $z_{s'}$. Show that

$$\hat{s} = \bar{s} - zz_{s'}[\overline{s^2} - (\bar{s})^2]f(r) + \cdots$$

and interpret the result. (Page 360.)

19–9. Derive Eq. (19–70) for $\overline{s^2} - \bar{s}^2$ from $\overline{s^n} = \sum_s s^n H_s \lambda^s / \sum_s H_s \lambda^s$. (Page 361.)

19–10. In Eq. (19–74), put $a = 60$ A, $\overline{z^2} = 16$, $T = 298.1$°K, $\epsilon = 78.5$, and determine the approximate concentration in moles·liter^{-1} of a 1–1 electrolyte (outside solution) at which the two terms making up B_2' have the same magnitude. (Page 362.)

19–11. Obtain expansions in powers of m_2 for $\bar{v}_1$, $\bar{H}_1$ and $\bar{S}_1$ in a binary solution at p, T, m_2. (Page 366.)

19–12. Verify Eq. (19–102) for Π/kT. (Page 368.)

19–13. Discuss the problem of an isomeric chemical equilibrium between two solutes in a very dilute solution at p and T (solutes = A, B; solvent = 1; use $m_A = \overline{N}_A/N_1$ and $m_B = \overline{N}_B/N_1$ as composition variables).

19–14. Find an expression for k_β, the Henry's law constant, in a dilute binary solution (β = solute, α = solvent), from Eq. (17–21).

19–15. Translate Eqs. (15–40) through (15–49) into corresponding equations for a binary solute in the McMillan-Mayer theory.

19–16. Show, from the general B_2' referred to just preceding Eq. (19–74), that

$$\frac{\partial B_2'}{\partial \ln \lambda} = \frac{\bar{z}z[\overline{s^2} - (\bar{s})^2]}{\sum} - \frac{\alpha z\overline{z}^2[z(\overline{s^3} - \overline{s^2}\bar{s}) + 2z_0(\overline{s^2} - (\bar{s})^2)]}{4(1 + \kappa a)^2 \sum}.$$

Supplementary Reading

McMillan, W. G., and Mayer, J. E., *J. Chem. Phys.* **13**, 276 (1945).
S. M., Chapter 6.

CHAPTER 20

THEORY OF CONCENTRATED SOLUTIONS

The methods employed in the previous chapter made possible a rigorous formulation of the theory of dilute solutions. But these methods cannot be extended in a practical way to concentrated solutions. If a phase separation occurs in the solution at higher concentrations, the methods of Chapter 19 are not applicable even in principle beyond the concentration of phase separation (because of divergence of the series expansions). In view of the fact that we are interested in the entire concentration range here, approximate theories must be introduced.

This situation is completely analogous to that in gases and liquids: the imperfect-gas theory of Chapter 15 had to be discarded in treating the theory of liquids (Chapter 16).

It is particularly appropriate to give only a brief introductory treatment in this chapter because of the existence of the two recent and excellent monographs on this subject by Guggenheim and by Prigogine (see Supplementary Reading list). In particular, these books should be consulted for detailed comparisons between theory and experiment. Section 20–1 on the lattice (or "strictly regular") theory of solutions may serve as an introduction to Guggenheim's book, and Sections 20–2 to 20–4 are related in the same way to Prigogine's book.

We study only binary solutions in Sections 20–1 to 20–3. But the methods discussed can all be extended to any number of components.

20–1 Lattice theory of solutions. In this section we consider a model of a binary solution which is sufficiently idealized so that we can take over the results of Chapter 14 on lattice statistics with only notational changes. The model is more nearly appropriate for a solid solution, but it is usually applied to liquid solutions.

The system is a condensed, incompressible solution containing N_A and N_B molecules of the two components at temperature T. The molecules occupy sites of a regular lattice; there are no vacant sites. Each site has c nearest-neighbor sites. The lattice is rigid, that is, it has a fixed volume per site. The volume V is then not an independent thermodynamic variable, since it is simply proportional to $N_A + N_B = M$ (total number of sites). Hence, this model omits all p-V effects. The molecules are spherical or effectively spherical and the two species are of approximately the same size (otherwise they would not be interchangeable on the same lattice).

Each molecule vibrates about a lattice site with a (three-dimensional) partition function $q_A(T)$ or $q_B(T)$, independent of the state of occupation of neighboring sites. If the molecules are polyatomic, q_A and q_B include the rotational and internal vibrational degrees of freedom. We take into account nearest-neighbor interactions: the pair interaction energies are denoted by w_{AA}, w_{AB}, and w_{BB}. That is, $w_{AA} = u_{AA}(a)$, etc., where a is the nearest-neighbor distance. We treat the w's as constants, though we could consider them functions of temperature. Usually all the w's have negative values.

The thermodynamic equations for the Helmholtz free energy are

$$dA = -S\,dT + \mu_A\,dN_A + \mu_B\,dN_B, \tag{20-1}$$

$$A = \mu_A N_A + \mu_B N_B. \tag{20-2}$$

If we rewrite Eq. (20–1) as

$$dA = -S\,dT + \mu_B\,d(N_A + N_B) + (\mu_A - \mu_B)\,dN_A, \tag{20-3}$$

we have the analog, term by term, of the equation

$$dA = -S\,dT - \Phi\,dM + \mu\,dN \tag{20-4}$$

for a lattice gas, where we have made the arbitrary association: occupied site $\leftrightarrow$ species A.

The canonical ensemble partition function for the solution is

$$Q(N_A, N_B, T) = q_A(T)^{N_A} q_B(T)^{N_B} \sum_{N_{AB}} g(N_A, N_A + N_B, N_{AB}) e^{-W/kT}, \tag{20-5}$$

where

$$W = N_{AA}w_{AA} + N_{AB}w_{AB} + N_{BB}w_{BB}.$$

The notation is essentially the same as that in Eq. (14–20); the function g is that of Chapter 14. We introduce [see Eqs. (14–19), (14–27), and (14–28)]

$$w = w_{AA} + w_{BB} - 2w_{AB}, \tag{20-6}$$

$$cN_A = 2N_{AA} + N_{AB}, \tag{20-7}$$

$$cN_B = 2N_{BB} + N_{AB}, \tag{20-8}$$

into Eq. (20–5) and obtain

$$Q = (q_A e^{-cw_{AA}/2kT})^{N_A} (q_B e^{-cw_{BB}/2kT})^{N_B} \sum_{N_{AB}} g x^{N_{AB}}, \tag{20-9}$$

where $x = e^{w/2kT}$. The sum is now exactly the sum which occurs in lattice-gas theory [see, for example, Eq. (14–29)]; we denote it by $\sum$ (a function of N_A, M, and T), below.

Incidentally, we note that in this model pure A [put $N_B = 0$ in Eq. (20–9)] has the partition function

$$Q(N_A, T) = (q_A e^{-cw_{AA}/2kT})^{N_A},$$

as in an Einstein crystal. If the solution is supposed to be a liquid, then q_A should include a communal entropy factor e [see Eq. (16–22) and Section 20–2]. The interaction potential energy in pure A, relative to infinite separation as zero, is seen to be $cw_{AA}N_A/2$, as expected. The chemical potential μ_A and vapor pressure p_A^0 of pure A are

$$\frac{\mu_A}{kT} = -\left(\frac{\partial \ln Q}{\partial N_A}\right)_T = -\ln q_A e^{-cw_{AA}/2kT}$$

$$= \frac{\mu_A(\text{gas})}{kT} = \frac{\mu_A^0(T)}{kT} + \ln p_A^0, \tag{20–10}$$

assuming the vapor is an ideal gas. There is, of course, a similar equation for component B.

Ideal solution. The first special case of Eq. (20–9) that we consider is $w = 0$; that is, $w_{AA} + w_{BB} = 2w_{AB}$. This means that, energetically, A and B molecules "like" the opposite species as well as their own species; or, that there is no energy change when an A molecule which is completely surrounded (nearest neighbors) by A molecules and a B which is surrounded by B's exchange places with each other. In Eq. (20–9), $x = 1$ and, as in Eq. (14–4),

$$\sum = \frac{(N_A + N_B)!}{N_A!N_B!}. \tag{20–11}$$

This corresponds to the ideal lattice statistics of Chapter 7. We can now put Eq. (20–11) for $\sum$ in Eq. (20–9) and find the thermodynamic functions of interest. For example,

$$\frac{\mu_A}{kT} = -\left(\frac{\partial \ln Q}{\partial N_A}\right)_{N_B,T} = \ln x_A - \ln q_A e^{-cw_{AA}/2kT}$$

$$= \frac{\mu_A(\text{gas})}{kT} = \frac{\mu_A^0(T)}{kT} + \ln p_A,$$

where x_A (not to be confused with x) is the mole fraction $N_A/(N_A + N_B)$ of A in the solution, and where p_A is the partial pressure of A above the

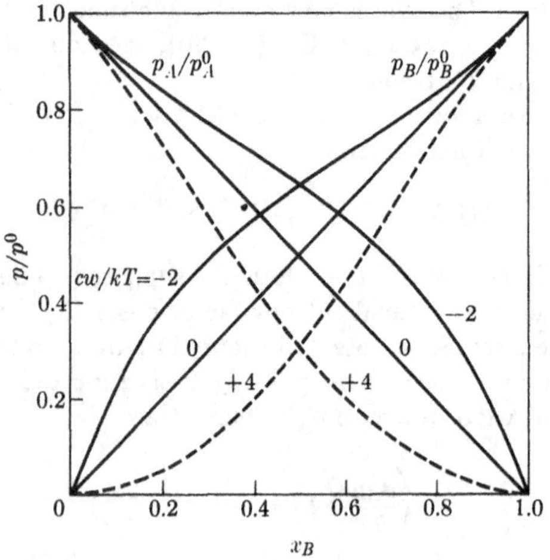

FIG. 20–1. Vapor pressure curves for Bragg-Williams binary solution.

solution. Using Eq. (20–10),

$$\frac{p_A}{p_A^0} = x_A. \tag{20–12}$$

Thus with this model, Raoult's law is obeyed over the whole composition range (Fig. 20–1) when $w = 0$. It should be recalled that, besides $w = 0$, we are also imposing the implicit condition that A and B molecules be near enough in size to fit into the same lattice. In thermodynamics, a solution exhibiting the behavior (20–12) is called "ideal."

Component B also follows Raoult's law, of course. This is, in fact, a purely thermodynamic consequence of Eq. (20–12) for A (Problem 20–1).

From the standard canonical ensemble equations (Section 1–4), we find for the entropy

$$S = N_A\left(kT\frac{d\ln q_A}{dT} + k\ln q_A\right) + N_B\left(kT\frac{d\ln q_B}{dT} + k\ln q_B\right) + k\ln\frac{(N_A+N_B)!}{N_A!N_B!}. \tag{20–13}$$

The last term arises from the configurational degeneracy. If we define ΔS_m ("entropy of mixing") as the entropy change in the process

$$\left.\begin{array}{c} N_A \text{ molecules of pure } A \text{ at } T \\ + \\ N_B \text{ molecules of pure } B \text{ at } T \end{array}\right\} \rightarrow \begin{array}{c}\text{solution with} \\ N_A,\ N_B,\ T,\end{array}$$

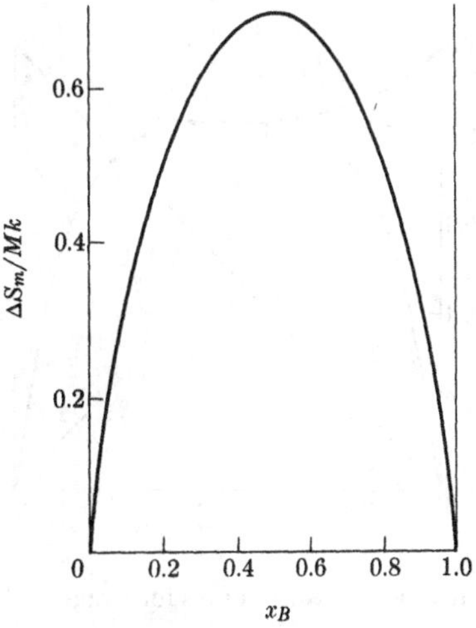

FIG. 20–2. Entropy of mixing in an ideal solution as a function of mole fraction of component B.

then

$$\frac{\Delta S_m}{Mk} = -x_A \ln x_A - x_B \ln x_B \geq 0. \tag{20–14}$$

In ordinary solution thermodynamics, "mixing" is defined at constant p and T, but we need not specify p above because of the absence of p-V effects in this model. The function (20–14) is plotted in Fig. 20–2. We find also

$$\frac{\Delta A_m}{MkT} = -\frac{\Delta S_m}{Mk}, \tag{20–15}$$

$$\Delta E_m = 0. \tag{20–16}$$

Bragg-Williams approximation. We recall that in this approximation (Section 14–4) the molecules are assumed to be distributed among sites in a random fashion, despite molecular interactions. Equation (20–9) becomes in this case

$$Q = (q_A e^{-cw_{AA}/2kT})^{N_A}(q_B e^{-cw_{BB}/2kT})^{N_B} \frac{(N_A + N_B)!x^{\overline{N}_{AB}}}{N_A!N_B!}, \tag{20–17}$$

where $\overline{N}_{AB} = cN_AN_B/M$. Then

$$\frac{\mu_A}{kT} = -\left(\frac{\partial \ln Q}{\partial N_A}\right)_{N_B,T} = \ln x_A - \ln q_A e^{-cw_{AA}/2kT} - \frac{cw(1 - x_A)^2}{2kT},$$

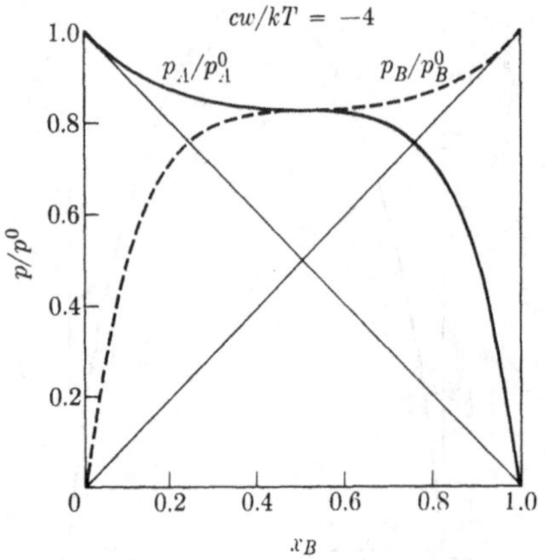

FIG. 20–3. Critical vapor pressure curves for Bragg-Williams binary solution.

and therefore

$$\frac{p_A}{p_A^0} = x_A e^{-cw(1-x_A)^2/2kT}, \tag{20–18}$$

with an analogous equation for p_B/p_B^0 (Problem 20–1). We plot Eq. (20–18) in Figs. 20–1, 20–3, and 20–4 for various choices of cw/kT.

In Fig. 20–1, the value $cw/kT = 0$ leads to ideal behavior, as already mentioned. If we take w negative, say $cw/kT = -2$, this means, according to Eq. (20–6), that $w_{AA} + w_{BB}$ is more negative than $2w_{AB}$ (AA and BB pairs are more stable, energetically, than AB pairs). This gives so-called "positive" deviations from Raoult's law, as shown in the figure: the presence of B molecules increases the "escaping tendency" of A molecules, and vice versa, compared to ideal behavior, because of the "dislike" of A and B for each other relative to A for A and B for B. When w is positive, we have, on the other hand, "negative" deviations from ideality (Fig. 20–1, $cw/kT = 4$).

We found that $cw/kT = -4$ leads to critical behavior in a lattice gas (Section 14–4). The same is true here, as should be expected, and as can be seen in Fig. 20–3. When $cw/kT < -4$, the two components are no longer miscible in all proportions; over part of the composition range we get a separation of the solution into two phases of different composition (Fig. 20–4). In one phase the majority of molecules are of type A and in the other, of type B. We can understand this as follows: if AA and BB interactions are sufficiently favored over AB interactions (or if the temperature is low enough), the system splits into two solutions with excess A in one and excess B in the other in order to have the advantage of a larger

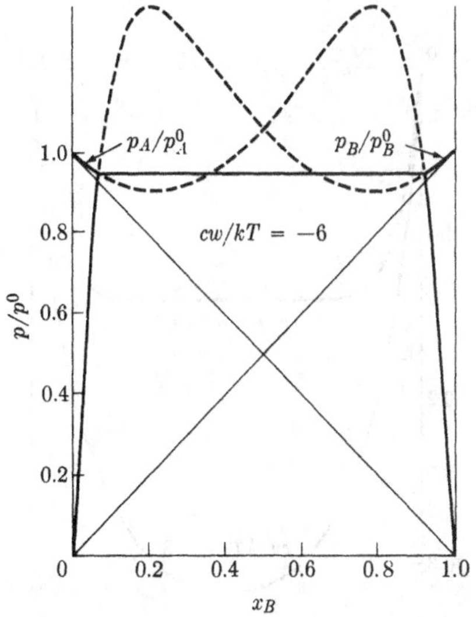

FIG. 20–4. Vapor pressure curves for Bragg-Williams binary solution showing phase separation.

number of AA and BB interactions. The location of the horizontal line (stable equilibrium curve) is fixed in Fig. 20–4 by the thermodynamic requirement that the chemical potential (or vapor pressure) of each component be the same, at equilibrium, in the two solutions of different composition.

It will already be apparent to the reader that this very simple (lattice, Bragg-Williams) theory of solutions predicts correctly some of the most important qualitative features observed experimentally with binary solutions. Of course, the theory cannot be expected to be satisfactory in a quantitative way.

We can also easily derive from the Bragg-Williams Q (Eq. 20–17) the following "mixing" properties (Problem 20–2):

$$\frac{\Delta A_m}{MkT} = x_A \ln x_A + x_B \ln x_B - \frac{cw}{2kT} x_A x_B, \tag{20–19}$$

$$\frac{\Delta S_m}{Mk} = -x_A \ln x_A - x_B \ln x_B, \tag{20–20}$$

$$\frac{\Delta E_m}{MkT} = -\frac{cw}{2kT} x_A x_B. \tag{20–21}$$

The entropy of mixing is the same as for an ideal solution. This is a con-

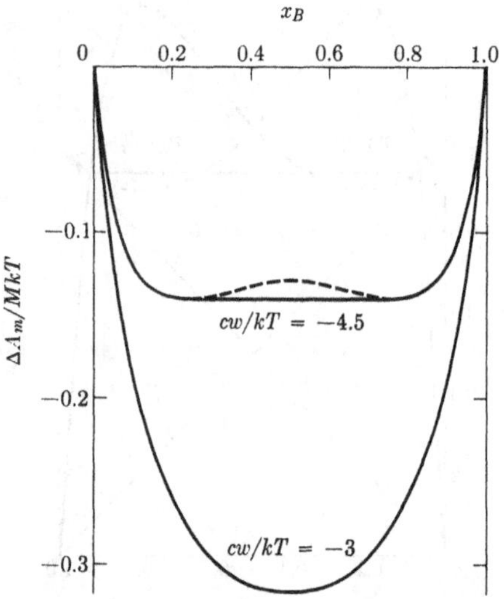

FIG. 20–5. Free energy of mixing for Bragg-Williams binary solution at temperatures above and below the critical temperature ($cw/kT_c = -4$).

sequence of the "random distribution" assumption in the Bragg-Williams theory. If w is negative (AA, BB more stable than AB), the energy of mixing is positive, as expected. The free energy of mixing is plotted in Fig. 20–5 for $cw/kT = -3$ and $cw/kT = -4.5$ (phase separation). The horizontal line (mixture of two solutions of different composition) is the stable equilibrium curve. This curve has a lower free energy than the dashed metastable curve.

It is useful for many purposes in the study of solutions to define "excess" quantities relative to the values the quantities would have if the pure components at a given pressure and temperature formed an *ideal* solution (at the same pressure and temperature). For example, for the entropy,

$$\Delta S_e = S - S\text{ (ideal)} = \Delta S_m - \Delta S_m\text{ (ideal)}. \qquad (20\text{–}22)$$

Thus, in the Bragg-Williams approximation,

$$\frac{\Delta A_e}{MkT} = -\frac{cw}{2kT}x_A x_B, \qquad (20\text{–}23)$$

$$\Delta S_e = 0, \qquad (20\text{–}24)$$

$$\Delta E_e = \Delta A_e. \qquad (20\text{–}25)$$

In experimental work it is in general found that ΔE_e and $T\,\Delta S_e$ are of the same order of magnitude. The fact that the present approximation predicts $\Delta S_e = 0$ is therefore a serious fault. In the next subsection we find that higher approximations (lattice model) do not improve this situation appreciably.

Other criticisms of the lattice model, which, incidentally, are also independent of the order of approximation used in treating the model, are the omission of p-V effects and the predicted symmetry in various properties about $x_A = \frac{1}{2}$, which is not always observed experimentally.

Quasi-chemical approximation. It is somewhat more troublesome to rederive binary solution equations in this approximation, so instead we take over results from Section 14–5 (lattice gas in the same approximation). To do this we take advantage of the fact that Q_{gas} in Eq. (14–29) and Q_{solution} in Eq. (20–9) contain the same sum $\sum$. From Eqs. (20–3), (20–4), and (14–62) we have

$$\frac{\mu_B}{kT} = -\left(\frac{\partial \ln Q_{\text{sol}}}{\partial M}\right)_{N_A,T} = -\ln q_B e^{-cw_{BB}/2kT} - \frac{\Phi}{kT}$$

$$= -\ln q_B e^{-cw_{BB}/2kT} + \ln x_B - \frac{c}{2}\ln\frac{(\beta+1)x_B}{\beta - 1 + 2x_B}, \tag{20–26}$$

where

$$\beta = [1 - 4x_A x_B(1 - e^{-w/kT})]^{1/2}.$$

Then

$$\frac{p_B}{p_B^0} = x_B\left[\frac{\beta - 1 + 2x_B}{(\beta+1)x_B}\right]^{c/2}. \tag{20–27}$$

Equations for μ_A and p_A may now be written down by symmetry considerations. The qualitative behavior of the vapor pressure equations is the same as in Figs. 20–1, 20–3, and 20–4. The critical temperature is given by Eq. (14–63).

The free energy is

$$\frac{A}{MkT} = \frac{x_A\mu_A}{kT} + \frac{x_B\mu_B}{kT}.$$

From Eq. (20–26) and its counterpart for component A, we have then

$$\frac{\Delta A_e}{MkT} = \frac{c}{2}\left[x_A \ln\frac{\beta - 1 + 2x_A}{(\beta+1)x_A} + x_B \ln\frac{\beta - 1 + 2x_B}{(\beta+1)x_B}\right]. \tag{20–28}$$

The total nearest-neighbor ("configurational") interaction energy, E_{config}, is, from Eqs. (20–6) through (20–8),

$$E_{\text{config}} = N_{AA}^* w_{AA} + N_{AB}^* w_{AB} + N_{BB}^* w_{BB}$$

$$= -\frac{N_{AB}^* w}{2} + \frac{cw_{AA}N_A}{2} + \frac{cw_{BB}N_B}{2},$$

where

$$\frac{N_{AB}^*}{cM} = \frac{2x_Ax_B}{\beta + 1}.$$

Then

$$\frac{\Delta E_m}{MkT} = \frac{\Delta E_e}{MkT} = -\frac{N_{AB}^* w}{2MkT} = -\frac{cwx_Ax_B}{kT(\beta + 1)}. \tag{20-29}$$

The expression for $T\,\Delta S_e = \Delta E_e - \Delta A_e$ then follows from Eqs. (20–28) and (20–29).

The significance of the results for the excess functions can best be seen by converting them into (high-temperature) power series in cw/kT. We find (Problem 20–3)

$$\frac{\Delta A_e}{MkT} = -\frac{x_Ax_B}{2}\left(\frac{cw}{kT}\right) - \frac{x_A^2x_B^2}{4c}\left(\frac{cw}{kT}\right)^2 + \cdots, \tag{20-30}$$

$$\frac{\Delta E_e}{MkT} = -\frac{x_Ax_B}{2}\left(\frac{cw}{kT}\right) - \frac{x_A^2x_B^2}{2c}\left(\frac{cw}{kT}\right)^2 + \cdots, \tag{20-31}$$

$$\frac{\Delta S_e}{Mk} = \frac{\Delta E_e}{MkT} - \frac{\Delta A_e}{MkT} = -\frac{x_A^2x_B^2}{4c}\left(\frac{cw}{kT}\right)^2 + \cdots. \tag{20-32}$$

Actually these series are exact, for a lattice model, as far as they go (in fact the quasi-chemical equations also give the exact cubic terms†). The linear terms are just the Bragg-Williams results (see Problem 14–17); they are associated with random mixing ($T \rightarrow \infty$). The quadratic terms are the first corrections for nonrandom mixing. Note that ΔS_e is always negative, as should be expected *in this model* from any reduction in randomness. But experimentally, ΔS_e *is sometimes positive.*

To get an idea of the orders of magnitudes involved in the above equations, let us take a numerical example: $x_A = x_B = 0.5$, $c = 10$ (a good average for a liquid), and $cw/kT = -4$ (a rather low temperature). Then

$$\frac{\Delta A_e}{MkT} = \frac{1}{2} - \frac{1}{40}, \qquad \frac{\Delta E_e}{MkT} = \frac{1}{2} - \frac{1}{20}, \qquad \frac{\Delta S_e}{MkT} = -\frac{1}{40}.$$

We see that the correction for nonrandomness is very small. Incidentally, the ideal mixing terms that have been subtracted out here [Eqs. (20–14) and (20–15)] are ln 2 = 0.69 for the entropy and −0.69 for the free energy. The main conclusion we reach is that improvement of the Bragg-Williams approximation does not provide an excess entropy term $T\,\Delta S_e$ of the same magnitude as ΔE_e, as found experimentally. This difficulty,

† See Guggenheim, pp. 62–70.

as well as the fact that ΔS_e is always negative, is therefore a property of the lattice model itself and not of approximations introduced in deriving thermodynamic functions from the model.

However, it should not be concluded from the above discussion that nonrandom mixing corrections are *never* important. For example, the location of the critical point is quite sensitive to such corrections [see Eq. (14–63), *et seq.*].

Because of the small correction found for ΔS_e in Eq. (20–32), the approximation of random mixing can appropriately be introduced, for molecules of like size, in lattice or cell solution theories that are otherwise fairly sophisticated (e.g., Section 20–2). Random mixing should not be assumed in such theories, however, if the molecules are very different in size (say have more than a 25% difference in diameters).

Solid solutions and alloys. The lattice theory of solutions has been applied recently by Halsey *et al.*, in considerable detail, to solid binary solutions of the rare gases.* One would expect that this kind of system would put the lattice theory in the best possible light. But, at least for argon-krypton solutions, the conclusion is reached that the experimental results cannot be explained by the present lattice model. We shall not attempt a comparison between experiment and the above equations because w was treated in Halsey's work as temperature-dependent. Negative values of w and critical solution phenomena are observed.

Another example of the application of lattice theory to binary solid solutions is found in the so-called order-disorder transition† in alloys (e.g., β brass, an alloy of copper and zinc). In this case w is positive and phase separation does not occur. However, a higher-order transition ("order-disorder") is observed, for example, in the heat capacity as a function of temperature at constant composition (say $x_A = 0.5$). The experimental heat capacity curves resemble the square lattice curve in Fig. 14–3. Qualitatively, the nature of the transition is the following. Suppose the lattice is simple cubic and $x_A = 0.5$, for ease of discussion. At low temperatures, because of the greater energetic stability of AB pairs relative to AA and BB, the molecules will tend to arrange themselves alternately in the lattice ($\cdots ABAB \cdots$), as for example in a crystal of NaCl. There is also a second and equivalent way to do this ($\cdots BABA \cdots$). In either case there is long-range configurational order of the type found in crystals but not in liquids (or in a solid solution when w

* J. F. Welling and G. D. Halsey, Jr., *J. Chem. Phys.* **30,** 1514 (1959); *J. Phys. Chem.* **62,** 752 (1958); J. H. Singleton and G. D. Halsey, Jr., *J. Phys. Chem.* **58,** 1011 (1954); M. Freeman and G. D. Halsey, Jr., *J. Phys. Chem.* **60,** 1119 (1956).

† See Guggenheim, Chapter 7, for details.

is negative). The extent of long-range order varies with temperature and completely disappears above a critical temperature T_c. Hence the term "order-disorder transition." It can be shown rigorously, by a symmetry argument,* that the heat capacity as a function of temperature, of a lattice model with $x_A = 0.5$, is the same for a given magnitude of w, whether w is negative (phase separation below T_c: $\cdots AAA \cdots BBB \cdots$) or positive (long-range order below T_c: $\cdots ABAB \cdots$). See, for example, Fig. 14–3. It follows from this that the temperature T_c of the transition is the same for the two systems (w positive or negative). For a square lattice, $e^{-|w|/2kT_c} = \sqrt{2} - 1$ exactly (Section 14–3), but the exact theoretical critical temperature is not known for any three-dimensional lattice.

Experimental heat capacity curves for binary solutions with negative w, at fixed (critical) composition, have been obtained recently.† They resemble Fig. 14–3, as expected.

20–2 Cell theories of binary solutions. We saw in the preceding section that our earlier study (Chapter 14) of lattice problems could easily be applied to binary solutions. Our object here is very similar: to show how the LJD cell theory of liquids (Section 16–2) can be extended to include binary solutions.

Our starting point is a modification of Eq. (20–5). We assume that the molecules are similar in size and that we have a lattice with every site (cell) occupied by an A or a B. The nearest-neighbor number is c and the nearest-neighbor distance is a. In view of our findings in Section 20–1, we assume further that we have a *random* distribution of molecules among sites. Then we write

$$Q = q_A^{N_A} q_B^{N_B} \frac{(N_A + N_B)!}{N_A! N_B!} e^{-\overline{W}/kT}, \qquad (20\text{–}33)$$

where $\overline{W}$ is the average configurational potential energy, calculated on a random-distribution basis, with each molecule at the center of its cell. That is,

$$\overline{W} = \overline{N}_{AA} w_{AA} + \overline{N}_{AB} w_{AB} + \overline{N}_{BB} w_{BB}$$

$$= \frac{Mc}{2} (x_A^2 w_{AA} + 2x_A x_B w_{AB} + x_B^2 w_{BB}). \qquad (20\text{–}34)$$

So far we seem to have just Eq. (20–17) for the Bragg-Williams lattice

---

* See S. M., Chapter 7.

† G. Jura, D. Fraga, G. Maki, and J. H. Hildebrand, *Proc. Nat. Acad. Sci.* **39**, 19 (1953).

theory. But now we introduce a refinement, which brings in the LJD theory. We evaluate q_A and q_B explicitly by a cell calculation of the LJD type, assuming that the c neighbors of a given molecule, smeared uniformly over the surface of a sphere of radius a, consist of cx_A A molecules and cx_B B molecules (again using the random distribution assumption). Thus q_A and q_B are now not only functions of T but also of $v = V/M$ and x_A (or x_B). Unlike the lattice theory, the present treatment does not make the solution incompressible, and V is an independent thermodynamic variable as it is in real solutions. V is related to a by Eq. (16–24): $a^3 = \gamma v$, where ordinarily we take $\gamma = \sqrt{2}$ (and $c = 12$). The thermodynamic equation for the Helmholtz free energy is

$$dA = -S\,dT - p\,dV + \mu_A\,dN_A + \mu_B\,dN_B, \tag{20–35}$$

$$A = -kT \ln Q.$$

Let us now begin to change notation to conform with Section 16–2. For each type of pair interaction, we use the Lennard-Jones potential (16–28) with parameters ϵ_{AA}, r^*_{AA}, ϵ_{AB}, r^*_{AB}, etc. Incidentally, in numerical calculations, in the absence of other information, one usually takes (Eq. 15–53)

$$\epsilon_{AB} = (\epsilon_{AA}\epsilon_{BB})^{1/2},$$

$$r^*_{AB} = \frac{r^*_{AA} + r^*_{BB}}{2}.$$

We assume A and B are both monatomic, so that

$$q_A = \frac{v_f^A e}{\Lambda_A^3}, \qquad q_B = \frac{v_f^B e}{\Lambda_B^3}, \tag{20–36}$$

as in Eq. (16–22). Polyatomic molecules with effective spherical symmetry can be treated simply by changing the definition of $\Lambda(T)$ (Problem 20–4). The factor e is included, rather arbitrarily, for a liquid (see Section 16–2). Actually, this factor or any other constant factor will have no effect on the equation of state, on excess or mixing properties, or on p_A/p_A^0, etc. The free volume for an A molecule is (Eq. 16–25)

$$v_f^A(v, x_A, T) = \int_\Delta e^{-\psi_A/kT} 4\pi r^2\,dr, \tag{20–37}$$

$$\psi_A = \varphi_A(\mathbf{r}) - \varphi_A(0),$$

where $\varphi_A(\mathbf{r})$ is the potential energy of interaction of an A molecule in its cell at $\mathbf{r}$ with its cx_A A neighbors and cx_B B neighbors. Instead of

Eq. (16–27), we have

$$\varphi_A(\mathrm{r}) = \frac{c}{2}\int_0^{\pi} [x_A u_{AA}(R) + x_B u_{AB}(R)] \sin\theta \, d\theta. \tag{20–38}$$

On writing out $x_A u_{AA} + x_B u_{AB}$ in detail, using the Lennard-Jones potential, we see that this quantity is equivalent to a single potential $u_A(R)$ with average parameters ϵ_A and r_A^*, which are functions of composition, determined by

$$\begin{aligned} \epsilon_A r_A^{*6} &= x_A \epsilon_{AA} r_{AA}^{*6} + x_B \epsilon_{AB} r_{AB}^{*6}, \\ \epsilon_A r_A^{*12} &= x_A \epsilon_{AA} r_{AA}^{*12} + x_B \epsilon_{AB} r_{AB}^{*12}. \end{aligned} \tag{20–39}$$

These equations can easily be solved for ϵ_A and r_A^{*6}. Thus ψ_A, $\varphi_A(0)$, and v_f^A depend in exactly the same way on ϵ_A and $v_A^* = r_A^{*3}/\gamma$ as do ψ, $\varphi(0)$, and v_f on ϵ and v^* in Eqs. (16–34) through (16–36). For example, if $g(v^*/v, \epsilon/kT)$ is the function in Eq. (16–36), then

$$v_f^A = 2\pi a^3 g(v_A^*/v, \epsilon_A/kT),$$

where v_A^* and ϵ_A are functions of x_A. Completely analogous results are of course found for v_f^B, etc. This means that numerical tables computed for the one-component liquid LJD theory can be employed here without alteration.

For the interaction potential energy in the LJD notation, we have

$$\overline{W} = \frac{N_A \varphi_A(0)}{2} + \frac{N_B \varphi_B(0)}{2}, \tag{20–40}$$

where, from Eq. (20–38),

$$\begin{aligned} \varphi_A(0) &= c[x_A u_{AA}(a) + x_B u_{AB}(a)], \\ \varphi_B(0) &= c[x_A u_{AB}(a) + x_B u_{BB}(a)]. \end{aligned}$$

It is easy to see that Eqs. (20–34) and (20-40) are the same, since $w_{AA} = u_{AA}(a)$, etc.

The random-distribution assumption can be pushed one step further in this theory, to simplify matters a little more. Here we consider all cells as equivalent (as they are, on the average) and use a *single* free volume v_f based on a potential energy that is averaged (random distribution) not only over the two kinds of nearest neighbors but also over the two kinds of occupation of the cell itself (A or B). Thus, in Eqs. (20–33) and (20–36), instead of $(v_f^A)^{N_A}(v_f^B)^{N_B}$, we have v_f^M, where

$$v_f = \int_\Delta e^{-\psi/kT} 4\pi r^2 \, dr, \tag{20–41}$$

$$\psi = \varphi(\mathrm{r}) - \varphi(0),$$

and

$$\varphi(\mathbf{r}) = \frac{c}{2}\int_0^\pi [x_A^2 u_{AA}(R) + 2x_A x_B u_{AB}(R) + x_B^2 u_{BB}(R)] \sin\theta \, d\theta. \quad (20\text{–}42)$$

If we replace the expression in square brackets by a single Lennard-Jones potential $u(R)$ with average parameters ϵ and r^* (functions of composition), we find that ϵ and r^* are determined by the equations

$$\begin{aligned} \epsilon r^{*6} &= x_A^2 \epsilon_{AA} r_{AA}^{*6} + 2x_A x_B \epsilon_{AB} r_{AB}^{*6} + x_B^2 \epsilon_{BB} r_{BB}^{*6}, \\ \epsilon r^{*12} &= x_A^2 \epsilon_{AA} r_{AA}^{*12} + 2x_A x_B \epsilon_{AB} r_{AB}^{*12} + x_B^2 \epsilon_{BB} r_{BB}^{*12}, \end{aligned} \quad (20\text{–}43)$$

which are again easy to solve. In this case, then, we get the LJD functions (only one set this time) ψ, $\varphi(0)$, and v_f in the same notation as in Section 16–2, but we have to keep in mind the fact that ϵ and r^* (or v^*) are functions of composition.

The interaction potential energy is

$$\overline{W} = \frac{M\varphi(0)}{2}, \quad (20\text{–}44)$$

where, from Eq. (20–42),

$$\varphi(0) = c[x_A^2 u_{AA}(a) + 2x_A x_B u_{AB}(a) + x_B^2 u_{BB}(a)].$$

This is the same $\overline{W}$ as in Eq. (20–40).

From Eqs. (20–33), (20–36), and (20–44), the canonical ensemble partition function in the single free-volume approximation is

$$Q = \frac{(v_f e)^M (N_A + N_B)! e^{-M\varphi(0)/2kT}}{\Lambda_A^{3N_A} \Lambda_B^{3N_B} N_A! N_B!}, \quad (20\text{–}45)$$

or

$$\frac{A}{MkT} = x_A \ln \Lambda_A^3 + x_B \ln \Lambda_B^3 - \ln v_f e + x_A \ln x_A + x_B \ln x_B + \frac{\varphi(0)}{2kT}, \quad (20\text{–}46)$$

where v_f is a function of v, T, and x_A, and $\varphi(0)$ is a function of v and x_A. The equation of state is found from

$$-p = \left(\frac{\partial A}{\partial V}\right)_{T,N_A,N_B} = \left(\frac{\partial A}{\partial V}\right)_{T,M,x_A}. \quad (20\text{–}47)$$

Since v_f and $\varphi(0)$ are the same functions as in the LJD theory [except for the dependence of ϵ and v^* on x_A, which, however is constant in the differentiation (20–47)], we get the *same* equation of state (but with ϵ and v^* functions of x_A), (16–37). We are usually interested in solution behavior at low and constant pressure. In computational work, $p = 0$ is often chosen, since for a condensed phase most properties are relatively insensitive to p. If we choose T and take $p = 0$ (or $p =$ any constant), Eq. (16–37) then provides $v = V/M$ as a function of x_A.

Now A/MkT in Eq. (20–46) is a function of v, T, and x_A. If we want p, T, and x_A as independent variables, v can be determined as a function of these variables from the equation of state, (16–37). Therefore we may consider v in Eq. (20–46) to be replaced by $v(p, T, x_A)$, so that A/MkT becomes a function of p, T, and x_A.

We know from thermodynamics that for an ideal solution (defined by stating that the fugacities f_A and f_B obey Raoult's law), the quantities ΔV_m, ΔE_m, and ΔH_m are all zero, for mixing at constant p and T, while ΔS_m is given by Eq. (20–14). These properties require, in the present theory, that for an ideal solution

$$\frac{A_{\text{ideal}}}{MkT} = x_A \ln \Lambda_A^3 + x_B \ln \Lambda_B^3 - x_A \ln v_f(x_A = 1)e - x_B \ln v_f(x_B = 1)e$$
$$+ x_A \ln x_A + x_B \ln x_B + \frac{x_A\varphi(0, x_A = 1) + x_B\varphi(0, x_B = 1)}{2}. \qquad (20\text{–}48)$$

The quantities labeled $x_A = 1$ and $x_B = 1$ refer to pure liquid A and B, and the independent variables are p, T, and x_A, as explained above. Hence v is in general different in the terms labeled $x_A = 1$ and $x_B = 1$ (i.e., pure A and B will usually have different molar volumes at the same pressure). Equation (20–48) necessarily gives

$$\frac{\Delta A_m \text{ (ideal)}}{MkT} = x_A \ln x_A + x_B \ln x_B,$$

as required. From Eqs. (20–46) and (20–48), we have for the excess Helmholtz free energy,

$$\frac{\Delta A_e}{MkT} = x_A \ln v_f(x_A = 1) + x_B \ln v_f(x_B = 1) - \ln v_f(x_A)$$
$$+ \frac{\varphi(0, x_A) - x_A\varphi(0, x_A = 1) - x_B\varphi(0, x_B = 1)}{2kT}, \qquad (20\text{–}49)$$

where p, T, and x_A are independent variables and v is in general different in the terms labeled $x_A = 1$, $x_B = 1$, and x_A. All terms in Eq. (20–49) contribute to ΔE_e, but only the v_f terms contribute to ΔS_e (Problem 20–5). The φ terms are essentially equivalent to Eq. (20–23) in the Bragg-Williams lattice theory (Problem 20–6), but are not exactly the same because the nearest-neighbor distance a is different in pure A, pure B, and in the solution at the same pressure. It is clear from the form of the v_f terms that ΔS_e might be positive or negative.

We shall not pursue this subject further in any detail. It is sufficient to say that Salsburg and Kirkwood* have made detailed calculations based on Eq. (20–45), but with the added refinement of including three shells of neighbors in the calculations instead of only one. These authors have compared their results with a number of experimental systems. A very considerable improvement over the lattice theory is found, and complete qualitative agreement with experiment is achieved. In particular, ΔS_e is calculated to be of the correct order of magnitude, though always somewhat too small. The theory gives the correct sign for ΔS_e (usually positive), unlike the lattice theory which always has $\Delta S_e < 0$. Both ΔV_e and ΔH_e are in general too large (but the lattice theory does not even consider the volume: $\Delta V_e = 0$ always). Usually experimental values of ΔH_e and ΔV_e have the same sign. We shall see that first-order conformal solution theory (Section 20–4) predicts that ΔH_e and ΔV_e *always* have the same sign. Recent experimental cases in which these two quantities have opposite signs have been observed (e.g., neopentane and carbon tetrachloride). The cell theory under discussion here correctly predicts this sign reversal in such cases.

20–3 Random-mixing, corresponding-states theory. The treatment in the preceding section has one serious source of error, which is easy to remedy: it is based on LJD theoretical functions for a pure liquid. If we set for ourselves the less ambitious goal of predicting properties of the solution *from those of the pure liquids* rather than from first principles, then instead of using LJD functions for the pure liquid state, we may use experimental (corresponding-states) functions. In this section we show briefly how this can be done. Otherwise we employ the methods (random mixing, average ϵ and r^*) of Section 20–2. As we should expect, since experimental information is built into the "theory," agreement with experimental data on solutions is better, quantitatively, than with use of the LJD theory.†

The method of this section was suggested independently by Brown,‡ Prigogine *et al.*,‡ and Scott.‡

---

* Z. W. SALSBURG and J. G. KIRKWOOD, *J. Chem. Phys.* **20,** 1538 (1952); **21,** 2169 (1953). See also I. PRIGOGINE and G. GARIKIAN, *Physica* **16,** 239 (1950); I. PRIGOGINE and V. MATHOT, *J. Chem. Phys.* **20,** 49 (1952); J. A. POPLE, *Trans. Faraday Soc.* **49,** 591 (1953).

† See Prigogine, Chapter 11, for a detailed summary.

‡ W. B. BROWN, *Phil. Trans. Roy. Soc. (London)* **250,** 175, 221 (1957); I. PRIGOGINE, A. BELLEMANS, and A. ENGLERT-CHWOLES, *J. Chem. Phys.* **24,** 518 (1956); R. L. SCOTT, *J. Chem. Phys.* **25,** 193 (1956).

We recall [Eqs. (20–33) through (20–40)] that in extending the LJD theory to mixtures we passed from the one-component equation

$$Q = \left(\frac{v_f e e^{-\varphi(0)/2kT}}{\Lambda^3}\right)^N \tag{20–50}$$

to the two-component equation

$$Q = \left(\frac{v_f^A e e^{-\varphi_A(0)/2kT}}{\Lambda_A^3}\right)^{N_A} \left(\frac{v_f^B e e^{-\varphi_B(0)/2kT}}{\Lambda_B^3}\right)^{N_B} \frac{(N_A + N_B)!}{N_A! N_B!}, \tag{20–51}$$

where v_f^A and $\varphi_A(0)$ depend on the average parameters ϵ_A and r_A^* determined by Eqs. (20–39). Here the procedure is identical except that the starting point is the *experimental* corresponding-states equation (16–45). That is, the one-component equation is, instead of (20–50),

$$Q = \left[\frac{r^{*3}\Psi(\epsilon/kT, v/r^{*3})}{\Lambda^3}\right]^N,$$

and the two-component equation is assumed to be

$$Q = \left[\frac{r_A^{*3}\Psi(\epsilon_A/kT, v/r_A^{*3})}{\Lambda_A^3}\right]^{N_A} \left[\frac{r_B^{*3}\Psi(\epsilon_B/kT, v/r_B^{*3})}{\Lambda_B^3}\right]^{N_B} \frac{(N_A + N_B)!}{N_A! N_B!}, \tag{20–52}$$

where ϵ_A, r_A^*, ϵ_B, and r_B^* are average values still determined by Eqs. (20–39). These equations refer specifically to monatomic molecules, but if we change the usual definition of $\Lambda(T)$ (see Problem 20–4), they also apply to polyatomic molecules (with effective spherical symmetry) that obey the law of corresponding states.

Just as we used a single v_f and $\varphi(0)$ in Eq. (20–45), we can simplify (20–52) by writing

$$Q = \frac{[r^{*3}\Psi(\epsilon/kT, v/r^{*3})]^M (N_A + N_B)!}{\Lambda_A^{3N_A}\Lambda_B^{3N_B} N_A! N_B!}, \tag{20–53}$$

where ϵ and r^* are averages found from Eqs. (20–43).

We shall not extend this discussion any further. Prigogine† (Supplementary Reading list) goes into the subject in great detail. It will be obvious (Problem 20–7) from the LJD equations (20–46) through (20–49) how to compute the equation of state, ΔA_e, etc. In fact, the LJD equations may be considered a special case of the present discussion in which an *approximate theoretical* function is used for Ψ instead of an *empirical* one.

† Chapters 9 and 10.

20-4 Conformal solution theory. Equation (20-53) is an approximate equation applicable to a binary mixture of liquids, each of which obeys the law of corresponding states. In writing Eq. (20-53) we have assumed a random distribution of the two components in: (a) the configurational factor $M!/N_A!N_B!$; (b) the occupation of any given element of volume [this is implied in the use of a single Ψ—see the discussion preceding Eq. (20-41)]; and (c) the occupation of an element of volume in the neighborhood of a fixed molecule of either component [this is implied in the use of average parameters r^* and ϵ from Eq. (20-43), based on a random distribution]. Now there would *really* be random mixing and these assumptions would be *exact* if both components had the same interaction parameters (e.g., an isotopic mixture), say ϵ_{00} and r^*_{00}. Similarly, in the lattice model, the random mixing assumed in the Bragg-Williams approximation is exact if $w = 0$. Actually, the assumption of random mixing in the lattice model leads to the *exact first-order or linear terms* if we use expansions in powers of w/kT as in Eqs. (20-30) through (20-32). This is not a special situation, but is in fact a property of perturbation methods generally (e.g., in quantum mechanics): exact first-order terms are obtained by averaging with *unperturbed* (in our case, random mixing) weights. Whereas the expansions were in powers of w/kT in the lattice model, here we should use expansions in powers of $\epsilon_{AA} - \epsilon_{00}$, $\epsilon_{AB} - \epsilon_{00}$, etc., or the equivalent, since the unperturbed state (with random mixing) corresponds to these quantities having the value zero. We therefore reach the following conclusion: use of the random-distribution assumptions listed above and Eq. (20-53) will lead to an *exact* first-order (in $\epsilon_{AA} - \epsilon_{00}$, etc., or equivalent) solution theory for molecules obeying the law of corresponding states. This means that the components must have very nearly the same interaction parameters ϵ and r^*. An equivalent statement is that it must be possible to choose parameters ϵ_{00} and r^*_{00} of a reference component (hypothetical, or, more often, one of the components in the solution) such that $\epsilon_{AA} - \epsilon_{00}$, etc., are all small.

This general approach is referred to as conformal solution theory and is due to Longuet-Higgins.† The argument we use here (for continuity and simplicity) is somewhat different from that of Longuet-Higgins. It is based on Eq. (20-53) as the starting point and is therefore restricted to first-order effects. Second-order conformal solution theory has been discussed in detail by Brown.‡

Without complication we can easily generalize Eq. (20-53) in two ways: (a) we consider any number of components; and (b) instead of restrict-

† H. C. Longuet-Higgins, *Proc. Roy. Soc.* **205A**, 247 (1951). For a review, see Prigogine, Chapter 4.

‡ W. B. Brown, *Proc. Roy. Soc.* **240A**, 561 (1957).

ing ourselves to the Lennard-Jones potential, we use (Eq. 15–33) $u(r) = \epsilon h(r/r^*)$ for each pair interaction.

For the reference component,

$$u_{00} = \epsilon_{00} h\left(\frac{r}{r_{00}^*}\right),$$

and for the interaction between a molecule of component i and one of component j,

$$u_{ij} = \epsilon_{ij} h\left(\frac{r}{r_{ij}^*}\right). \tag{20–54}$$

Following Longuet-Higgins, we define f_{ij} [not to be confused with f_{ij} in Eq. (15–38)] and g_{ij} by

$$\epsilon_{ij} = f_{ij}\epsilon_{00}, \qquad r_{ij}^* = \frac{r_{00}^*}{g_{ij}}, \tag{20–55}$$

and we shall use expansions in powers of $f_{ij} - 1$ and $g_{ij} - 1$ (the reference component has $f_{00} = g_{00} = 1$).

Let us rewrite Eq. (20–53) for a multicomponent system and at the same time omit the factors $\Lambda_i^{3N_i}$, calling the remaining partition function the configurational partition function Q_c:

$$Q_c(\mathbf{N}, V, T) = \frac{[r^{*3}\Psi(\epsilon/kT, v/r^{*3})]^M M!}{\prod_i N_i!}, \tag{20–56}$$

where $M = \sum_i N_i$, $v = V/M$, r^* and ϵ are (random) average parameters, and Ψ is an empirical corresponding-states function applicable to each pure component. For example, for pure s,

$$Q_{cs} = \left[r_{ss}^{*3}\Psi\left(\frac{\epsilon_{ss}}{kT}, \frac{v}{r_{ss}^{*3}}\right)\right]^{N_s}. \tag{20–57}$$

For a hypothetical (isotopic) reference solution with all parameters the same as ϵ_{00} and r_{00}^*,

$$Q_c^0(\mathbf{N}, V, T) = \frac{[r_{00}^{*3}\Psi(\epsilon_{00}/kT, \ v/r_{00}^{*3})]^M M!}{\prod_i N_i!}. \tag{20–58}$$

At this point, let us calculate the first-order random average parameters r^* and ϵ for use in Eq. (20–56). We have (x = mole fraction)

$$r^* = \sum_{ij} x_i x_j r_{ij}^* = \sum_{ij} x_i x_j \frac{r_{00}^*}{g_{ij}},$$

or

$$r^* - r^*_{00} = r^*_{00} \sum_{ij} x_i x_j \left(\frac{1}{g_{ij}} - 1\right).$$

Thus to first-order terms in $g_{ij} - 1$,

$$r^* = r^*_{00}\left[1 - \sum_{ij} x_i x_j (g_{ij} - 1)\right]$$

and

$$r^{*3} = r^{*3}_{00}\left[1 - 3\sum_{ij} x_i x_j (g_{ij} - 1)\right]. \tag{20-59}$$

Similarly,

$$\epsilon = \epsilon_{00}\left[1 + \sum_{ij} x_i x_j (f_{ij} - 1)\right]. \tag{20-60}$$

We notice, from Eq. (20-58), that

$$Q_c^0(\mathbf{N}, V', T') = \frac{[r^{*3}_{00}\Psi(\epsilon/kT, v/r^{*3})]^M M!}{\prod_i N_i!},$$

where $V' = Vr^{*3}_{00}/r^{*3}$ and $T' = T\epsilon_{00}/\epsilon$. Comparison with Eq. (20-56) then gives

$$Q_c(\mathbf{N}, V, T) = \left(\frac{r^*}{r^*_{00}}\right)^{3M} Q_c^0(\mathbf{N}, V', T'). \tag{20-61}$$

Next, we want to expand $\ln Q_c^0(\mathbf{N}, V', T')$ in powers of the $g_{ij} - 1$ and $f_{ij} - 1$. This, together with Eq. (20-59), will give us $\ln Q_c(\mathbf{N}, V, T)$ (that is, the logarithm of the configurational partition function of the actual solution of interest), expanded in powers of the $g_{ij} - 1$ and $f_{ij} - 1$. To linear terms,

$$\begin{aligned}\ln Q_c^0(\mathbf{N}, V', T') = \ln Q_c^0(\mathbf{N}, V, T) &+ \sum_{ij}\left[\left(\frac{\partial \ln Q_c^0(\mathbf{N}, V', T')}{\partial g_{ij}}\right)_{\substack{\mathbf{N},V,T\\ f=g=1}} (g_{ij} - 1)\right.\\ &+ \left.\left(\frac{\partial \ln Q_c^0(\mathbf{N}, V', T')}{\partial f_{ij}}\right)_{\substack{\mathbf{N},V,T\\ f=g=1}} (f_{ij} - 1)\right],\end{aligned} \tag{20-62}$$

where

$$\frac{\partial \ln Q_c^0(\mathbf{N}, V', T')}{\partial g_{ij}} = \frac{\partial \ln Q_c^0}{\partial V'}\frac{\partial V'}{\partial g_{ij}} = \frac{3pV}{kT} x_i x_j \xrightarrow[f,g\to 1]{} \frac{3p^0 V x_i x_j}{kT}$$

and we have used Eq. (20–59) in differentiating V'. The pressure $p^0(M, V, T)$ is the pressure of the reference solution (or of the pure reference component at the same M, V, and T, since the intermolecular forces are the same in either case). Similarly, from Eq. (20–60),

$$\frac{\partial \ln Q_c^0(\mathbf{N}, V', T')}{\partial f_{ij}} = \frac{\partial \ln Q_c^0}{\partial T'} \frac{\partial T'}{\partial f_{ij}} \xrightarrow[f,g \to 1]{} - \frac{E_c^0 x_i x_j}{kT},$$

where $E_c^0(M, V, T)$ is the configurational energy (intermolecular potential energy) of the reference solution (or of the pure reference component at M, V, and T). Then Eq. (20–62) becomes

$$\ln Q_c^0(\mathbf{N}, V', T') = \ln Q_c^0(\mathbf{N}, V, T) + \sum_{ij} x_i x_j \left[\frac{3p^0 V}{kT} (g_{ij} - 1) - \frac{E_c^0}{kT} (f_{ij} - 1)\right] + \cdots.$$

Finally, we put this result and Eq. (20–59) in Eq. (20–61) to obtain

$$A_c(\mathbf{N}, V, T) - A_c^0(\mathbf{N}, V, T) = \sum_{ij} x_i x_j [E_c^0(f_{ij} - 1) + 3(MkT - p^0 V)(g_{ij} - 1)] + \cdots, \quad (20\text{–}63)$$

where $A_c = -kT \ln Q_c$ is the configurational Helmholtz free energy. This exact and elegant result expresses A_c for the solution in terms of the intermolecular interaction parameters f_{ij} and g_{ij} and of the properties of the pure reference component E_c^0, p^0, and

$$A_{c,\text{1-comp}}^0(M, V, T) = A_{c,\text{sol}}^0(\mathbf{N}, V, T) - MkT \sum_i x_i \ln x_i. \quad (20\text{–}64)$$

The A_c^0 in Eq. (20–63) refers, of course, to the solution; the last term in Eq. (20–64) is the ideal entropy of mixing term. These properties of the pure reference component are determined by ϵ_{00}, r_{00}^*, and the empirical law of corresponding states.

The independent variables $\mathbf{N}$, p, T and the function F_c are more convenient than $\mathbf{N}$, V, T and the function A_c, so we proceed now to make this change. Let p be the pressure on the actual solution when its volume is V and let $V^\square$ be the volume of the actual solution when its pressure is p^0. Now let us abbreviate Eq. (20–63) by the notation

$$A_c(p, V) - A_c^0(p^0, V) = \sum (p^0, V). \quad (20\text{–}65)$$

What we want (since we keep the same reference solution at p^0, V) is the corresponding equation for

$$F_c(p^0, V^\square) - F_c^0(p^0, V),$$

where

$$F_c(p^0, V^\square) = A_c(p^0, V^\square) + p^0V^\square, \tag{20–66}$$

$$F_c(p^0, V) = A_c(p^0, V) + p^0V. \tag{20–67}$$

If we integrate $\partial A_c/\partial V = -p$ from p^0, $V^\square$ to p, V, we obtain, to the linear term in $V - V^\square$,

$$A_c(p, V) - A_c(p^0, V^\square) = -p^0(V - V^\square). \tag{20–68}$$

From Eqs. (20–65) through (20–68) we find, then, that

$$F_c(\mathbf{N}, p^0, T) - F_c^0(\mathbf{N}, p^0, T)$$
$$= \sum_{ij} x_i x_j [E_c^0(f_{ij} - 1) + 3(MkT - p^0V)(g_{ij} - 1)] + \cdots, \tag{20–69}$$

where the right side is the same as in Eqs. (20–63) and (20–65). In Eq. (20–69), however, we regard V and E_c^0 as functions of M, p^0, and T.

Our next task is to derive an equation for ΔF_e, which, incidentally, is the same as $\Delta F_{c(\text{excess})}$, since nonconfigurational contributions are the same in F and F_{ideal} and therefore cancel. For N_i molecules of a single component i at p^0 and T, Eq. (20–69) reads

$$F_{ci}(N_i, p^0, T) - F_{ci}^0(N_i, p^0, T)$$
$$= E_{ci}^0(f_{ii} - 1) + 3(N_i kT - p^0 V_i)(g_{ii} - 1) + \cdots, \tag{20–70}$$

where F_{ci}^0, E_{ci}^0, and V_i all refer to N_i molecules of the reference component at p^0 and T. If the pure components mixed to form an ideal solution, we would have

$$F_c(\mathbf{N}, p^0, T)\ (\text{ideal}) = \sum_i F_{ci}(N_i, p^0, T)$$
$$+ MkT \sum_i x_i \ln x_i. \tag{20–71}$$

Since the reference solution is ideal, we also have

$$F_c^0(\mathbf{N}, p^0, T) = \sum_i F_{ci}^0(N_i, p^0, T)$$
$$+ MkT \sum_i x_i \ln x_i. \tag{20–72}$$

Finally, then, from Eqs. (20–69) through (20–72),

$$\begin{aligned}\Delta F_e = \Delta F_{c(\text{excess})} &= F_c(\mathbf{N}, p^0, T) - F_c(\mathbf{N}, p^0, T) \text{ (ideal)} \\ &= \sum_{ij} x_i x_j [E_c^0(f_{ij} - 1) + 3(MkT - p^0 V)(g_{ij} - 1)] \\ &\quad - \sum_i [E_{ci}^0(f_{ii} - 1) + 3(N_i kT - p^0 V_i)(g_{ii} - 1)] + \cdots .\end{aligned}$$

But $E_{ci}^0 = x_i E_c^0$ and $V_i = x_i V$ (the reference solution is ideal), so that this last equation reduces to (put $x_i = x_i \sum_j x_j$)

$$\begin{aligned}\Delta F_e &= E_c^0 \sum_{i<j} x_i x_j (2f_{ij} - f_{ii} - f_{jj}) \\ &\quad + 3(MkT - p^0 V) \sum_{i<j} x_i x_j (2g_{ij} - g_{ii} - g_{jj}) + \cdots . \end{aligned} \tag{20–73}$$

This is the central result of (first-order) conformal solution theory. If we make the presumably excellent approximation that

$$r_{ij}^* = \frac{(r_{ii}^* + r_{jj}^*)}{2},$$

or (to first order)

$$g_{ij} = \frac{(g_{ii} + g_{jj})}{2},$$

Eq. (20–73) simplifies to

$$\Delta F_e(\mathbf{N}, p^0, T) = E_c^0(M, p^0, T) \sum_{i<j} x_i x_j d_{ij}, \tag{20–74}$$

where

$$d_{ij} = 2f_{ij} - f_{ii} - f_{jj}.$$

Other excess quantities follow immediately from Eq. (20–74):

$$\Delta S_e = -\left(\frac{\partial \Delta F_e}{\partial T}\right)_{p^0, \mathbf{N}} = -\left(\frac{\partial E_c^0}{\partial T}\right)_{p^0, M} \sum_{i<j} x_i x_j d_{ij}, \tag{20–75}$$

$$\Delta H_e = \Delta F_e + T \Delta S_e = \left(E_c^0 - T \frac{\partial E_c^0}{\partial T}\right) \sum_{i<j} x_i x_j d_{ij}, \tag{20–76}$$

$$\Delta V_e = \left(\frac{\partial \Delta F_e}{\partial p^0}\right)_{T, \mathbf{N}} = \left(\frac{\partial E_c^0}{\partial p^0}\right)_{T, M} \sum_{i<j} x_i x_j d_{ij}. \tag{20–77}$$

Note that for binary solutions the sum $\sum$ in the above equations reduces to a single term, $x_1x_2\,d_{12}$, symmetrical about $x_1 = \frac{1}{2}$, as in the Bragg-Williams theory (Eq. 20–23).

At ordinary pressures, E_c^0 is negative (E_c^0 is practically equal to the negative of the energy of vaporization of the reference component at T), as are also all the other coefficients of $\sum$ in Eqs. (20–75) through (20–77): an increase in T expands the liquid, increases the distance between neighbors, and hence makes E_c^0 less negative; an increase in p^0 contracts the liquid and makes E_c^0 more negative; both terms in the coefficient of $\sum$ in Eq. (20–76) are negative. The sum $\sum$ can be positive or negative, but is of course the same in all these equations. In a binary solution, d_{12} and $\sum$ are positive if 12 pairs are energetically more stable than 11 and 22 pairs [this corresponds to a positive w in Eq. (20–6)]. In this case (binary, $d_{12} > 0$), all the excess quantities are negative. In general (multicomponent), the excess quantities are all of the same sign, positive or negative, and in the ratio of the coefficients of $\sum$ for any composition.

The above predictions (and others) of the first-order conformal theory of solutions were extensively compared with experiment by Longuet-Higgins in 1951. The theory proved to be quite successful when applied to data on suitable binary systems available at that time. In particular, all the excess functions were found to have the same sign. However, more recently other "suitable" solutions, such as neopentane + carbon tetrachloride and methane + carbon monoxide, have been found experimentally to have positive ΔH_e and negative ΔV_e. We have already mentioned that this feature can be explained by the approximate "random" theories of Sections 20–2 and 20–3.

The conclusion we reach is that although first-order conformal solution theory is formally exact for almost ideal solutions of "corresponding-states" molecules, its range of applicability to real systems is quite limited. To extend the range of applicability, one must turn to the exact (and relatively complicated) second-order conformal solution theory or to the approximate theories of Sections 20–2 and 20–3.

As a final remark, we remind the reader that the "theories" discussed in Section 20–3 and in the present section are semiempirical in a sense, for they are based on experimental "corresponding-states" information about the pure components. This is justified in the absence of an exact theory of the liquid state. On the other hand, Section 20–2 is completely theoretical, but in it we must use an approximate (LJD) theory of liquids.

PROBLEMS

20–1. Use the Gibbs-Duhem equation to deduce an expression for p_B/p_B^0 from (a) Eq. (20–12) (ideal), and (b) Eq. (20–18) (Bragg-Williams). (Pages 374 and 376.)

20–2. Derive the Bragg-Williams "mixing" equations (20–19) through (20–21). (Page 377.)

20–3. Expand the quasi-chemical equations (20–28) and (20–29) in powers of w/kT to obtain the second-order results (20–30) through (20–32). (Page 380.)

20–4. Show how to redefine Λ in equations such as (20–36) so that it may refer to diatomic or polyatomic molecules instead of to monatomic molecules. (Page 383.)

20–5. Break up Eq. (20–49) for ΔA_e into separate expressions for ΔE_e and ΔS_e. (Page 386.)

20–6. Show that the φ terms in Eq. (20–49) (LJD theory) are essentially equivalent to Eq. (20–23) (Bragg-Williams). (Page 386.)

20–7. Discuss the equation of state and ΔA_e from Eq. (20–53). (Page 388.)

20–8. Develop a Bragg-Williams lattice theory of binary solutions, allowing each site to be occupied either by A, by B, *or be empty*. Take the lattice distance a = constant.

20–9. Derive expressions for μ_A from Eqs. (20–52) and (20–53). Will p_A/p_A^0 and p_B/p_B^0 have the usual symmetry about $x_A = \frac{1}{2}$?

20–10. Show from Eq. (20–74) that, for a binary conformal solution,

$$\mu_1 = (\text{function of } p^0, T) + kT \ln x_1 + \left(\frac{E_c^0 d_{12}}{N_1 + N_2}\right) x_2^2.$$

Compare this with the Bragg-Williams equation preceding (20–18).

20–11. Discuss the equation of state and critical behavior (gas-liquid and solution-solution) of a conformal solution.

20–12. In the application of Eq. (20–74) to a binary solution, ΔF_e should be the same irrespective of the arbitrary choice of component 1 or component 2 as reference component. Investigate this point.

20–13. What can be deduced about d_{12} if $\epsilon_{12} = (\epsilon_{11}\epsilon_{22})^{1/2}$?

20–14. In a one-component "corresponding-states" system, Q is given by the equation preceding Eq. (20–52). If we regard A as a function of T, V, N, ϵ, and r^{*3}, find the coefficients a_1 and a_2 in the equation

$$dA = -S\,dT - p\,dV + \mu\,dN + a_1\,d\epsilon + a_2\,dr^{*3}.$$

Derive an analogous equation for a conformal solution regarding A as a function of T, V, $\mathbf{N}$, $\mathbf{f}$, $\mathbf{g}$. Use this result to provide an alternative (and more elegant) derivation of Eq. (20–69) for F_c from Eq. (20–63) for A_c.

Supplementary Reading

Fowler and Guggenheim, Chapter 8.
Guggenheim, Chapters 1–4.
Hildebrand and Scott, Chapters 2, 3, 6–8.
Prigogine, Chapters 3, 4, 8–11.
S. M., Chapter 7.

CHAPTER 21

POLYMER AND POLYELECTROLYTE SOLUTIONS AND GELS

In Chapter 13 we considered the configuration of polymer molecules and rubber elasticity. Our object in the present chapter is to extend this discussion to polymer and polyelectrolyte solutions and gels. Since this is a relatively specialized topic in statistical mechanics, we shall confine ourselves in each section to the simplest possible analysis that brings out the essential features.

To present a unified discussion, we devote the first section to the Wall theory of rubber elasticity. This leads to the same length-force equation (13–57) as the (simplified) James-Guth argument of Section 13–3, but the method of Wall has the advantage that it has been exploited in the theory of polymer and polyelectrolyte gels (Sections 21–3 and 21–4). As far as the theory of rubber elasticity itself is concerned, the James-Guth theory (which we have not fully presented) is more detailed and fundamental.

21–1 Wall theory of rubber elasticity.* The reader should review Section 13–3 through Eq. (13–54) for general background and point of view.

We start with an isotropic cube of rubber with edge L_0, under no forces. On a molecular level, we assume that the sample is made up of a cross-linked network of N chains, all of the same length. But our results will turn out to be independent of chain length, so the assumption of uniformity of chain length is not really necessary. We assume further that the distribution of end-to-end lengths r of the N chains in the undeformed (L_0^3) network is gaussian and is in fact the same distribution that N *free* molecules would have (Chapter 13). Equation (13–40) is therefore applicable, and we rewrite it in terms of the components of r as

$$P_0(x, y, z)\,dx\,dy\,dz = \frac{\beta^3}{\pi^{3/2}}\,e^{-\beta^2(x^2+y^2+z^2)}\,dx\,dy\,dz\,, \qquad (21\text{–}1)$$

$$\beta^2 = \frac{3}{2\overline{r^2}}\cdot$$

---

* F. T. Wall, *J. Chem. Phys.* **10**, 132, 485 (1942); **11**, 527 (1943). See also Flory, Chapter 11.

The subscript zero refers to the undeformed state. Our final assumption is that if the cube is deformed from edges of length L_0 to edges of length L_x, L_y, L_z, then the end-to-end distribution becomes

$$P(x, y, z)\,dx\,dy\,dz = \frac{\beta^3}{\alpha_x\alpha_y\alpha_z\pi^{3/2}}\,e^{-\beta^2[(x^2/\alpha_x^2)+(y^2/\alpha_y^2)+(z^2/\alpha_z^2)]}\,dx\,dy\,dz, \quad (21\text{–}2)$$

where $\alpha_x = L_x/L_0$, etc. This is obviously true if each of the N chains is deformed in the same way (α_x, α_y, α_z) as the bulk sample itself (Problem 21–1), but this detailed an assumption is not needed [only the over-all distribution (21–2)].

In the undeformed state the chains are free to assume the unbiased "random walk" distribution (21–1). We may therefore anticipate that this distribution corresponds to the maximum possible entropy of the system, S_0. Any deformation of the sample will lead to a decrease in entropy because the chains are forced to assume a distribution [e.g., (21–2)] which is not "random" or "unbiased." Our next step is to compute the entropy difference $S(\alpha_x, \alpha_y, \alpha_z) - S_0$.

Let us divide the "end-to-end space" x, y, z [one end of the chain is chosen as origin, the other end is at x, y, z (Fig. 13–1)] into small elements of volume $d\mathbf{r}_i = dx_i\,dy_i\,dz_i$. Then $p_i = P_0(\mathbf{r}_i)\,d\mathbf{r}_i$ is the probability of a free chain having an end-to-end vector in $d\mathbf{r}_i$. The undeformed state of the network corresponds to the chains having their most probable end-to-end distribution, with $n_i = Np_i$ chains in $d\mathbf{r}_i$ for each i. The deformed state α_x, α_y, α_z, on the other hand, corresponds to a distribution which is *not* the most probable: there are, according to our assumption (21–2), $s_i = NP(\mathbf{r}_i)\,d\mathbf{r}_i$ chains in $d\mathbf{r}_i$, with $s_i \neq n_i$ unless $\alpha_x = \alpha_y = \alpha_z = 1$. We want to calculate the ratio of the probabilities of observing these two distributions, $\Omega(\alpha_x, \alpha_y, \alpha_z)/\Omega_0$, because $S - S_0 = k \ln (\Omega/\Omega_0)$.

The probability that any specific chain is in $d\mathbf{r}_i$ is p_i, and the probability that $n_i = Np_i$ specific chains are in $d\mathbf{r}_i$ is $p_i^{n_i}$. Therefore, for the undeformed state (most probable distribution),

$$\Omega_0 = \frac{N!}{\prod_i n_i!}\prod_i p_i^{n_i}, \quad (21\text{–}3)$$

where the combinatorial factor is inserted because it is immaterial which particular chains are in $d\mathbf{r}_i$, etc. Now if we place in each of the $d\mathbf{r}_i$, not the most probable number of chains n_i, but some other number s_i, then the corresponding Ω is

$$\Omega = \frac{N!}{\prod_i s_i!}\prod_i p_i^{s_i}, \quad (21\text{–}4)$$

where we expect $\Omega < \Omega_0$ (Problem 21–2). Equation (21–4) is a quite general expression for any set of numbers $s_1, s_2, \ldots$, but we are interested in the particular set $s_i = NP\,d\mathbf{r}_i$ from Eq. (21–2). On using

$$\sum_i p_i = 1, \qquad \sum_i s_i = \sum_i n_i = N,$$

we find from Eqs. (21–3) and (21–4),

$$\ln \frac{\Omega}{\Omega_0} = \sum_i s_i \ln \frac{n_i}{s_i}.$$

But

$$\ln \frac{n_i}{s_i} = \ln \frac{P_0(\mathbf{r}_i)}{P(\mathbf{r}_i)} = \ln \alpha_x\alpha_y\alpha_z - \beta^2\left[x_i^2\left(1 - \frac{1}{\alpha_x^2}\right) + y_i^2\left(1 - \frac{1}{\alpha_y^2}\right) + z_i^2\left(1 - \frac{1}{\alpha_z^2}\right)\right],$$

so that

$$\ln \frac{\Omega}{\Omega_0} = N \ln \alpha_x\alpha_y\alpha_z - \beta^2\left[\left(1 - \frac{1}{\alpha_x^2}\right)\sum_i s_i x_i^2 + \text{etc.}\right].$$

Now

$$\sum_i s_i x_i^2 = N \sum_i P(\mathbf{r}_i) x_i^2\, d\mathbf{r}_i = N \iiint_{-\infty}^{+\infty} x^2 P(x, y, z)\, dx\, dy\, dz = \frac{N\alpha_x^2}{2\beta^2},$$

and therefore

$$\Delta S = S(\alpha_x, \alpha_y, \alpha_z) - S_0 = k \ln \frac{\Omega}{\Omega_0} = Nk\left[\ln \alpha_x\alpha_y\alpha_z - \tfrac{1}{2}(\alpha_x^2 + \alpha_y^2 + \alpha_z^2 - 3)\right]. \tag{21–5}$$

It is easy to verify (Problem 21–3) that ΔS has its maximum value at $\alpha_x = \alpha_y = \alpha_z = 1$, as we have anticipated. We note that ΔS is independent of chain length; it depends only on the number of chains in the network and on the macroscopic deformation parameters α_x, α_y, α_z. Therefore, if the network has a distribution in chain lengths, we obtain a contribution to ΔS of the form (21–5) for each group of chains of the same length and a *total* ΔS still given by (21–5), where N is the *total* number of chains of all lengths.

We next introduce a correction into Eq. (21–5), suggested by Flory.* Actually, this correction does not affect the length-force equation for

---

* See Flory, p. 468.

rubber, derived below, but will influence our results beginning with Section 21–3. In deriving Eq. (21–5) we have assumed that the entropy of a network is the same as the entropy of the same number of polymer chains *not* involved in a network, but with the same end-to-end distribution. This neglects the fact that when a network is formed from chains, there is an entropy change which is volume-dependent, and therefore this network-formation entropy will in general be different at $\alpha_x, \alpha_y, \alpha_z$ than at $\alpha_x = \alpha_y = \alpha_z = 1$. A network of N chains has $N/2$ cross-links. Each cross-link is formed by chemical bond formation between a monomer on one chain and a monomer on another. If the second monomer must be within a small volume δV (a constant) around the first monomer in order for reaction to occur, the probability for cross-link formation is proportional to $\delta V/V$. Hence the probability of forming $N/2$ cross-links is proportional to $(\delta V/V)^{N/2}$, and the ratio of this probability at V to the probability at L_0^3 is $(\alpha_x\alpha_y\alpha_z)^{-N/2}$. This contributes an additional term to ΔS, $-(Nk/2) \ln (\alpha_x\alpha_y\alpha_z)$. The corrected form of Eq. (21–5) is thus

$$\Delta S = S(\alpha_x, \alpha_y, \alpha_z) - S_0 = \frac{Nk}{2} (\ln \alpha_x\alpha_y\alpha_z - \alpha_x^2 - \alpha_y^2 - \alpha_z^2 + 3). \tag{21–6}$$

We now consider the rubber elasticity problem specifically. Here, for stretching in the x-direction,

$$V = L_0^3 = \text{constant}, \qquad \alpha_x\alpha_y\alpha_z = 1,$$
$$L_x = L, \qquad \alpha_x = \alpha,$$
$$L_y = L_z, \qquad \alpha_y = \alpha_z = \frac{1}{\alpha^{1/2}}.$$

Therefore, Eq. (21–6) becomes

$$S(\alpha) - S(1) = \frac{Nk}{2}\left(3 - \alpha^2 - \frac{2}{\alpha}\right).$$

Then from Eq. (13–52),

$$\tau = -\frac{T}{L_0}\left(\frac{\partial S}{\partial \alpha}\right)_T = \frac{NkT}{L_0}\left(\alpha - \frac{1}{\alpha^2}\right). \tag{21–7}$$

This is just the same equation as (13–57), obtained from the James-Guth theory, with $C = L_0^{-2}$ (see also Fig. 13–4).

21–2 Flory-Huggins polymer solution theory.* We leave, temporarily, the discussion of polymer networks begun in the previous section and turn here to solutions of free polymer molecules (component 2) in a solvent

* See Flory, Chapters 12 and 13.

(component 1). In the next section we will then combine the results of the present and preceding sections to study the swelling of polymer networks (gels) in a solvent.

The theory to be presented here, due independently to Flory and Huggins, is a direct generalization of the Bragg-Williams approximation in the lattice model of binary solutions (Section 20–1). The thermodynamic equations of Section 20–1 are applicable without change. The Bragg-Williams theory is appropriate for solutions of molecules of approximately equal size—each site in the lattice can be occupied by either an A molecule or a B molecule. The essential difference in the present problem is that the polymer molecules are much larger (a factor of usually 10^3 or 10^4) than the solvent molecules. This leads to striking departures from the A–B symmetry we became accustomed to in Chapter 20. We still use a lattice model and assume random mixing, but whereas a solvent molecule occupies only one site in the lattice, the polymer molecule occupies M sites along a "random walk" (Fig. 21–1), where $M = O(10^3 \text{ or } 10^4)$. We assume that all the chains have the same length. Since M should be regarded as the ratio of the two molar volumes, it will be of the order of the number of monomers or of the number of statistical units in a polymer chain, but not usually exactly equal to either of these numbers.

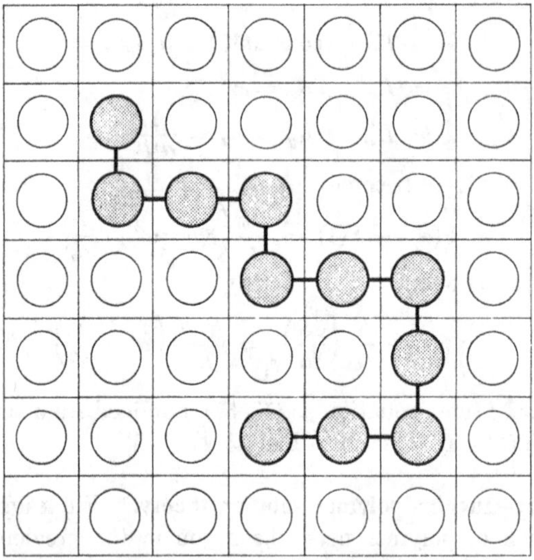

FIG. 21–1. Lattice model (schematic, in two dimensions here) for polymer molecule in a solution. Sites not occupied by polymer segments are occupied by solvent molecules (one per site).

Because of the great difference in size of the two molecules, the mole fractions x_1 and x_2 are no longer very useful. Instead, we use volume or site fractions:

$$\varphi_1 = \frac{N_1}{N_1 + MN_2}, \qquad \varphi_2 = \frac{MN_2}{N_1 + MN_2}, \qquad \varphi_1 + \varphi_2 = 1.$$

We shall sometimes denote the total number of sites by $M_0 = N_1 + MN_2$.

We investigate first the entropy of (random) mixing solvent and polymer molecules. Let $\Omega(N_1, N_2)$ be the number of possible configurations or arrangements of N_1, N_2 molecules on M_0 sites, and let $\Omega(0, N_2)$ be the number of possible configurations of N_2 polymer molecules on MN_2 sites (this refers to the pure polymer before mixing). Then, since $\Omega = 1$ for N_1 solvent molecules on N_1 sites (pure solvent), the desired entropy of mixing is

$$\Delta S_m = k \ln \frac{\Omega(N_1, N_2)}{\Omega(0, N_2)}. \tag{21-8}$$

We now find $\Omega(N_1, N_2)$.

The number $\Omega(N_1, N_2)$ is just equal to the number of ways of arranging N_2 polymer molecules on M_0 sites, for after we place the polymer molecules in the originally empty lattice, there is only one way to place the solvent molecules (i.e., we simply fill up all the remaining unoccupied sites). Imagine that we label the polymer molecules from 1 to N_2 and introduce them one at a time, in order, into the lattice. Let ω_i be the number of ways of putting the i-th polymer molecule into the lattice with $i - 1$ molecules already there (assumed to be arranged in an average, random distribution). Then the approximation to $\Omega(N_1, N_2)$ which we use is

$$\Omega(N_1, N_2) = \frac{1}{N_2!} \prod_{i=1}^{N_2} \omega_i. \tag{21-9}$$

The factor $(N_2!)^{-1}$ is inserted because we have treated the molecules as distinguishable in the product, whereas they are actually indistinguishable.

Next, we derive an expression for ω_{i+1}. With i polymer molecules already in the lattice, the fraction of sites filled is $f_i = Mi/M_0$. The first unit of the $i + 1$-th molecule can be placed in any one of the $M_0 - Mi$ vacant sites. The first unit has c nearest-neighbor sites, of which $c(1 - f_i)$ are empty (random distribution assumed). Therefore the number of possible locations for the second unit is $c(1 - f_i)$. Similarly, the third unit can go in $(c - 1)(1 - f_i)$ different places. At this point we make the approximation that units 4, 5, ..., M also each have $(c - 1)(1 - f_i)$ possibilities, though this is not quite correct (Huggins' treatment is more detailed here, but in view of the rather crude model, we omit this refinement). Multiplying all of these factors together, we have for ω_{i+1},

$$\omega_{i+1} = (M_0 - Mi)c(c-1)^{M-2}(1-f_i)^{M-1}$$
$$= (M_0 - Mi)^M \left(\frac{c-1}{M_0}\right)^{M-1},$$

where we have replaced c by $c - 1$ as a further approximation.

Now we will need

$$\ln \prod_{i=1}^{N_2} \omega_i = N_2(M-1)\ln\left(\frac{c-1}{M_0}\right) + M \sum_{i=0}^{N_2-1} \ln (M_0 - Mi). \quad (21\text{–}10)$$

We approximate the sum by an integral:

$$\sum \cong \int_0^{N_2} \ln (M_0 - Mi)\, di = \frac{1}{M}\int_{N_1}^{M_0} \ln u\, du$$
$$= \frac{1}{M}(M_0 \ln M_0 - M_0 - N_1 \ln N_1 + N_1). \quad (21\text{–}11)$$

We put Eqs. (21–10) and (21–11) in (21–9) and find

$$\ln \Omega(N_1, N_2) = -N_2 \ln N_2 + N_2 - N_1 \ln N_1 + N_1$$
$$+ M_0 \ln M_0 - M_0 + N_2(M-1)\ln\left(\frac{c-1}{M_0}\right).$$

From this result, we also have

$$\ln \Omega(0, N_2) = -N_2 \ln N_2 + N_2 + MN_2 \ln MN_2 - MN_2$$
$$+ N_2(M-1)\ln\left(\frac{c-1}{MN_2}\right).$$

Therefore, from Eq. (21–8),

$$\frac{\Delta S_m}{k} = -N_1 \ln \varphi_1 - N_2 \ln \varphi_2. \quad (21\text{–}12)$$

This very simple relation is the generalization of Eq. (20–20) to solutions of molecules of unequal size. Note that if we put $M = 1$ ($\varphi_1 = x_1, \varphi_2 = x_2$), we recover Eq. (20–20). Figure 21–2 shows $\Delta S_m/M_0k$ plotted against φ_2 with $M > 500$ (Problem 21–4). This should be compared with Fig. 20–2 for molecules of equal size.

Next, we calculate ΔE_m on the same random mixing (Bragg-Williams) basis. We let w_{11} be the interaction energy between nearest-neighbor solvent molecules, w_{22} between nearest-neighbor polymer units (not chemically bonded), and w_{12} between one solvent molecule and one poly-

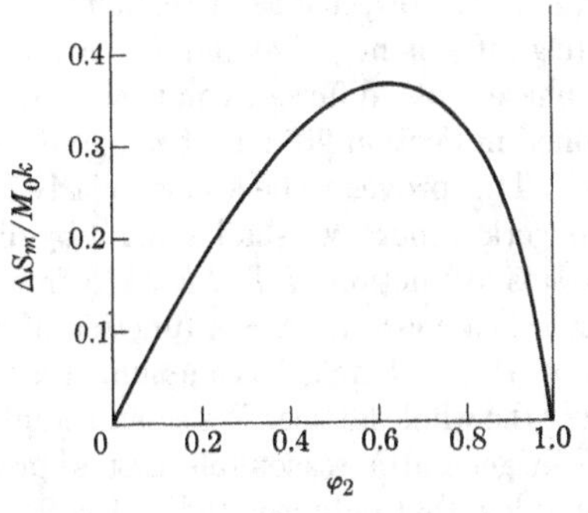

FIG. 21-2. Entropy of mixing as a function of volume fraction of polymer for $M > 500$ (the curve is independent of M for sufficiently large M).

mer unit. In the solution N_1, N_2, the probability that any site is occupied by a solvent molecule is φ_1 and by a polymer unit is φ_2. The average number of solvent molecules nearest neighbor to a polymer unit is $(c - 2)\varphi_1$, neglecting end-of-chain effects. Therefore the number $\overline{N}_{12}$ of 12 interactions is $\overline{N}_{12} = (c - 2)\varphi_1 M N_2$. Similarly, $\overline{N}_{11} = c\varphi_1 N_1/2$ and $\overline{N}_{22} = (c - 2)\varphi_2 M N_2/2$. In the pure polymer, $\varphi_2 = 1$ and $\overline{N}_{22} = (c - 2)MN_2/2$, while in the pure solvent, $\overline{N}_{11} = cN_1/2$. As a further simplification and approximation, we replace $c - 2$ by c in the above expressions, so that

$$\begin{aligned}\Delta E_m &= (c\varphi_1 M N_2 w_{12} + \tfrac{1}{2}c\varphi_1 N_1 w_{11} + \tfrac{1}{2}c\varphi_2 M N_2 w_{22}) \\ &\quad -(\tfrac{1}{2}cMN_2 w_{22} + \tfrac{1}{2}cN_1 w_{11}) \\ &= -\frac{cM_0\varphi_1\varphi_2 w}{2},\end{aligned} \tag{21–13}$$

where, as usual,

$$w = w_{11} + w_{22} - 2w_{12}.$$

Equation (21–13) is the generalization of Eq. (20–21); the two equations are the same when $M = 1$. We define the "mixing parameter" χ by $\chi = -cw/2kT$. Then

$$\frac{\Delta E_m}{kT} = \chi M_0 \varphi_1 \varphi_2. \tag{21–14}$$

This is often called the "van Laar heat of mixing" expression. When w is negative and χ positive, 11 and 22 neighbors are more stable than 12 neighbors. In this case ($\chi > 0$) the solvent is said to be a "poor solvent" (solvent and polymer molecules "dislike" each other). A "good" solvent has $\chi \leq 0$. We shall find below, just as in Section 20–1, that if χ is positive and large enough (e.g., if w is negative and T is low enough), the solution will split into two phases with different compositions.

We have mentioned in Section 20–1 that w may be regarded as a function of temperature. This proves to be a particularly useful generalization in polymer solution work, though we shall avoid it by discussing isothermal processes only. If w is a function of T, ΔE_m is a free energy rather than an energy. Even if w is a constant, χ is a function of T, of course.

Our derivations of ΔS_m and ΔE_m have assumed a random distribution of polymer units over the whole lattice. We should emphasize that although this assumption is in general a reasonable first approximation, it is not realistic when φ_2 is so low that only isolated molecules or pair, triplet, etc., interactions between otherwise isolated polymer molecules are involved. For in this case the concentration of polymer units will have a finite value within the space more or less occupied by polymer molecules but will be zero in the space between polymer molecules. Thus, for example, we cannot expect to get a reasonable osmotic pressure second virial coefficient from the present theory (see Section 21–6).

Finally, the free energy of mixing follows from Eqs. (21–12) and (21–14):

$$\frac{\Delta A_m}{kT} = \frac{\Delta E_m}{kT} - \frac{\Delta S_m}{k} = N_1 \ln \varphi_1 + N_2 \ln \varphi_2 + \chi M_0 \varphi_1 \varphi_2. \quad (21\text{–}15)$$

The form of this result, which should be compared with Eq. (20–19), suggests that it ought to be possible to derive it without using a lattice model, and indeed this has been done.*

With the free energy of mixing available, we can now easily find the chemical potentials μ_1 and μ_2 from the thermodynamic equation (20–1). Thus

$$\left(\frac{\partial \Delta A_m/kT}{\partial N_1}\right)_{N_2,T} = \frac{\mu_1(\varphi_2) - \mu_1(0)}{kT} = \ln \frac{p_1}{p_1^0}$$

$$= \ln (1 - \varphi_2) + \left(1 - \frac{1}{M}\right) \varphi_2 + \chi \varphi_2^2, \quad (21\text{–}16)$$

where $\mu_1(0)$ is the chemical potential of the pure solvent and p_1^0 is the vapor pressure over the pure solvent. Equation (21–16) reduces to Eq. (20–18) when $M = 1$, but otherwise there is a term in φ_2 that destroys the symmetry of Figs. 20–1, 20–3, and 20–4. When $w = 0$ in Eq. (20–18) (molecules of approximately equal size), Raoult's law is obeyed. But if we put

---

* J. H. Hildebrand, *J. Chem. Phys.* **15**, 225 (1947); H. C. Longuet-Higgins, *Faraday Soc. Disc.* **15**, 73 (1953).

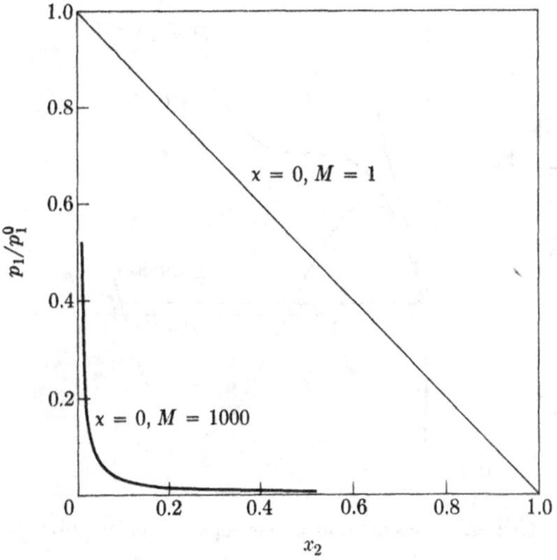

FIG. 21-3. Effect of molecular size of solute on vapor pressure of solvent when $\chi = 0$.

$\chi = 0$ in Eq. (21-16), there remain considerable deviations from Raoult's law (Fig. 21-3) that can only be associated with the great difference in molecular size of the two components.

For μ_2 we have

$$\left(\frac{\partial \Delta A_m/kT}{\partial N_2}\right)_{N_1,T} = \frac{\mu_2(\varphi_2) - \mu_2(1)}{kT}$$
$$= \ln \varphi_2 - (M-1)(1-\varphi_2) + \chi M(1-\varphi_2)^2. \quad (21\text{-}17)$$

Actually, Eq. (21-17) contains no new information. It follows from (21-16) and the Gibbs-Duhem equation or from Eqs. (20-2) and (21-16) (Problem 21-5).

Equation (21-16) is also essentially an equation for the osmotic pressure (2 = nondiffusible solute, 1 = diffusible solvent). For, with an incompressible solvent as in the present model, we have from thermodynamics

$$\frac{\mu_1(\varphi_2) - \mu_1(0)}{kT} = -\frac{\Pi v_1}{kT}, \quad (21\text{-}18)$$

where v_1 is the volume per molecule (i.e., volume per site) of pure solvent, a constant.

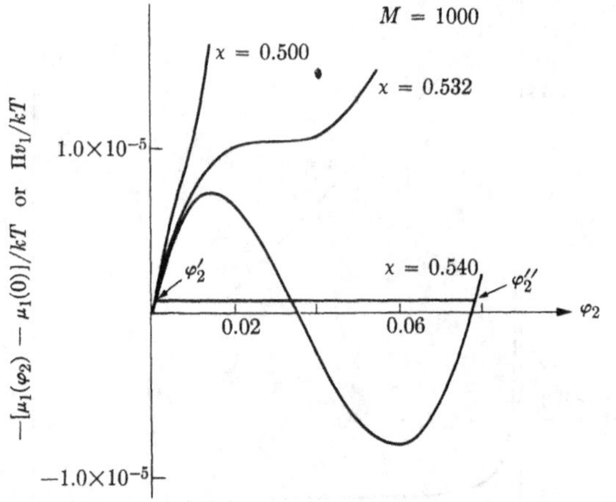

FIG. 21–4. Critical behavior and phase separation for polymer solution with $M = 1000$.

Figure 21–4 shows a plot of $-[\mu_1 - \mu_1(0)]/kT$ or $\Pi v_1/kT$ against φ_2 for $M = 1000$ and different positive values of χ (i.e., different temperatures for constant w). Critical behavior and phase separation are evident, as in Section 20–1, but here the curves are very unsymmetrical. The horizontal line can be located by use of the criteria that

$$\mu_1(\varphi_2') = \mu_1(\varphi_2''), \qquad \mu_2(\varphi_2') = \mu_2(\varphi_2''),$$

where φ_2' and $\varphi_2''(\varphi_2' < \varphi_2'')$ give the composition of the two phases in equilibrium. The first of these criteria is of course satisfied by any horizontal line; but the horizontal line has to be adjusted to satisfy the second one, using Eq. (21–17) for μ_2. Alternatively, if the plot in Fig. 21–4 is made against $1/\varphi_2$ instead of φ_2 (Π vs. $1/\varphi_2$ is analogous to p vs. v for a gas), the equal-area theorem can be used to locate the horizontal line (Problem 21–6).

To locate the critical point we put

$$\frac{\partial \mu_1}{\partial \varphi_2} = \frac{\partial^2 \mu_1}{\partial \varphi_2^2} = 0,$$

using Eq. (21–16), and find

$$-\frac{1}{1-\varphi_2} + \left(1 - \frac{1}{M}\right) + 2\chi\varphi_2 = 0,$$

$$-\frac{1}{(1-\varphi_2)^2} + 2\chi = 0.$$

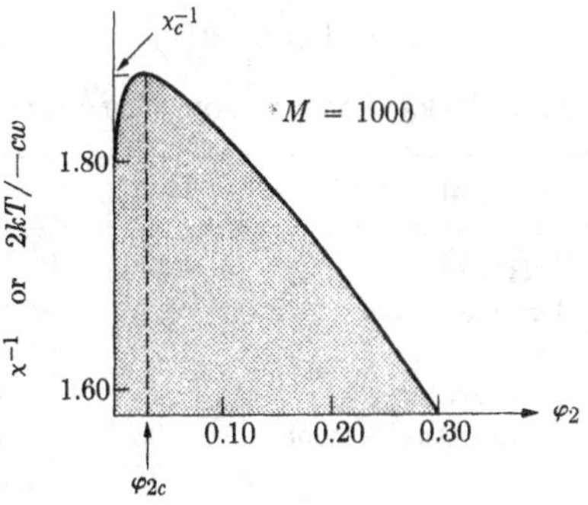

FIG. 21-5. Temperature-composition curve for polymer solution with $M = 1000$.

Eliminating 2χ, we get the critical value

$$\varphi_{2c} = \frac{1}{1 + M^{1/2}}. \tag{21-19}$$

Then

$$\chi_c = \frac{(1 + M^{1/2})^2}{2M}. \tag{21-20}$$

As $M \to \infty$, these relations become

$$\varphi_{2c} \to \frac{1}{M^{1/2}}, \qquad \chi_c \to \frac{1}{2} + \frac{1}{M^{1/2}} \to \frac{1}{2}.$$

For $M = 1000$, $\varphi_{2c} = 0.0307$ and $\chi_c = 0.532$. When $M = 1$, we get the same results as in Section 20-1 (Bragg-Williams).

If χ is positive and less than $\chi_c(\cong 0.5)$ or if χ is negative, the polymer and solvent are miscible in all proportions. But if $\chi > \chi_c$, phase separation occurs for some compositions. Since necessarily $\varphi_2' < \varphi_{2c}$, one phase is extremely dilute in polymer. If we vary χ by varying T, we obtain a temperature-composition curve, as in Fig. 21-5 for $M = 1000$. Note especially the marked asymmetry. In the shaded area of this figure, two solution phases are present (immiscible region); the miscible (one solution) region is outside the shaded area.

As an illustration, we give in Table 21-1 a few experimental values of χ for the polymer molecules of natural rubber.

A great deal of work has been done on checking the predictions of the Flory-Huggins theory against experimental results. The reader should consult the Supplementary Reading list for summarizing discussions. We shall merely state here that the theory proves to be completely satisfactory

TABLE 21-1

MIXING PARAMETER χ FOR NATURAL RUBBER

Solvent	t, °C	χ
CCl_4	15–20	0.28
Benzene	25	0.44
CS_2	27	0.49
Ethyl acetate	25	0.78
Methyl ethyl ketone	25	0.94
Acetone	25	1.37

in a qualitative or, indeed, semiquantitative way, but a number of detailed discrepancies have been uncovered which require refinement in the theory. There is no doubt, though, that this relatively simple theory contains the essential features which distinguish high polymer solutions from ordinary solutions of small molecules.

21–3 Swelling of polymer gels.* If a sample of free (uncross-linked) polymer molecules is put in contact with pure solvent at the same pressure, the polymer will take up solvent to form a solution. In fact, if a large amount of solvent is available, it will continue to enter the solution indefinitely ($\varphi_2 \rightarrow 0$) because the chemical potential of the solvent in the solution is always lower than in the pure solvent at the same pressure. However, we can stop the process after a finite amount of solvent has mixed with polymer by making the pressure on the solution higher than on the pure solvent (Δp = osmotic pressure = Π). If, on the other hand, a *network* of polymer molecules is put in contact with free solvent at the same pressure, solvent will be absorbed by the network forming a gel, but the process will stop after a finite amount of solvent absorption (without establishing a pressure difference) because the tendency of solvent molecules to mix with the polymer chains is resisted as a result of the fact that mixing entails stretching of the network in this case.

In this section we employ the concepts of Sections 21–1 and 21–2 to construct an approximate theory of polymer network swelling. Actually, it is easy to give a treatment that includes stretching as well as swelling: we shall therefore investigate, below, the equilibrium swelling of a network, under a pulling force τ in the x-direction, when in contact with solvent at the same pressure (Fig. 21–6).

In view of the semiquantitative success of the Wall and Flory-Huggins theories (the latter is more accurate than the former), we can anticipate

* See Flory, Chapter 13, for further details.

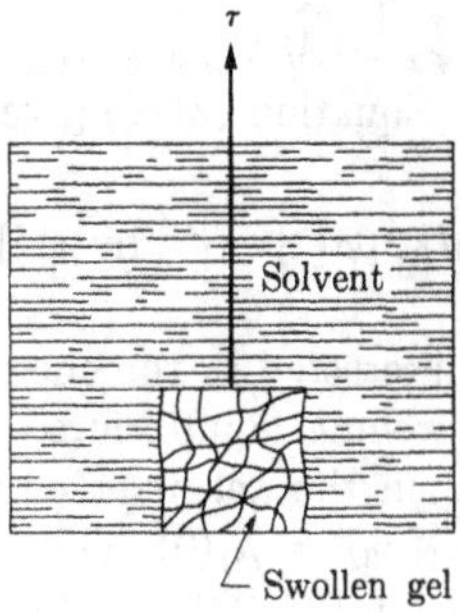

FIGURE 21-6

the same order of agreement with experiment here, for no new ingredients need to be added to the theory.

Consider the following mixing process. The initial state, with Helmholtz free energy A_0, is: (1) an undeformed polymer network (free of solvent) made up of N_2 chains, each with M units (of solvent size), with volume $V_0 = L_0^3 = MN_2v_1$; and (2) N_1 molecules of pure solvent with volume N_1v_1. The final state, with Helmholtz free energy A, is the above network, swollen with N_1 molecules of solvent, with volume (assumed additive) $V = (N_1 + MN_2)v_1$ and shape $L_x = L = \alpha L_0$, $L_y = L_z = (V/L)^{1/2}$. We also have the relations $V/V_0 = 1/\varphi_2$ (called the swelling ratio) and $\alpha_y = \alpha_z = 1/(\alpha\varphi_2)^{1/2}$.

The basic thermodynamic equation for the gel (swollen network) is

$$dA = -S\,dT + \tau\,dL + \mu_1\,dN_1 + \mu_2\,dN_2.$$

In practice, of course, one can vary T, L, and N_1 but not N_2.

The free energy change, $\Delta A = A - A_0$, in the above process is

$$\Delta A = \Delta A_m + \Delta A_d = \Delta A_m - T\,\Delta S_d, \tag{21-21}$$

where ΔA_m is the free energy of mixing network with solvent, and ΔS_d is the entropy of deforming the network. The implicit approximation is made here that the free energies of deformation and mixing are independent of each other. We obtain ΔA_m from Eq. (21-15), but we have to make an appropriate alteration to take care of the fact that here, instead of N_2 free polymer molecules, we have a single giant polymer molecule (the entire network). Thus the coefficient of $\ln \varphi_2$ in Eq. (21-15) is essentially zero. The volume fractions retain the same physical meaning, however, so we have

$$\frac{\Delta A_m}{kT} = N_1 \ln \varphi_1 + \chi M_0 \varphi_1 \varphi_2, \tag{21-22}$$

where, in the relation $\varphi_2 = MN_2/M_0$, N_2 now refers to the number of chains in the network. Equation (21–6) gives ΔS_d. Therefore,

$$\frac{\Delta A}{kT} = N_1 \ln \varphi_1 + \chi M_0 \varphi_1 \varphi_2 + \frac{N_2}{2}\left(\ln \varphi_2 + \alpha^2 + \frac{2}{\alpha\varphi_2} - 3\right). \quad (21\text{–}23)$$

We are particularly interested in the chemical potential of the solvent, for it determines the swelling equilibrium. When equilibrium has been reached (Fig. 21–6), μ_1 in the gel must equal μ_1 in the pure solvent in contact with the gel: $\mu_1(\varphi_2) = \mu_1(0)$. Therefore, we have for this equilibrium condition, from Eq. (21–23),

$$\frac{\mu_1(\varphi_2) - \mu_1(0)}{kT} = \left(\frac{\partial \Delta A/kT}{\partial N_1}\right)_{T,L,N_2} = \left(\frac{\partial \Delta A/kT}{\partial N_1}\right)_{T,\alpha,N_2}$$

$$= \ln(1 - \varphi_2) + \varphi_2 + \chi\varphi_2^2 + \frac{1}{M}\left(\frac{1}{\alpha} - \frac{\varphi_2}{2}\right) = 0. \quad (21\text{–}24)$$

Equation (21–24) determines the equilibrium swelling ratio $1/\varphi_2$ for a given extension α. One can give the following physical interpretation of this equation. The first three terms are equal to $-\Pi_{\text{net}}v_1/kT$ [see Eq. (21–18)], where Π_{net} is the osmotic pressure of the network. These terms lead to an expansion of the gel. (By themselves, they would equal zero and hence lead to equilibrium only when $\varphi_2 \to 0$.) The last two terms are equal to $-p_d v_1/kT$, where p_d is the pressure associated with deformation of the network. This follows from

$$p_d = -\frac{\partial \Delta A_d}{\partial V} = -\frac{1}{v_1}\frac{\partial \Delta A_d}{\partial N_1}.$$

Actually, p_d is negative and $-p_d$ is a force per unit area tending to contract the gel. When equilibrium is reached, these two opposing pressures just balance each other: $\Pi_{\text{net}} = -p_d$.

The special case of Eq. (21–24) which is of most interest is free (and isotropic) swelling of the gel in the solvent; that is, $\tau = 0$ and $V/V_0 = \alpha^3 = 1/\varphi_2$. Therefore we replace $1/\alpha$ in Eq. (21–24) by $\varphi_2^{1/3}$ and solve the resulting equation for φ_2. In a good (or not too poor) solvent and with ordinary values of M, the swelling ratio $1/\varphi_2$ will exceed 10, and $\varphi_2 < 0.1$. (Of course, if the swelling is too great, say $1/\varphi_2 > 30$, the gaussian distribution used in deriving the deformation terms begins to become inaccurate.) Therefore we expand the logarithm in Eq. (21–24) to obtain the approximate equation

$$\frac{1}{\varphi_2}\left(\frac{1}{\varphi_2^{2/3}} - \frac{1}{2}\right) = M\left(\frac{1}{2} - \chi\right). \quad (21\text{–}25)$$

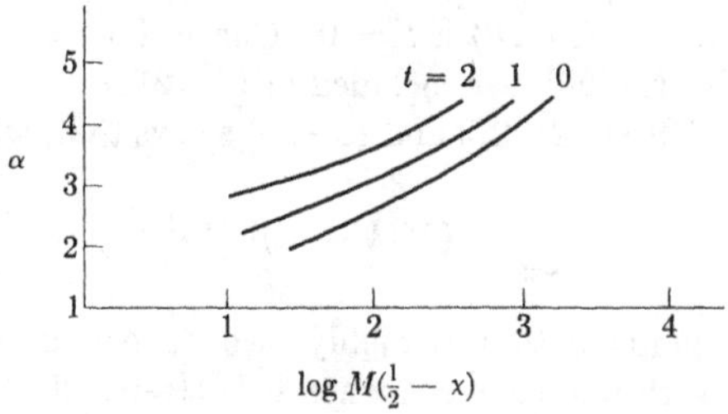

FIG. 21-7. Theoretical behavior of a stretched, swollen gel.

If we also neglect $\frac{1}{2}$ compared to $\varphi_2^{-2/3}$, we have

$$\left(\frac{1}{\varphi_2}\right)^{5/3} = \left(\frac{V}{V_0}\right)^{5/3} = M\left(\frac{1}{2} - \chi\right). \tag{21-26}$$

As a numerical example of this last equation, a swelling of $\alpha = 3$ and $1/\varphi_2 = \alpha^3 = 27$ is obtained if we take $\chi = 0.25$ and $M = 972$, or if we take $\chi = 0$ and $M = 486$. Figure 21-7 ($t = 0$ curve) shows a plot of α against $\log M(\frac{1}{2} - \chi)$, calculated from Eq. (21-25).

Equation (21-26) gives reasonable agreement with experiment for solvents leading to large swelling. The exponent 5/3 has been verified experimentally, and the values of χ calculated from swelling experiments and Eq. (21-26) agree fairly well with those found from vapor pressure measurements on the uncross-linked material using the Flory-Huggins theory.

Returning now to the more general situation in which the gel may be stretched, we calculate the length-force relation from

$$\tau = \left(\frac{\partial A}{\partial L}\right)_{T,N_1,N_2} = \frac{kT}{L_0}\left(\frac{\partial \Delta A/kT}{\partial \alpha}\right)_{T,N_1,N_2},$$

with the result

$$\frac{\tau L_0}{N_2 kT} \equiv t = \alpha - \frac{1}{\alpha^2 \varphi_2}. \tag{21-27}$$

For a given α, we find φ_2 from Eq. (21-24) and then t from Eq. (21-27). When $1/\varphi_2$ is large, we expand the logarithm and find (Problem 21-7) the generalization of Eq. (21-25),

$$\alpha^2(\alpha - t)[\alpha(\alpha - t) - \tfrac{1}{2}] = M(\tfrac{1}{2} - \chi). \tag{21-28}$$

If we also drop $\frac{1}{2}$ compared to α^2, we have

$$t = \alpha - \frac{[M(\frac{1}{2} - \chi)]^{1/2}}{\alpha^{3/2}}, \tag{21-29}$$

which reduces to Eq. (21–26) if $t = 0$. Curves for $t = 1$ and $t = 2$, calculated from Eq. (21–28), are included in Fig. 21–7.

Comparison of Eqs. (21–27) and (21–29) shows that, when $1/\varphi_2$ is large,

$$\frac{1}{\varphi_2} = [M(\tfrac{1}{2} - \chi)]^{1/2}\alpha^{1/2}. \tag{21–30}$$

This equation predicts that a highly swollen gel, in equilibrium with solvent, will absorb more solvent when it is stretched. This is confirmed experimentally.

21–4 Swelling of polyelectrolyte gels.* Suppose that along the molecular chains of the network studied in the preceding section there is a fixed charge ze on every xth† unit (on the average) formed by the dissociation of an ion (e.g., $—XNa \rightarrow —X^- + Na^+$). In this case the "solvent" is an electrolyte solution (usually aqueous), though we still treat it for simplicity as a one-component solvent in nonelectrostatic expressions. The gel now resembles the "inside" solution in a Donnan membrane equilibrium (Section 19–2), and the "solvent" in which the gel is immersed (Fig. 21–6) is the "outside" solution. The fixed ("solute") charges are nondiffusible (confined to the gel), but the electrolyte ions are diffusible. No membrane is necessary because the fixed charges are attached to the polymer chains.

The repulsion between the fixed charges furnishes an additional tendency for the gel to swell (beyond what one would calculate from Section 21–3). We give in this section the first-order theory of this effect. Later in the section, we indicate how the treatment can be generalized to take care of the case in which the charge on the polymer chains can vary in amount because of a dissociative or binding equilibrium (e.g., $—COOH \rightarrow COO^- + H^+$).

Aside from intrinsic interest, important applications of this work are to the electrostatic theory of muscle contraction and to ion exchange resins. Muscle contraction will be referred to again at the end of the section.

For simplicity we assume that although the fixed charges are attached to polymer chains, they are spatially distributed in the volume V occupied by the gel just as if they were free to move throughout V. Because of the Brownian motion of the polymer chains, this is not an unrealistic assumption. However, these charges are not and do not count as additional

---

* The simple theory in the present section is included in the following papers, where a more detailed treatment will be found: T. L. HILL, *J. Chem. Phys.* **20,** 1259 (1952); *Faraday Soc. Discussions* **13,** 132 (1953).

† This x has no relation to x or x_i in Chapter 20.

molecules in contributing, for example, to the osmotic pressure. We assume further that electrostatic effects are independent of mixing and deformation effects—we have merely to add appropriate electrostatic terms.

The simplest procedure for us is to work directly with Eq. (21–24) rather than with the free energy, because of information already available in Chapter 19. (The free-energy approach is the subject of Problem 21–8.) We have seen that the first three terms in this equation may be associated with the osmotic pressure Π_{net} of the chains of the network. Similarly, the additional tendency of the repulsion between fixed charges on the polymer chains to expand the gel may be attributed to the osmotic pressure Π_{el} of these "nondiffusible" or "solute" charges. We proceed next to evaluate Π_{el}.

We saw in Section 19–2 that the leading terms in the osmotic pressure for a system of charged ($\mathbf{z}\mathbf{e}$) solute molecules with number density ρ is

$$\frac{\Pi}{kT} = \rho + \left(\frac{\mathbf{z}^2}{2\sum}\right)\rho^2 + \cdots \tag{21–31}$$

$$\sum = \sum_k \rho_k^0 \mathbf{z}_k^2,$$

where ρ_k^0 in $\sum$ refers as usual to the kth diffusible ionic species in the outside solution. The second term in Eq. (21–31) is a consequence of the electrical neutrality in the outside solution (see Problem 19–5). (The second term can also be interpreted as due to the excess total concentration of diffusible ions, inside over outside; or as due to the repulsion between solute molecules.) In the present case, the fixed charges do not count as additional molecules contributing to Π, but they do produce a "neutrality" term. That is, in Eq. (21–31), since we are dealing with a single giant molecule (network), we let $\rho \to 0$, keeping the "solute" charge density $\rho\mathbf{z}\mathbf{e}$ constant [this resembles the argument leading to Eq. (21–22)]:

$$\frac{\Pi_{\text{el}}}{kT} = \frac{(\text{charge density})^2}{2\mathbf{e}^2\sum}, \tag{21–32}$$

The fixed charge density in the gel is

$$\frac{MN_2}{x}\cdot\frac{\mathbf{z}\mathbf{e}}{V} = \frac{MN_2\mathbf{z}\mathbf{e}}{xV_0}\cdot\frac{V_0}{V} = \frac{\mathbf{z}\mathbf{e}\varphi_2}{xv_1}.$$

The term to be added to Eq. (21–24) is $-\Pi_{\text{el}}v_1/kT$. Thus we have

$$\ln(1-\varphi_2) + \varphi_2 + \chi\varphi_2^2 + \frac{1}{M}\left(\frac{1}{\alpha} - \frac{\varphi_2}{2}\right) - \frac{\mathbf{z}^2\varphi_2^2}{2x^2v_1\sum} = 0. \tag{21–33}$$

The length-force relation, (21–27), is unaltered because Π_{el} depends only on the volume of the gel and not its shape.

It is clear from Eq. (21–33) that the effect of adding charges to the network, when an electrolyte solution is "solvent," is simply to lower the effective value of χ to

$$\chi - \frac{z^2}{2x^2 v_1 \sum} \cdot \tag{21–34}$$

This increases the degree of swelling, as expected. Figure 21–7 is applicable with the effective χ in (21–34) replacing χ itself. The electrostatic term in (21–34) is seen to be proportional to the square of the amount of charge placed on the gel ($\propto z^2/x^2$) and inversely proportional to the ionic strength of the electrolyte solution in which the gel is immersed. As a numerical example, take $|z| = 1$ and let the electrolyte be of type 1–1 with number density ρ^0. In an aqueous solution, $\rho^0 v_1 = c^0/55.5$, where c^0 is the molar concentration of electrolyte. Thus (21–34) becomes, in this special case,

$$\chi - \frac{55.5}{4x^2 c^0} \cdot \tag{21–35}$$

Now χ is of order unity, often less than 0.5. If we take $c^0 = 0.1$ mole·liter^{-1}, and $x = 10$, the electrostatic term in (21–35) is 1.39; if $x = (1000)^{1/2} = 31.6$, it is 0.139.

When $\tau = 0$ (free swelling) and the degree of swelling is large, Eqs. (21–26) and (21–35) give us

$$\frac{V(z = 1)}{V(z = 0)} = \left[\frac{\frac{1}{2} - \chi + (13.9/x^2 c^0)}{\frac{1}{2} - \chi}\right]^{3/5} \tag{21–36}$$

for the ratio of gel volumes with and without fixed charges on the gel. For example, if $\chi = 0.25$ and $c^0 = 0.15$ mole·liter^{-1},

$$\frac{V(1)}{V(0)} = 9.4 \qquad \text{when} \quad x = 3,$$

$$\frac{V(1)}{V(0)} = 2.5 \qquad \text{when} \quad x = 10.$$

Because of the enhanced swelling in polyelectrolyte gels, the gaussian distribution is often inadequate if an accurate theory is desired, and must be replaced by the Langevin function (Eq. 13–20).

Next, we consider a polyelectrolyte gel with an ionic dissociation (or binding) equilibrium. Let N be the number of ions bound to the network (each such ion has a charge $m\mathbf{z}\mathbf{e}$) and let μ be the chemical potential of the bound ions. The total number of sites for binding is B. These are distributed uniformly over the chains of the network, and are all equivalent. The partition function of a bound ion is q. A term $\mu\, dN$ has to be added to the thermodynamic expression for dA. When no ions are bound, the network is assumed to have a total fixed charge of $MN_2\mathbf{z}\mathbf{e}/x$. This becomes $n\mathbf{z}\mathbf{e}$ when N ions are bound, where

$$n = \frac{N_2M}{x} + Nm.$$

We have to extend Eq. (21–21) for ΔA. The initial state, with Helmholtz free energy A_0, has $N = 0$. The final state, with free energy A, has N bound ions. Thus, assuming additivity of all contributions to ΔA,

$$\Delta A = A - A_0 = \Delta A_m - T\,\Delta S_d + \Delta A_{\text{el}} + \Delta A_{\text{ads}}, \tag{21–37}$$

where ΔA_{el} is the electrostatic free energy (the work necessary to charge up the network in the presence of electrolyte), which we have to deduce from Π_{el}, and where ΔA_{ads} is (Eq. 7–5)

$$\Delta A_{\text{ads}} = -kT\,[B \ln B - N \ln N - (B - N) \ln (B - N) + N \ln q(T)]. \tag{21–38}$$

Now we consider ΔA_{el}. First we note that n can be written as

$$n = \frac{N_2M}{x}\left[1 + \theta m\left(\frac{xB}{N_2M}\right)\right].$$

The quantity xB/N_2M, a constant, is the number of ion binding sites on the polymer chains (B) divided by the number of fixed charges on the chains (N_2M/x). As usual, $\theta = N/B$, the fraction of binding sites occupied. Thus the last term in Eq. (21–33) has to be modified to read

$$-\frac{\Pi_{\text{el}}v_1}{kT} = -\frac{\mathbf{z}^2\varphi_2^2[1 + \theta m(xB/N_2M)]^2}{2x^2v_1\sum}. \tag{21–39}$$

Now, as in Eq. (21–24), we have

$$\left(\frac{\partial\,\Delta A_{\text{el}}/kT}{\partial N_1}\right)_{T,\alpha,N_2,N} = -\frac{\Pi_{\text{el}}v_1}{kT}, \tag{21–40}$$

so we can obtain ΔA_{el} from Eqs. (21–39) and (21–40) by integrating with respect to N_1, which enters only in $\varphi_2 = N_2M/(N_1 + N_2M)$. We choose as limits N_1 and $N_1 = \infty$. When $N_1 = \infty$, $V = \infty$ and $\Delta A_{\text{el}} = 0$ (the work of charging is obviously zero). We find easily (see also Problem 21–8)

$$\frac{\Delta A_{\text{el}}}{kT} = \frac{z^2 n^2}{2V\Sigma}. \tag{21–41}$$

It has already been mentioned that the swelling equilibrium equation, (21–33), is modified by use of (21–39) as the electrostatic term. The length-force relation is still (21–27). We must, finally, derive the binding equilibrium equation from

$$\frac{\mu}{kT} = \left(\frac{\partial A/kT}{\partial N}\right)_{N_1,N_2,L,T} = \left(\frac{\partial\, \Delta A_{\text{el}}/kT}{\partial N}\right)_{N_1,N_2,T} + \left(\frac{\partial\, \Delta A_{\text{ads}}/kT}{\partial N}\right)_T$$

$$= \frac{z^2 nm}{V\Sigma} + \ln\frac{\theta}{(1-\theta)q}.$$

We can rewrite this as

$$\lambda q = \frac{\theta}{1-\theta}\exp\left\{\frac{z^2 m\varphi_2[1 + \theta m(xB/N_2M)]}{xv_1\Sigma}\right\}. \tag{21–42}$$

The quantity λq is equal to the outside concentration of the ions in the binding equilibrium, multiplied by a constant (Eq. 7–9).

The three basic equations referred to in the above paragraph determine the properties of the system. For example, consider the free swelling ($\tau = 0$) of a polyelectrolyte gel with a binding equilibrium. The parameters χ, M, z, m, x, B/N_2M, v_1, and Σ all have to be assigned. From Eq. (21–27), $\alpha^{-1} = \varphi_2^{1/3}$, which is put in Eq. (21–33) [modified by (21–39)] to eliminate α^{-1}. Then, if we assign a value to φ_2, θ can be calculated from this equation. Finally, the pair of values φ_2, θ is put in Eq. (21–42) and λq is calculated (Problem 21–9).

A theory of polyelectrolyte gels, in a somewhat more sophisticated form than above, has been developed and tested experimentally by A. Katchalsky and his collaborators.* Again the theory proves to be adequate for semiquantitative purposes.

Important early work in this field was done by Hermans and Overbeek.† Perhaps the most elaborate unified theory of polyelectrolyte solutions and gels at present available is that due to Harris and Rice.‡

---

* See, for example, A. Katchalsky, *J. Polymer Sci.* **7**, 393 (1951); A. Katchalsky, S. Lifson, and H. Eisenberg, *ibid*, **7**, 571 (1951; A. Katchalsky and I. Michaeli, *ibid.* **15**, 69 (1955); A. Katchalsky and M. Zwick, *ibid.* **16**, 221 (1955).

† J. J. Hermans and J. Th. G. Overbeek, *Rec. trav. chim.* **67**, 761 (1948).

‡ F. E. Harris and S. A. Rice, *J. Phys. Chem.* **58**, 725, 733 (1954); *J. Chem. Phys.* **25**, 955 (1956); S. A. Rice and F. E. Harris, *J. Chem. Phys.* **24**, 326, 336 (1956); *Z. physik. Chem.* (*Frankfort*) **8**, 207 (1956).

Finally, we mention briefly an interesting possibility in the application of the above theory to muscle action. It can be shown* that a polymer or polyelectrolyte gel, constrained to a constant radius and with χ sufficiently larger than 0.5, will exhibit a first-order phase transition in its length-force curve. This means, for example, that the gel may undergo a very large change in length ("contraction") at constant force as a consequence of a very small change in some parameter, such as ionic strength or the concentration of an ion or molecule being bound on the polymer chains. This type of precipitous or "razor-edge" behavior, observed in all first-order transitions, seems characteristic of muscle action. Other first-order phase transitions which have also been mentioned as possibilities in this connection are the *two-dimensional* α-β transition* (Section 14–2) and the crystalline-amorphous transition in polymers.†

21–5 Isolated polymer or polyelectrolyte molecules in solution.‡ We emphasized in Section 21–2 that the Flory-Huggins solution theory was not suitable for extremely dilute polymer solutions because the theory assumed a uniform distribution of polymer units or segments throughout the volume of the solution. In this section we give an approximate treatment designed to take care of the limit of *infinite dilution*. In this limit we need consider only a single polymer ($z = 0$) or polyelectrolyte molecule immersed in the solvent. The polymer segments are confined to the region occupied by the molecule (Fig. 21–8). Between molecules are large spaces of pure solvent. In the next section we treat pair interactions between polymer molecules (which determine, for example, the osmotic pressure second virial coefficient).

The point of view we adopt here resembles that of Chapter 13 in that we consider a single polymer molecule and the solvent it encloses as a thermodynamic system. Specifically, we treat the single molecule as if it were a spherical gel with uniform polymer density. This allows us to take over, with little change, the equations of the preceding two sections on macroscopic gels.

In a good, or not too poor, solvent, or if charged, the polymer molecule will expand over a considerable volume of solvent (in which the polymer volume fraction might be, say, 0.01 or less). The mixing and electrostatic free energies are responsible for this, just as in Section 21–4. However,

---

* T. L. Hill, *J. Chem. Phys.* **20,** 1259 (1952); *Faraday Soc. Disc.* **13,** 132 (1953).

† P. J. Flory, *Science,* **124,** 53 (1956).

‡ See Flory, Chapter 14, for the polymer problem and T. L. Hill, *J. Chem. Phys.* **20,** 1173 (1952) for the polyelectrolyte problem.

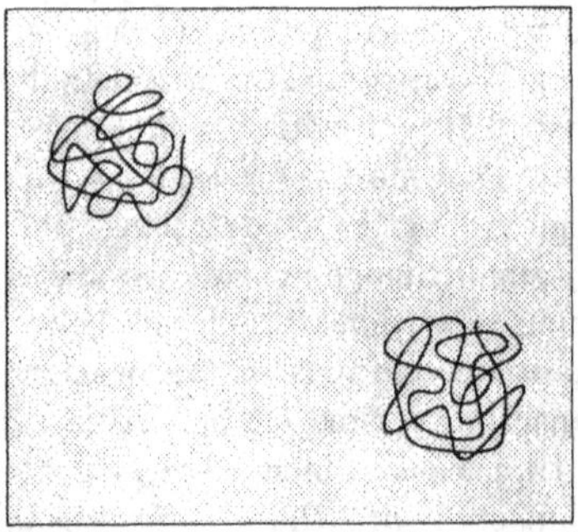

FIG. 21–8. Isolated polymer molecules in a very dilute solution.

the tendency to expand is resisted by the low probability of highly extended molecular configurations (Chapter 13). This is a deformation free energy effect, just as in previous sections.

When we treat a single polymer molecule in the solvent as a uniform "gel," we introduce two new approximations. First, the polymer units have approximately a gaussian distribution rather than a uniform distribution about the center of the sphere. We shall replace this gaussian distribution by a uniform equivalent sphere. Second, especially in connection with ΔA_{el}, there are surface effects which may become significant for a relatively small sphere but which are negligible for macroscopic gels. The criterion, as far as electrostatic effects are concerned, is whether or not the radius of the sphere is large compared to the Debye-Hückel $1/\kappa$. If the electrostatic forces are large enough, these surface effects will, in fact, lead to nonspherical (e.g., ellipsoidal) shapes for the molecule.* In an extreme case, the molecule will become an extended rod in order to reduce ΔA_{el}. We shall not pursue these shape changes here, however.

Flory† has shown that, in the present problem, the actual approximately gaussian distribution of polymer units about the center of the polymer molecule in a "free" (or random walk) polymer chain can be replaced by an effective uniform distribution of polymer units over a sphere of diameter equal to the root mean-square end-to-end distance $(\overline{r^2})^{1/2} = M'^{1/2}a'$ (Eq. 13–45), where a' is the length of a statistical unit and M' is the number of statistical units in the chain. This is an intuitively reasonable result. When the chain is expanded (or contracted) in each dimension by a factor

* T. L. HILL, *J. Chem. Phys.* **20**, 1173 (1952).

† P. J. FLORY, *J. Chem. Phys.* **17**, 303 (1949).

α relative to the random walk state, the radius of the effective sphere becomes

$$R(\alpha) = \frac{\alpha(\overline{r^2})^{1/2}}{2} = \frac{\alpha M'^{1/2}a'}{2}$$

and the volume

$$V(\alpha) = \frac{4}{3}\pi R^3 = \frac{\pi\alpha^3 M'^{3/2}a'^3}{6} = C\alpha^3, \tag{21-43}$$

which defines C (the volume of the sphere when $\alpha = 1$).

We use the basic equation

$$\Delta A = A - A_0 = \Delta A_m - T\,\Delta S_d + \Delta A_{\text{el}}, \tag{21-44}$$

where A is the free energy of the single polymer or polyelectrolyte chain of M units and of the solvent in the volume $V(\alpha) = v_1(N_1 + M)$. Here A_0 is the free energy of the initial (before mixing) pure solvent and of the (discharged) polymer chain in the random walk state, $\alpha = 1$ (i.e., in a sample of undeformed and uncross-linked polymer chains). We retain the notation $V_0 = Mv_1$ and $\varphi_2 = V_0/V = M/(N_1 + M)$, where now φ_2 refers to the interior of a single polymer molecule and *not* to the whole solution (for which $\varphi_2 \to 0$). The free energy of mixing is given by Eq. (21–22), which we rewrite as

$$\frac{\Delta A_m}{kT} = \frac{V - V_0}{v_1}\ln\frac{V - V_0}{V} + \frac{\chi V_0(V - V_0)}{v_1 V}\cdot \tag{21-45}$$

We get ΔS_d from Eq. (21–5) [the correction in Eq. (21–6) is omitted since there are no cross-links here], with $N = 1$:

$$\frac{\Delta S_d}{k} = \ln\alpha^3 - \frac{3}{2}(\alpha^2 - 1) = \ln\frac{V}{C} - \frac{3}{2}\left[\left(\frac{V}{C}\right)^{2/3} - 1\right]\cdot \tag{21-46}$$

Finally, if we are concerned with a polyelectrolyte molecule, we have from Eq. (21–41),

$$\frac{\Delta A_{\text{el}}}{kT} = \frac{z^2M^2}{2x^2V\sum}\cdot \tag{21-47}$$

Substitution of Eqs. (21–45) through (21–47) in Eq. (21–44) gives ΔA. We find the equilibrium degree of swelling by setting $v_1\,\partial\,\Delta A/\partial V = 0$, which is obviously equivalent to Eq. (21–24). We obtain in this way

$$\ln(1 - \varphi_2) + \varphi_2 + \left(\chi - \frac{z^2}{2x^2v_1\sum}\right)\varphi_2^2 + \frac{v_1}{C}\left(\frac{1}{\alpha} - \frac{1}{\alpha^3}\right) = 0, \tag{21-48}$$

just as in Eq. (21–33), except for the deformation terms.

The volume fraction of polymer, φ_2, will be smaller here than in the corresponding swollen network; for in a network, when $\alpha = 1$ (random walk state), $\varphi_2 = 1$, but here, when $\alpha = 1$, φ_2 is already small, say, 0.01 (Problem 21–10). The reason for this is that a random walk configuration is a rather open one, and in the present problem the open space is filled with solvent rather than other chains of a network. The relation between φ_2 and α is $\varphi_2 = Mv_1/C\alpha^3$.

Therefore it will almost always be legitimate to expand the logarithm in Eq. (21–48), with the result

$$\alpha^5 - \alpha^3 = \left(\frac{1}{2} + \frac{\mathbf{z}^2}{2x^2 v_1 \Sigma} - \chi\right)\frac{V_0^2}{Cv_1}. \tag{21–49}$$

If we put $Ma = M'a'$ (a is the length of a polymer unit with volume v_1) and $v_1 \cong a^3$, the coefficient in Eq. (21–49) becomes

$$\frac{V_0^2}{Cv_1} = \frac{6}{\pi}\left(\frac{a}{a'}\right) M'^{1/2}.$$

In a typical case, $a' \cong 3a$, so that

$$\alpha^5 - \alpha^3 \cong \left(1 + \frac{\mathbf{z}^2}{x^2 v_1 \Sigma} - 2\chi\right)\frac{M'^{1/2}}{3}. \tag{21–50}$$

If the polymer is not charged ($\mathbf{z} = 0$) and $\chi = 0$ (good solvent), and

if $M' = 900$, then $\alpha = 1.7$,

if $M' = 10{,}000$, then $\alpha = 2.1$.

If $\chi = \frac{1}{2}$, $\alpha = 1$. For this value of χ, the tendency of the molecule to swell because of the entropy of mixing with solvent is just balanced by the tendency of the molecule to compress itself because of the poor solvent. The result is a molecule in the random walk state ($\alpha = 1$).

With a polyelectrolyte, the coefficient of $M'^{1/2}/3$ can easily be 3 [see Eq. (21–35)]. In this particular case,

if $M' = 900$, then $\alpha = 2.1$,

if $M' = 10{,}000$, then $\alpha = 2.6$.

Equation (21–49) has been subjected to extensive experimental checking (see Flory, Chapter 14) using, especially, frictional properties of the polymer molecules. The general conclusion is, again, that the theory is fairly successful but has some shortcomings, as should be expected. The electrostatic term, in particular, seems to predict too large an effect.

It is easy to extend the above theory to a polyelectrolyte molecule with an ionic binding equilibrium (see Section 21-4), but we leave this as an exercise for the reader (Problem 21-11).

21-6 Second virial coefficient in polymer and polyelectrolyte solutions.† We have so far combined the basic ideas of the Wall and Flory-Huggins theories to treat macroscopic polymer (and polyelectrolyte) gels and also infinitely dilute solutions in which a single polymer molecule in a solvent is treated essentially as a very small gel. In this section we make a straightforward extension of the considerations of the preceding section to a slightly more concentrated polymer solution. Specifically, we deduce the second virial coefficient B_2^* (Eq. 19-37) from a potential of mean force $w(r)$ calculated as the work necessary to bring two isolated polymer molecules together from $r = \infty$ to r.

The argument we use is based on the uniform effective sphere model of the preceding section. This model is quite sufficient to bring out the essential points. Refinements will be found in the references already given.

Consider first two isolated uncharged ($z = 0$) polymer molecules in a solvent, each with radius R (Section 21-5). If the distance r between centers is greater than $2R$, there is no overlapping of the spheres and no interaction between the molecules. Thus $w(r) = 0$ for $r > 2R$. If $r < 2R$, the spheres overlap. We assume, as a first approximation, that the overlapping does not distort the molecular configuration or the distribution of polymer segments. If V' is the volume of overlap of the two molecules, we find from elementary geometrical considerations (overlap of two spheres),

$$V'(r) = \frac{2\pi}{3}\left(2R^3 - \frac{3R^2r}{2} + \frac{r^3}{8}\right). \qquad (21\text{-}51)$$

Also, if the polymer volume fraction in each of the isolated spheres is φ_2, it is $2\varphi_2$ in V'.

The above discussion also applies to polyelectrolyte molecules, but this statement requires some special comment. In general, we would expect that two spherical polyelectrolyte molecules of radius R and fixed charge zeM/x would have $w(r) > 0$ for $r > 2R$, even though electrolyte solution can penetrate inside the spheres. This will be true for small polyelectrolyte molecules.‡ But implicit in the model adopted in the preceding and present

---

† See FLORY, Chapter 12; also B. H. ZIMM, *J. Chem. Phys.* **14,** 164 (1946) and T. A. OROFINO and P. J. FLORY, *J. Chem. Phys.* **26,** 1067 (1957); *J. Phys. Chem.* **63,** 283 (1959).

‡ See, for example, F. T. WALL, *et al.*, *J. Chem. Phys.* **26,** 114 (1957); **31,** 1640 (1959); T. L. HILL, *J. Am. Chem. Soc.* **78,** 1577 (1956); **80,** 3241 (1958).

sections is the assumption that a polyelectrolyte molecule is large enough to be considered to have the properties of a very small macroscopic gel. As far as electrostatic effects go, this condition is equivalent to $R \gg 1/\kappa$. In this case, the fixed charges will be virtually neutralized by the diffusible ions of the electrolyte. The electrostatic potential will be constant and the net charge density will be zero both inside and outside the sphere (the potential has a different value inside and out), except for a relatively unimportant boundary region at the edge of the sphere with thickness of order $1/\kappa$. Thus, if $R \gg 1/\kappa$, two polyelectrolyte molecules with $r > 2R$ "see" each other as essentially electrically neutral bodies, and hence $w(r) = 0$. Consequently, we can use the same kind of argument for both polymer and polyelectrolyte molecules.

In a typical numerical example, $1/\kappa$ might be 20 A (Eq. 18–25) and

$$R = \frac{\alpha M'^{1/2} a'}{2} = \frac{2 \times 30 \times 12}{2} = 350 \text{ A}.$$

In this connection we might recall Eq. (19–43) for the second virial coefficient of large spheres with a surface charge and in the presence of nonpenetrating electrolyte. The "neutrality" term in that equation is proportional to the square of the net charge ($z\epsilon$) in the region $r \leq a$ (all of the charge happens to be on the surface). In the present problem, the corresponding term in B_2^* vanishes because the net charge in $r \leq 2R$ is essentially zero.

We return now to the overlap volume V' when $r < 2R$. When we bring two molecules together from infinity to r, the only change occurs in the two volumes V' destined to overlap. Thus we can represent the process by:

$$\underset{(1)}{2\,[\text{volume } V' \text{ with } \varphi_2]} \rightarrow$$

$$\underset{(2)}{[\text{volume } V' \text{ with } 2\varphi_2]} +$$

$$\underset{(0)}{[\text{volume } V' \text{ with } \varphi_2 = 0]}. \tag{21–52}$$

The free energy change in this process is

$$\Delta A_{\text{overlap}} = w(r) = \Delta A^{(2)} + \Delta A^{(0)} - 2\,\Delta A^{(1)}, \tag{21–53}$$

where $\Delta A^{(n)}$ is to be calculated as in Eq. (21–44). A helpful simplification is that $\Delta S_d^{(0)} = 0$, $\Delta S_d^{(2)} = 2\,\Delta S_d^{(1)}$, and hence there is no deformation contribution to ΔA_{ov}. This is a consequence of our assumption that overlapping is "additive," i.e., without distortion of polymer configuration.

Let us put Eq. (21–44) in the form

$$\frac{\Delta A^{(n)}}{kT} = N_1^{(n)} \ln (1 - \varphi_2^{(n)}) + \frac{\chi V' \varphi_2^{(n)}(1 - \varphi_2^{(n)})}{v_1} - \frac{\Delta S_d^{(n)}}{k} + \frac{z^2 (\varphi_2^{(n)})^2 V'}{2x^2 v_1^2 \Sigma}, \tag{21–54}$$

where

$$\varphi_2^{(0)} = 0, \qquad \varphi_2^{(1)} = \varphi_2, \qquad \varphi_2^{(2)} = 2\varphi_2,$$

and

$$N_1^{(n)} = \frac{(1 - \varphi_2^{(n)}) V'}{v_1}.$$

Substitution of Eq. (21–54) into Eq. (21–53), with expansion of the logarithm, leads to

$$\frac{w(r)}{kT} = \frac{\varphi_2^2}{v_1}\left(1 - 2\chi + \frac{z^2}{x^2 v_1 \Sigma}\right) V'(r) \qquad (r \leq 2R), \tag{21–55}$$

where $V'(r)$ is given by Eq. (21–51) (Problem 21–12). The same electrostatic term as in Eqs. (21–33) (polyelectrolyte gel) and (21–49) (single polyelectrolyte molecule) enters here and lowers the effective value of χ. When $z = 0$, $w(r)$ has the sign of $1 - 2\chi$. Thus, in a good solvent (e.g., $\chi = 0$), $w(r) > 0$ and work has to be done to force the molecules to overlap. This is because the overlapping process, (21–52), involves a partial unmixing of polymer and solvent. In a poor solvent with $\chi > 0.5$, $w(r)$ is negative: the molecules "want" to overlap to increase the number of polymer-polymer and solvent-solvent interactions relative to polymer-solvent interactions. When $\chi = \frac{1}{2}$, these mixing and interaction effects just balance, and $w(r) = 0$ for all r. That is, polymer molecules interpenetrate without effective interaction. This of course leads to a vanishing second virial coefficient, $B_2^* = 0$.

Next, we use the $w(r)/kT$ of Eq. (21–55) to calculate B_2^* from Eq. (19–37). Because the integration cannot be carried out analytically, we resort to an expansion of $e^{-w/kT}$:

$$B_2^* = -2\pi \int_0^{2R} \left[-\frac{w}{kT} + \frac{1}{2}\left(\frac{w}{kT}\right)^2 - \cdots \right] r^2\, dr. \tag{21–56}$$

The contribution from $r > 2R$ is zero because $w = 0$ in this range. We find from Eqs. (21–51), (21–55), and (21–56), after elementary integrations,

$$B_2^* = M^2 v_1\left(\frac{1}{2} + \frac{z^2}{2x^2 v_1 \sum} - \chi\right) \times \left\{1 - \frac{0.324[\frac{1}{2} + (z^2/2x^2 v_1 \sum) - \chi]V_0^2}{v_1 V} + \cdots\right\}. \quad (21\text{–}57)$$

This is the same as Flory's result [Eq. (75′) of his Chapter 12], derived from a gaussian instead of uniform segment distribution, except that 0.324 is replaced by 0.354. Using $V = C\alpha^3$ and Eq. (21–49), Eq. (21–57) becomes

$$B_2^* = C\alpha^3(\alpha^2 - 1)[1 - 0.324(\alpha^2 - 1) + \cdots], \quad (21\text{–}58)$$

where, it will be recalled, C is the volume of the effective isolated polymer sphere (Eq. 21–43) when $\alpha = 1$. Equation (21–58) expresses B_2^* in terms of the swelling parameter α of an isolated polymer molecule, which in turn is determined by Eq. (21–49). When the quantity in large parentheses in Eqs. (21–49) and (21–57) is zero, the isolated molecule is in the "random walk state," $\alpha = 1$, and $B_2^* = 0$. When $B_2^* \neq 0$, B_2^* has the sign of the quantity in large parentheses in Eq. (21–57), as expected from our discussion of Eq. (21–55). Convergence of Eq. (21–58) is rapid only when α is near unity, that is, when the quantity in large parentheses in Eq. (21–57) is small.

Incidentally, the leading term in Eq. (21–57) is the same as the second virial coefficient obtained (a) from the Flory-Huggins theory (Problem 21–13), and (b) as a first approximation by Zimm, using the McMillan-Mayer theory.

Orofino and Flory have tested a somewhat refined version of this theory against extensive experimental results on both polymers and polyelectrolytes. The agreement between theory and experiment is very satisfactory for polymers but less so for polyelectrolytes. Again the predicted electrostatic effects are too large.

Problems

21–1. If Eq. (21–1) gives the end-to-end distribution in an undeformed sample of rubber, show that Eq. (21–2) gives the distribution in a deformed sample, assuming each chain is deformed in the same way ($\alpha_x, \alpha_y, \alpha_z$) as the bulk sample. (Page 399.)

21–2. Use the method of undetermined multipliers to prove that the maximum Ω in Eq. (21–4) for given N and $\mathbf{p}$ is obtained when $s_i = Np_i$. (Page 399.)

21–3. Show that ΔS in Eq. (21–5) has its maximum value when $\alpha_x = \alpha_y = \alpha_z = 1$. (Page 400.)

21–4. Make a rough plot of φ_2 vs. x_2 and $\Delta S_m/M_0k$ vs. x_2 for $M = 1000$. (Page 404.)

21–5. Use Eqs. (21–16) and (21–17) for μ_1 and μ_2, and $A = N_1\mu_1 + N_2\mu_2$, to check Eq. (21–15) for ΔA_m. (Page 407.)

21–6. Show that use of the equal-area theorem on a plot of Π vs. $1/\varphi_2$ to locate the two-phase equilibrium point in a polymer solution is equivalent to using the equality of the two chemical potentials in the two phases. (Page 408.)

21–7. Derive Eq. (21–28), for the relation between α and t in a polymer gel, from Eqs. (21–24) and (21–27) by expanding $\ln(1 - \varphi_2)$. (Page 413.)

21–8. In a macroscopic polyelectrolyte gel the total charge density can be written (compare the Poisson-Boltzmann equation of Chapter 18)

$$\frac{n\mathbf{z}e}{V} + \sum_j \rho_j^0 z_j e e^{-z_j e\psi/kT},$$

where ψ is the electrostatic potential in the gel (relative to the outside as zero). Linearize the exponential, solve for ψ, and deduce ΔA_{el} in Eq. (21–41) by a charging process. (Page 418.)

21–9. Consider the free swelling of a gel with, say, —COOH groups which dissociate into $—COO^-$ and H^+. That is, the gel has fixed negative charges on the $—COO^-$ groups and can bind H^+, one for every $—COO^-$. Calculate and plot the swelling ratio $1/\varphi_2$ as a function of $1 - \theta$, the fraction of dissociated —COOH groups (as, for example, in a titration of the gel with base). This particular case implies $\mathbf{z} = -1$, $m = -1$, and $xB/N_2M = 1$. Assume the electrolyte is of type $1 - 1$, with concentration $c^0 = 0.15$ mole·liter^{-1}. Also, take $x = 10$, $\chi = 0.25$, $M = 400$, and $v_1 = 18$ cm^3·mole^{-1}. In Eq. (21–33), expand the logarithm and drop the term $\varphi_2/2$ compared with $1/\alpha = \varphi_2^{-1/3}$. Also, calculate and plot the titration curve, $1 - \theta$ against λq. Repeat the calculation with $c^0 = 0.05$ mole·liter^{-1}. (Page 418.)

21–10. Show from Eq. (21–43) that φ_2 is of order 10^{-2} when $\alpha = 1$, in a typical case. (Page 422.)

21–11. Make the necessary modifications in Section 21–5 when the polyelectrolyte molecule is involved in an ionic dissociation or binding equilibrium. (Page 423.)

21–12. Make a rough plot of the function $V'(r)$ between $r = 0$ and $r = 2R$. (Page 425.)

21–13. Show that the osmotic pressure second virial coefficient obtained from the Flory-Huggins theory agrees with the leading term in Eq. (21–57). (Page 426.)

21–14. Modify the Flory-Huggins theory for a heterogeneous polymer (mixture of polymer molecular weights).

21–15. Investigate whether the Flory-Huggins type of argument can be applied to long rigid rods (instead of random coils).

21–16. Extend the Flory-Huggins theory to polyelectrolyte molecules. If the polyelectrolyte molecules are involved in an ionic binding equilibrium, what happens to the titration curve (adsorption isotherm) when a phase separation occurs?

21–17. Discuss qualitatively some of the errors made in assuming additivity of the free energy terms in Eq. (21–37).

21–18. Beginning with Section 21–4, the solvent (electrolyte solution) has several components, but we treat it as a single component in nonelectrostatic expressions. Investigate the legitimacy of this procedure as far as the entropy of mixing solvent with polymer is concerned.

21–19. According to Section 21–5, how does the radius R of the effective sphere vary with the molecular weight of the polymer?

21–20. Deduce one more term in the expansion of B_2^* in Eq. (21–57). Compare with Flory [Eq. (75′), Chapter 12].

Supplementary Reading

FLORY, Chapters 11–14.

GUGGENHEIM, Chapters 10–12.

HILDEBRAND and SCOTT, Chapter 20.

MILLER, A. R., *Theory of Solutions of High Polymers*. Oxford: University Press, 1948.

TOMPA, H., *Polymer Solutions*. London: Butterworths, 1956.

TRELOAR, L. R. G., *Physics of Rubber Elasticity*. Oxford: University Press, 1958. Chapters 4–7.

Part IV

Quantum Statistics

CHAPTER 22

QUANTUM STATISTICS

Although the argument in the first few chapters of the book is based on quantum mechanics, up to now we have bypassed all topics which involve any but the most elementary and familiar quantum-mechanical ideas (energy levels, degeneracy, etc.). This has been done for the benefit of those readers with practically no background in quantum mechanics. In this chapter we consider a few of the more fundamental and simple problems in quantum statistics. Even here the demands on the reader's knowledge of quantum mechanics are very modest—in keeping with the introductory nature of the present work.

Sections 22–1 to 22–4 are concerned with the quantum statistics of one-component systems of indistinguishable particles without interparticle interactions. (We use the term "particle" here and below to refer to fundamental particles, atoms, or molecules.) In Sections 22–5 to 22–7 molecular interactions are allowed. This makes the problem very involved, so our treatment is strictly introductory. Section 22–8 is concerned with the special topic of dilute symmetrical diatomic gases at low temperatures (e.g., ortho- and parahydrogen).

22–1 Introduction to Fermi-Dirac and Bose-Einstein statistics. The problem we consider here is the same as that of Section 3–2 but now we give a general treatment. We are concerned with a macroscopic system of identical and indistinguishable particles, which do not exert forces on each other, in a volume V (or area $\mathcal{A}$, etc.) and at temperature T. The possible energy eigenvalues for a single particle in V are denoted by $\epsilon_1, \epsilon_2, \epsilon_3, \ldots$ (functions of V), and the associated energy eigenfunctions are $\psi_1, \psi_2, \psi_3, \ldots$. For simplicity of notation, an energy level with degeneracy ω is listed here ω times (i.e., these are energy states, not levels). In Section 3–2 we limited ourselves to special conditions such that the number of energy states of the above type (available to any one particle) is very much larger than the number (N) of particles in the system. When this is the case, we found (Eq. 3–10) that the canonical ensemble partition function of the system is simply

$$Q = \frac{1}{N!} q^N, \tag{22–1}$$

where

$$q = \sum_j e^{-\epsilon_j/kT}. \tag{22–2}$$

When we remove the above restriction, as we do here, the situation is more complicated. In particular, because of symmetry restrictions on the wave functions of the system, the particles are not independent of each other (even though interparticle interactions are absent): Q is no longer essentially given by a product of (independent) q's, q^N; indeed, the single particle partition function q no longer arises as a significant quantity. But the general results we shall find will reduce to Eqs. (22–1) and (22–2) as a limiting case, as expected.

There is no way of detecting experimentally an exchange, one for the other, of two identical particles in a system. Let ψ be the wave function representing the state of a system of N identical and indistinguishable particles. Then we can easily deduce* from the above-mentioned experimental fact that ψ may behave in one of two ways: when the coordinates of two particles are exchanged in the function ψ, ψ may remain unchanged or ψ may change sign. We say that ψ is *symmetrical* in the coordinates of identical particles in the former case and *antisymmetrical* in the latter case. One of the fundamental postulates or principles of quantum mechanics is that (a) the only states available (or "accessible") to real systems of indistinguishable particles are those represented by wave functions which are either symmetrical or antisymmetrical, and (b) particles with half-integral spin (e.g., electrons, protons, neutrons) have antisymmetrical wave functions, while those with integral spin (e.g., photons) have symmetrical wave functions. Nuclei, atoms, ions, and molecules made up of an odd number of electrons, protons, and neutrons (e.g., D, He^3) are in the antisymmetrical class, and those with an even number (e.g., H, H_2, D_2, He^4) are in the symmetrical class. The above restrictions on accessibility are applicable whether or not there are interparticle forces. They are also applicable to multicomponent systems in an obvious way (Problem 22–1).

The canonical ensemble partition function is in any case (interparticle forces, multicomponent systems)

$$Q = \sum_j e^{-E_j/kT}, \tag{22–3}$$

where the E_j are eigenvalues of the Hamiltonian operator $\mathcal{H}$ for the system. But we now understand explicitly that we are to include in the sum only those energy eigenstates associated with energy eigenfunctions with the correct symmetry properties. This was implicit in our discussion in Chapter 1, but the point was not emphasized. In other words, the sum is over all *accessible* energy states; those states which are inaccessible because of symmetry (or other) restrictions are omitted from the sum.

---

* See, for example, L. D. LANDAU and E. M. LIFSHITZ, *Quantum Mechanics*. Reading, Mass.: Addison-Wesley, 1958; p. 204.

In a one-component system, we include in the sum either all the symmetrical or all the antisymmetrical states, depending on the kind of particle. It is customary to say that *Bose-Einstein statistics* is involved in the former case and *Fermi-Dirac* in the latter.

We now return to the special case of a one-component system of noninteracting particles. It will no doubt be helpful to the reader if we illustrate the above considerations on symmetry restrictions by an explicit simple example. Consider a system with only three particles, $N = 3$; call them a, b, and c. For simplicity, suppose that the (nondegenerate) single-particle energy eigenvalues $\epsilon_1, \epsilon_2, \epsilon_3, \ldots$, are proportional to the quantum number: $\epsilon_j = j\delta$, where δ is a constant (as for a harmonic oscillator with proper choice of energy zero). Let us use the canonical ensemble partition function in the form (Eq. 1–37) of a sum over energy levels instead of states:

$$Q = \sum_i \Omega_i e^{-E_i/kT}. \tag{22–4}$$

Consider, say, the system energy level $E = 9\delta$. That is, 9 units (1 unit $= \delta$) of energy are to be divided up among the three particles. The possibilities are listed in Table 22–1 (Problem 22–2). The numbers under a, b, and c are particle quantum numbers j for the respective particles; the sum of the j values is 9 in every case ($E = 9\delta$). The first row corresponds to the system energy eigenfunction $\psi = \psi_7(a)\,\psi_1(b)\,\psi_1(c)$, where (a) refers to the coordinates of particle a, etc. This function satisfies the Schrödinger equation for the system, $\mathcal{H}\psi = 9\delta\psi$. The number 3 under "permutations" refers to the fact that the first row might have read 1, 7, 1 or 1, 1, 7 as well as 7, 1, 1 [the first of these corresponds to the system energy eigenfunction $\psi_1(a)\,\psi_7(b)\,\psi_1(c)$, etc.]. In the second row, there are six different functions possible for the set of quantum numbers 6, 2, 1;

TABLE 22–1

SYSTEM OF THREE PARTICLES

Row	a	b	c	Permutations
1.	7	1	1	3
2.	6	2	1	6
3.	5	3	1	6
4.	5	2	2	3
5.	4	4	1	3
6.	4	3	2	6
7.	3	3	3	1

etc. So far we have ignored symmetry. If we continue to do so, and this would be correct *if the particles are distinguishable* (as in an Einstein crystal, for example), we deduce, by adding up the column of numbers under "permutations," that there are 28 different wave functions which satisfy the equation $\mathcal{H}\psi = 9\delta\psi$. That is, the system energy level $E = 9\delta$ has a degeneracy $\Omega_D = 28$ ($D =$ "distinguishable"). Thus one term in Eq. (22–4), for this simple example, is $28e^{-9\delta/kT}$.

Suppose the three particles obey Fermi-Dirac statistics (e.g., they might be electrons). If we run through the list of 28 energy eigenfunctions referred to above, we find that *none* of them is antisymmetrical. For example,

$$\psi_6(a)\,\psi_2(b)\,\psi_1(c) \neq -\psi_6(b)\,\psi_2(a)\,\psi_1(c),$$

where we have exchanged the coordinates of particles a and b. But we have not really exhausted all the possibilities. Any linear combination of two or more of the 28 functions is also a solution of the (linear) equation $\mathcal{H}\psi = 9\delta\psi$. On examining linear combinations, we find, for example, that the function

$$\psi = C \begin{vmatrix} \psi_6(a) & \psi_6(b) & \psi_6(c) \\ \psi_2(a) & \psi_2(b) & \psi_2(c) \\ \psi_1(a) & \psi_1(b) & \psi_1(c) \end{vmatrix}, \tag{22–5}$$

where C is a normalization constant, meets our requirements. This function satisfies the equation $\mathcal{H}\psi = 9\delta\psi$ because it is a linear combination of the 6 functions corresponding to row 2 in Table 22–1 (as can be seen on expanding the determinant). Also, it is an antisymmetrical wave function, for to exchange, say, (a) and (b), amounts to exchanging two columns in the determinant. But it is well known that the exchange of two columns changes the sign of a determinant. In this way, (22–5), we can construct one antisymmetrical energy eigenfunction for the system from each of rows 2, 3, and 6 in Table 22–1 (the quantum numbers are all different in each case). No such function can be formed from any of the other rows, however. For example, the determinant constructed from the set of quantum numbers 7, 1, 1 has two rows exactly alike and therefore is identically equal to zero. This will be the case *whenever two or more particles are in the same quantum state.* Further inspection reveals no other antisymmetrical linear combinations of the 28 functions (except linear combinations of the three determinants already found). Therefore, in this example, there are 3 antisymmetrical system energy states belonging to the eigenvalue $E = 9\delta$; that is, $\Omega_{FD} = 3$. Thus one term in Eq. (22–4), for a Fermi-Dirac system, is $3e^{-9\delta/kT}$.

The above argument can obviously be generalized to a system containing any number N of noninteracting particles. We reach the general conclusion that a quantum state of a Fermi-Dirac system in which two or more particles are in the same particle quantum state is inaccessible. An alternative statement is that, in Fermi-Dirac statistics, each particle quantum state can be "occupied" by either zero or one particle at a time. This is just the *Pauli exclusion principle* (which forms the basis of the periodic table when applied to electrons in atoms). For example, when the above system of 3 particles is in the state (22–5), particle states 1, 2, and 6 are occupied (we cannot say by which particle), while states 3, 4, 5, 7, 8, . . . are unoccupied.

Next, we consider the Bose-Einstein case (e.g., the three particles might be He^4 atoms). We examine the 28 wave functions above and find that only the last one,

$$\psi = \psi_3(a)\,\psi_3(b)\,\psi_3(c),$$

is symmetrical. But, if we allow linear combinations, we find that in fact one symmetrical wave function can be formed from *each* of the rows in Table 22–1. All these functions can be represented in the form of a modified determinant; for example, from row 1 (Problem 22–3),

$$\psi = C' \begin{vmatrix} \psi_7(a) & \psi_7(b) & \psi_7(c) \\ \psi_1(a) & \psi_1(b) & \psi_1(c) \\ \psi_1(a) & \psi_1(b) & \psi_1(c) \end{vmatrix}' , \qquad (22\text{–}6)$$

where the prime on the determinant means that in expanding the determinant *all signs are to be taken positive*. This scheme gives, for row 7,

$$\psi = 6C'\psi_3(a)\,\psi_3(b)\,\psi_3(c),$$

etc. Thus, in this example, $\Omega_{\text{BE}} = 7$, and one term in Eq. (22–4) is $7e^{-9\delta/kT}$.

From the above simple example we can make the following general statement: in Bose-Einstein statistics (symmetrical wave functions), the number of particles in a particle quantum state is unrestricted; the number can be $0, 1, 2, 3, \ldots, N$ (whereas in Fermi-Dirac statistics only 0, 1 are possible).

There is another important general deduction we can make by inspection of Table 22–1. We note that, in this example,

$$\Omega_{\text{BE}} \geq \frac{\Omega_D}{N!} \geq \Omega_{\text{FD}}. \qquad (22\text{–}7)$$

It is easy to see that this will always be true. Each Fermi-Dirac state (rows 2, 3, and 6 in Table 22–1) corresponds to $N!$ "distinguishable" states. But there may be some sets of particle quantum numbers (rows 1, 4, 5, and 7 in Table 22–1) which give "distinguishable" states but not a Fermi-Dirac state. Therefore $\Omega_D \geq N!\Omega_{\text{FD}}$. Also, each set of particle quantum numbers gives one Bose-Einstein state. The maximum number of "distinguishable" states for one set of quantum numbers is $N!$ (all quantum numbers different). Therefore $\Omega_D \leq N!\Omega_{\text{BE}}$.

It is also clear that if E is sufficiently large (high temperature), the number of available particle quantum states can be very large compared with the number of particles N (see Sections 3–2 and 4–1). In this case the vast majority of sets of particle quantum numbers will have all quantum numbers different (i.e., because of the large excess of particle quantum states, only infrequently will two particles be in the same quantum state, even if allowed). Thus, for large values of E, the equalities in (22–7) are approached:

$$\Omega_{\text{BE}} \rightarrow \frac{\Omega_D}{N!} \leftarrow \Omega_{\text{FD}} \qquad (E \rightarrow \infty). \tag{22–8}$$

Hence, the two kinds of statistics give the same limiting result. This limiting type of statistics (treating the particles as distinguishable to get Ω_D, and correcting with $N!$) is referred to as "classical" or "Boltzmann." We have been using Boltzmann statistics in most of the book, since the limiting condition is achieved in the usual problems of interest.

The canonical ensemble expression corresponding to Eq. (22–8) is

$$\sum_i \Omega_i^{\text{BE}} e^{-E_i/kT} \rightarrow \frac{1}{N!} \sum_i \Omega_i^D e^{-E_i/kT} \leftarrow \sum_i \Omega_i^{\text{FD}} e^{-E_i/kT}. \tag{22–9}$$

Actually, this relation and (22–8) are not restricted to systems of noninteracting particles (see Sections 6–2 and 22–6).

We turn now to the explicit problem of calculating thermodynamic functions for a one-component system of noninteracting indistinguishable particles. The customary treatment* is based on Ω and the microcanonical ensemble. Table 22–1 makes it clear how one is to proceed in principle. For given N and E, Ω_{BE} is the total number of different ways of dividing E up among the N particles, treating the particles as indistinguishable and without restricting the number of particles in any one particle quantum (energy) state. For Ω_{FD}, however, we have the limitation that not more than one particle can be in a particle quantum state (hence $\Omega_{\text{FD}} \leq \Omega_{\text{BE}}$). In practical cases, N and E are of course very large (macro-

---

* See, for example, Mayer and Mayer, Chapter 5.

scopic). The restriction of fixed values for N and E requires use of undetermined multipliers (Problem 22–4), or the method of steepest descents. A simpler procedure is to employ the grand partition function (μ, V, T specified). This avoids the restraints of constant N and E.

In Chapter 7 [see, especially, Eqs. (7–26), (7–31), and (7–34), and the discussion following Eq. (7–27)], we found that a grand partition function ξ for a single subsystem was useful. Specifically, we considered a system in which the adsorption sites were distinguishable and the particles indistinguishable. In Eq. (7–31) (Langmuir adsorption), the number of molecules per site was restricted to zero or one, while in Eq. (7–34) (B.E.T. adsorption) the number of molecules per site was not restricted. The situation here is very similar. A particle quantum state with the particles occupying it is a subsystem in the present problem. The particle quantum states are distinguishable, but the particles are indistinguishable. In Fermi-Dirac statistics, the subsystem has $s = 0$ or 1, and in Bose-Einstein statistics $s = 0, 1, 2, \ldots$.

In *Fermi-Dirac statistics*, then,

$$\Xi(\mu, V, T) = \prod_j \xi_j(V, T), \qquad \xi_j = 1 + e^{-\epsilon_j(V)/kT}\lambda, \qquad \lambda = e^{\mu/kT}. \tag{22–10}$$

The first term in ξ_j is associated with $s_j = 0$, and the second with $s_j = 1$, where $e^{-\epsilon_j/kT}$ [corresponding to q in Eq. (7–31)] is the partition function of one particle in the energy state ϵ_j. The mean number of particles in

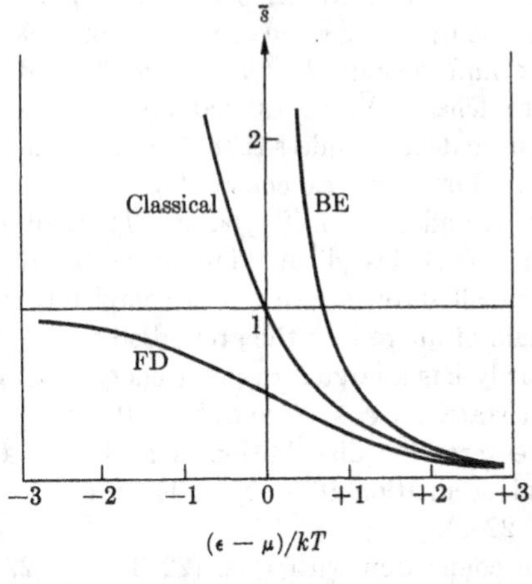

Fig. 22–1. Bose-Einstein, Fermi-Dirac, and classical (Boltzmann) distributions: mean number ($\bar{s}$) of particles in a quantum state as a function of the energy (ϵ) of the state.

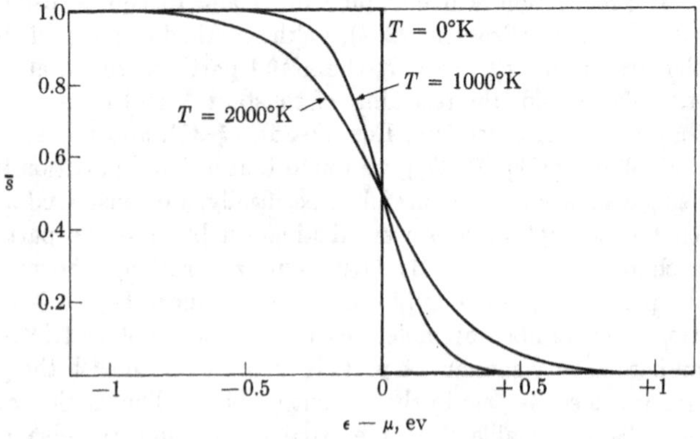

FIG. 22–2. Fermi-Dirac distribution at different temperatures.

the system is

$$\overline{N} = \lambda \left(\frac{\partial \ln \Xi}{\partial \lambda} \right)_{V,T} = \sum_j \lambda \left(\frac{\partial \ln \xi_j}{\partial \lambda} \right)_{V,T} = \sum_j \bar{s}_j, \tag{22–11}$$

where

$$\bar{s}_j = \lambda \left(\frac{\partial \ln \xi_j}{\partial \lambda} \right)_{V,T} = \frac{e^{-\epsilon_j/kT}\lambda}{1 + e^{-\epsilon_j/kT}\lambda} = \frac{1}{e^{(\epsilon_j - \mu)/kT} + 1}. \tag{22–12}$$

This is the Fermi-Dirac distribution law (see Fig. 22–1).

Equation (22–12) gives the mean number of molecules in the state ϵ_j. For any finite ϵ_j and constant T: if $\mu \to -\infty$, $\bar{s}_j \to 0$ (therefore the system approaches zero density, $\overline{N}/V \to 0$); and if $\mu \to +\infty$, $\bar{s}_j \to 1$ (the system approaches infinite density unless there are only a finite number of particle energy states). For finite ϵ_j, constant μ, and $T \to 0$ (see Fig. 22–2): $\bar{s}_j \to 0$ if $\epsilon_j > \mu$ and $\bar{s}_j \to 1$ if $\epsilon_j < \mu$. Thus, at the absolute zero of temperature, the particles fill up (one per state) all energy states below $\epsilon = \mu$ and leave all states above $\epsilon = \mu$ completely unoccupied. Because the particles cannot all go into the ground state at $T = 0$, a Fermi-Dirac system necessarily has a large zero-point energy. (If μ is chosen less than ϵ_1, the ground state, all $\bar{s}_j = 0$ and $\overline{N} = 0$ at $T = 0$.) For T slightly above absolute zero, the distribution is no longer a step function, and there is partial occupation ($0 < \bar{s}_j < 1$) of states in the neighborhood of $\epsilon = \mu$ (Fig. 22–2).

We noted in connection with Eqs. (22–8) and (22–9) that both kinds of quantum statistics are expected to go over into classical statistics if the number of available quantum states greatly exceeds the number of particles, N. In the present context and notation, this means $\bar{s}_j \to 0$

for all energy states (i.e., $s_j = 0$ almost always, but occasionally $s_j = 1$). We see from Eq. (22-12) that this condition is achieved if $\lambda \to 0$ (e.g., $\overline{N}/V \to 0$ with T constant or $T \to +\infty$ with $\overline{N}/V$ constant). Equation (22-12) becomes, for small λ,

$$\bar{s}_j = \lambda e^{-\epsilon_j/kT} \qquad (\lambda \to 0), \tag{22-13}$$

or

$$\frac{\bar{s}_j}{\overline{N}} = \frac{e^{-\epsilon_j/kT}}{q}, \qquad q = \sum_j e^{-\epsilon_j/kT} \qquad (\lambda \to 0), \tag{22-14}$$

where λ has been eliminated ($\lambda = \overline{N}/q$) by summing (22-13) over j. The quantity $\bar{s}_j/\overline{N}$ is the fraction of molecules in state j, or the probability of any one particular molecule being in state j. Equation (22-14) is the Boltzmann distribution, the same as Eq. (3-35), as should be expected. We observe that q and the Boltzmann probability $e^{-\epsilon_j/kT}/q$ arise *only in the limit* $\lambda \to 0$. Figure 22-1 includes the classical or Boltzmann case, (22-13).

In *Bose-Einstein statistics*,

$$\Xi(\mu, V, T) = \prod_j \xi_j(V, T), \tag{22-15}$$

$$\xi_j = 1 + e^{-\epsilon_j(V)/kT}\lambda + e^{-2\epsilon_j(V)/kT}\lambda^2 + \cdots = \frac{1}{1 - e^{-\epsilon_j/kT}\lambda}. \tag{22-16}$$

The successive terms in the above series are for $s_j = 0, 1, 2, \ldots$, the number of molecules in state j, with energies $0, \epsilon_j, 2\epsilon_j, \ldots$, and partition functions $1, e^{-\epsilon_j/kT}, e^{-2\epsilon_j/kT}$, etc. The series converges only if $e^{-\epsilon_j/kT}\lambda < 1$. Since this condition must hold for all states j, we have that μ must be less than the ground state energy ϵ_1 (usually taken as zero for convenience). There is no corresponding restriction on values of μ in Fermi-Dirac statistics. Equation (22-11) for $\overline{N}$ still applies but here we find

$$\bar{s}_j = \lambda\left(\frac{\partial \ln \xi_j}{\partial \lambda}\right)_{V,T} = \frac{1}{e^{(\epsilon_j-\mu)/kT} - 1}. \tag{22-17}$$

This is the Bose-Einstein distribution law, which differs from the Fermi-Dirac law only by the sign in front of the unity in the denominator. Figure 22-1 shows that for a given $\mu < \epsilon_1$, each energy state ϵ_j has a mean population which decreases in the order BE, Class, FD. The result BE > FD is to be expected simply because of the restriction $s_j = 0$ or 1 for FD. Relative to the classical case, the larger (smaller) number of particles in the system for given μ in the Bose-Einstein (Fermi-Dirac) case corresponds to an effective attraction (repulsion) between the particles. We shall see this also in Eqs. (22-38) and (22-76).

In a Bose-Einstein system, for any $\mu < \epsilon_1$, when $T \to 0$ all $\bar{s}_j \to 0$ and $\overline{N} \to 0$. However, if μ is only infinitesimally less than ϵ_1 and $T \to 0$, then we still have $\bar{s}_j \to 0$ for all $j > 1$, but $\bar{s}_1$ can be made as large as one pleases. Thus if we put $\epsilon_1 - \mu = kT/\overline{N}$,

$$\bar{s}_1 = \frac{1}{e^{1/\overline{N}} - 1} = \overline{N} \qquad (T \to 0). \tag{22–18}$$

In other words, in the limit $T \to 0$, the only thermodynamically significant ($\overline{N} > 0$) value of μ is $\mu = \epsilon_1$; also, in this limit, all particles are in the ground state (ϵ_1) and all excited states ($j > 1$) are unoccupied. This is in distinct contrast to the behavior of a Fermi-Dirac system at $T = 0$.

The low-temperature behavior for either kind of statistics also follows from Fig. 22–1 if we keep in mind that as $T \to 0$ the horizontal energy scale in the figure shrinks proportionally.

If we let $\lambda \to 0$ in Eq. (22–17), we have

$$\bar{s}_j \to \frac{1}{e^{\epsilon_j/kT}/\lambda} = \lambda e^{-\epsilon_j/kT} \qquad (\lambda \to 0). \tag{22–19}$$

This is again the classical or Boltzmann distribution, (22–13), as it should be. Each $\bar{s}_j \ll 1$: usually $s_j = 0$, occasionally $s_j = 1$, and only very rarely is $s_j > 1$.

Finally, let us indicate in a formal way how the various thermodynamic functions are obtained in either kind of quantum statistics. The variables μ, V, and T are independent and specified in advance. Then, from Eq. (22–11), the mean number of particles in the system is (upper sign FD; lower sign BE)

$$\overline{N} = \sum_j \frac{1}{e^{(\epsilon_j - \mu)/kT} \pm 1}. \tag{22–20}$$

The average energy is obviously

$$\overline{E} = \sum_j \epsilon_j \bar{s}_j = \sum_j \frac{\epsilon_j}{e^{(\epsilon_j - \mu)/kT} \pm 1}. \tag{22–21}$$

Also, for the equation of state, we have

$$pV = kT \ln \Xi = \pm kT \sum_j \ln [1 \pm e^{(\mu - \epsilon_j)/kT}]. \tag{22–22}$$

In these equations the ϵ_j are functions of V (or of the external variables of the problem). All other thermodynamic functions can now be found from μ, V, T, $\overline{N}$, $\overline{E}$, and p (Problem 22–5).

We consider the three most important special cases in the following sections.

22-2 Ideal Fermi-Dirac gas; electrons in metals. The particular model we consider here is a noninteracting gas of Fermi-Dirac particles in a box of volume V (ideal Fermi-Dirac gas). The energy levels ϵ_j are then those of a particle in a box, Eq. (4-2). Because the energy levels are extremely close together, for macroscopic V, we can treat them as continuous. Equation (4-9) gives the number of translational states between ϵ and $\epsilon + d\epsilon$.

For concreteness, we discuss one particular example—the electrons in a metal. In this case there is an electron spin of one-half (in units of $h/2\pi$) and degeneracy of two. We therefore multiply Eq. (4-9) by two and have

$$\omega(\epsilon)\, d\epsilon = 4\pi \left(\frac{2m}{h^2}\right)^{3/2} V\epsilon^{1/2}\, d\epsilon. \tag{22-23}$$

If an electron at rest in the gas phase is used to locate the zero of energy, then each electron in the metal is assumed to move in a constant potential field φ (compare φ in Chapter 16). This potential is negative (of the order of several electron volts), owing to the attraction between an electron and the lattice of positive ions, and goes to zero at the surface of the metal.

It is at first glance quite surprising that a noninteracting gas of electrons is of any use at all as a model for the electrons in a metal. The fact is that the deductions from this model are in good qualitative (not quantitative) agreement with experiment (heat capacity, electrical and thermal conductivity, thermionic emission, etc.). The essential reasons for this agreement are: (1) each electron interacts with the other electrons and the lattice of positive ions by long range coulomb interactions which change only relatively slowly with the position of the electron (hence the use of $\varphi =$ constant); and (2) as we shall see below, quantum effects are extreme in this case, and these (rather than interaction) dominate in the behavior of the electrons.

For simplicity, from here on we take (but see Problem 22-6) the zero of energy as the ground state in the metal, $\epsilon_1 = 0$ (instead of $\epsilon_1 = \varphi$). Then Eq. (22-20) becomes

$$\overline{N} = 4\pi \left(\frac{2m}{h^2}\right)^{3/2} V \int_0^\infty \frac{\epsilon^{1/2}\, d\epsilon}{e^{(\epsilon-\mu)/kT} + 1}\cdot \tag{22-24}$$

Let us check immediately to see if the classical limit (omit the $+1$) can be used. This requires $\lambda \ll 1$. We can use the classical (ideal gas) λ for this test, Eq. (4-23) (the electron spin introduces a factor of two, which is unimportant for the moment). That is, if $\lambda_{\text{class}} \ll 1$ then $\lambda = \lambda_{\text{class}}$. Now the condition $\lambda_{\text{class}} \ll 1$ from Eq. (4-23) is just the condition $\Lambda^3 N/V \ll 1$ in Eq. (4-6). Let us take $m =$ electron mass, $T = 300°\text{K}$, and N/V corresponding to $10\ \text{cm}^3\cdot\text{mole}^{-1}$ for the metal and one free electron per metal atom. We find

$$\frac{\Lambda^3 N}{V} = 4800,$$

which is large, rather than small, compared with unity (see also Table 4–1). Thus quantum effects are large here; this is because of the high free-electron density and, especially, low mass. (The term "strong degeneracy" is often used in both kinds of statistics to refer to the case $\lambda_{\text{class}} \gg 1$; "weak degeneracy" means almost classical, $\lambda_{\text{class}} < 1$.) One can use the classical limit of Eq. (22–24) only above about $T = 10^{5}$°K (Problem 22–7), at which temperature, however, the metal would be vaporized.

Thus, from the point of view of the electrons in a metal, room temperature is itself a relatively *low temperature* (see Fig. 22–2). In fact, equations valid in the limit $T \to 0$ give numerical results that are quite accurate (as we shall confirm below) at room temperature. Therefore we begin by investigating the low-temperature limit.

When $T = 0$, $\bar{s} = 1$ for $\epsilon < \mu$ and $\bar{s} = 0$ for $\epsilon > \mu$, as already pointed out. Equation (22–24) becomes, then,

$$\overline{N} = 4\pi \left(\frac{2m}{h^2}\right)^{3/2} V \int_0^{\mu} \epsilon^{1/2}\, d\epsilon = \frac{8\pi}{3} \left(\frac{2m\mu}{h^2}\right)^{3/2} V. \tag{22–25}$$

The number of electrons that can be accommodated by the system is clearly just equal to the number of quantum states with energy μ or less [compare Eq. (4–5)]. Or, if we regard $\overline{N}/V$ as given, all energy states up to the so-called Fermi energy,

$$\mu = \frac{h^2}{8m} \left(\frac{3\overline{N}}{\pi V}\right)^{2/3}, \tag{22–26}$$

are filled, and all above μ are empty. The Fermi energy is of the order of several electron volts (Problem 22–8).

The (kinetic) energy at $T = 0$ is

$$\overline{E} = \sum_j \epsilon_j \bar{s}_j = 4\pi \left(\frac{2m}{h^2}\right)^{3/2} V \int_0^{\mu} \epsilon^{3/2}\, d\epsilon$$

$$= \frac{3\overline{N}\mu}{5}. \tag{22–27}$$

That is, the average energy (ground state as zero) per particle is $3\mu/5$. This is the zero-point energy of the system. It is large because of the Pauli exclusion principle. If we use Eq. (22–26) to eliminate μ from Eq. (22–27), we have $\overline{E}$ as a function of $\overline{N}$ and V. We see, then, that $C_V = 0$ $(T \to 0)$. An early problem in the (classical) theory of metals was to try to reconcile the fact that although "free" electrons exist as far as thermal and electrical conductivity are concerned, they still do not contribute significantly to the heat capacity (the expected value was $3k/2$ per electron). We have here the quantum-mechanical explanation: at ordinary temperatures the vast majority of electrons are in the filled states, $\epsilon < \mu$; the only electrons that have higher energy levels available (i.e., unoccupied) to which they can jump (and absorb heat) are those near $\epsilon = \mu$. The first nonzero term in C_V, expanded in powers of T, is given below in Eq. (22–31).

We can find the pressure from Eq. (22–22). When $\mu > \epsilon$ (and $T \to 0$), it is permissible to drop the unity. When $\mu < \epsilon$, we drop the exponential. Thus

$$pV = kT \int_0^{\mu} \left(\frac{\mu - \epsilon}{kT}\right) \omega(\epsilon)\, d\epsilon = \overline{N}\mu - \overline{E}$$

$$= \tfrac{2}{5}\overline{N}\mu = \tfrac{2}{3}\overline{E}. \tag{22–28}$$

The electrons in a metal exert a very high pressure (even at $T = 0$, because of the large zero-point kinetic energy), of the order of 10^5 atm. This pressure is balanced, however, by a potential energy contribution from $\partial\varphi/\partial V$, which we have not included explicitly.

We digress here to note that the result $pV = 2\overline{E}/3$ is rather general: it holds for any noninteracting (ideal) gas of particles in a three-dimensional box, irrespective of statistics or temperature. To see this in a formal way, put $\omega(\epsilon)\, d\epsilon = C\epsilon^{1/2}\, d\epsilon$ (Eq. 4–9) in Eqs. (22–21) and (22–22):

$$\overline{E} = C \int_0^{\infty} \frac{\epsilon^{3/2} d\epsilon}{e^{(\epsilon-\mu)/kT} \pm 1},$$

$$pV = \pm kTC \int_0^{\infty} \epsilon^{1/2} \ln\left[1 \pm e^{(\mu-\epsilon)/kT}\right] d\epsilon.$$

An explicit expression for C is not needed here. An integration of the pV equation by parts leads immediately to $pV = 2\overline{E}/3$. The physical explanation of this general result has already been given in connection with Eq. (4–19).

For the entropy of a Fermi-Dirac gas at $T = 0$, we have

$$S = 0, \tag{22–29}$$

since there is only one way ($\Omega = 1$) to put $\overline{N}$ indistinguishable particles in the lowest $\overline{N}$ energy states, one per state. For $T > 0$, S is proportional to T [see Eq. (22–31)].

The above equations for the low-temperature limit are the most important for this simple model of a metal. To find the next term in expansions in power of T requires straightforward but quite lengthy mathematical manipulations, which we omit because the results obtained are not sufficiently important for our purposes. (We shall see below that the corrections found are rather negligible.) Let us simply state, as a typical example, the result for the energy:

$$E = \frac{3}{5} N\mu_0 \left[1 + \frac{5\pi^2}{12} \left(\frac{kT}{\mu_0} \right)^2 + \cdots \right], \tag{22-30}$$

where μ_0 is the Fermi energy:

$$\mu_0 = \frac{h^2}{8m} \left(\frac{3N}{\pi V} \right)^{2/3}.$$

This gives E as a function of the independent variables N, V, T (closed, isothermal system). Since μ_0 is of the order of electron volts and 1 ev corresponds to kT with $T = 1.16 \times 10^4$°K, we see that the correction in Eq. (22–30) at, say, $T = 300$°K, is of the order of only 0.1%.

Other thermodynamic functions follow easily from Eq. (22–30) (Problem 22–9). For example,

$$C_V = \left(\frac{\partial E}{\partial T} \right)_{N,V} = Nk \frac{\pi^2}{2} \frac{kT}{\mu_0} + \cdots. \tag{22-31}$$

This predicts that C_V is proportional to T, a conclusion which has been verified at low temperatures (where the Debye T^3 heat capacity of the metal—arising from vibration of positive ions—goes rapidly to zero). Also, integration of C_V/T leads to $S \propto T$. At room temperature, Eq. (22–31) gives, in agreement with experiment, a heat capacity which is of the order of only one percent of (a) the classical value for free electrons, or (b) the vibrational heat capacity of the positive ions (Chapter 5). Figure 22–3 shows the heat capacity of an ideal Fermi-Dirac gas over the whole temperature range.

A much more detailed account of this subject is available in a number of places. See especially Mayer and Mayer, Chapter 16.

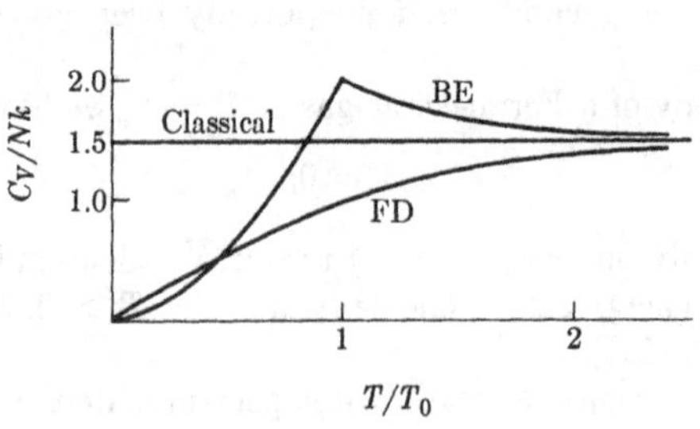

FIG. 22–3. Heat capacity of ideal Bose-Einstein and Fermi-Dirac gases. T_0 is defined in both cases by Eq. (22–45).

22-3 Ideal Bose-Einstein gas; helium. We saw in the preceding section that very strong quantum effects occur in an electron gas even at ordinary temperatures. In this section we shall be concerned with a gas of non-interacting Bose-Einstein molecules (ideal Bose-Einstein gas). The mass of a particle (molecule) here is several thousand times the mass of an electron. Hence quantum effects are much smaller: they are absent at room temperature and become very strong only at very low temperatures (a few degrees absolute). We found, for example, in Table 4-1 that $\lambda_{\text{class}} = \Lambda^3 N/V = 1.5$ for helium at the liquid density and 4.2°K. The condition for classical behavior, it will be recalled, is $\lambda_{\text{class}} \ll 1$. We conclude from the above remarks that in the present problem (and of course also for a Fermi-Dirac gas of *molecules*) we have to consider both weak and strong quantum effects. Low-temperature expressions (strong degeneracy) are not valid over the whole range of practical interest, as is the case with electrons in a metal.

The specific example we choose is helium. Helium is almost the lightest gas available and it has the weakest known intermolecular forces. Hence, with helium, quantum effects have the best possible chance to make themselves evident in the gas at low temperatures without intermolecular forces confusing the issue. Unfortunately, it turns out that the extremely interesting condensation properties of helium, observed experimentally, are due to an inextricable mixture of quantum and intermolecular force effects. But there is little doubt that the "condensation" in an *ideal* Bose-Einstein gas, which we shall study below, is intimately related to the actual behavior of helium at low temperatures, though it furnishes only part of the story.

Let us begin by investigating the region of slight degeneracy (low density or high temperature). This is most simply done using essentially the grand-partition-function, virial-expansion method of imperfect gas theory (Chapter 15). The question of the range of convergence of the series will be examined later. We shall write explicit expressions only through the second virial coefficient and related terms, but it is easy to extend the various series further [see, for example, Problem 22-10 and Eq. (22-89)]. Up to the second virial coefficient there is no inconvenience in including both Fermi-Dirac (upper sign) and Bose-Einstein (lower sign) cases in the same equations, so we may as well do this.

We begin with

$$\begin{aligned} pV &= kT \ln \Xi \\ &= \pm 2\pi kT \left(\frac{2m}{h^2}\right)^{3/2} V \int_0^\infty \epsilon^{1/2} \ln\left(1 \pm e^{-\epsilon/kT}\lambda\right) d\epsilon. \qquad (23\text{-}32) \end{aligned}$$

For small λ we can expand the logarithm and integrate term by term:

$$pV = \mp 2\pi kT\left(\frac{2m}{h^2}\right)^{3/2} V\sum_{j\geq 1}\frac{(\mp 1)^j\lambda^j}{j}\int_0^\infty \epsilon^{1/2}e^{-j\epsilon/kT}\,d\epsilon,$$

or

$$\frac{p}{kT} = \mp\frac{1}{\Lambda^3}\sum_{j\geq 1}\frac{(\mp 1)^j\lambda^j}{j^{5/2}}. \tag{22–33}$$

In the limit of zero density ($\lambda \to 0$), $p/kT = \lambda/\Lambda^3 = \rho$ (classical ideal gas; Chapter 4). As usual, we define the activity z so that $z \to \rho$ when $\rho \to 0$. That is, $z \equiv \lambda/\Lambda^3$. If we replace λ by z in Eq. (22–33), we have

$$\frac{p}{kT} = \sum_{j\geq 1} b_j(T)z^j, \tag{22–34}$$

where

$$b_j(T) = \frac{(\mp 1)^{j-1}\Lambda^{3(j-1)}}{j^{5/2}}. \tag{22–35}$$

Equation (22–34) is in the same form as Eq. (15–7) of imperfect gas theory. While those departures from the ideal gas law ($p/kT = z = \rho$) that were discussed in Section 15–2 were due to intermolecular forces in a classical gas, here they are due to quantum effects. Because Eq. (22–34) is a special case of Eq. (15–7), we can use the general methods of Chapter 15 without modification: Eqs. (15–7) through (15–21) are all applicable if we take Eq. (22–35) for the b_j. Also, we have

$$\frac{\overline{E}}{\overline{N}kT} = \frac{3}{2}\frac{p}{\rho kT} = \frac{3}{2}\left[1 + \sum_{k\geq 1} B_{k+1}(T)\rho^k\right] \tag{22–36}$$

and (see also Problem 22–11)

$$\frac{S}{\overline{N}k} = \frac{\overline{E}}{\overline{N}kT} + \frac{p}{\rho kT} - \frac{\mu}{kT}$$
$$= \frac{5}{2} - \ln\Lambda^3\rho + \sum_{k\geq 1}\left(\frac{5}{2} - \frac{k+1}{k}\right)B_{k+1}\rho^k. \tag{22–37}$$

Next, we list a few explicit equations (weak degeneracy) through the second virial coefficient, using $B_2 = -b_2 = \pm\Lambda^3/2^{5/2}$ (upper sign, Fermi-Dirac):

$$\frac{p}{\rho kT} = 1 \pm \frac{\Lambda^3}{2^{5/2}}\rho + \cdots, \tag{22–38}$$

$$\frac{\mu}{kT} = \ln\Lambda^3\rho \pm \frac{\Lambda^3}{2^{3/2}}\rho + \cdots, \tag{22–39}$$

$$\frac{\overline{E}}{\overline{N}kT} = \frac{3}{2}\left(1 \pm \frac{\Lambda^3}{2^{5/2}}\rho + \cdots\right), \tag{22–40}$$

$$\frac{S}{\overline{N}k} = \frac{5}{2} - \ln \Lambda^3\rho \pm \frac{\Lambda^3}{2^{7/2}}\rho + \cdots, \tag{22–41}$$

$$\frac{C_V}{\overline{N}k} = \frac{3}{2}\left(1 \mp \frac{\Lambda^3}{2^{7/2}}\rho + \cdots\right). \tag{22–42}$$

The following comments on these equations may be of interest. (a) The sign of B_2 corresponds to a "repulsion" between Fermi-Dirac molecules and an "attraction" between Bose-Einstein molecules. (b) The magnitude of the purely quantum second virial coefficient above is $B_2^Q = O(\Lambda^3)$. The magnitude of a classical B_2 arising from intermolecular forces is $B_2^F = O(1/\rho_{\text{cond}})$, where ρ_{cond} is the number density of a condensed phase. The ratio of the two magnitudes is then $B_2^Q/B_2^F = O(\Lambda^3\rho_{\text{cond}})$. But this is just the order of the quantity calculated in Table 4–1; hence, we have a verification of the fact that in helium, at about 4°K, quantum and intermolecular effects both contribute significantly to B_2. (c) If intermolecular forces are suddenly "switched on" in a classical monatomic ideal gas (ρ and T constant), the entropy *decreases* whether the forces are repulsive or attractive (Problem 15–11). But note in Eq. (22–41) that although a Bose-Einstein gas has a lower entropy than a classical gas, a Fermi-Dirac gas has a *higher* entropy. This shows that the analogy between quantum effects and intermolecular forces referred to in (a) cannot be pushed too far. (d) These series may be regarded not only as in powers of ρ (coefficients are functions of T) but also as in powers of $\Lambda^3\rho$ (coefficients are constants), powers of $T^{-3/2}$ (coefficients are functions of ρ), or powers of h^3 (coefficients are functions of ρ and T). (e) Equation (22–42) informs us that the C_V curves in Fig. 22–3 deviate from the classical value at high temperature by a quantity proportional to $T^{-3/2}$.

The above series are valid for small λ. Now let us examine what happens, in Bose-Einstein statistics (the complications we find below do not arise in Fermi-Dirac statistics) when we take larger values of λ. First, we note, from Eq. (22–17), that the number of molecules in the ground state ($\epsilon_1 = 0$) is $\bar{s}_1 = \lambda/(1-\lambda)$ and hence that λ can never be greater than unity (μ cannot be positive). We are interested, then, in the range $0 \le \lambda \le 1$. Next, let us look at the integral form of the equation for $\overline{N}$:

$$\overline{N} = 2\pi\left(\frac{2m}{h^2}\right)^{3/2} V\int_0^\infty \frac{\epsilon^{1/2}\,d\epsilon}{e^{\epsilon/kT}\lambda^{-1} - 1}.$$

We multiply numerator and denominator of the integrand by $e^{-\epsilon/kT}\lambda$, expand $(1 - e^{-\epsilon/kT}\lambda)^{-1}$, integrate term by term, and find

$$\overline{N} = \left(\frac{2\pi mkT}{h^2}\right)^{3/2} VF_{3/2}(\alpha), \tag{22-43}$$

where

$$\alpha = -\frac{\mu}{kT}, \qquad e^{-\alpha} = \lambda,$$

and the function $F_\sigma(\alpha)$ is defined by

$$\begin{aligned} F_\sigma(\alpha) &= \sum_{j\geq 1} j^{-\sigma}e^{-j\alpha}, \\ F_\sigma(0) &= \sum_{j\geq 1} j^{-\sigma} = \zeta(\sigma), \end{aligned} \tag{22-44}$$

where $\zeta(\sigma)$ is the Riemann ζ-function. Since μ cannot be positive, α cannot be negative. As λ increases from 0 to 1, α decreases from $+\infty$ to 0. In order to study increasing degeneracy, then, we can start with large α and let α decrease toward zero. In this process the function $F_{3/2}(\alpha)$ increases monotonically (something like $e^{-\alpha}$, but more rapidly for small α) from a value zero at $\alpha = +\infty$ to a value $\zeta(3/2) = 2.612$ at $\alpha = 0$. Thus, for given V and T, $\overline{N}$ increases from zero at $\alpha = +\infty$ to a *maximum* value $2.612V/\Lambda^3$ at $\alpha = 0$. But this result is inconsistent with the relation $\bar{s}_1 = \lambda/(1-\lambda)$, which implies that $\bar{s}_1$ and hence $\overline{N}$ (since $\overline{N} \geq \bar{s}_1$) can be made arbitrarily large by taking λ arbitrarily close to unity. The reason for the discrepancy is that in converting the sum (22–20) to an integral, we have incorrectly omitted a contribution from the ground state, because the weight $\omega(\epsilon) \propto \epsilon^{1/2}$ is zero at $\epsilon = 0$.

Let us now change our point of view a little and suppose that $N/V = \rho$ is fixed, and that Eq. (22–43) determines α as a function of T. This equation is valid so long as α is significantly greater than zero or λ is significantly less than unity, for in this case the number of molecules in the ground state, $\lambda/(1-\lambda)$, is of negligible order compared with N. For example, if $\alpha = 10^{-6}$, $\lambda = 1 - 10^{-6}$, and $\bar{s}_1 = 10^6$, which is negligible in relation to N. At high temperatures, α is large and Eq. (22–43) is equivalent to the series (22–39) (lower sign). As T decreases, α decreases until the limiting value $\alpha = 0+$ (or $\mu = 0-$) is reached at a special temperature $T_0(\rho)$ defined by the equation

$$\left(\frac{h^2}{2\pi mkT_0}\right)^{3/2} = \frac{\zeta(3/2)}{\rho} = \frac{2.612}{\rho}. \tag{22-45}$$

At this temperature there is apparently some kind of singularity. Hence, for given ρ, Eq. (22–43) and the series (22–38) through (22–42) are valid for $T > T_0$, but not for $T < T_0$. Alternatively, we can say that, for given T, these series are valid for $\rho < \rho_0(T)$ but not for $\rho > \rho_0$, where ρ_0 is defined by

$$\left(\frac{h^2}{2\pi mkT}\right)^{3/2} = \frac{2.612}{\rho_0}. \qquad (22\text{–}46)$$

The numerical value of T_0 turns out to be 3.13°K at the density of liquid helium (27.6 $\text{cm}^3\cdot\text{mole}^{-1}$). This is of the same order of magnitude as the temperature of the (higher-order) phase transition in helium (referred to again below), 2.19°K. Closer agreement is not to be expected, since intermolecular forces are being neglected in the present discussion.

For $T < T_0$, and N and V fixed, the right side of Eq. (22–43) gives the number of molecules in excited states. Then N is equal to this number plus the number $\bar{s}_1$ in the ground state:

$$N = \frac{1}{e^{\alpha} - 1} + \left(\frac{2\pi mkT}{h^2}\right)^{3/2} VF_{3/2}(\alpha).$$

This equation has a solution for α if α is only slightly greater than zero:

$$N = \frac{1}{\alpha} + \left(\frac{2\pi mkT}{h^2}\right)^{3/2} VF_{3/2}(0+) = \frac{1}{\alpha} + N\left[\frac{T}{T_0(\rho)}\right]^{3/2},$$

where we have eliminated V by use of Eq. (22–45). Then

$$\frac{1}{\alpha} = -\frac{kT}{\mu} = \bar{s}_1 = N\left[1 - \left(\frac{T}{T_0}\right)^{3/2}\right] \qquad (T < T_0). \qquad (22\text{–}47)$$

This verifies that α and μ are essentially zero in the whole range $0 \leq T \leq T_0$: $\alpha = O(N^{-1})$ and $-\mu = O(kT/N)$. At $T \cong 0$, $-\mu = kT/N$, in agree-

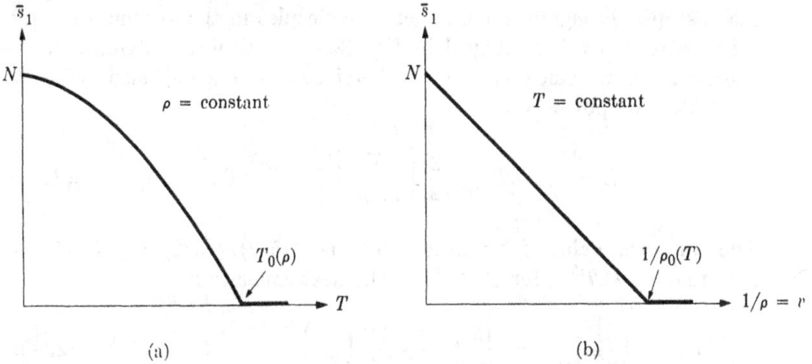

FIG. 22–4. Number of molecules in the ground state as a function of (a) temperature (density constant) and (b) volume per molecule (temperature constant), for an ideal Bose-Einstein gas.

ment with Eq. (22–18). Figure 22–4(a) shows how the number of molecules $\bar{s}_1$ in the ground state varies with T. Above T_0, $\bar{s}_1$ is a thermodynamically negligible number, but starting suddenly at T_0 (as T is decreased), there is a "condensation," called the "Bose-Einstein condensation," of a macroscopic number of molecules into the ground state. At $T = 0$, all molecules are in the ground state: $\bar{s}_1 = N$.

Alternatively, for $\rho > \rho_0(T)$ with T constant, we have from Eqs. (22–45) and (22–46),

$$\bar{s}_1 = N\left(1 - \frac{\rho_0}{\rho}\right) \qquad (\rho > \rho_0), \tag{22–48}$$

as shown in Fig. 22–4(b).

We shall return to the Bose-Einstein condensation following investigation of other thermodynamic properties. The equation for the energy of a Bose-Einstein gas is

$$E = 2\pi\left(\frac{2m}{h^2}\right)^{3/2} V\int_0^\infty \frac{\epsilon^{3/2}\,d\epsilon}{e^{\epsilon/kT}\lambda^{-1} - 1}$$

$$= \frac{3}{2}kT\left(\frac{2\pi mkT}{h^2}\right)^{3/2} VF_{5/2}(\alpha) \tag{22–49}$$

$$= \frac{3}{2}NkT\frac{F_{5/2}(\alpha)}{F_{3/2}(\alpha)}. \tag{22–50}$$

For $T > T_0$ (ρ constant) or $\rho < \rho_0$ (T constant), Eq. (22–50) is equivalent to Eq. (22–40) [where α has been eliminated as an independent variable in favor of ρ, using essentially Eq. (22–39)]. For $T < T_0$, the macroscopically significant number of molecules in the ground state [not taken care of by the integral in Eq. (22–49)] do not contribute to the energy anyhow, since $\epsilon_1 = 0$. Therefore Eq. (22–49) also holds for $T < T_0$ if we put $\alpha = 0+$:

$$E = \frac{3}{2}NkT\frac{\zeta(5/2)}{\zeta(3/2)}\left[\frac{T}{T_0(\rho)}\right]^{3/2} \qquad (T < T_0). \tag{22–51}$$

The numerical value of $\zeta(5/2)$ is 1.342, and $\zeta(5/2)/\zeta(3/2) = 0.5134$. We note that $E \propto T^{5/2}$, for $T < T_0$. The heat capacity is

$$C_V = \left(\frac{\partial E}{\partial T}\right)_{N,V} = \frac{15}{4}Nk\frac{\zeta(5/2)}{\zeta(3/2)}\left(\frac{T}{T_0}\right)^{3/2} \qquad (T < T_0), \tag{22–52}$$

or $C_V \propto T^{3/2}$. The complete C_V from Eqs. (22–42) and (22–52) is shown in Fig. 22–3. There is no discontinuity in C_V at T_0, but there is a dis-

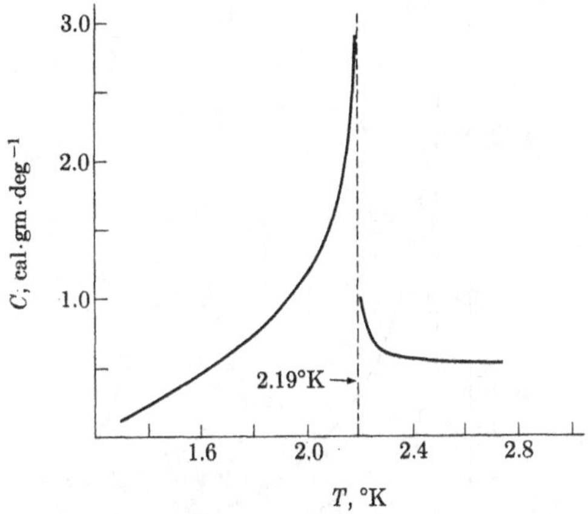

FIG. 22-5. Experimental heat capacity of liquid helium under its own vapor pressure.

continuity in $\partial C_V/\partial T$. The experimental heat capacity of liquid helium under its own vapor pressure (neither C_V nor C_p) is shown in Fig. 22-5, with a singularity at 2.19°K. Because of the shape of this curve, the singularity is often referred to as a λ-transition. Experimental curves of this kind are also found in the order-disorder transition in alloys (Section 20-1).

The equations for pV follow immediately from $pV = (2/3)E$. We already have the case $T > T_0$ or $\rho < \rho_0$ in Eq. (22-38). This can be deduced from Eqs. (22-50) for E and (22-39) for α (or μ). For $T < T_0$ or $\rho > \rho_0$, we have from Eq. (22-51),

$$pV = NkT\,\frac{\zeta(5/2)}{\zeta(3/2)}\left(\frac{T}{T_0}\right)^{3/2} \qquad (T < T_0), \qquad (22\text{-}53)$$

or

$$\frac{p}{kT} = \frac{\zeta(5/2)}{\zeta(3/2)}\,\rho_0(T) \qquad (\rho > \rho_0). \qquad (22\text{-}54)$$

Equation (22-54) is particularly important, for it states that if T is held constant, p remains constant in value for $\rho > \rho_0$. The p-v isotherms are shown in Fig. 22-6.

It should now be clear that, *thermodynamically*, the Bose-Einstein condensation is a first-order phase transition (but the experimental transition in helium is not first order). The two phases in equilibrium are the condensed phase with T, $1/\rho = v = 0$ and the dilute phase with T, $v = 1/\rho_0(T)$. The two phases have the same pressure (the vapor pressure),

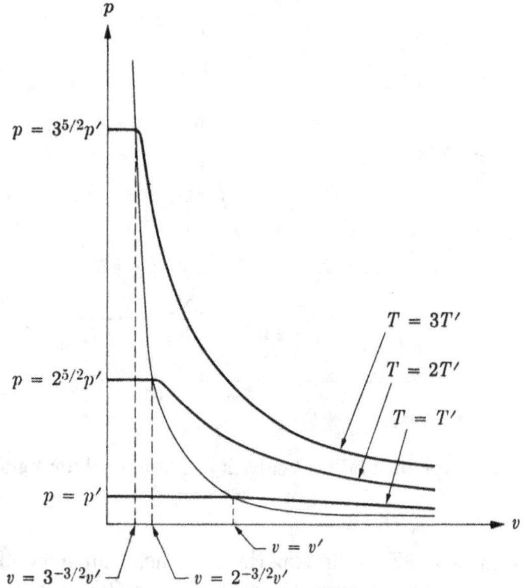

FIG. 22–6. Pressure-volume isotherms for an ideal Bose-Einstein gas.

$p = 0.514kT\rho_0(T)$, and the same chemical potential, $\mu = 0$. However, this first-order phase transition is very unusual on the *molecular* level. When $0 < v < 1/\rho_0(T)$, that is, in the flat portion of a p-v isotherm (Fig. 22–6), the system has a uniform macroscopic density $1/v$ (instead of the customary two different densities). As it is usually stated, the condensation occurs in momentum space rather than in coordinate space: the condensed phase consists of molecules with zero energy and momentum, and macroscopic de Broglie wavelength. (Actually a zero-point energy and momentum exist for a molecule in a box, but they are completely negligible in magnitude since V is macroscopic.)

If $T < T_0$ or $\rho > \rho_0$, the entropy is given by

$$\begin{aligned} S &= \frac{E + pV - N\mu}{T} \\ &= \frac{E + pV}{T} = \frac{H}{T} = \frac{5}{3}\frac{E}{T} \\ &= \frac{5}{2}Nk\,\frac{\zeta(5/2)}{\zeta(3/2)}\left[\frac{T}{T_0(\rho)}\right]^{3/2} \qquad (T < T_0). \end{aligned} \tag{22–55}$$

Hence $S \propto T^{3/2}$ and $S = 0$ at $T = 0$ (all molecules are in the ground state). For the heat of the phase transition, per molecule, we then find [replace $(T/T_0)^{3/2}$ by ρ_0/ρ in Eq. (22-55)]

$$\frac{\Delta H}{N} = \frac{T\,\Delta S}{N} = \frac{T(S_{\text{dil}} - S_{\text{cond}})}{N}$$

$$= \frac{T[S(\rho = \rho_0) - S(\rho = \infty)]}{N} = \frac{5}{2}\,kT\,\frac{\zeta(5/2)}{\zeta(3\text{-}2)}. \qquad (22\text{-}56)$$

For further details the reader is advised to consult the treatment by F. London (Supplementary Reading list).

22-4 Blackbody radiation (photon gas). Here we give a brief derivation of the thermodynamic properties of electromagnetic radiation in thermal equilibrium, which we regard as a gas of photons. We consider the photons to be in a container of volume V with heat insulating and perfectly reflecting inside walls. Thus the system is isolated and has a definite energy E. Because photons do not interact with each other, a very small blackbody is assumed to be present in the container to absorb and emit photons, thus making thermal equilibrium possible. Because photons absorbed at one frequency might be emitted at another, and in different numbers to conserve energy, the total number N of photons in the system is not fixed, even though the system is isolated.

There are two essential respects, then, in which a photon gas differs from a gas of particles with nonzero rest mass (e.g., electrons or helium atoms): (1) the photons do not interact with each other, so that in the photon gas *no approximation* is made by neglecting interparticle interactions; and (2) the number of particles in an isolated system is not conserved.

Thermodynamically, the system is characterized by E and V only (not E, V, and N, as usual for an isolated system).

Photons have spin one (units of $h/2\pi$) and hence obey Bose-Einstein statistics. The spin degeneracy is two (corresponding to two independent directions of polarization); it is not three, as would be the case for particles with spin one and nonzero rest mass.

Electrons also have a spin degeneracy of two, so we might expect to be able to take over Eq. (22-23) for $\omega(\epsilon)\,d\epsilon$, the number of energy eigenstates between ϵ and $\epsilon + d\epsilon$, for a single particle in V. However, because of their zero rest mass, Eq. (22-23) is not applicable to photons. Instead, we write this equation in terms of the absolute value of the momentum ($\epsilon = p^2/2m$, $p \geq 0$, since the energy is all kinetic):

$$\frac{8\pi V p^2\,dp}{h^3}$$

is the number of quantum states with p between p and $p + dp$. (Note that we use p for both momentum and pressure in this section.) This expression does not involve the mass. In applying it to photons, we use the de Broglie relation,

$$p = \frac{h}{\lambda} = \frac{h\nu}{c}, \tag{22-57}$$

to again change variables, this time from p to ν. This gives

$$G(\nu)\, d\nu = \frac{8\pi V \nu^2\, d\nu}{c^3} \tag{22-58}$$

as the number of quantum states with frequencies between ν and $\nu + d\nu$. Note that Eq. (22–58) agrees with the "transverse part" of Eq. (VI–30) for $g(\nu)\, d\nu$ in Debye's continuum model for a crystal. In fact a *wave* argument of the type presented in Appendix VI is ordinarily used to derive Eq. (22–58). We have, instead, taken the *particle* point of view. This illustrates the fact that photons may be viewed either as particles or waves.

We now wish to apply the relations (22–20) through (22–22) for N, E, and pV to the special case of a photon gas. We use the lower sign for Bose-Einstein statistics, take Eq. (22–58) as the weighting function in converting the sums to integrals, and put $\epsilon = h\nu$ for the energy of a photon. Also, because of the fact that this (isolated) system is defined thermodynamically by E and V only, we can make a special assignment of μ at the outset. We first consider this point.

We have already encountered other isolated systems in which the numbers of molecules are not fixed: systems with chemical equilibria. Let us review this topic briefly (see Chapter 10, especially Section 10–2). Suppose we have a two-component system characterized by E, V, N_A, and N_B. Then a catalyst is added making possible a chemical reaction between A and B, say $A \rightleftarrows B$. Thermodynamically, this reduces the number of independent components by one, from two to one. The equilibrium point can be located by, for example, maximizing S with respect to N_A holding E, V, and $N_A + N_B$ constant, minimizing F (or A) with respect to N_A holding p (or V), T, and $N_A + N_B$ constant, etc. In any case, the equilibrium condition $\Delta\mu = \mu_B - \mu_A = 0$ is found. The situation in a photon gas is rather similar. The small blackbody catalyzes the "reaction" $nA \rightleftarrows mA$ (n and m arbitrary integers, A = photon), which leads to a number of independent components one less than would otherwise be the case (i.e., the number of independent components here is zero instead of one). The equilibrium point is located by maximizing S with respect to N holding E and V constant, minimizing F (or A) with respect to N holding p (or V) and T constant, etc. In any case, all the relations

$$\left(\frac{\partial S}{\partial N}\right)_{E,V} = -\frac{\mu}{T} = 0, \qquad \left(\frac{\partial F}{\partial N}\right)_{p,T} = \left(\frac{\partial A}{\partial N}\right)_{V,T} = \mu = 0, \qquad \text{etc.,}$$

give $\mu = 0$. We also get $\mu = 0$ from

$$\Delta\mu = m\mu - n\mu = (m - n)\mu = 0.$$

In summary: because the number N of photons is not conserved in an isolated system, N must be regarded as a parameter and not as an independent thermodynamic variable; location of the equilibrium point with respect to the value of the parameter leads to the condition $\mu = 0$.

Whereas a molecular Bose-Einstein gas has $\mu = 0$ for $T < T_0$, a photon gas has $\mu = 0$ for all temperatures.

The remainder of the analysis is very straightforward. Equations (22–20) through (22–22) become

$$N = \frac{8\pi V}{c^3} \int_0^\infty \frac{\nu^2\, d\nu}{e^{h\nu/kT} - 1}, \tag{22–59}$$

$$E = \frac{8\pi V h}{c^3} \int_0^\infty \frac{\nu^3\, d\nu}{e^{h\nu/kT} - 1}, \tag{22–60}$$

$$pV = -\frac{8\pi V kT}{c^3} \int_0^\infty \nu^2 \ln\left(1 - e^{-h\nu/kT}\right) d\nu. \tag{22–61}$$

The integrand in Eq. (22–60) is essentially the blackbody energy distribution function, discovered at first empirically by Planck in 1900, which provided the starting point for the whole development of quantum theory. If we write this integrand as

$$\frac{\nu^3}{e^{h\nu/kT} - 1} = \frac{\nu^3 e^{-h\nu/kT}}{1 - e^{-h\nu/kT}} = \nu^3 \sum_{n\geq 1} (e^{-h\nu/kT})^n,$$

we find

$$E = \frac{48\pi V h}{c^3}\left(\frac{kT}{h}\right)^4 \zeta(4).$$

Since $\zeta(4) = \pi^4/90$, we have

$$E = \frac{8\pi^5 V (kT)^4}{15(hc)^3}. \tag{22–62}$$

Note that E is a function of V and T only; or T is a function of E/V only. The fact that the total energy of the radiation is proportional to T^4 is the well-known Stefan-Boltzmann law. The heat capacity is

$$C_V = \left(\frac{\partial E}{\partial T}\right)_V = \frac{32\pi^5 V k}{15}\left(\frac{kT}{hc}\right)^3. \tag{22–63}$$

The temperature dependence of E and C_V is the same as that of a crystal at low temperatures, as we should expect from Sections 5–3 and 5–4 and the discussion following Eq. (22–58). The resemblance between Eqs. (5–33) and (22–60) for E should also be noted. Because of the relation between these two problems, the acoustical quanta in a crystal are often regarded as forming a *phonon* gas in analogy to a photon gas.

Integration of Eq. (22–60) by parts leads to

$$pV = \frac{E}{3}. \tag{22–64}$$

The reason why we get $pV = E/3$ here and not $pV = 2E/3$, as in earlier sections, is that this is an extreme relativistic system ($\epsilon = cp$ instead of $p^2/2m$, p = momentum). Since E/V is a function of T only, the pressure is a function of T only, as in a first-order phase transition. This result might have been expected since μ = constant (=0). Equation (22–54) for the pressure in the Bose-Einstein condensation is similar.

The radiation pressure $p = E/3V$ is completely negligible in magnitude except at extreme temperatures. For example (Problem 22–12), $p = 1/5$ atm at 10^{5}°K (atomic bomb explosion).

We handle Eq. (22–59) for N just as we did Eq. (22–60) for E and find for the equilibrium number of photons

$$N = 16\pi\zeta(3)V\left(\frac{kT}{hc}\right)^3, \tag{22–65}$$

where $\zeta(3) = 1.202$. Thus N is determined by V and T (or E and V, etc.). We can combine Eqs. (22–62), (22–64), and (22–65) to obtain

$$E = \frac{\pi^4 NkT}{30\zeta(3)} = 2.701NkT, \tag{22–66}$$

$$pV = \frac{E}{3} = 0.900\,NkT. \tag{22–67}$$

The entropy follows from

$$N\mu = 0 = E - TS + pV,$$

$$S = \frac{E + pV}{T} = \frac{4E}{3T} = \frac{32\pi^5 Vk(kT)^3}{45(hc)^3}. \tag{22–68}$$

Thus $S \propto T^3$ and $S = 0$ at $T = 0$.

22–5 Quantum statistics with intermolecular interactions. This is a very involved subject, so we must confine ourselves to a brief introduction to it. Further details will be found in S. M., Sections 8–11 and 16, and, especially, in Hirschfelder, *et al.*, Chapters 1, 2, and 6. Many of the contributions in this field are due to J. de Boer and his collaborators. Because of their complexity, we omit entirely any discussion of the many current attempts being made to attack the liquid helium and other quantum fluid problems exactly.

The method of Sections 22–1 to 22–4 breaks down when intermolecular interactions are present, for the energy eigenvalues (eigenfunctions) of the whole system can no longer be written simply as a sum (product) of the energy eigenvalues (eigenfunctions) of the individual molecules in the system. We have to return instead to the completely general equation (22–3),

$$Q = \sum_j e^{-E_j/kT}, \tag{22–69}$$

where the sum is over all energy states accessible to the particular system under consideration.

There are alternative but equivalent forms of Eq. (22–69) that are often useful. Let ψ_j be the normalized energy eigenfunction belonging to the eigenvalue E_j. That is, $\mathcal{H}\psi_j(q) = E_j\psi_j(q)$, where $\mathcal{H}$ is the Hamiltonian operator for the system, $\mathcal{H} = \mathcal{H}$ (kinetic energy operator) $+ U(q)$ (potential energy), and q represents all the coordinates. Then we can write

$$\begin{aligned} Q &= \sum_j e^{-E_j/kT} \int \psi_j^*(q)\psi_j(q)\, dq \\ &= \sum_j \int \psi_j^*(q) e^{-E_j/kT} \psi_j(q)\, dq. \end{aligned} \tag{22–70}$$

Let us define the operator $e^{-\mathcal{H}/kT}$ by

$$e^{-\mathcal{H}/kT} = 1 - \frac{\mathcal{H}}{kT} + \frac{1}{2}\frac{\mathcal{H}^2}{(kT)^2} - \cdots.$$

Then, since,

$$\mathcal{H}^m\psi_j(q) = E_j^m\, \psi_j(q) \qquad (m = 1, 2, \ldots),$$

we have

$$e^{-\mathcal{H}/kT}\psi_j(q) = e^{-E_j/kT}\psi_j(q),$$

and Eq. (22–70) becomes

$$Q = \sum_j \int \psi_j^*(q) e^{-\mathcal{H}/kT} \psi_j(q)\, dq. \tag{22–71}$$

Actually, there is a more general form of Eq. (22–71):

$$Q = \sum_n \int \varphi_n^*(q) e^{-\mathcal{H}/kT} \varphi_n(q)\, dq, \tag{22–72}$$

where the φ_n are any complete set of orthonormal functions. To prove this, we expand φ_n in the set of functions ψ_j:

$$\varphi_n = \sum_j A_{nj}\psi_j,$$

$$A_{nj} = \int \psi_j^* \varphi_n \, dq.$$

Substitution of this expansion for φ_n in Eq. (22–72) gives

$$Q = \sum_j e^{-E_j/kT} \sum_n A_{nj}^* A_{nj}$$

$$= \sum_j e^{-E_j/kT}.$$

The last step can be justified by expanding the function ψ_j in the set φ_n and then forming the integral $\int \psi_j^* \psi_j \, dq = 1$. Thus both Eqs. (22–71) and (22–72) are equivalent to Eq. (22–69) and hence are equivalent to each other. Equation (22–71) is the special case of (22–72) obtained when we choose the energy eigenfunctions for the set φ_n.

One quite general application of Eq. (22–72) is the following: it allows us to write the quantum-mechanical analog of the classical statistical-mechanical quantity $e^{-U/kT}$ (recall that $e^{-U/kT}$ is proportional to the configurational probability density). Consider, for example, a one-component system of monatomic molecules. In classical mechanics (Eq. 6–21)

$$Q = \frac{1}{N!\Lambda^{3N}} \int e^{-U(q)/kT} \, dq. \tag{22–73}$$

Equation (22–72) also involves an integral over dq. Comparison of (22–72) and (22–73) shows that the analog of $e^{-U/kT}$ is

$$[S] = N!\Lambda^{3N} \sum_n \varphi_n^*(q) e^{-\mathcal{H}/kT} \varphi_n(q) \tag{22–74}$$

$$= N!\Lambda^{3N} \sum_j \psi_j^*(q)\psi_j(q) e^{-E_j/kT}. \tag{22–75}$$

We call $[S]$ the "Slater sum." We can easily verify that $[S]$ has the same physical significance as $e^{-U/kT}$ by noting that

$$\frac{e^{-E_j/kT}}{Q} \cdot \psi_j^*(q)\psi_j(q) \, dq$$

is the probability of observing the system in the energy state E_j and also in a configuration between q and $q + dq$. If we sum this quantity over j, we have the probability of observing a configuration between q and $q + dq$, irrespective of j. This sum over j is essentially that in (22-75).

As a very simple example of Eq. (22-75) we can calculate $[S]$ for two noninteracting particles in a one-dimensional box of length L. We find (Problem 22-13)

$$[S] = 1 \mp e^{-2\pi(x_1-x_2)^2/\Lambda^2} \quad \text{(upper sign FD)}, \tag{22-76}$$

where x_1 and x_2 are the coordinates of the two particles. A negligible term in $(x_1 + x_2)^2$ has been dropped here. Clearly, there is an effective interaction between the particles when their distance apart is of order Λ (thermal de Broglie wavelength) or less, an attraction in the Bose-Einstein case and a repulsion in the Fermi-Dirac case. The leading term (unity) is the classical value (h, $\Lambda \to 0$) for noninteracting molecules.

For two particles in three dimensions (see Hirschfelder, *et al.*, p. 402),

$$[S] = 1 \mp e^{-2\pi r_{12}^2/\Lambda^2} \quad \text{(upper sign FD)}. \tag{22-77}$$

In view of Eqs. (22-72) and (22-74) this gives

$$\begin{aligned} Q_2 &= \frac{1}{2\Lambda^6} \int [S]\, d\mathbf{r}_1\, d\mathbf{r}_2 \\ &= \frac{V^2}{2\Lambda^6} \mp \frac{V}{2^{5/2}\Lambda^3}. \end{aligned} \tag{22-78}$$

The equations of Section 15-1 then lead to the second virial coefficient

$$B_2 = \pm \frac{\Lambda^3}{2^{5/2}}, \tag{22-79}$$

in agreement with Eq. (22-38).

In general, we can write for monatomic molecules [compare Eq. (6-21)]

$$Q_N = \frac{Z_N}{N!\Lambda^{3N}},$$

where

$$Z_N = \int [S]\, d\mathbf{r}_1 \dots d\mathbf{r}_N. \tag{22-80}$$

This is the quantum-mechanical configuration integral. Even if U_N is pairwise additive, in general the quantity $-kT \ln [S]$ (the quantum analog of U_N in Z_N) will not be. Hence quantum analogs of equations such as (15-30) for B_3, B_4, etc., do not exist (but we recall that the equations of Section 15-1 are general and apply to the quantum case).

It may be instructive if we also evaluate Q_2 (Eq. 22-78) for two noninteracting molecules in V using Eq. (22-69) directly. We are seeking just first-order quantum effects; for this purpose, we can replace sums by integrals. First, we have from the energy-level expression (4–2) that Q_1 is given by (this is Problem 4–2)

$$Q_1 = \sum_{l_x,}\sum_{l_y,}\sum_{l_z \geq 1} \exp\left[-(l_x^2 + l_y^2 + l_z^2)h^2/8mV^{2/3}kT\right]$$

$$= \iiint_0^\infty \exp[\;]\, dl_x\, dl_y\, dl_z = \frac{V}{\Lambda^3}.$$

Now, to start on Q_2 we take the corresponding sixfold sum over all possible values of l_{x1}, l_{y1}, l_{z1}, l_{x2}, l_{y2}, and l_{z2}:

$$\sum_{l_{x1},\ldots,l_{z2} \geq 1} \cdots \sum \exp\left[-(l_{x1}^2 + \cdots + l_{z2}^2)h^2/8mV^{2/3}kT\right].$$

The value of this sum (from integration) is obviously $(V/\Lambda^3)^2$. In this sum we have counted, for example, the states

$$l_{x1}, \ldots, l_{z2} = 4, 9, 8, 13, 3, 3$$

and

$$l_{x1}, \ldots, l_{z2} = 13, 3, 3, 4, 9, 8$$

as separate states (with the same energy). Because of indistinguishability, this is wrong, and we divide the above sum by two, giving $V^2/2\Lambda^6$. But this correction is not proper for those states in which

$$l_{x1}, l_{y1}, l_{z1} = l_{x2}, l_{y2}, l_{z2};$$

for example,

$$l_{x1}, \ldots, l_{z2} = 4, 9, 8, 4, 9, 8.$$

Such states should not be included at all in Fermi-Dirac statistics, and their contribution should not have been divided by two in Bose-Einstein statistics (because each occurs only once in the sixfold sum). We can compensate for these errors by adding an appropriate term. Thus for Q_2 we have (upper sign Fermi-Dirac)

$$Q_2 = \frac{V^2}{2\Lambda^6} \mp \frac{1}{2}\sum_{l_{x1},}\sum_{l_{y1},}\sum_{l_{z1} \geq 1} \exp\left[-2(l_{x1}^2 + l_{y1}^2 + l_{z1}^2)h^2/8mV^{2/3}kT\right]. \tag{22–81}$$

On replacing the sums by integrals, we again get Eq. (22–78).

The calculation of the quantum-mechanical second virial coefficient for interacting molecules (e.g., a Lennard-Jones 6–12 interaction potential) has been carried out for several monatomic gases, but the procedure is too complicated to discuss here (see Hirschfelder, *et al.*, Chapter 6). At low temperatures, Q_2 in the form $\sum_j e^{-E_j/kT}$ was the starting point. At higher temperatures, Q_2 in the form of an integral of the Slater sum (Eq. 22–80), using momentum eigenfunctions for the φ_n, was the starting point. Figure 15–1 contains curves for the theoretical (high-temperature method) and experimental second virial coefficients of H_2 and He, using the Lennard-Jones potential with values of the parameters ϵ and r^* chosen to give best fit (these are the values in Table IV–1). Agreement is excellent. The curves for H_2 and He depart significantly from the classical curve.

At high temperatures (slight quantum effects) the explicit equation for B_2 is found to be (upper sign FD)

$$B_2(T) = -2\pi \int_0^\infty [e^{-u(r)/kT} - 1]r^2\,dr \pm \frac{\Lambda^3}{2^{5/2}} + \frac{h^2}{24\pi m(kT)^3} \int_0^\infty e^{-u(r)/kT} \left(\frac{du}{dr}\right)^2 r^2\,dr + O(h^4/m^2). \tag{22–82}$$

The first term is the classical B_2 (Eq. 15–24), and the second term (of order $h^3/m^{3/2}$) is the quantum B_2 for noninteracting molecules (Eq. 22–79).

It is appropriate at this point to discuss the modification in the law of corresponding states (Sections 15–2 and 16–4) that must be made for quantum fluids. We consider, as usual, monatomic gases with a pair interaction potential of the form $u(r) = \epsilon h(r/r^*)$ (this h is an arbitrary function and not Planck's constant). The Lennard-Jones 6–12 potential is an example. The virial expansion can be written

$$\frac{pv}{kT} = 1 + \frac{B_2(T)}{r^{*3}}\left(\frac{r^{*3}}{v}\right) + \frac{B_3(T)}{r^{*6}}\left(\frac{r^{*3}}{v}\right)^2 + \cdots. \tag{22–83}$$

In the classical case, we have already seen that B_2/r^{*3}, B_3/r^{*6}, etc., are universal functions of kT/ϵ only, and hence pv/kT is a universal function of kT/ϵ and v/r^{*3} (law of corresponding states). In quantum statistics the thermodynamic expansion (22–83) of course still holds, but we observe on inspection of Eq. (22–82) that B_2/r^{*3} is now a function (different in Fermi-Dirac and Bose-Einstein statistics) of kT/ϵ and *also* of the quantum parameter $h/(m\epsilon)^{1/2}r^*$, where $h/(m\epsilon)^{1/2}$ is essentially the de Broglie wavelength for a molecule with energy ϵ. This is true as well for the higher virial coefficients B_3/r^{*6}, etc. (as can be verified by dimensional analysis). Therefore pv/kT is a function, different in the two statistics, of the reduced thermodynamic variables kT/ϵ and v/r^{*3}, and also of the reduced quantum parameter $h/(m\epsilon)^{1/2}r^*$. These two functions approach each other and become the universal classical function (Section 16–4) in the limit $h/(m\epsilon)^{1/2}r^* \rightarrow 0$. The B_2/r^{*3} curves for H_2 and He (both Bose-Einstein) in Fig. 15–1 do not coincide because they have different values of the quantum parameter (Problem 22–14). In fact, the three curves in Fig. 15–1 may be regarded as a Bose-Einstein family of curves corresponding to three different choices of the quantum parameter.

Hirschfelder *et al.* give many more details about the quantum-mechanical law of corresponding states and its applications.

The next section contains another application of the quantum equation for Q, Eq. (22–72).

22–6 The factors h^n and $N!$ in classical statistics. In Chapters 3, 4, and 6 we found it necessary to introduce the factors h^n (n degrees of freedom) and $N!$ (N molecules) into the classical canonical ensemble partition function [see, for example, Eq. (6–25)]. In Chapter 6, by appealing to special cases, we justified the use of h^n as the volume in classical phase space corresponding to a single quantum state. Here we give an argument, due to Kirkwood, which provides a general justification for this. The appropriateness of the factor $N!$ has already been demonstrated adequately in Sections 3–2, 6–2, and 22–1. In the case of $N!$ there can be no doubt about the correct quantitative expression [whereas with h^n we might have expected, say, $(h/4\pi)^n$, etc.]. In fact, we can deduce the factor $N!$ simply by combining the quantum-mechanical idea of indistinguishability of identical particles with the classical phase integral (Eq. 6–18). But Kirkwood's method can be extended (we omit this here) to provide a *formal* verification of the $N!$ correction.

We consider a system of N point particles with a potential energy $U(\mathbf{r})$, where $\mathbf{r}$ stands for $\mathbf{r}_1, \ldots, \mathbf{r}_N$. We use cartesian coordinates, even though for some systems other coordinates would be more natural. For example, the system might consist of two atoms, $N = 2$, forming a diatomic molecule, in which case the rotational (θ, φ), vibrational (r_{12}), and translational (x, y, z of center of mass) coordinates would be the conventional choice; but instead we employ $\mathbf{r}_1 = x_1, y_1, z_1$ and $\mathbf{r}_2 = x_2, y_2, z_2$. Our procedure is to obtain the desired result concerning h^n with cartesian coordinates and then use the classical mechanical theorem, (VII–11), that an element of volume in phase space is invariant under a canonical transformation $(q, p \rightarrow q', p')$, to embrace *any* choice of coordinates.

We start with Eq. (22–72), and use for the $\varphi_n(\mathbf{r})$ the normalized momentum eigenfunctions

$$\varphi(\mathbf{r}; \mathbf{p}) = \frac{1}{h^{3N/2}} e^{2\pi i \mathbf{p}\cdot\mathbf{r}/h}. \tag{22–84}$$

The normalization is such that if we expand the arbitrary function $\psi(\mathbf{r})$

in momentum eigenfunctions,

$$\psi(\mathbf{r}) = \frac{1}{h^{3N/2}} \int A(\mathbf{p}) e^{2\pi i \, \mathbf{p}\cdot\mathbf{r}/h} \, d\mathbf{p},$$

then by the inversion formula for the Fourier transform,

$$A(\mathbf{p}) = \frac{1}{h^{3N/2}} \int \psi(\mathbf{r}) e^{-2\pi i \, \mathbf{p}\cdot\mathbf{r}/h} \, d\mathbf{r}.$$

Whereas $\psi^*\psi$ is the probability density for $\mathbf{r}$, A^*A is the probability density for $\mathbf{p}$. Temporarily we ignore restrictions on the symmetry of wave functions. Substitution of Eq. (22-84) in Eq. (22-72) gives

$$Q = \frac{1}{h^{3N}} \int_{\mathbf{p}} \int_{\mathbf{r}} e^{-2\pi i \, \mathbf{p}\cdot\mathbf{r}/h} e^{-\mathcal{H}/kT} e^{2\pi i \, \mathbf{p}\cdot\mathbf{r}/h} \, d\mathbf{r} \, d\mathbf{p}. \qquad (22\text{-}85)$$

It should be apparent that momentum eigenfunctions were chosen for the φ_n in Eq. (22-72) because we now have, in (22-85), Q expressed as an integral over $\mathbf{r}$ and $\mathbf{p}$—just as for the *classical* Q. Next, we define a function $w(\mathbf{r}, \mathbf{p}, T)$ by the equation

$$e^{-\mathcal{H}/kT} e^{2\pi i \, \mathbf{p}\cdot\mathbf{r}/h} = e^{-H/kT} e^{2\pi i \, \mathbf{p}\cdot\mathbf{r}/h} w(\mathbf{r}, \mathbf{p}, T). \qquad (22\text{-}86)$$

If we put this in Eq. (22-85), we obtain

$$Q = \frac{1}{h^{3N}} \int_{\mathbf{p}} \int_{\mathbf{r}} e^{-H/kT} w(\mathbf{r}, \mathbf{p}, T) \, d\mathbf{r} \, d\mathbf{p}. \qquad (22\text{-}87)$$

Now all we have to show is that in the classical limit ($h \to 0$), $w \to 1$.

To do this, we go back to Eq. (22-86) defining w. For simplicity of notation, let us consider a one-dimensional (x) system here. Because we are using cartesian coordinates, the extension to any number of variables is very straightforward. The reader should verify this using two variables (x and y or x_1 and x_2). With the aid of the relations

$$e^{-\mathcal{H}/kT} = 1 - \frac{1}{kT}\mathcal{H} + \frac{1}{2!(kT)^2}\mathcal{H}^2 - \cdots,$$

$$\mathcal{H} = -\frac{h^2}{8\pi^2 m}\frac{d^2}{dx^2} + U(x),$$

$$\frac{d^{2n} e^{2\pi i p x/h}}{dx^{2n}} = \left(-\frac{4\pi^2 p^2}{h^2}\right)^n e^{2\pi i p x/h},$$

the left side of Eq. (22–86) becomes

$$e^{-\mathcal{H}/kT}e^{2\pi ipx/h} = e^{2\pi ipx/h}\left[1 - \frac{1}{kT}\left(\frac{p^2}{2m} + U\right) + \frac{1}{2!(kT)^2}\left(\frac{p^2}{2m} + U\right)^2 + \cdots\right]$$

$$+ \text{ terms of order } h^2, h^4, \ldots .$$

In the classical limit ($h \to 0$), the terms of order h^2, h^4, etc., drop out, and there remains

$$e^{-\mathcal{H}/kT}e^{2\pi ipx/h} \to e^{2\pi ipx/h}e^{-H/kT} \qquad (h \to 0).$$

Therefore, from Eq. (22–86),

$$w \to 1 \qquad \text{as} \qquad h \to 0.$$

Kirkwood used a more elegant argument based on the so-called Bloch differential equation, but the result is the same.

On referring back to Eq. (22–87), we see, then, that the factor h^{-3N} in front of the classical phase integral is indeed correct, at least with cartesian coordinates. But in view of Eq. (VII–11), we can change variables from $\mathbf{r}$, $\mathbf{p}$ to *any* set $\mathbf{q}'$, $\mathbf{p}'$ and still have

$$Q = \frac{1}{h^{3N}}\int_{\mathbf{p}'}\int_{\mathbf{q}'} e^{-H/kT}\, d\mathbf{q}'\, d\mathbf{p}' \qquad (h \to 0). \tag{22–88}$$

This is what we set out to prove.

We observe that Eq. (22–88) does not have the desired factor $(N!)^{-1}$ in front of the integral signs. The reason is that we have ignored symmetry restrictions on the wave functions [Eq. (22–88) is correct as it stands for a system of distinguishable particles—an Einstein crystal, for example]. Actually, the same kind of argument as above can be carried through, including symmetry effects (indistinguishable particles). But the argument is too sophisticated for the present text and we omit it (see S. M., pp. 85–89). The result is that h^{3N} in Eq. (22–88) is replaced by $h^{3N}N!$.

22–7 Free-volume theories of quantum liquids. In Chapter 16 we discussed several approximate theories of classical liquids. Similar approximations have been applied to quantum liquids. Because the theoretical foundation is especially unsatisfactory when quantum effects are included in these theories, we shall merely sketch here what is perhaps the simplest procedure in converting a classical free-volume theory into a quantum theory. For further discussion of this general subject, see Prigogine (Chapter 18) and Band (Chapter 8).

To be specific, we consider a Bose-Einstein system. The same kind of argument will obviously apply to a Fermi-Dirac system. In Sections 16-1 and 16-2, the transition from an ideal classical gas (no intermolecular interactions) to a real classical fluid was made by passing from the ideal gas equation (16-15),

$$-\frac{A}{NkT} = \ln \frac{ve}{\Lambda^3},$$

to the "real fluid" equation (16-22),

$$-\frac{A}{NkT} = \ln \frac{v_f(v, T)e}{\Lambda^3} - \frac{\varphi(v)}{2kT}.$$

That is, we replace v by the free volume v_f, and we add a potential energy term $-\varphi(v)/2kT$. We now follow exactly the same formal procedure for a Bose-Einstein fluid,* and for the same physical reasons. Thus, we first write the expression for the Helmholtz free energy of an ideal (no interactions) Bose-Einstein gas, from Eqs. (22-40), (22-41), (22-51), and (22-55):

$$-\frac{A}{NkT} = \ln \frac{ve}{\Lambda^3} + 0.4618 \frac{v_0(T)}{v} + 0.0112 \left[\frac{v_0(T)}{v}\right]^2 + 0.00065 \left[\frac{v_0(T)}{v}\right]^3 + \cdots, \qquad v > v_0(T) \quad (22\text{-}89)$$

$$= 0.5134 \frac{v}{v_0(T)}, \qquad v < v_0(T), \quad (22\text{-}90)$$

where we have included additional terms (see Problem 22-10) in the $v > v_0$ series and (Eq. 22-46)

$$v_0(T) \equiv \frac{1}{\rho_0(T)} = \frac{\Lambda^3}{2.612}.$$

Next, we replace v by $v_f(v, T)$ and add the potential energy term $-\varphi(v)/2kT$:

$$-\frac{A}{NkT} = \ln \frac{v_f e}{\Lambda^3} + 0.4618 \frac{v_0}{v_f} + 0.0112 \left(\frac{v_0}{v_f}\right)^2 + \cdots - \frac{\varphi}{2kT} \qquad (v_f > v_0) \quad (22\text{-}91)$$

$$= 0.5134 \frac{v_f}{v_0} - \frac{\varphi}{2kT} \qquad (v_f < v_0). \quad (22\text{-}92)$$

Equations (22-91) and (22-92) now determine all the properties of the fluid once we adopt some definite model which provides v_f and φ.

---

* T. L. Hill, *J. Phys. Chem.* **51**, 1219 (1947).

The simplest example is the van der Waals model [Eqs. (16–4), (16–5), and (16–23)]:

$$v_f = v - b, \varphi = -\frac{2a_v}{v}; \qquad b = \frac{2\pi r^{*3}}{3}, a_v = \frac{2\pi\epsilon r^{*3}}{3}. \tag{22–93}$$

If we substitute these expressions for v_f and φ in Eqs. (22–91) and (22–92), and then use $p = -(\partial A/\partial V)_{N,T}$, we find for the equation of state

$$\frac{p}{kT} = \frac{1}{(v-b)}\left[1 - 0.4618\frac{v_0}{v-b} - 0.0225\left(\frac{v_0}{v-b}\right)^2 - \cdots\right] - \frac{a_v}{v^2kT} \qquad (v > v_0 + b) \tag{22–94}$$

$$= \frac{0.5134}{v_0} - \frac{a_v}{v^2kT} \qquad (b < v < v_0 + b) \tag{22–95}$$

$$= +\infty \qquad (v < b). \tag{22–96}$$

It is easy to verify (Problem 22–15) that pv/kT is a universal function of kT/ϵ, v/r^{*3}, and $h/(m\epsilon)^{1/2}r^*$, and hence that the quantum-mechanical law of corresponding states is obeyed (Section 22–5). It is also apparent that the above very simple theory gives for the second virial coefficient (upper sign Fermi-Dirac)

$$B_2(T) = b - \frac{a_v}{kT} \pm \frac{\Lambda^3}{2^{5/2}}, \tag{22–97}$$

which should be compared with Eq. (22–82).

We leave it as an exercise for the reader (Problem 22–16) to deduce further properties of the van der Waals-Bose-Einstein fluid.

22–8 Gas of symmetrical diatomic molecules at low temperatures. In this section we consider a rather different type of application of quantum symmetry restrictions. A symmetrical (or homonuclear, which is a less confusing term in the present context) diatomic molecule is made up of two identical nuclei and one or more electrons. It is impossible, experimentally, to distinguish one of these nuclei from the other (when the molecule rotates), just as two identical atoms in a box are indistinguishable. A wave function representing the state of a system containing homonuclear diatomic molecules must therefore be symmetrical or antisymmetrical (depending on the nucleus) in the exchange of the two nuclei of any homonuclear diatomic molecule. The wave function must also be antisymmetrical in the exchange of any two electrons. A similar restriction obviously applies to more complicated molecules with equivalent and identical atoms, such as CH_4, CDH_3, etc., but we limit ourselves to the diatomic problem.

Clearly, there is no analogous symmetry restriction for the nuclei of heteronuclear diatomic molecules. Nothing essential needs to be added to the discussion of this case already given in Chapter 8.

Let us investigate a one-component homonuclear diatomic gas at sufficiently low density (Eq. 4–6) so that classical statistics can be used for the *translational* motion. Then, as in Eq. (22–1), $Q = q^N/N!$, where q is the partition function $\sum_j e^{-\epsilon_j/kT}$ for a single molecule in a box of volume V. In this special case we need consider only the energy eigenvalues ϵ_j and eigenfunctions ψ_j of a single molecule. We are to include in the q-sum only those molecular energy eigenstates with proper symmetry (e.g., states which are antisymmetrical in the exchange of the two nuclei for H_2 and symmetrical for D_2, since a hydrogen nucleus contains an odd number of nucleons and a deuterium nucleus an even number).

We have already seen in Section 8–3 that the above symmetry restrictions can be ignored and a classical treatment of rotation can be used if $T \gg \Theta_r$. Our problem here is to provide a formulation which is suitable at low temperatures, that is, when $T \not\gg \Theta_r$ and the classical approach breaks down. In order to reach the range of temperatures at which quantum effects would show up, very low temperatures are needed (see Table 8–1) except for H_2 ($\Theta_r = 85.4°K$) and D_2 ($\Theta_r = 42.7°K$).

As in Eqs. (4–41) and (8–1) through (8–3) we write, for a homonuclear diatomic molecule,

$$H = H_t + H_r + H_v + H_e + H_n,$$

$$\epsilon = \epsilon_t + \epsilon_r + \epsilon_v + \epsilon_e + \epsilon_n,$$

$$\psi = \psi_t \psi_r \psi_v \psi_e \psi_n.$$

We consider only the ground electronic and nuclear states, for reasons already discussed in Chapters 4 and 8. We now examine the symmetry of ψ to see which energy eigenstates are to be included in the q-sum. The ground state electronic function ψ_e is necessarily antisymmetrical in the electrons. Usually (for example, H_2), the ground state ψ_e is also symmetrical in the nuclei: this is the only case we consider. The function ψ_t (which depends only on the coordinates of the center of mass) is not affected by exchanging nuclei, nor is ψ_v (which depends only on the internuclear distance). Hence both of these functions are symmetrical in the nuclei. Therefore ψ has the nuclear symmetry of the product $\psi_r\psi_n$. If the nuclei have an odd (even) number of nucleons ("mass number"), ψ and $\psi_r\psi_n$ must be antisymmetrical (symmetrical) in exchange of the nuclei.

Let s_n be the nuclear spin (in units of $h/2\pi$). For example, $s_n = \frac{1}{2}$ for H, $s_n = 1$ for D, $s_n = 0$ for He^4, etc. Then the nuclear ground state degeneracy of each nucleus is $\omega_{n1} = 2s_n + 1$. Let the nuclear ground state energy eigenfunctions be $\psi_1, \psi_2, \ldots, \psi_{\omega_{n1}}$. If we denote the two (identical) nuclei by A and B, a function of the form $\psi_1(A)\psi_2(B) - \psi_1(B)\psi_2(A)$ is a possible antisymmetrical nuclear function ψ_n for the diatomic molecule (if A and B are interchanged, the function changes sign). Such a function exists for each pair of different numbers 12, 13, $\ldots$, $1\omega_{n1}$, etc. Altogether there are $\omega_{n1}(\omega_{n1} - 1)/2$ such pairs and therefore this number of antisymmetrical nuclear states for the molecule.

The function $\psi_1(A)\psi_2(B) + \psi_1(B)\psi_2(A)$, obtained simply by changing the sign in an antisymmetrical function, is symmetrical in the exchange of the two nuclei A and B. Other symmetrical functions are $\psi_1(A)\psi_1(B)$, $\psi_2(A)\psi_2(B)$, etc. Thus there are a total of

$$\frac{\omega_{n1}(\omega_{n1} - 1)}{2} + \omega_{n1} = \frac{\omega_{n1}(\omega_{n1} + 1)}{2}$$

symmetrical nuclear states.

The total number of nuclear states for the molecule is

$$\frac{\omega_{n1}(\omega_{n1} - 1)}{2} + \frac{\omega_{n1}(\omega_{n1} + 1)}{2} = \omega_{n1}^2,$$

as one should expect.

Now we turn to ψ_r. Aside from normalization, we have*

$$\psi_r(\theta, \varphi) = e^{im\varphi}P_j^{|m|}(\cos\theta),$$

in rather standard notation. The associated Legendre functions can be defined by

$$P_j^{|m|}(x) = (1 - x^2)^{|m|/2}\frac{d^{|m|}P_j(x)}{dx^{|m|}},$$

$$P_j(x) = \frac{1}{2^j j!}\frac{d^j(x^2 - 1)^j}{dx^j}.$$

If we interchange nuclei, θ becomes $\pi - \theta$ and φ becomes $\varphi + \pi$. Thus $x = \cos\theta$ becomes $-x$. Then we note that

$$P_j(-x) = (-1)^j P_j(x)$$

and

$$P_j^{|m|}(-x) = (-1)^{j-|m|}P_j^{|m|}(x).$$

---

* See any text on quantum mechanics.

Also,

$$e^{im(\varphi+\pi)} = e^{im\pi}e^{im\varphi} = (-1)^{|m|}e^{im\varphi}.$$

Therefore, finally,

$$\begin{aligned}\psi_r(\pi - \theta, \varphi + \pi) &= (-1)^{j-|m|}(-1)^{|m|}\psi_r(\theta, \varphi)\\ &= (-1)^j\psi_r(\theta, \varphi).\end{aligned}$$

We conclude that ψ_r is antisymmetrical in exchange of the nuclei if j is odd and symmetrical if j is even.

Putting these results together, we have the following situation. If the nuclei of the homonuclear diatomic molecule contain an odd (even) number of nucleons, ψ and $\psi_r\psi_n$ must be antisymmetrical (symmetrical) in exchange of the nuclei. For rotational states with j even, ψ_r is symmetrical, and hence only antisymmetrical (symmetrical) nuclear states are accessible—in order to make the product $\psi_r\psi_n$ antisymmetrical (symmetrical). On the other hand, for j odd, ψ_r is antisymmetrical, and only symmetrical (antisymmetrical) nuclear states are accessible in order to make $\psi_r\psi_n$ antisymmetrical (symmetrical). We therefore have

$$q = q_t q_v q_e q_{rn}, \tag{22-98}$$

with [see Eq. (8-23)]

$$\begin{aligned}q_{rn} = \frac{\omega_{n1}(\omega_{n1} - 1)}{2} &\sum_{j=0,2,\ldots} (2j + 1)e^{-j(j+1)\Theta_r/T}\\ &+ \frac{\omega_{n1}(\omega_{n1} + 1)}{2} \sum_{j=1,3,\ldots} (2j + 1)e^{-j(j+1)\Theta_r/T} \quad \text{(odd mass no.)},\end{aligned} \tag{22-99}$$

$$q_{rn} = \frac{\omega_{n1}(\omega_{n1} + 1)}{2} \sum_{j=0,2,\ldots} + \frac{\omega_{n1}(\omega_{n1} - 1)}{2} \sum_{j=1,3,\ldots} \quad \text{(even mass no.)}, \tag{22-100}$$

where q_{rn} is the combined rotational and nuclear partition function which cannot here be factored into q_r and q_n.

If $s_n = 0$ (which occurs only with even mass number), as for example in O_2^{16}, $\omega_{n1} = 1$ and

$$q_{rn} = \sum_{j=0,2,\ldots} .$$

In this case odd rotational states do not appear at all. This has been verified spectroscopically.

For a heteronuclear diatomic molecule XY (see Section 8–3), the equation corresponding to Eqs. (22–99) and (22–100) is

$$q_{rn} = q_n q_r,$$

$$q_n = \omega_{n1}^{X} \omega_{n1}^{Y}, \qquad q_r = \sum_{j=0,1,2,\ldots} \qquad (22\text{–}101)$$

At high temperatures (Eq. 8–25), Eqs. (22–99) and (22–100) both reduce to

$$q_{rn} = \left[\frac{\omega_{n1}(\omega_{n1} - 1)}{2} + \frac{\omega_{n1}(\omega_{n1} + 1)}{2}\right]\frac{1}{2}\frac{T}{\Theta_r}$$

$$= \frac{\omega_{n1}^2 T}{2\Theta_r} \quad \text{(homonuclear)}, \qquad (22\text{–}102)$$

in agreement with Eq. (8–29) with $\sigma = 2$. In the same limit, Eq. (22–101) becomes

$$q_{rn} = \frac{\omega_{n1}^{X} \omega_{n1}^{Y} T}{\Theta_r} \qquad \text{(heteronuclear)}. \qquad (22\text{–}103)$$

When the classical forms for q_{rn}, (22–102) and (22–103), can be used (which is almost always the case), q_n is separable and constant (the product of the two nuclear spin weights) and is ordinarily omitted, as already explained in Sections 2–4 and 4–4.

We confine ourselves in the remainder of this section to the special case of hydrogen gas (see also Problem 22–17 concerning deuterium gas, D_2). A hydrogen nucleus has odd mass number (one proton) and $s_n = \frac{1}{2}$, $\omega_{n1} = 2$ (spin "up" or "down"). Therefore Eq. (22–99) becomes

$$q_{rn} = 1 \sum_{j=0,2,\ldots} + 3 \sum_{j=1,3,\ldots} . \qquad (22\text{–}104)$$

TABLE 22–2

EQUILIBRIUM PERCENT p-H_2 IN H_2 GAS

T, °K	% p-H_2
0	100
15	99.989
30	96.95
50	76.8
100	38.5
200	25.953
298.1	25.074

Hydrogen molecules in even rotational states ($j = 0, 2, \ldots$) are conventionally referred to as parahydrogen, or p-H_2, and those in odd rotational states are called orthohydrogen, o-H_2 (Problem 22-18).

The thermodynamic properties of hydrogen gas, at equilibrium, are determined by $Q = q^N/N!$ and $q = q_t q_v q_e q_{rn}$, with q_{rn} given by Eq. (22-104). At the temperatures of interest here (10°K-300°K), translation is classical, while the vibrational and electronic degrees of freedom are completely unexcited.

The ratio of the number of ortho- to parahydrogen molecules, at equilibrium, is obviously

$$\frac{N_o}{N_p} = \frac{3\sum_{j=1,3,\ldots}}{1\sum_{j=0,2,\ldots}}.$$

The high-temperature limit is $N_o/N_p \to 3$, and the low-temperature limit is

$$\frac{N_o}{N_p} \to \frac{3(3e^{-2\Theta_r/T} + \cdots)}{1(1 + 5e^{-6\Theta_r/T} + \cdots)} \to 0.$$

That is, the equilibrium mixture approaches 100% para ($j = 0$) as $T \to 0$ (Table 22-2; Problem 22-19).

The rotational-nuclear contribution to the heat capacity, C_{Vrn}, can be calculated from Eq. (22-104) after two differentiations with respect to temperature [see, for example, Eqs. (8-31) and (8-33)]. The curve obtained is labeled e-H_2 (e for equilibrium) in Fig. 22-7. Also included in

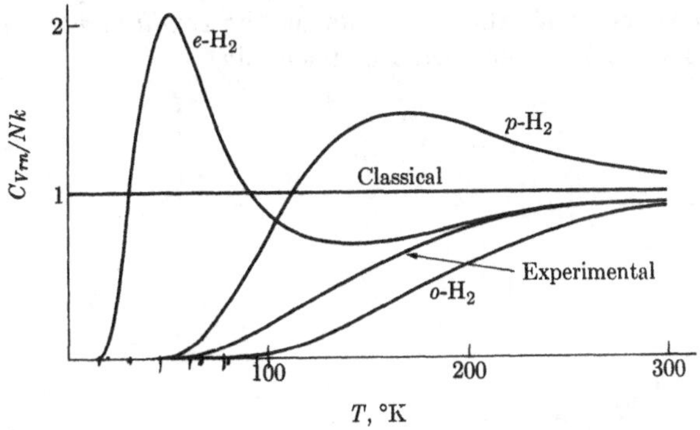

FIG. 22-7. Rotational-nuclear contribution to the heat capacity for o-H_2, p-H_2, e-H_2 (equilibrium mixture at each temperature) and $\frac{3}{4}$ o-H_2 (labeled "experimental").

the figure are curves for pure p-H_2 and pure o-H_2, calculated from

$$\text{para:} \quad q_{rn} = 1 \sum_{j=0,2,\ldots}$$

$$\text{ortho:} \quad q_{rn} = 3 \sum_{j=1,3,\ldots} .$$

The experimental curve for H_2, shown in the figure, does not agree with any of these! This interesting puzzle was resolved by Dennison, who realized that, in the absence of a catalyst, the half-life for the interconversion between ortho and para states is very long compared with the time of an experiment. Therefore when ordinary hydrogen gas is cooled down from room temperature for low-temperature heat-capacity measurements, the high-temperature composition ($\frac{3}{4}$ o-H_2) obtains even at low temperatures, instead of the equilibrium composition (Table 22–2). The gas is *not* in a state of complete equilibrium (as assumed in Table 22–2), but is in a state of "frozen" metastable equilibrium with $\frac{3}{4}$ o-H_2. (See Sections 2–3 and 2–4 in this connection.) If we calculate a heat-capacity curve based on $\frac{3}{4}$ o-H_2 at all temperatures, using

$$C_{Vrn} = \tfrac{3}{4}C_{Vrn}\,(\text{ortho}) + \tfrac{1}{4}C_{Vrn}\,(\text{para})$$

and the ortho and para curves already included in Fig. 22–7, we find a theoretical curve that is in excellent agreement with the experimental curve shown in the figure.

The above analysis is elegantly confirmed by an experiment due to Bonhoeffer and Harteck: active charcoal is found to be a catalyst (the mechanism involves adsorbed free H atoms) for the ortho-para conversion; in the presence of a catalyst, points on the *equilibrium curve* (labeled e-H_2) in Fig. 22–7 are obtained experimentally.

PROBLEMS

22–1. Explain how symmetry restrictions on accessible quantum states apply to multicomponent systems. (Page 432.)

22–2. Construct a table analogous to Table 22–1 for a system of four particles with a total of 12 energy "units." (Page 433.)

22–3. Expand the determinant in Eq. (22–6), and show that the wave function ψ is symmetrical. (Page 435.)

22–4. Use the microcanonical ensemble and the method of undetermined multipliers to derive the Fermi-Dirac distribution law, (22–12). To avoid applying the Stirling approximation to small numbers, the quantum states should be treated in large groups with essentially the same energy. (Page 437.)

22–5. Show that the entropy of an ideal quantum gas can be written in the form (upper sign Fermi-Dirac)

$$S = -k \sum_j [\bar{s}_j \ln \bar{s}_j \pm (1 \mp \bar{s}_j) \ln (1 \mp \bar{s}_j)].$$

(Page 440.)

22–6. Derive an equation for the equilibrium vapor pressure of electrons in a metal assuming the gas phase is very dilute (classical statistics; ignore space charge) and using the low temperature limit for μ in the metal. Note that it is necessary to introduce the potential φ in this problem. (Page 441.)

22–7. Show that $\lambda_{\text{class}} \ll 1$ for the electrons in a typical metal only above about 10^5°K. (Page 442.)

22–8. Calculate a numerical value (in ev) for the Fermi energy μ_0 of a typical metal. Show in an energy-level diagram the relation between μ_0 and φ (electron at rest in gas phase = zero of energy). (Page 442.)

22–9. Use Eqs. (22–30) and (22–31) for an electron gas to derive equations for the thermodynamic functions S, pV, and $N\mu$. (Page 444.)

22–10. Extend the series (22–38) through (22–42) one more term for the Bose-Einstein case. (Page 445.)

22–11. Derive the equation

$$\frac{S}{Nk} = \frac{5}{2} - \ln \Lambda^3 \rho - \sum_{k \geq 1} \left(B_{k+1} + T \frac{dB_{k+1}}{dT} \right) \frac{\rho^k}{k}$$

for a monatomic gas from the virial expansion and show its equivalence with Eq. (22–37) for an ideal quantum gas. (Page 446.)

22–12. Show that the radiation pressure $p = E/3V$ is about 1/5 atm at 10^5°K. (Page 456.)

22–13. Use Eq. (22–75) to derive Eq. (22–76) for the Slater sum of two non-interacting particles in a one-dimensional box. Note that $N!\Lambda^{3N}$ in Eq. (22–75) becomes $2!\Lambda^2$ in this case. (Page 459.)

22–14. Use the values of ϵ and r^* in Table IV–1 to calculate the quantum parameter $h/(m\epsilon)^{1/2}r^*$ for H_2, He^3, He^4, A, and Xe. (Page 462.)

22–15. Show that the van der Waals-Bose-Einstein equation of state, (22–94) through (22–96), obeys the quantum-mechanical law of corresponding states. (Page 466.)

22–16. Investigate the nature of the phase transition and $C_V(T)$ for the van der Waals-Bose-Einstein fluid. (Page 466.)

*22–17. Give a discussion of the D_2 case analogous to that in the text for H_2. (Page 470.)

22–18. Write out the four ground state nuclear wave functions for H_2. Classify them as ortho and para and as symmetrical or antisymmetrical in exchange of the nuclei. (Page 471.)

22–19. Calculate the equilibrium percent of p-H_2 in H_2 gas at 30°K. (Page 471.)

22–20. Investigate the properties of one- and two-dimensional ideal Bose-Einstein gases.

22–21. Investigate the properties of the following ideal two component systems: BE–BE, BE–FD, FD–FD.

22–22. Derive an expression for the radial distribution function in an ideal quantum gas in the limit $\rho \to 0$.

22–23. (a) Investigate the fluctuation in N in an isolated photon gas (E and V given). Compare Section 10–3. (b) Consider the same question for an isothermal system (T and V given).

22–24. Derive equations for the fluctuation in s_j in systems of noninteracting Bose-Einstein and Fermi-Dirac particles. Compare Section 7–2.

22–25. Verify that the Clausius-Clapeyron equation, $dp/dT = \Delta S/\Delta V$ for a first-order phase transition, where p is the vapor pressure, is satisfied in the ideal Bose-Einstein condensation.

22–26. Consider the following *hypothetical* system with hybrid "quantum" statistics. The particles do not interact in the usual sense, and any number of particles can be in a single quantum state. But when s particles are in the same state ϵ, there is a new kind of "interaction" energy, $(s - 1)w$, which favors or disfavors more than one particle occupying the same state. The s particles contribute a total of $s\epsilon + (s - 1)w$ to the energy. In Bose-Einstein statistics, $w = 0$; in Fermi-Dirac statistics, $w = +\infty$. Consider the general case, for a gas, where w is finite (positive or negative) and investigate the thermodynamic properties of the system. Use the grand partition function method of Eqs. (22–10) and (22–15). Note the formal resemblance to B.E.T. adsorption theory (Eq. 7–36).

*22–27. Derive equations for the asymptotic low-temperature heat capacity C_{Vrn} for o-H_2 and p-H_2. Calculate C_{Vrn} at 60°K in the two cases.

Supplementary Reading

Band, Chapters 8, 10 and 11.
Fowler and Guggenheim, Chapters 1 and 11.
Hirschfelder, Curtiss, and Bird, Chapter 6.
Kittel, Part 1.
Landau and Lifshitz, Chapters 5–7, Section 33.
London, F., *Superfluids*, Vol. II. New York: Wiley, 1954.
Mayer and Mayer, Chapters 2, 5, 7 and 16.
Prigogine, Chapters 18 and 19.
Schrödinger, Chapters 7–9.
Slater, Chapters 5, 19, 27–29.
S. M., Chapters 2 and 3.
ter Haar, Chapters 3, 4, 7–10.

APPENDIX I

NATURAL CONSTANTS

Quantity	Symbol	Value
Avogadro's number	N_0	6.02486×10^{23} mole^{-1}
Velocity of light	c	2.997930×10^{10} cm·sec^{-1}
Electronic charge	**e**	4.80286×10^{-10} esu
Electron rest mass		9.1083×10^{-28} gm
Planck's constant	h	6.62517×10^{-27} erg·sec
Mass of hydrogen atom		1.67330×10^{-24} gm
Mass of unit atomic weight		1.65979×10^{-24} gm
Mass of proton		1.67239×10^{-24} gm
Boltzmann's constant	k	1.38044×10^{-16} erg·deg^{-1}
Gas constant	$R = N_0k$	8.31696×10^{7} erg·deg^{-1}·mole^{-1}
Temperature scales		0°C = 273.15°K

1 cal = 4.184 joules = 4.184×10^7 ergs
1 ev = 1.60206×10^{-12} erg = 23.0693 kcal·mole^{-1}
1 atm = 1.0133×10^6 dynes·cm^{-2} = 760 mm Hg

APPENDIX II

MAXIMUM-TERM METHOD

In statistical mechanics we frequently use the approximation of replacing the logarithm of a sum by the logarithm of the largest term in the sum. Offhand this may sound unreasonable, but in practice, under suitable conditions, the "approximation" is in fact not an approximation at all—to terms with an order of magnitude significant in thermodynamics.

As a typical illustration, let us examine in some detail a sum which is simple but which has the same form and qualitative properties as the sums encountered in actual problems. Consider

$$\sum = \sum_{N=0}^{M} t_N, \tag{1}$$

where

$$t_N = \frac{M!x^N}{N!(M-N)!} \tag{2}$$

Here $x = O(1)$ and $M = O(10^{20})$, say. $\sum$ might be, for example, a grand partition function, N the (fluctuating) number of molecules in the system, and M the maximum number of molecules in the system. As another example, $\sum$ resembles $\sum_n \Omega_t(n)$ in Eq. (1–4), if we put $N = n_1$, $M - N = n_2$, $M = \mathfrak{N}$, and $x = 1$. Here the limit $\mathfrak{N}$ or $M \to \infty$ is involved.

First we note that the summation in Eq. (1) can be carried out immediately and exactly to give

$$\sum = (1 + x)^M,$$

or

$$\ln \sum = M \ln (1 + x). \tag{3}$$

It is always the logarithm of such a sum, rather than the sum itself, that is related to a thermodynamic function (see, for example, Section 1–7) and is therefore of interest to us. Next, let us find the value of N, N^*, which gives the largest term in the sum. Since the largest t_N is also the largest $\ln t_N$, we can locate N^*, for convenience, by reference to $\ln t_N$ rather than to t_N. N^* and $M - N^*$ will be of the same order of magnitude as M, that is, very large numbers. Hence we can use, without detectable error, the well-known Stirling approximation for the logarithm of the factorial of a very large number y:

$$\ln y! = \sum_{i=1}^{y} \ln i \cong \int_1^y \ln i \, di$$

$$\cong y \ln y - y. \tag{4}$$

Then

$$\ln t_N = M \ln M - N \ln N - (M - N) \ln (M - N) + N \ln x, \tag{5}$$

$$\frac{\partial \ln t_N}{\partial N} = 0 = -\ln N + \ln (M - N) + \ln x,$$

or

$$\frac{N^*}{M - N^*} = x, \qquad N^* = \frac{xM}{1 + x}. \tag{6}$$

To find the logarithm of the maximum term, we put the expression for N^* in Eq. (6) in place of N in Eq. (5) and find

$$\ln t_{N^*} = M \ln (1 + x). \tag{7}$$

Thus, to terms of order M, and these are the only terms of thermodynamic significance, the logarithm of the complete sum $\sum$ (Eq. 3) has the same value as the logarithm of just the largest term (Eq. 7).

To understand more clearly what is behind this superficially surprising agreement, let us pursue our example a little further. Consider the Taylor expansion of $\ln t_N$ about $N = N^*$:

$$\ln t_N = \ln t_{N^*} + \frac{1}{2}\left(\frac{\partial^2 \ln t}{\partial N^2}\right)_{N=N^*}(N - N^*)^2 + \cdots.$$

The linear term in $N - N^*$ is missing because the first derivative is zero at $N = N^*$. From Eq. (5) we have

$$\frac{\partial^2 \ln t_N}{\partial N^2} = -\frac{1}{N} - \frac{1}{M - N}.$$

Then using Eq. (6),

$$\ln t_N = \ln t_{N^*} - \frac{(1 + x)^2 (N - N^*)^2}{2xM} + \cdots,$$

or

$$t_N = t_{N^*}\, e^{-(N-N^*)^2/2\sigma^2}, \tag{8}$$

where the standard deviation σ is

$$\sigma = \frac{x^{1/2} M^{1/2}}{1 + x}. \tag{9}$$

Thus t_N is a *very* sharp gaussian curve, centered about N^*, with a standard deviation which is extremely small relative to the most probable value,

$$\frac{\sigma}{N^*} = \frac{1}{x^{1/2}M^{1/2}} = O(M^{-1/2}) = O(10^{-10}) \tag{10}$$

if $M = O(10^{20})$. To within the accuracy of terms retained above, $\overline{N} = N^*$. Higher terms would introduce a discrepancy between $\overline{N}$ and N^* of completely negligible order.

The gaussian function t_N falls off essentially to zero a few standard deviations on either side of $N = N^*$. Hence the number of terms which contribute significantly to $\sum$ is a few times σ, or $O(M^{1/2})$; all other terms can be ignored. The order of magnitude of each of these terms is the same as that of t_{N^*}, namely, $O(e^M)$. Hence

$$\sum = O(e^M)O(M^{1/2})$$

and

$$\ln \sum = O(M) + O(\ln M) = \ln t_{N^*} + O(\ln M). \tag{11}$$

But, because of the magnitude of M, the term $O(\ln M)$ in this equation is completely negligible compared with the other term. Thus, even though $\sum$ is, say, 10^{10} times larger than t_{N^*}, the difference between $\ln \sum$ and $\ln t_{N^*}$ is inconsequential [as we have already verified in Eqs. (3) and (7)].

As a final comment, we emphasize that all the significant terms in the sum, referred to above, have values of N that differ negligibly from N^*. That is, N may differ from the mean by $O(\sigma)$ or $O(M^{1/2})$, but *relative* to the value of the mean itself, this is only $O(M^{-1/2})$, as in Eq. (10).

APPENDIX III

METHOD OF UNDETERMINED MULTIPLIERS

If we wish to locate the maximum (or minimum) in, say, the function $F(x, y, z)$, we require that at the maximum, by definition,

$$dF = \frac{\partial F}{\partial x}\,dx + \frac{\partial F}{\partial y}\,dy + \frac{\partial F}{\partial z}\,dz = 0 \tag{1}$$

for *any* choices of dx, dy, and dz. That is, dx, dy, and dz are independent variations. This implies that

$$\frac{\partial F}{\partial x} = 0, \qquad \frac{\partial F}{\partial y} = 0, \qquad \frac{\partial F}{\partial z} = 0 \tag{2}$$

separately, in order to satisfy Eq. (1) in general. The relations (2) give us three equations to solve for the three unknown values of x, y, and z at the maximum.

Now suppose we want to locate the maximum in $F(x, y, z)$, not including all possible values of x, y, z as above, but only those values that satisfy some relation or condition, say, $G(x, y, z) = c$, where c is a constant. In this case, at the *conditional* maximum, located, say, at x_0, y_0, z_0, we still have the relation $dF = 0$, as in Eq. (1), but now dx, dy, and dz are not all independent. They are, in fact, interrelated by the requirement

$$dG = \frac{\partial G}{\partial x}\,dx + \frac{\partial G}{\partial y}\,dy + \frac{\partial G}{\partial z}\,dz = 0, \tag{3}$$

which must also be satisfied at the conditional maximum, since $G(x, y, z)$ is held equal to a constant.

It may occur to the reader that there are two rather obvious ways to proceed. One is to solve, say, for z as a function of x and y in the equation $G(x, y, z) = c$, use this expression to eliminate z from $F(x, y, z)$, and finally maximize F with respect to x and y, as in Eq. (2). Actually, we have already used this method in Appendix II. The problem there is equivalent to finding the maximum term in the sum

$$\Sigma = \sum_{N,P=0}^{M} \frac{(N + P)!x^N}{N!P!},$$

subject to the condition or restraint $N + P = M$. If we use this condition to eliminate P, we obtain Eq. (II–1). With the condition already contained in Eq. (II–1), we find the absolute maximum of $\ln t_N$ rather than a conditional maximum.

The other method is to solve Eq. (3) for, say, dz in terms of dx and dy, put this expression in place of dz in Eq. (1), and then set the resulting coefficients of dx and dy equal to zero separately, since now dx and dy are independent. In this second method, a third relation is needed to give three equations in the three unknowns, and this can be obtained by eliminating either dx or dy instead of dz from Eq. (1). These two methods are necessarily equivalent to each other (and also to the method to be described below). However, they are not easily generalized, so we pass on to a third procedure, due to Lagrange.

At the conditional maximum, $dG = 0$ and also $\alpha\, dG = 0$, where α is *any* constant. Furthermore, $dF = 0$ and therefore $dF - \alpha\, dG = 0$. That is,

$$\left(\frac{\partial F}{\partial x} - \alpha \frac{\partial G}{\partial x}\right) dx + \left(\frac{\partial F}{\partial y} - \alpha \frac{\partial G}{\partial y}\right) dy + \left(\frac{\partial F}{\partial z} - \alpha \frac{\partial G}{\partial z}\right) dz = 0. \quad (4)$$

Only two of the variations dx, dy, and dz are independent here, since Eq. (1) or (3) must also be satisfied. It should be understood that the derivatives in Eq. (4) [and in (1) and (3)] are all to be evaluated at the maximum, x_0, y_0, z_0. Now Eq. (4) holds for *any* value of α. We are therefore free to make any choice of α we please, and we choose

$$\alpha = \frac{(\partial F/\partial z)_{x_0,y_0,z_0}}{(\partial G/\partial z)_{x_0,y_0,z_0}}\,. \quad (5)$$

Of course at this stage x_0, y_0, z_0 are not known, so the actual value of α is still undetermined. With this particular choice for α, the last term in Eq. (4) drops out. The remaining two terms contain dx and dy only. Since dx and dy can be varied independently [for any dx and dy, dz is fixed by Eq. (1) or (3)], and since the sum of the two terms must always be zero, we have, as in Eq. (2),

$$\frac{\partial F}{\partial x} - \alpha \frac{\partial G}{\partial x} = 0, \quad (6)$$

$$\frac{\partial F}{\partial y} - \alpha \frac{\partial G}{\partial y} = 0, \quad (7)$$

where α is given by Eq. (5) and the derivatives are evaluated at x_0, y_0, z_0.

In practice, since x_0, y_0, z_0 are not known in advance, the procedure is to regard Eqs. (5) through (7),

$$\frac{\partial F}{\partial x} - \alpha \frac{\partial G}{\partial x} = \frac{\partial F}{\partial y} - \alpha \frac{\partial G}{\partial y} = \frac{\partial F}{\partial z} - \alpha \frac{\partial G}{\partial z} = 0, \quad (8)$$

as three equations in three unknowns, x_0, y_0, z_0. Of course, the solution of Eqs. (8) will depend on α, so we actually obtain as roots from (8), $x_0(\alpha)$, $y_0(\alpha)$, and $z_0(\alpha)$. But then α can be evaluated by substitution in $G(x, y, z) = c$. That is, α follows from

$$G(x_0(\alpha), y_0(\alpha), z_0(\alpha)) = c.$$

If there are two conditions on the maximum, say $G_1(x, y, z) = c_1$ and $G_2(x, y, z) = c_2$, the three relations of the type

$$\frac{\partial F}{\partial x} - \alpha\frac{\partial G_1}{\partial x} - \beta\frac{\partial G_2}{\partial x} = 0$$

must be solved for $x_0(\alpha, \beta)$, $y_0(\alpha, \beta)$, and $z_0(\alpha, \beta)$. We evaluate the two undetermined multipliers α and β by substitution of x_0, y_0, and z_0 into $G_1 = c_1$ and $G_2 = c_2$, giving two equations in the two unknowns, α and β.

The method can easily be generalized to an indefinite number of independent variables $x, y, z, \ldots$ and conditions $G_1 = c_1$, $G_2 = c_2, \ldots$. Of course, the number of conditions must be less than the number of independent variables.

APPENDIX IV

THE LENNARD-JONES POTENTIAL

A pair of spherically symmetrical (or effectively spherical) and chemically saturated molecules can be shown by a quantum-mechanical calculation to attract each other with a force falling off with distance as $1/r^7$, for sufficiently large r. The potential of this force varies, then, as $1/r^6$. This is the "London dispersion" or "van der Waals attractive" force. It arises from an attraction between mutually induced fluctuating dipoles in the two molecules. When r is small, the two molecules repel each other strongly because of internuclear repulsion and overlap of electronic shells. A quantum-mechanical calculation of this latter effect is impossibly difficult in general. A useful approximation is to represent the repulsive contribution to the potential by a term in $1/r^{12}$.

The complete potential (called a "Lennard-Jones 6–12 potential") then has the form

$$u(r) = -2\epsilon\left(\frac{r^*}{r}\right)^6 + \epsilon\left(\frac{r^*}{r}\right)^{12} \tag{1}$$

$$= 4\epsilon\left[-\left(\frac{r_0}{r}\right)^6 + \left(\frac{r_0}{r}\right)^{12}\right], \tag{2}$$

where $r^* = 2^{1/6}r_0$. Clearly $u(r)$ has a minimum ($u = -\epsilon$) at $r = r^*$ and $u = 0$ at $r = r_0$. Figure IV–1 shows the behavior of the function $u(r)$.

A quantum-mechanical investigation of the interaction between *three* molecules shows that, except when the molecules all repel each other strongly (i.e., at small distances—which would be important only at very

TABLE IV–1

LENNARD-JONES PARAMETERS

	ϵ/k, °K	r^*, A	r_0, A
He	10.22	2.869	2.556
H_2	37.00	3.287	2.928
Ne	34.9	3.12	2.78
A	119.8	3.822	3.405
Kr	171	4.04	3.60
Xe	221	4.60	4.10
N_2	95.05	4.151	3.698
O_2	118	3.88	3.46
CH_4	148.2	4.285	3.817

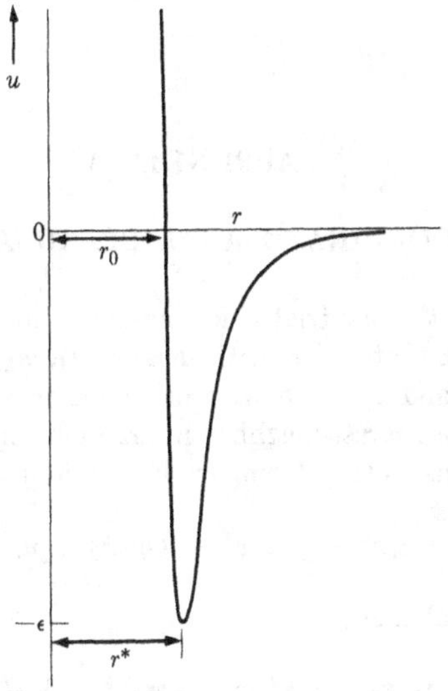

FIG. IV–1. Lennard-Jones 6–12 intermolecular pair potential.

high temperatures and pressures), the total interaction potential can be considered to be a sum of pairwise interactions, as in Eq. (1), to an excellent approximation. This approximation is quite generally used in condensed systems.

In principle, the parameters ϵ and r^* can be obtained for a given molecule from a quantum-mechanical calculation, as already indicated, but in practice they must be deduced from indirect experimental evidence. The second virial coefficient is the most useful source. Table IV–1 contains values of ϵ and r^* for a number of molecules, taken from Hirschfelder, Curtiss, and Bird. This book contains a vast amount of information about intermolecular forces. The reader interested in further details should consult it.

APPENDIX V

NORMAL COORDINATE ANALYSIS IN A SPECIAL CASE

Consider a hypothetical triatomic, one-dimensional molecule with atoms 1, 2, 3 arranged in that order, with masses m_1, m_2, and m_3, and with coordinates x_1, x_2, and x_3 (each with origin taken at the equilibrium position). We assume nearest-neighbor interactions only. The Hamiltonian, for small vibrations, is then [compare Eq. (5–51)]

$$H = \tfrac{1}{2}m_1\dot{x}_1^2 + \tfrac{1}{2}m_2\dot{x}_2^2 + \tfrac{1}{2}m_3\dot{x}_3^2 + u_e(x_1, x_2, x_3),$$

where the potential energy is

$$u_e(x_1, x_2, x_3) = \tfrac{1}{2}f_1(x_2 - x_1)^2 + \tfrac{1}{2}f_3(x_3 - x_2)^2. \tag{1}$$

The potential energy $u_e(0)$ (i.e., the potential energy at $x_1 = x_2 = x_3 = 0$) is taken as the zero of energy here for convenience. The f's are force constants. Our object is to find a new set of "normal" coordinates ξ_1, ξ_2, ξ_3, linearly related to x_1, x_2, x_3, as in Eq. (5–23), and having the property that the motion of the system can be decomposed into independent harmonic oscillations ("normal modes of vibration") of ξ_1, ξ_2, ξ_3, as in Eqs. (5–25) and (5–26). We might note that the linear combinations $x_2 - x_1$ and $x_3 - x_2$ put the potential energy, but not the kinetic energy, in the form of a sum of squares.

The inverse of the linear transformation (5–23) gives each x_i as a linear combination of ξ_1, ξ_2, and ξ_3. If, in the classical motion, only ξ_1 is excited (i.e., $\xi_2 = \xi_3 = 0$ and ξ_1 varies sinusoidally with time, with frequency ν_1), then the inverse transformation simplifies to each x_i being proportional to ξ_1. Because of this proportionality, in the normal mode associated with ξ_1, each x_i must also oscillate with frequency ν_1, and furthermore when $\xi_1 = 0$, each $x_i = 0$. (That is, in a pure normal mode, all atoms pass through their respective origins at the same time.)

Still referring to the normal mode $\xi_1(t)$, if atom 1 is to undergo simple harmonic motion with frequency ν_1, its equation of motion must satisfy

$$m_1\ddot{x}_1 = -4\pi^2 m_1\nu_1^2 x_1, \tag{2}$$

since integration of this equation gives the required form

$$x_1 = C_1 \sin(2\pi\nu_1 t + C_2), \tag{3}$$

where C_1 and C_2 are constants. Another condition on the equation of motion of atom 1 is that

$$m_1\ddot{x}_1 = -\frac{\partial u_e}{\partial x_1} = f_1(x_2 - x_1). \tag{4}$$

That is,

$$f_1(x_2 - x_1) = -4\pi^2 m_1 \nu_1^2 x_1. \tag{5}$$

We can easily obtain equations similar to (5) from $\ddot{x}_2$ and $\ddot{x}_3$, also for the frequency ν_1. Then in these three equations we can just as well put ν_2 or ν_3 as ν_1.

The three equations of type (5) for any one of the frequencies ν_1, ν_2, or ν_3 can be written

$$\begin{aligned} (f_1 - 4\pi^2 m_1\nu^2)x_1 - f_1x_2 \qquad\qquad\qquad &= 0, \\ -f_1x_1 + (f_1 + f_3 - 4\pi^2 m_2\nu^2)x_2 - f_3x_3 &= 0, \\ -f_3x_2 + (f_3 - 4\pi^2 m_3\nu^2)x_3 &= 0. \end{aligned} \tag{6}$$

These are three linear homogeneous equations in x_1, x_2, x_3, which have a nontrivial solution only if the determinant of the coefficients vanishes:

$$\begin{vmatrix} f_1 - m_1\gamma & -f_1 & 0 \\ -f_1 & f_1 + f_3 - m_2\gamma & -f_3 \\ 0 & -f_3 & f_3 - m_3\gamma \end{vmatrix} = 0, \tag{7}$$

where $\gamma = 4\pi^2\nu^2$. The three roots of this cubic equation in γ give the three possible normal vibration frequencies, $\gamma_1 = 4\pi^2\nu_1^2$, etc. The roots are

$$\begin{aligned} \gamma_1 &= 0, \\ \gamma_2 + \gamma_3 &= f_1\left(\frac{m_1 + m_2}{m_1m_2}\right) + f_3\left(\frac{m_2 + m_3}{m_2m_3}\right), \\ \gamma_2\gamma_3 &= f_1f_3\left(\frac{m_1 + m_2 + m_3}{m_1m_2m_3}\right). \end{aligned} \tag{8}$$

To study the motion associated with each γ_i, we substitute the value of γ_i back in Eqs. (6). If we put $\gamma_1 = 0$, we find $x_1 = x_2 = x_3$. This, together with $\nu_1 = 0$, obviously corresponds to the translation of the whole molecule with interatomic distances fixed at the equilibrium distances. At this point we should note that there is a total of three degrees of freedom in this system. Therefore we should anticipate one translational degree (center of mass) and two internal vibrational degrees. Here we see that the translational motion (the same is true for rotation in two or three dimensions) drops naturally out of the normal coordinate analysis as a special mode with zero frequency.

To investigate ξ_2 and ξ_3, let us simplify matters by taking $m_1 = m_2 = m_3 = m$ and $f_1 = f_3 = f$. Then

$$\gamma_2 = \frac{f}{m}, \qquad \gamma_3 = \frac{3f}{m},$$

$$\nu_2 = \frac{1}{2\pi}\sqrt{\frac{f}{m}}, \qquad \nu_3 = \frac{1}{2\pi}\sqrt{\frac{3f}{m}}. \tag{9}$$

If we put γ_2 in Eqs. (6), we find $x_2 = 0$ and $x_1 = -x_3$, corresponding to the motion

ξ_2:  . (10)

Substitution of γ_3 in Eqs. (6) gives $x_2 = -2x_1 = -2x_3$, corresponding to

ξ_3:  . (11)

In both (10) and (11) the center of mass stays fixed.

In the motion corresponding to each of the three pure normal modes, we have from the above:

Mode 1	Mode 2	Mode 3
$x_1 = C_1\xi_1,$	$x_1 = C_2\xi_2,$	$x_1 = -C_3\xi_3,$
$x_2 = C_1\xi_1,$	$x_2 = 0,$	$x_2 = 2C_3\xi_3,$
$x_3 = C_1\xi_1,$	$x_3 = -C_2\xi_2,$	$x_3 = -C_3\xi_3,$

or, in general,

$$\begin{aligned} x_1 &= C_1\xi_1 + C_2\xi_2 - \tfrac{1}{2}C_3\xi_3, \\ x_2 &= C_1\xi_1 \qquad\qquad + C_3\xi_3, \\ x_3 &= C_1\xi_1 - C_2\xi_2 - \tfrac{1}{2}C_3\xi_3, \end{aligned} \tag{12}$$

where the C_i are arbitrary. Equations (12) are the inverse transformation referred to above, i.e., inverse to (5–23). If we substitute Eqs. (12) in Eq. (1) we find

$$\begin{aligned} H = {} & \tfrac{1}{2}(3mC_1^2\dot{\xi}_1^2 + 2mC_2^2\dot{\xi}_2^2 + \tfrac{3}{2}mC_3^2\dot{\xi}_3^2) \\ & + \tfrac{1}{2}(2fC_2^2\xi_2^2 + \tfrac{9}{2}fC_3^2\xi_3^2), \end{aligned} \tag{13}$$

which is indeed separable, as in Eq. (5–25). If we put the coefficients of Eq. (13) into Eq. (5–26), we recover $\nu_1 = 0$ and Eqs. (9) for ν_2 and ν_3.

From Eqs. (12) we find for the transformation of type (5–23).

$$\begin{aligned} 3C_1\xi_1 &= x_1 + x_2 + x_3, \\ 2C_2\xi_2 &= x_1 - x_3, \\ 3C_3\xi_3 &= -x_1 + 2x_2 - x_3. \end{aligned} \tag{14}$$

Eqs. (14) should be compared with (10) and (11).

APPENDIX VI

VIBRATIONAL FREQUENCY DISTRIBUTION IN A SOLID CONTINUUM

The object here is to find the function $g(\nu)$, where $g(\nu)\,d\nu$ is the number of normal modes of vibration with frequencies between ν and $\nu + d\nu$ in a monatomic crystal regarded as a continuum. This problem arises in Debye's theory of the thermodynamic properties of a crystal (Chapter 5).

One-dimensional solid. We consider first a one-dimensional crystal of length L and consisting of N atoms. We treat the crystal as a continuum with only longitudinal (along the x-axis) displacements allowed. A point at x in the unstrained solid is displaced to $x + u(x, t)$ during the propagation of an elastic wave. The partial differential equation for the displacement $u(x, t)$ is

$$\frac{\partial^2 u}{\partial x^2} = \frac{1}{v_1^2}\frac{\partial^2 u}{\partial t^2}, \tag{1}$$

where v_1 is the velocity of the wave. The equation for a traveling wave with frequency ν is, in complex form,

$$u = Ae^{2\pi i(\nu t - \alpha x)}. \tag{2}$$

Equation (2) satisfies Eq. (1) if $\alpha = \pm\nu/v_1$. That is,

$$u = Ae^{2\pi i\nu[t-(x/v_1)]}, \tag{3}$$

or

$$u = Ae^{2\pi i\nu[t+(x/v_1)]}. \tag{4}$$

Equation (3) represents a wave traveling to the right, and Eq. (4) a wave traveling to the left. The wavelength in either case is clearly

$$\lambda = \frac{v_1}{\nu}. \tag{5}$$

The normal modes of vibration in a crystal or polyatomic molecule (Appendix V) obviously correspond to standing rather than traveling "waves." Therefore we are primarily interested in standing waves here. By taking linear combinations of the sine and cosine parts of Eqs. (3)

and (4), with proper choices of coefficients, standing waves of the forms

$$\sin 2\pi\nu t \sin \frac{2\pi x}{\lambda}, \tag{6}$$

$$\cos 2\pi\nu t \sin \frac{2\pi x}{\lambda}, \tag{7}$$

$$\sin 2\pi\nu t \cos \frac{2\pi x}{\lambda}, \tag{8}$$

$$\cos 2\pi\nu t \cos \frac{2\pi x}{\lambda}, \tag{9}$$

may be obtained.

The precise choice of boundary conditions for a large crystal is immaterial. We assume the solid is held rigidly at the two ends: $u = 0$ at $x = 0$ and $x = L$. The standing waves (6) and (7) satisfy this requirement at $x = 0$ and also at $x = L$, provided

$$\frac{2L}{\lambda} = n \qquad (n = 1, 2, 3, \ldots). \tag{10}$$

The allowed wavelengths and frequencies are therefore

$$\lambda_n = \frac{2L}{n}, \qquad \nu_n = \frac{v_1 n}{2L} \qquad (n = 1, 2, 3, \ldots, N), \tag{11}$$

where N is written as the maximum value of n since there are really only N vibrational degrees of freedom, despite the fact that we are using a continuum model. In a true continuum, there would be no lower limit to λ ($n \to \infty$) and an infinite number of normal modes. If the nearest-neighbor distance is a, $L = Na$ and the shortest wavelength allowed is $\lambda_N = 2a$. Correspondingly, the highest frequency ν_m is

$$\nu_m = \nu_N = \frac{v_1}{2a}. \tag{12}$$

Since ν_n depends linearly on n in Eq. (11), it is apparent that $g(\nu)$ in this case is just a constant:

$$\int_0^{\nu_m} g\, d\nu = g\nu_m = N, \tag{13}$$

$$g = \frac{N}{\nu_m} = \frac{2aN}{v_1} = \frac{2L}{v_1}. \tag{14}$$

From the theory of elasticity, v_1 can be expressed in terms of the elastic properties of the one-dimensional solid. But from a molecular point of view it is more satisfying to relate v_1 to the intermolecular potential energy, atomic mass, and nearest-neighbor distance in the crystal. This connection can be established (see Section 5–4) by identifying the low-frequency asymptotic behavior of the exact $g(\nu)$ from molecular theory with the g obtained here by elastic continuum theory.

Three-dimensional isotropic solid. Let the solid contain N atoms and be in the shape of a rectangular parallelepiped with edges of lengths L_x, L_y, L_z. A point at x, y, z in the unstrained solid is in general displaced to $x + u_x$, $y + u_y$, $z + u_z$ during the propagation of an elastic wave. If the most general partial differential equation for wave propagation in an isotropic continuous solid, with coefficients determined by the elastic constants, is investigated, it is found* that there are two different possible velocities of propagation, v_{3l} for longitudinal waves (one mode of vibration for a given wavelength, in the direction of propagation) and v_{3t} for transverse waves (two modes of vibration for a given wavelength, perpendicular to the direction of propagation):

$$v_{3l} = \left[\frac{3(1 - \sigma)}{\chi\rho(1 + \sigma)}\right]^{1/2}, \tag{15}$$

$$v_{3t} = \left[\frac{3(1 - 2\sigma)}{2\chi\rho(1 + \sigma)}\right]^{1/2}, \tag{16}$$

where χ is the volume compressibility, σ is Poisson's ratio, and ρ the density. The longitudinal wave is the sound wave. We are therefore interested in the separate equations

$$\nabla^2 u_l = \frac{1}{v_{3l}^2} \frac{\partial^2 u_l}{\partial t^2}, \tag{17}$$

$$\nabla^2 u_t = \frac{1}{v_{3t}^2} \frac{\partial^2 u_t}{\partial t^2}, \tag{18}$$

where u_l and u_t represent the amplitudes of the displacements in the two cases. It will be convenient to use

$$\nabla^2 u = \frac{1}{v_3^2} \frac{\partial^2 u}{\partial t^2} \tag{19}$$

to stand for either Eq. (17) or Eq. (18).

---

* See, for example, A. E. H. Love, *Mathematical Theory of Elasticity*, pp. 297–302.

The expression

$$u = Ae^{2\pi i(\nu t - \boldsymbol{\alpha}\cdot\mathbf{r})}$$

$$= Ae^{2\pi i[\nu t - (\alpha_x x + \alpha_y y + \alpha_z z)]} \tag{20}$$

represents a traveling plane wave of frequency ν with wave front normal to $\boldsymbol{\alpha}$. This follows because

$$\alpha_x x + \alpha_y y + \alpha_z z = c,$$

where c is a constant, is the equation of a plane normal to $\boldsymbol{\alpha}$. Equation (20) satisfies Eq. (19) provided that ν and v_3 are related by

$$\frac{\nu^2}{v_3^2} = \alpha_x^2 + \alpha_y^2 + \alpha_z^2. \tag{21}$$

From Eq. (20), we see that the wavelength λ is the perpendicular distance between, say, the planes $\mathbf{a}\cdot\mathbf{r} = 0$ and $\mathbf{a}\cdot\mathbf{r} = 1$. Then $|\boldsymbol{\alpha}|\lambda = 1$, or $\lambda = 1/|\boldsymbol{\alpha}|$. From Eq. (21),

$$\lambda = \frac{v_3}{\nu}. \tag{22}$$

If we define

$$l_x = \frac{\alpha_x}{|\boldsymbol{\alpha}|}, \qquad l_y = \frac{\alpha_y}{|\boldsymbol{\alpha}|}, \qquad l_z = \frac{\alpha_z}{|\boldsymbol{\alpha}|}, \tag{23}$$

where

$$l_x^2 + l_y^2 + l_z^2 = 1, \tag{24}$$

Eq. (20) becomes

$$u = A \exp\left\{2\pi i\nu\left[t - \frac{(l_x x + l_y y + l_z z)}{v_3}\right]\right\}. \tag{25}$$

Actually, we can write $\pm l_x$, $\pm l_y$, and $\pm l_z$ in Eq. (25) and still have a traveling wave with velocity v_3, frequency ν, and wavelength $\lambda = 1/|\boldsymbol{\alpha}|$, which is a solution of Eq. (19). By proper linear combinations of these eight waves traveling in different directions we can find standing waves as in Eqs. (6) through (9). For example,

$$\left.\begin{matrix}\sin 2\pi\nu t\\ \cos 2\pi\nu t\end{matrix}\right\} \sin\frac{2\pi l_x x}{\lambda}\sin\frac{2\pi l_y y}{\lambda}\sin\frac{2\pi l_z z}{\lambda}. \tag{26}$$

For boundary conditions, we assume the solid to be held rigidly ($u = 0$) on the surfaces $x = 0$, $x = L_x$, $y = 0$, $y = L_y$, $z = 0$, $z = L_z$.

Equation (26) is a standing wave which satisfies these requirements provided that

$$\frac{2l_xL_x}{\lambda} = n_x, \qquad \frac{2l_yL_y}{\lambda} = n_y, \qquad \frac{2l_zL_z}{\lambda} = n_z \qquad (n_x, n_y, n_z = 1, 2, \ldots). \tag{27}$$

We can eliminate l_x, l_y, and l_z by substituting Eqs. (27) into Eq. (24):

$$\nu_{n_xn_yn_z} = \frac{v_3}{\lambda_{n_xn_yn_z}} = \frac{v_3}{2}\left(\frac{n_x^2}{L_x^2} + \frac{n_y^2}{L_y^2} + \frac{n_z^2}{L_z^2}\right)^{1/2}. \tag{28}$$

Equation (28) gives the wavelengths and frequencies of the normal modes of vibration. For each set of integers n_x, n_y, n_z, there are two transverse modes and one longitudinal mode with the same wavelength but with different frequencies (since v_{3t} and v_{3l} are different).

Each point n_x, n_y, n_z (all integers) in the positive octant of n_x, n_y, n_z space gives a possible frequency $\nu_{n_xn_yn_z}$. Further, there is one such point per unit volume of this space. The number of frequencies $F(\nu)$ with $0 \leq \nu_{n_xn_yn_z} \leq \nu$ is therefore equal to one-eighth the volume of the ellipsoid [from Eq. (28)]

$$1 = \frac{n_x^2}{(2\nu L_x/v_3)^2} + \frac{n_y^2}{(2\nu L_y/v_3)^2} + \frac{n_z^2}{(2\nu L_z/v_3)^2}. \tag{29}$$

That is,

$$F(\nu) = \frac{1}{8}\cdot\frac{4\pi}{3}\left(\frac{2\nu}{v_3}\right)^3 L_xL_yL_z = \frac{4\pi\nu^3V}{3v_3^3}.$$

The number of frequencies between ν and $\nu + d\nu$ is then

$$\frac{dF}{d\nu}\,d\nu = \frac{4\pi\nu^2V}{v_3^3}\,d\nu.$$

Finally, for the total density of vibrational modes,

$$g(\nu) = 4\pi V\left(\frac{1}{v_{3l}^3} + \frac{2}{v_{3t}^3}\right)\nu^2. \tag{30}$$

Thus, in a three-dimensional solid, $g(\nu) \propto \nu^2$. Note the resemblance of the above argument to that in Section 4–1.

To limit the total number of vibrational modes to $3N$, we have

$$\frac{12\pi V}{\bar{v}_3^3}\int_0^{\nu_m} \nu^2\,d\nu = 3N, \tag{31}$$

where

$$\frac{3}{\bar{v}_3^3} \equiv \frac{1}{v_{3l}^3} + \frac{2}{v_{3t}^3} \tag{32}$$

and $g(\nu) = 0$ for $\nu > \nu_m$. That is, ν_m is the maximum frequency. From Eq. (31),

$$\nu_m = \left(\frac{3N}{4\pi V}\right)^{1/3} \bar{v}_3 \tag{33}$$

and

$$g(\nu) = \frac{9N\nu^2}{\nu_m^3}. \tag{34}$$

Two-dimensional isotropic solid. By the methods of Eqs. (17) through (29), we find

$$F(\nu) = \frac{1}{4} \cdot \pi \left(\frac{2\nu}{v_2}\right)^2 L_x L_y$$

$$= \frac{\pi \nu^2 \mathcal{A}}{v_2^2},$$

$$\frac{dF}{d\nu} = \frac{2\pi\nu\mathcal{A}}{v_2^2},$$

where $\mathcal{A}$ is the area, and

$$g(\nu) = 2\pi\mathcal{A}\left(\frac{1}{v_{2l}^2} + \frac{1}{v_{2t}^2}\right)\nu. \tag{35}$$

That is, $g(\nu) \propto \nu$ in two dimensions. In summary, for n dimensions, $g(\nu) \propto \nu^{n-1}$.

APPENDIX VII

GENERALIZED COORDINATES

We merely summarize here a few definitions and properties in classical mechanics needed in the text. For details and proofs the reader should consult Tolman or Goldstein.*

Consider a conservative mechanical system of particles with masses $m_1, m_2, \ldots$, cartesian coordinates $x_1, y_1, z_1, x_2, y_2, z_2, \ldots$, and potential energy $U(x_1, y_1, z_1, x_2, y_2, z_2, \ldots)$. The kinetic energy is

$$K = \tfrac{1}{2}m_1(\dot{x}_1^2 + \dot{y}_1^2 + \dot{z}_1^2) + \tfrac{1}{2}m_2(\dot{x}_2^2 + \dot{y}_2^2 + \dot{z}_2^2) + \cdots, \qquad (1)$$

and the total energy

$$E = \text{constant} = U(x_1, y_1, z_1, \ldots) + K(\dot{x}_1, \dot{y}_1, \dot{z}_1, \ldots). \qquad (2)$$

The equations of motion in Newtonian form are

$$\frac{d}{dt}(m_k\dot{x}_k) = -\frac{\partial U}{\partial x_k}, \quad \frac{d}{dt}(m_k\dot{y}_k) = -\frac{\partial U}{\partial y_k}, \quad \frac{d}{dt}(m_k\dot{z}_k) = -\frac{\partial U}{\partial z_k},$$
$$k = 1, 2, 3, \ldots. \qquad (3)$$

The Lagrangian function L is defined by $L = K - U$. The equations of motion in Lagrangian form, which we shall not use, are

$$\frac{d}{dt}\frac{\partial L}{\partial \dot{q}_i} - \frac{\partial L}{\partial q_i} = 0 \qquad (i = 1, 2, \ldots), \qquad (4)$$

where here the "generalized" coordinates (i.e., not necessarily cartesian coordinates) are denoted by $q_1, q_2, \ldots$ ($3N$ q's for N particles). In general, each q may be a function of all the cartesian coordinates.

It will be observed that Eqs. (3) are very unsymmetrical in x_k and $\dot{x}_k$, etc., and that Eqs. (4) are similarly unsymmetrical in q_i and $\dot{q}_i$. There is a considerable simplification and symmetrization in classical mechanics and classical statistical mechanics if, for a given choice of coordinates $q_1, q_2, \ldots$, we regard the energy as a function not of the q's (which may or may not be cartesian coordinates) and $\dot{q}$'s but of the q's and p's, where p_i is defined in general by

$$p_i = \frac{\partial L}{\partial \dot{q}_i} \qquad (i = 1, 2, \ldots). \qquad (5)$$

---

* H. Goldstein, *Classical Mechanics*. Reading, Mass.: Addison-Wesley, 1950.

In conservative systems, $U = U(q)$ and $K = K(q, \dot{q})$, so that for our purposes

$$p_i = \frac{\partial K}{\partial \dot{q}_i} \qquad (i = 1, 2, \ldots). \tag{6}$$

The variable p_i is called the momentum conjugate to q_i. The energy expressed as a function of the q's and p's is called the Hamiltonian function, $H(q, p) = E =$ constant. The equations of motion can now be put in the very symmetrical (i.e., with respect to q and p) Hamiltonian form

$$\frac{dq_i}{dt} = \frac{\partial H}{\partial p_i}, \qquad \frac{dp_i}{dt} = -\frac{\partial H}{\partial q_i} \qquad (i = 1, 2, \ldots). \tag{7}$$

As a very simple example, let the q's be cartesian coordinates. Then from Eqs. (1) and (6),

$$p_{xk} = \frac{\partial K}{\partial \dot{x}_k} = m_k \dot{x}_k, \qquad \text{etc.,} \tag{8}$$

and

$$H(q, p) = U(q) + \frac{1}{2m_1}(p_{x1}^2 + p_{y1}^2 + p_{z1}^2) + \frac{1}{2m_2}(p_{x2}^2 + p_{y2}^2 + p_{z2}^2) + \cdots. \tag{9}$$

The conjugate momenta are just the ordinary components of momentum, in this special case. Then Eqs. (7) become

$$\frac{dx_k}{dt} = \frac{p_{xk}}{m_k}, \qquad \frac{dp_{xk}}{dt} = -\frac{\partial U}{\partial x_k}, \ldots, \tag{10}$$

in agreement with Eqs. (3) and (8).

An important property of the conjugate variables q, p, which we shall need, is the following. If we change coordinates from the arbitrary set $q_1, q_2, \ldots$ to a new arbitrary set $q_1', q_2', \ldots$, and also change conjugate momenta from $p_1, p_2, \ldots$ to $p_1', p_2', \ldots$ [where the p' are defined in terms of q' by Eq. (6)], then the element of volume in q, p space is equal to the element of volume in q', p' space:

$$dq_1\, dq_2 \ldots dp_1\, dp_2 \ldots = dq_1'\, dq_2' \ldots dp_1'\, dp_2' \ldots. \tag{11}$$

In other words, the Jacobian of the transformation $q, p \rightarrow q', p'$ is unity.

INDEX

INDEX

A CATALOG OF SELECTED

DOVER BOOKS

IN SCIENCE AND MATHEMATICS

Astronomy

BURNHAM'S CELESTIAL HANDBOOK, Robert Burnham, Jr. Thorough guide to the stars beyond our solar system. Exhaustive treatment. Alphabetical by constellation: Andromeda to Cetus in Vol. 1; Chamaeleon to Orion in Vol. 2; and Pavo to Vulpecula in Vol. 3. Hundreds of illustrations. Index in Vol. 3. 2,000pp. 6⅛ x 9¼.
Vol. I: 0-486-23567-X
Vol. II: 0-486-23568-8
Vol. III: 0-486-23673-0

EXPLORING THE MOON THROUGH BINOCULARS AND SMALL TELESCOPES, Ernest H. Cherrington, Jr. Informative, profusely illustrated guide to locating and identifying craters, rills, seas, mountains, other lunar features. Newly revised and updated with special section of new photos. Over 100 photos and diagrams. 240pp. 8¼ x 11. 0-486-24491-1

THE EXTRATERRESTRIAL LIFE DEBATE, 1750–1900, Michael J. Crowe. First detailed, scholarly study in English of the many ideas that developed from 1750 to 1900 regarding the existence of intelligent extraterrestrial life. Examines ideas of Kant, Herschel, Voltaire, Percival Lowell, many other scientists and thinkers. 16 illustrations. 704pp. 5⅜ x 8½. 0-486-40675-X

THEORIES OF THE WORLD FROM ANTIQUITY TO THE COPERNICAN REVOLUTION, Michael J. Crowe. Newly revised edition of an accessible, enlightening book recreates the change from an earth-centered to a sun-centered conception of the solar system. 242pp. 5⅜ x 8½. 0-486-41444-2

A HISTORY OF ASTRONOMY, A. Pannekoek. Well-balanced, carefully reasoned study covers such topics as Ptolemaic theory, work of Copernicus, Kepler, Newton, Eddington's work on stars, much more. Illustrated. References. 521pp. 5⅜ x 8½.
0-486-65994-1

A COMPLETE MANUAL OF AMATEUR ASTRONOMY: TOOLS AND TECHNIQUES FOR ASTRONOMICAL OBSERVATIONS, P. Clay Sherrod with Thomas L. Koed. Concise, highly readable book discusses: selecting, setting up and maintaining a telescope; amateur studies of the sun; lunar topography and occultations; observations of Mars, Jupiter, Saturn, the minor planets and the stars; an introduction to photoelectric photometry; more. 1981 ed. 124 figures. 25 halftones. 37 tables. 335pp. 6½ x 9¼. 0-486-40675-X

AMATEUR ASTRONOMER'S HANDBOOK, J. B. Sidgwick. Timeless, comprehensive coverage of telescopes, mirrors, lenses, mountings, telescope drives, micrometers, spectroscopes, more. 189 illustrations. 576pp. 5⅝ x 8¼. (Available in U.S. only.)
0-486-24034-7

STARS AND RELATIVITY, Ya. B. Zel'dovich and I. D. Novikov. Vol. 1 of *Relativistic Astrophysics* by famed Russian scientists. General relativity, properties of matter under astrophysical conditions, stars, and stellar systems. Deep physical insights, clear presentation. 1971 edition. References. 544pp. 5⅝ x 8¼. 0-486-69424-0

Chemistry

THE SCEPTICAL CHYMIST: THE CLASSIC 1661 TEXT, Robert Boyle. Boyle defines the term "element," asserting that all natural phenomena can be explained by the motion and organization of primary particles. 1911 ed. viii+232pp. 5⅜ x 8½.
0-486-42825-7

RADIOACTIVE SUBSTANCES, Marie Curie. Here is the celebrated scientist's doctoral thesis, the prelude to her receipt of the 1903 Nobel Prize. Curie discusses establishing atomic character of radioactivity found in compounds of uranium and thorium; extraction from pitchblende of polonium and radium; isolation of pure radium chloride; determination of atomic weight of radium; plus electric, photographic, luminous, heat, color effects of radioactivity. ii+94pp. 5⅜ x 8½. 0-486-42550-9

CHEMICAL MAGIC, Leonard A. Ford. Second Edition, Revised by E. Winston Grundmeier. Over 100 unusual stunts demonstrating cold fire, dust explosions, much more. Text explains scientific principles and stresses safety precautions. 128pp. 5⅜ x 8½. 0-486-67628-5

THE DEVELOPMENT OF MODERN CHEMISTRY, Aaron J. Ihde. Authoritative history of chemistry from ancient Greek theory to 20th-century innovation. Covers major chemists and their discoveries. 209 illustrations. 14 tables. Bibliographies. Indices. Appendices. 851pp. 5⅜ x 8½. 0-486-64235-6

CATALYSIS IN CHEMISTRY AND ENZYMOLOGY, William P. Jencks. Exceptionally clear coverage of mechanisms for catalysis, forces in aqueous solution, carbonyl- and acyl-group reactions, practical kinetics, more. 864pp. 5⅜ x 8½.
0-486-65460-5

ELEMENTS OF CHEMISTRY, Antoine Lavoisier. Monumental classic by founder of modern chemistry in remarkable reprint of rare 1790 Kerr translation. A must for every student of chemistry or the history of science. 539pp. 5⅜ x 8½. 0-486-64624-6

THE HISTORICAL BACKGROUND OF CHEMISTRY, Henry M. Leicester. Evolution of ideas, not individual biography. Concentrates on formulation of a coherent set of chemical laws. 260pp. 5⅜ x 8½. 0-486-61053-5

A SHORT HISTORY OF CHEMISTRY, J. R. Partington. Classic exposition explores origins of chemistry, alchemy, early medical chemistry, nature of atmosphere, theory of valency, laws and structure of atomic theory, much more. 428pp. 5⅜ x 8½. (Available in U.S. only.) 0-486-65977-1

GENERAL CHEMISTRY, Linus Pauling. Revised 3rd edition of classic first-year text by Nobel laureate. Atomic and molecular structure, quantum mechanics, statistical mechanics, thermodynamics correlated with descriptive chemistry. Problems. 992pp. 5⅜ x 8½. 0-486-65622-5

FROM ALCHEMY TO CHEMISTRY, John Read. Broad, humanistic treatment focuses on great figures of chemistry and ideas that revolutionized the science. 50 illustrations. 240pp. 5⅜ x 8½. 0-486-28690-8

Engineering

DE RE METALLICA, Georgius Agricola. The famous Hoover translation of greatest treatise on technological chemistry, engineering, geology, mining of early modern times (1556). All 289 original woodcuts. 638pp. 6¾ x 11. 0-486-60006-8

FUNDAMENTALS OF ASTRODYNAMICS, Roger Bate et al. Modern approach developed by U.S. Air Force Academy. Designed as a first course. Problems, exercises. Numerous illustrations. 455pp. 5⅜ x 8½. 0-486-60061-0

DYNAMICS OF FLUIDS IN POROUS MEDIA, Jacob Bear. For advanced students of ground water hydrology, soil mechanics and physics, drainage and irrigation engineering and more. 335 illustrations. Exercises, with answers. 784pp. 6⅛ x 9¼. 0-486-65675-6

THEORY OF VISCOELASTICITY (Second Edition), Richard M. Christensen. Complete consistent description of the linear theory of the viscoelastic behavior of materials. Problem-solving techniques discussed. 1982 edition. 29 figures. xiv+364pp. 6⅛ x 9¼. 0-486-42880-X

MECHANICS, J. P. Den Hartog. A classic introductory text or refresher. Hundreds of applications and design problems illuminate fundamentals of trusses, loaded beams and cables, etc. 334 answered problems. 462pp. 5⅜ x 8½. 0-486-60754-2

MECHANICAL VIBRATIONS, J. P. Den Hartog. Classic textbook offers lucid explanations and illustrative models, applying theories of vibrations to a variety of practical industrial engineering problems. Numerous figures. 233 problems, solutions. Appendix. Index. Preface. 436pp. 5⅜ x 8½. 0-486-64785-4

STRENGTH OF MATERIALS, J. P. Den Hartog. Full, clear treatment of basic material (tension, torsion, bending, etc.) plus advanced material on engineering methods, applications. 350 answered problems. 323pp. 5⅜ x 8½. 0-486-60755-0

A HISTORY OF MECHANICS, René Dugas. Monumental study of mechanical principles from antiquity to quantum mechanics. Contributions of ancient Greeks, Galileo, Leonardo, Kepler, Lagrange, many others. 671pp. 5⅜ x 8½. 0-486-65632-2

STABILITY THEORY AND ITS APPLICATIONS TO STRUCTURAL MECHANICS, Clive L. Dym. Self-contained text focuses on Koiter postbuckling analyses, with mathematical notions of stability of motion. Basing minimum energy principles for static stability upon dynamic concepts of stability of motion, it develops asymptotic buckling and postbuckling analyses from potential energy considerations, with applications to columns, plates, and arches. 1974 ed. 208pp. 5⅜ x 8½. 0-486-42541-X

METAL FATIGUE, N. E. Frost, K. J. Marsh, and L. P. Pook. Definitive, clearly written, and well-illustrated volume addresses all aspects of the subject, from the historical development of understanding metal fatigue to vital concepts of the cyclic stress that causes a crack to grow. Includes 7 appendixes. 544pp. 5⅜ x 8½. 0-486-40927-9

ROCKETS, Robert Goddard. Two of the most significant publications in the history of rocketry and jet propulsion: "A Method of Reaching Extreme Altitudes" (1919) and "Liquid Propellant Rocket Development" (1936). 128pp. 5⅜ x 8½. 0-486-42537-1

STATISTICAL MECHANICS: PRINCIPLES AND APPLICATIONS, Terrell L. Hill. Standard text covers fundamentals of statistical mechanics, applications to fluctuation theory, imperfect gases, distribution functions, more. 448pp. 5⅜ x 8½. 0-486-65390-0

ENGINEERING AND TECHNOLOGY 1650–1750: ILLUSTRATIONS AND TEXTS FROM ORIGINAL SOURCES, Martin Jensen. Highly readable text with more than 200 contemporary drawings and detailed engravings of engineering projects dealing with surveying, leveling, materials, hand tools, lifting equipment, transport and erection, piling, bailing, water supply, hydraulic engineering, and more. Among the specific projects outlined-transporting a 50-ton stone to the Louvre, erecting an obelisk, building timber locks, and dredging canals. 207pp. 8⅜ x 11¼. 0-486-42232-1

THE VARIATIONAL PRINCIPLES OF MECHANICS, Cornelius Lanczos. Graduate level coverage of calculus of variations, equations of motion, relativistic mechanics, more. First inexpensive paperbound edition of classic treatise. Index. Bibliography. 418pp. 5⅜ x 8½. 0-486-65067-7

PROTECTION OF ELECTRONIC CIRCUITS FROM OVERVOLTAGES, Ronald B. Standler. Five-part treatment presents practical rules and strategies for circuits designed to protect electronic systems from damage by transient overvoltages. 1989 ed. xxiv+434pp. 6⅛ x 9¼. 0-486-42552-5

ROTARY WING AERODYNAMICS, W. Z. Stepniewski. Clear, concise text covers aerodynamic phenomena of the rotor and offers guidelines for helicopter performance evaluation. Originally prepared for NASA. 537 figures. 640pp. 6⅛ x 9¼. 0-486-64647-5

INTRODUCTION TO SPACE DYNAMICS, William Tyrrell Thomson. Comprehensive, classic introduction to space-flight engineering for advanced undergraduate and graduate students. Includes vector algebra, kinematics, transformation of coordinates. Bibliography. Index. 352pp. 5⅜ x 8½. 0-486-65113-4

HISTORY OF STRENGTH OF MATERIALS, Stephen P. Timoshenko. Excellent historical survey of the strength of materials with many references to the theories of elasticity and structure. 245 figures. 452pp. 5⅜ x 8½. 0-486-61187-6

ANALYTICAL FRACTURE MECHANICS, David J. Unger. Self-contained text supplements standard fracture mechanics texts by focusing on analytical methods for determining crack-tip stress and strain fields. 336pp. 6⅛ x 9¼. 0-486-41737-9

STATISTICAL MECHANICS OF ELASTICITY, J. H. Weiner. Advanced, self-contained treatment illustrates general principles and elastic behavior of solids. Part 1, based on classical mechanics, studies thermoelastic behavior of crystalline and polymeric solids. Part 2, based on quantum mechanics, focuses on interatomic force laws, behavior of solids, and thermally activated processes. For students of physics and chemistry and for polymer physicists. 1983 ed. 96 figures. 496pp. 5⅜ x 8½. 0-486-42260-7

Mathematics

FUNCTIONAL ANALYSIS (Second Corrected Edition), George Bachman and Lawrence Narici. Excellent treatment of subject geared toward students with background in linear algebra, advanced calculus, physics and engineering. Text covers introduction to inner-product spaces, normed, metric spaces, and topological spaces; complete orthonormal sets, the Hahn-Banach Theorem and its consequences, and many other related subjects. 1966 ed. 544pp. 6⅛ x 9¼. 0-486-40251-7

ASYMPTOTIC EXPANSIONS OF INTEGRALS, Norman Bleistein & Richard A. Handelsman. Best introduction to important field with applications in a variety of scientific disciplines. New preface. Problems. Diagrams. Tables. Bibliography. Index. 448pp. 5⅜ x 8½. 0-486-65082-0

VECTOR AND TENSOR ANALYSIS WITH APPLICATIONS, A. I. Borisenko and I. E. Tarapov. Concise introduction. Worked-out problems, solutions, exercises. 257pp. 5⅜ x 8¼. 0-486-63833-2

AN INTRODUCTION TO ORDINARY DIFFERENTIAL EQUATIONS, Earl A. Coddington. A thorough and systematic first course in elementary differential equations for undergraduates in mathematics and science, with many exercises and problems (with answers). Index. 304pp. 5⅜ x 8½. 0-486-65942-9

FOURIER SERIES AND ORTHOGONAL FUNCTIONS, Harry F. Davis. An incisive text combining theory and practical example to introduce Fourier series, orthogonal functions and applications of the Fourier method to boundary-value problems. 570 exercises. Answers and notes. 416pp. 5⅜ x 8½. 0-486-65973-9

COMPUTABILITY AND UNSOLVABILITY, Martin Davis. Classic graduate-level introduction to theory of computability, usually referred to as theory of recurrent functions. New preface and appendix. 288pp. 5⅜ x 8½. 0-486-61471-9

ASYMPTOTIC METHODS IN ANALYSIS, N. G. de Bruijn. An inexpensive, comprehensive guide to asymptotic methods–the pioneering work that teaches by explaining worked examples in detail. Index. 224pp. 5⅜ x 8½ 0-486-64221-6

APPLIED COMPLEX VARIABLES, John W. Dettman. Step-by-step coverage of fundamentals of analytic function theory–plus lucid exposition of five important applications: Potential Theory; Ordinary Differential Equations; Fourier Transforms; Laplace Transforms; Asymptotic Expansions. 66 figures. Exercises at chapter ends. 512pp. 5⅜ x 8½. 0-486-64670-X

INTRODUCTION TO LINEAR ALGEBRA AND DIFFERENTIAL EQUATIONS, John W. Dettman. Excellent text covers complex numbers, determinants, orthonormal bases, Laplace transforms, much more. Exercises with solutions. Undergraduate level. 416pp. 5⅜ x 8½. 0-486-65191-6

RIEMANN'S ZETA FUNCTION, H. M. Edwards. Superb, high-level study of landmark 1859 publication entitled "On the Number of Primes Less Than a Given Magnitude" traces developments in mathematical theory that it inspired. xiv+315pp. 5⅜ x 8½. 0-486-41740-9

INTRODUCTORY REAL ANALYSIS, A.N. Kolmogorov, S. V. Fomin. Translated by Richard A. Silverman. Self-contained, evenly paced introduction to real and functional analysis. Some 350 problems. 403pp. 5⅜ x 8½. 0-486-61226-0

APPLIED ANALYSIS, Cornelius Lanczos. Classic work on analysis and design of finite processes for approximating solution of analytical problems. Algebraic equations, matrices, harmonic analysis, quadrature methods, much more. 559pp. 5⅜ x 8½. 0-486-65656-X

AN INTRODUCTION TO ALGEBRAIC STRUCTURES, Joseph Landin. Superb self-contained text covers "abstract algebra": sets and numbers, theory of groups, theory of rings, much more. Numerous well-chosen examples, exercises. 247pp. 5⅜ x 8½. 0-486-65940-2

QUALITATIVE THEORY OF DIFFERENTIAL EQUATIONS, V. V. Nemytskii and V.V. Stepanov. Classic graduate-level text by two prominent Soviet mathematicians covers classical differential equations as well as topological dynamics and ergodic theory. Bibliographies. 523pp. 5⅜ x 8½. 0-486-65954-2

THEORY OF MATRICES, Sam Perlis. Outstanding text covering rank, nonsingularity and inverses in connection with the development of canonical matrices under the relation of equivalence, and without the intervention of determinants. Includes exercises. 237pp. 5⅜ x 8½. 0-486-66810-X

INTRODUCTION TO ANALYSIS, Maxwell Rosenlicht. Unusually clear, accessible coverage of set theory, real number system, metric spaces, continuous functions, Riemann integration, multiple integrals, more. Wide range of problems. Undergraduate level. Bibliography. 254pp. 5⅜ x 8½. 0-486-65038-3

MODERN NONLINEAR EQUATIONS, Thomas L. Saaty. Emphasizes practical solution of problems; covers seven types of equations. ". . . a welcome contribution to the existing literature...."–*Math Reviews*. 490pp. 5⅜ x 8½. 0-486-64232-1

MATRICES AND LINEAR ALGEBRA, Hans Schneider and George Phillip Barker. Basic textbook covers theory of matrices and its applications to systems of linear equations and related topics such as determinants, eigenvalues and differential equations. Numerous exercises. 432pp. 5⅜ x 8½. 0-486-66014-1

LINEAR ALGEBRA, Georgi E. Shilov. Determinants, linear spaces, matrix algebras, similar topics. For advanced undergraduates, graduates. Silverman translation. 387pp. 5⅜ x 8½. 0-486-63518-X

ELEMENTS OF REAL ANALYSIS, David A. Sprecher. Classic text covers fundamental concepts, real number system, point sets, functions of a real variable, Fourier series, much more. Over 500 exercises. 352pp. 5⅜ x 8½. 0-486-65385-4

SET THEORY AND LOGIC, Robert R. Stoll. Lucid introduction to unified theory of mathematical concepts. Set theory and logic seen as tools for conceptual understanding of real number system. 496pp. 5⅜ x 8¼. 0-486-63829-4

TENSOR CALCULUS, J.L. Synge and A. Schild. Widely used introductory text covers spaces and tensors, basic operations in Riemannian space, non-Riemannian spaces, etc. 324pp. 5⅜ x 8¼. 0-486-63612-7

ORDINARY DIFFERENTIAL EQUATIONS, Morris Tenenbaum and Harry Pollard. Exhaustive survey of ordinary differential equations for undergraduates in mathematics, engineering, science. Thorough analysis of theorems. Diagrams. Bibliography. Index. 818pp. 5⅜ x 8½. 0-486-64940-7

INTEGRAL EQUATIONS, F. G. Tricomi. Authoritative, well-written treatment of extremely useful mathematical tool with wide applications. Volterra Equations, Fredholm Equations, much more. Advanced undergraduate to graduate level. Exercises. Bibliography. 238pp. 5⅜ x 8½. 0-486-64828-1

FOURIER SERIES, Georgi P. Tolstov. Translated by Richard A. Silverman. A valuable addition to the literature on the subject, moving clearly from subject to subject and theorem to theorem. 107 problems, answers. 336pp. 5⅜ x 8½. 0-486-63317-9

INTRODUCTION TO MATHEMATICAL THINKING, Friedrich Waismann. Examinations of arithmetic, geometry, and theory of integers; rational and natural numbers; complete induction; limit and point of accumulation; remarkable curves; complex and hypercomplex numbers, more. 1959 ed. 27 figures. xii+260pp. 5⅜ x 8½. 0-486-63317-9

POPULAR LECTURES ON MATHEMATICAL LOGIC, Hao Wang. Noted logician's lucid treatment of historical developments, set theory, model theory, recursion theory and constructivism, proof theory, more. 3 appendixes. Bibliography. 1981 edition. ix + 283pp. 5⅜ x 8½. 0-486-67632-3

CALCULUS OF VARIATIONS, Robert Weinstock. Basic introduction covering isoperimetric problems, theory of elasticity, quantum mechanics, electrostatics, etc. Exercises throughout. 326pp. 5⅜ x 8½. 0-486-63069-2

THE CONTINUUM: A CRITICAL EXAMINATION OF THE FOUNDATION OF ANALYSIS, Hermann Weyl. Classic of 20th-century foundational research deals with the conceptual problem posed by the continuum. 156pp. 5⅜ x 8½. 0-486-67982-9

CHALLENGING MATHEMATICAL PROBLEMS WITH ELEMENTARY SOLUTIONS, A. M. Yaglom and I. M. Yaglom. Over 170 challenging problems on probability theory, combinatorial analysis, points and lines, topology, convex polygons, many other topics. Solutions. Total of 445pp. 5⅜ x 8½. Two-vol. set.
Vol. I: 0-486-65536-9 Vol. II: 0-486-65537-7

INTRODUCTION TO PARTIAL DIFFERENTIAL EQUATIONS WITH APPLICATIONS, E. C. Zachmanoglou and Dale W. Thoe. Essentials of partial differential equations applied to common problems in engineering and the physical sciences. Problems and answers. 416pp. 5⅜ x 8½. 0-486-65251-3

THE THEORY OF GROUPS, Hans J. Zassenhaus. Well-written graduate-level text acquaints reader with group-theoretic methods and demonstrates their usefulness in mathematics. Axioms, the calculus of complexes, homomorphic mapping, p-group theory, more. 276pp. 5⅜ x 8½. 0-486-40922-8

Math–Decision Theory, Statistics, Probability

ELEMENTARY DECISION THEORY, Herman Chernoff and Lincoln E. Moses. Clear introduction to statistics and statistical theory covers data processing, probability and random variables, testing hypotheses, much more. Exercises. 364pp. 5⅜ x 8½. 0-486-65218-1

STATISTICS MANUAL, Edwin L. Crow et al. Comprehensive, practical collection of classical and modern methods prepared by U.S. Naval Ordnance Test Station. Stress on use. Basics of statistics assumed. 288pp. 5⅜ x 8½. 0-486-60599-X

SOME THEORY OF SAMPLING, William Edwards Deming. Analysis of the problems, theory and design of sampling techniques for social scientists, industrial managers and others who find statistics important at work. 61 tables. 90 figures. xvii +602pp. 5⅜ x 8½. 0-486-64684-X

LINEAR PROGRAMMING AND ECONOMIC ANALYSIS, Robert Dorfman, Paul A. Samuelson and Robert M. Solow. First comprehensive treatment of linear programming in standard economic analysis. Game theory, modern welfare economics, Leontief input-output, more. 525pp. 5⅜ x 8½. 0-486-65491-5

PROBABILITY: AN INTRODUCTION, Samuel Goldberg. Excellent basic text covers set theory, probability theory for finite sample spaces, binomial theorem, much more. 360 problems. Bibliographies. 322pp. 5⅜ x 8½. 0-486-65252-1

GAMES AND DECISIONS: INTRODUCTION AND CRITICAL SURVEY, R. Duncan Luce and Howard Raiffa. Superb nontechnical introduction to game theory, primarily applied to social sciences. Utility theory, zero-sum games, n-person games, decision-making, much more. Bibliography. 509pp. 5⅜ x 8½. 0-486-65943-7

INTRODUCTION TO THE THEORY OF GAMES, J. C. C. McKinsey. This comprehensive overview of the mathematical theory of games illustrates applications to situations involving conflicts of interest, including economic, social, political, and military contexts. Appropriate for advanced undergraduate and graduate courses; advanced calculus a prerequisite. 1952 ed. x+372pp. 5⅜ x 8½. 0-486-42811-7

FIFTY CHALLENGING PROBLEMS IN PROBABILITY WITH SOLUTIONS, Frederick Mosteller. Remarkable puzzlers, graded in difficulty, illustrate elementary and advanced aspects of probability. Detailed solutions. 88pp. 5⅜ x 8½. 65355-2

PROBABILITY THEORY: A CONCISE COURSE, Y. A. Rozanov. Highly readable, self-contained introduction covers combination of events, dependent events, Bernoulli trials, etc. 148pp. 5⅜ x 8¼. 0-486-63544-9

STATISTICAL METHOD FROM THE VIEWPOINT OF QUALITY CONTROL, Walter A. Shewhart. Important text explains regulation of variables, uses of statistical control to achieve quality control in industry, agriculture, other areas. 192pp. 5⅜ x 8½. 0-486-65232-7

Math–Geometry and Topology

ELEMENTARY CONCEPTS OF TOPOLOGY, Paul Alexandroff. Elegant, intuitive approach to topology from set-theoretic topology to Betti groups; how concepts of topology are useful in math and physics. 25 figures. 57pp. 5⅜ x 8½. 0-486-60747-X

COMBINATORIAL TOPOLOGY, P. S. Alexandrov. Clearly written, well-organized, three-part text begins by dealing with certain classic problems without using the formal techniques of homology theory and advances to the central concept, the Betti groups. Numerous detailed examples. 654pp. 5⅜ x 8½. 0-486-40179-0

EXPERIMENTS IN TOPOLOGY, Stephen Barr. Classic, lively explanation of one of the byways of mathematics. Klein bottles, Moebius strips, projective planes, map coloring, problem of the Koenigsberg bridges, much more, described with clarity and wit. 43 figures. 210pp. 5⅜ x 8½. 0-486-25933-1

THE GEOMETRY OF RENÉ DESCARTES, René Descartes. The great work founded analytical geometry. Original French text, Descartes's own diagrams, together with definitive Smith-Latham translation. 244pp. 5⅜ x 8½. 0-486-60068-8

EUCLIDEAN GEOMETRY AND TRANSFORMATIONS, Clayton W. Dodge. This introduction to Euclidean geometry emphasizes transformations, particularly isometries and similarities. Suitable for undergraduate courses, it includes numerous examples, many with detailed answers. 1972 ed. viii+296pp. 6⅛ x 9¼. 0-486-43476-1

PRACTICAL CONIC SECTIONS: THE GEOMETRIC PROPERTIES OF ELLIPSES, PARABOLAS AND HYPERBOLAS, J. W. Downs. This text shows how to create ellipses, parabolas, and hyperbolas. It also presents historical background on their ancient origins and describes the reflective properties and roles of curves in design applications. 1993 ed. 98 figures. xii+100pp. 6½ x 9¼. 0-486-42876-1

THE THIRTEEN BOOKS OF EUCLID'S ELEMENTS, translated with introduction and commentary by Sir Thomas L. Heath. Definitive edition. Textual and linguistic notes, mathematical analysis. 2,500 years of critical commentary. Unabridged. 1,414pp. 5⅜ x 8½. Three-vol. set.
Vol. I: 0-486-60088-2 Vol. II: 0-486-60089-0 Vol. III: 0-486-60090-4

SPACE AND GEOMETRY: IN THE LIGHT OF PHYSIOLOGICAL, PSYCHOLOGICAL AND PHYSICAL INQUIRY, Ernst Mach. Three essays by an eminent philosopher and scientist explore the nature, origin, and development of our concepts of space, with a distinctness and precision suitable for undergraduate students and other readers. 1906 ed. vi+148pp. 5⅜ x 8½. 0-486-43909-7

GEOMETRY OF COMPLEX NUMBERS, Hans Schwerdtfeger. Illuminating, widely praised book on analytic geometry of circles, the Moebius transformation, and two-dimensional non-Euclidean geometries. 200pp. 5⅝ x 8¼. 0-486-63830-8

DIFFERENTIAL GEOMETRY, Heinrich W. Guggenheimer. Local differential geometry as an application of advanced calculus and linear algebra. Curvature, transformation groups, surfaces, more. Exercises. 62 figures. 378pp. 5⅜ x 8½. 0-486-63433-7

History of Math

THE WORKS OF ARCHIMEDES, Archimedes (T. L. Heath, ed.). Topics include the famous problems of the ratio of the areas of a cylinder and an inscribed sphere; the measurement of a circle; the properties of conoids, spheroids, and spirals; and the quadrature of the parabola. Informative introduction. clxxxvi+326pp. 5⅜ x 8½.
0-486-42084-1

A SHORT ACCOUNT OF THE HISTORY OF MATHEMATICS, W. W. Rouse Ball. One of clearest, most authoritative surveys from the Egyptians and Phoenicians through 19th-century figures such as Grassman, Galois, Riemann. Fourth edition. 522pp. 5⅜ x 8½. 0-486-20630-0

THE HISTORY OF THE CALCULUS AND ITS CONCEPTUAL DEVELOPMENT, Carl B. Boyer. Origins in antiquity, medieval contributions, work of Newton, Leibniz, rigorous formulation. Treatment is verbal. 346pp. 5⅜ x 8½. 0-486-60509-4

THE HISTORICAL ROOTS OF ELEMENTARY MATHEMATICS, Lucas N. H. Bunt, Phillip S. Jones, and Jack D. Bedient. Fundamental underpinnings of modern arithmetic, algebra, geometry and number systems derived from ancient civilizations. 320pp. 5⅜ x 8½. 0-486-25563-8

A HISTORY OF MATHEMATICAL NOTATIONS, Florian Cajori. This classic study notes the first appearance of a mathematical symbol and its origin, the competition it encountered, its spread among writers in different countries, its rise to popularity, its eventual decline or ultimate survival. Original 1929 two-volume edition presented here in one volume. xxviii+820pp. 5⅜ x 8½. 0-486-67766-4

GAMES, GODS & GAMBLING: A HISTORY OF PROBABILITY AND STATISTICAL IDEAS, F. N. David. Episodes from the lives of Galileo, Fermat, Pascal, and others illustrate this fascinating account of the roots of mathematics. Features thought-provoking references to classics, archaeology, biography, poetry. 1962 edition. 304pp. 5⅜ x 8½. (Available in U.S. only.) 0-486-40023-9

OF MEN AND NUMBERS: THE STORY OF THE GREAT MATHEMATICIANS, Jane Muir. Fascinating accounts of the lives and accomplishments of history's greatest mathematical minds–Pythagoras, Descartes, Euler, Pascal, Cantor, many more. Anecdotal, illuminating. 30 diagrams. Bibliography. 256pp. 5⅜ x 8½. 0-486-28973-7

HISTORY OF MATHEMATICS, David E. Smith. Nontechnical survey from ancient Greece and Orient to late 19th century; evolution of arithmetic, geometry, trigonometry, calculating devices, algebra, the calculus. 362 illustrations. 1,355pp. 5⅜ x 8½. Two-vol. set. Vol. I: 0-486-20429-4 Vol. II: 0-486-20430-8

A CONCISE HISTORY OF MATHEMATICS, Dirk J. Struik. The best brief history of mathematics. Stresses origins and covers every major figure from ancient Near East to 19th century. 41 illustrations. 195pp. 5⅜ x 8½. 0-486-60255-9

Physics

OPTICAL RESONANCE AND TWO-LEVEL ATOMS, L. Allen and J. H. Eberly. Clear, comprehensive introduction to basic principles behind all quantum optical resonance phenomena. 53 illustrations. Preface. Index. 256pp. 5⅜ x 8½. 0-486-65533-4

QUANTUM THEORY, David Bohm. This advanced undergraduate-level text presents the quantum theory in terms of qualitative and imaginative concepts, followed by specific applications worked out in mathematical detail. Preface. Index. 655pp. 5⅜ x 8½. 0-486-65969-0

ATOMIC PHYSICS (8th EDITION), Max Born. Nobel laureate's lucid treatment of kinetic theory of gases, elementary particles, nuclear atom, wave-corpuscles, atomic structure and spectral lines, much more. Over 40 appendices, bibliography. 495pp. 5⅜ x 8½. 0-486-65984-4

A SOPHISTICATE'S PRIMER OF RELATIVITY, P. W. Bridgman. Geared toward readers already acquainted with special relativity, this book transcends the view of theory as a working tool to answer natural questions: What is a frame of reference? What is a "law of nature"? What is the role of the "observer"? Extensive treatment, written in terms accessible to those without a scientific background. 1983 ed. xlviii+172pp. 5⅜ x 8½. 0-486-42549-5

AN INTRODUCTION TO HAMILTONIAN OPTICS, H. A. Buchdahl. Detailed account of the Hamiltonian treatment of aberration theory in geometrical optics. Many classes of optical systems defined in terms of the symmetries they possess. Problems with detailed solutions. 1970 edition. xv + 360pp. 5⅜ x 8½. 0-486-67597-1

PRIMER OF QUANTUM MECHANICS, Marvin Chester. Introductory text examines the classical quantum bead on a track: its state and representations; operator eigenvalues; harmonic oscillator and bound bead in a symmetric force field; and bead in a spherical shell. Other topics include spin, matrices, and the structure of quantum mechanics; the simplest atom; indistinguishable particles; and stationary-state perturbation theory. 1992 ed. xiv+314pp. 6⅛ x 9¼. 0-486-42878-8

LECTURES ON QUANTUM MECHANICS, Paul A. M. Dirac. Four concise, brilliant lectures on mathematical methods in quantum mechanics from Nobel Prize-winning quantum pioneer build on idea of visualizing quantum theory through the use of classical mechanics. 96pp. 5⅜ x 8½. 0-486-41713-1

THIRTY YEARS THAT SHOOK PHYSICS: THE STORY OF QUANTUM THEORY, George Gamow. Lucid, accessible introduction to influential theory of energy and matter. Careful explanations of Dirac's anti-particles, Bohr's model of the atom, much more. 12 plates. Numerous drawings. 240pp. 5⅜ x 8½. 0-486-24895-X

ELECTRONIC STRUCTURE AND THE PROPERTIES OF SOLIDS: THE PHYSICS OF THE CHEMICAL BOND, Walter A. Harrison. Innovative text offers basic understanding of the electronic structure of covalent and ionic solids, simple metals, transition metals and their compounds. Problems. 1980 edition. 582pp. 6⅛ x 9¼. 0-486-66021-4